UAV or Drones for Remote Sensing Applications

UAV or Drones for Remote Sensing Applications

Volume 2

Special Issue Editors

Felipe Gonzalez Toro
Antonios Tsourdos

MDPI • Basel • Beijing • Wuhan • Barcelona • Belgrade

MDPI

Special Issue Editors

Felipe Gonzalez Toro
Queensland University of Technology
Australia

Antonios Tsourdos
Cranfield University
UK

Editorial Office
MDPI
St. Alban-Anlage 66
Basel, Switzerland

This is a reprint of articles from the Special Issue published online in the open access journal *Sensors* (ISSN 1424-8220) from 2017 to 2018 (available at: http://www.mdpi.com/journal/sensors/special_issues/UAV_drones_remote_sensing)

For citation purposes, cite each article independently as indicated on the article page online and as indicated below:

LastName, A.A.; LastName, B.B.; LastName, C.C. Article Title. *Journal Name* **Year**, *Article Number, Page Range.*

Volume 2
ISBN 978-3-03897-111-5 (Pbk)
ISBN 978-3-03897-112-2 (PDF)

Volume 1–2
ISBN 978-3-03897-113-9 (Pbk)
ISBN 978-3-03897-114-6 (PDF)

Contents

About the Special Issue Editors

Felipe Gonzalez Toro, Associate Professor at the Science and Engineering Faculty, Queensland University of Technology (Australia), with a passion for innovation in the fields of aerial robotics and automation and remote sensing. He creates and uses aerial robots, drones or UAVs that possess a high level of cognition using efficient on-board computer algorithms and advanced optimization and game theory approaches that assist us to understand and improve our physical and natural world. Dr. Gonzalez leads the UAVs-based remote sensing research at QUT. As of 2017, he has published nearly 120 peer reviewed papers. To date, Dr. Gonzalez has been awarded \$10.1M in chief investigator/partner investigator grants. This grant income represents a mixture of sole investigator funding, international, multidisciplinary collaborative grants and funding from industry. He is also a Chartered Professional Engineer, Engineers Australia—National Professional Engineers Register (NPER), a member of the Royal Aeronautical Society (RAeS), The IEEE, American Institute of Aeronautics and Astronautics (AIAA) and holder of a current Australian Private Pilot Licence (CASA PPL).

Antonios Tsourdos obtained a MEng on Electronic, Control and Systems Engineering, from the University of Sheffield (1995), an MSc on Systems Engineering from Cardiff University (1996) and a PhD on Nonlinear Robust Autopilot Design and Analysis from Cranfield University (1999). He joined the Cranfield University in 1999 as lecturer, was appointed Head of the Centre of Autonomous and Cyber-Physical Systems in 2007 and Professor of Autonomous Systems and Control in 2009 and Director of Research—Aerospace, Transport and Manufacturing in 2015. Professor Tsourdos was a member of the Team Stellar, the winning team for the UK MoD Grand Challenge (2008) and the IET Innovation Award (Category Team, 2009). Professor Tsourdos is an editorial board member of: Proceedings of the IMechE Part G: *Journal of Aerospace Engineering, IEEE Transactions of Aerospace and Electronic Systems, Aerospace Science & Technology, International Journal of Systems Science, Systems Science & Control Engineering* and the *International Journal of Aeronautical and Space Sciences*. Professor Tsourdos is Chair of the IFAC Technical Committee on Aerospace Control, a member of the IFAC Technical Committee on Networked Systems, Discrete Event and Hybrid Systems, and Intelligent Autonomous Vehicles. Professor Tsourdos is also a member of the AIAA Technical Committee on Guidance, Control and Navigation; AIAA Unmanned Systems Program Committee; IEEE Control System Society Technical Committee on Aerospace Control (TCAC) and IET Robotics & Mechatronics Executive Team.

Preface to "UAV or Drones for Remote Sensing Applications"

The rapid development and growth of unmanned aerial vehicles (UAVs) as a remote sensing platform, as well as advances in the miniaturization of instrumentation and data systems, have resulted in an increasing uptake of this technology in the environmental and remote sensing science communities. Although tough regulations across the globe may still limit the broader use of UAVs, their use in precision agriculture, ecology, atmospheric research, disaster response biosecurity, ecological and reef monitoring, forestry, fire monitoring, quick response measurements for emergency disaster, Earth science research, volcanic gas sampling, monitoring of gas pipelines, mining plumes, humanitarian observations and biological/chemo-sensing tasks continues to increase.

This Special Issue provides a forum for high-quality peer-reviewed papers that broaden the awareness and understanding of UAV developments, applications of UAVs for remote sensing, and associated developments in sensor technology, data processing and communications, and UAV system design and sensing capabilities.

This topic encompasses many algorithms and process flows and tools, including: robust vehicle detection in aerial images based on cascaded convolutional neural networks; a stereo dual-channel dynamic programming algorithm for UAV image stitching, as well as seamline determination based on PKGC segmentation for remote sensing image mosaicking; the implementation of an IMU-aided image stacking algorithm in digital cameras; the study of multispectral characteristics at different observation angles, rapid three-dimensional reconstruction for image sequence acquired from UAV cameras; comparisons of Riegl Ricopter UAV Lidar-derived canopy height and DBH with terrestrial Lidar; vision-based target finding and inspection of a ground target using a multirotor UAV system; a localization framework for real-time UAV autonomous landing using an on-ground deployed visual approach; curvature continuous and bounded path planning for fixed-wing UAVs; the calculation and identification of the aerodynamic parameters for small-scaled fixed-wing UAVs

Several wildfire and agricultural applications of UAVS including: deep learning-based wildfire identification in UAV imagery; postfire vegetation survey campaigns; secure utilization of beacons; and UAVS used in emergency response systems for building fire hazards; observing spring and fall phenology in a deciduous forest with aerial drone imagery; the design and testing of a UAV mapping system for agricultural field surveying; artificial neural network to predict vine water status spatial variability using multispectral information obtained from an unmanned aerial vehicle; automatic hotspot and sun glint detection in UAV multispectral images obtained via uncooled thermal camera calibration and optimization of the photogrammetry process for UAV applications in agriculture; olive yield forecast tool based on the tree canopy geometry using UAS imagery; spatial scale gap filling downscaling method for applications in precision agriculture; automatic co-registration algorithm to remove canopy shaded pixels in UAV-borne thermal images to improve the estimation of crop water stress on vineyards; methodologies for improving plant pest surveillance in vineyards and crops using UAV-based hyperspectral and spatial data; UAV-assisted dynamic clustering of wireless sensors and networks for crop health monitoring.

Several applications of UAVS in the fields of environment and conservation including the following: the automatic detection of pre-existing termite mounds through UAS and hyperspectral imagery; aerial mapping of forests affected by pathogens using UAVs; hyperspectral sensors and artificial i ntelligence; c oral r eef and coral bleaching monitoring; invasive grass and vegetation surveys in remote arid lands. UAVs are also utilized in many other applications: vicarious calibration of SUAS microbolometer temperature imagery for the estimation of radiometric land surface temperature; the documentation of hiking trails in alpine areas; the detection of nuclear sources by UAV teleoperation using a visuo-haptic augmented reality interface; the design of a UAV-embedded microphone array system for sound source localization in outdoor environments; the monitoring of concentrated solar power plants, accuracy analysis of a dam model from drone surveys, mobile sensing and actuation infrastructure, UAV-based frameworks for river hydromorphological characterization; online aerial terrain mapping for ground robot navigation.

Felipe Gonzalez Toro, Antonios Tsourdos

Special Issue Editors

sensors

MDPI

Article

Automatic Coregistration Algorithm to Remove Canopy Shaded Pixels in UAV-Borne Thermal Images to Improve the Estimation of Crop Water Stress Index of a Drip-Irrigated Cabernet Sauvignon Vineyard

Tomas Poblete [1], Samuel Ortega-Farías [1,2,*] and Dongryeol Ryu [3]

[1]   Centro de Investigación y Transferencia en Riego y Agroclimatología (CITRA), Universidad de Talca, Casilla 747, Talca 3460000, Chile; totopoblete@gmail.com
[2]   Research Program on Adaptation of Agriculture to Climate Change (A2C2), Universidad de Talca, Casilla 747, Talca 3460000, Chile
[3]   Department of Infrastructure Engineering, The University of Melbourne, Parkville 3010, Australia; dryu@unimelb.edu.au
*   Correspondence: sortega@utalca.cl

Received: 28 November 2017; Accepted: 25 January 2018; Published: 30 January 2018

Abstract: Water stress caused by water scarcity has a negative impact on the wine industry. Several strategies have been implemented for optimizing water application in vineyards. In this regard, midday stem water potential (SWP) and thermal infrared (TIR) imaging for crop water stress index (CWSI) have been used to assess plant water stress on a vine-by-vine basis without considering the spatial variability. Unmanned Aerial Vehicle (UAV)-borne TIR images are used to assess the canopy temperature variability within vineyards that can be related to the vine water status. Nevertheless, when aerial TIR images are captured over canopy, internal shadow canopy pixels cannot be detected, leading to mixed information that negatively impacts the relationship between CWSI and SWP. This study proposes a methodology for automatic coregistration of thermal and multispectral images (ranging between 490 and 900 nm) obtained from a UAV to remove shadow canopy pixels using a modified scale invariant feature transformation (SIFT) computer vision algorithm and Kmeans++ clustering. Our results indicate that our proposed methodology improves the relationship between CWSI and SWP when shadow canopy pixels are removed from a drip-irrigated Cabernet Sauvignon vineyard. In particular, the coefficient of determination (R^2) increased from 0.64 to 0.77. In addition, values of the root mean square error (RMSE) and standard error (SE) decreased from 0.2 to 0.1 MPa and 0.24 to 0.16 MPa, respectively. Finally, this study shows that the negative effect of shadow canopy pixels was higher in those vines with water stress compared with well-watered vines.

Keywords: multispectral and thermal automatic coregistration; shadow removal; crop water stress index (CWSI); UAV; midday stem water potential

1. Introduction

Water availability is a critical limiting factor in the agricultural industry; therefore, a wide range of new technologies and strategies have been adopted to optimize the agricultural water consumption [1–4]. Granier et al. [5] argued that the measurements of physiological parameters can provide better information about the whole-plant-level water use with changing atmospheric water demands. For example, the water potential has been used to characterize the plant water stress and to schedule irrigation in vineyards [6–8], as well as for nuts trees [9,10], and olive trees [11,12]. However, water potential is typically measured on a plant-by-plant basis leading to high costs and requiring a considerable time when these measurements are extended to cover a large area [13,14]. This limitation has motivated the development of cost- and time-effective alternatives to evaluate plant water status.

Multispectral imagery to capture images at the leaf and canopy levels has been proposed as an effective tool for agricultural applications [15] to indirectly and remotely assess plant water status. For example, Rapaport et al. [16] reported that estimating the water balance index (WABI-2) using visible (538 nm) and short-wave infrared (1500 nm) spectrum is a good indicator of water stress in grapevines. Rallo, et al. [17] suggested that spectral information between the near infrared (NIR) (750 nm) and short-wave infrared (SWIR) (1550 nm) ranges can improve the prediction of leaf water potential. In addition, Pôças et al. in [18,19] showed that the wavelength information of visible (VIS) and NIR spectra can be used to predict water status. Poblete, et al. [20] suggested that artificial neural networks using information obtained from 500 to 800 nm could be used to predict the stem water potential (SWP) spatial variability in vineyards.

Furthermore, the Crop Water Stress Index (CWSI) derived from the radiometric temperature of a plant canopy measured using thermal infrared (TIR) sensors has been suggested as a reliable tool to assess water stress [21–25] showing good correlations with ground measurements of water potential. However, as in the case of ground-based water potential measurements, when large crop areas are to be assessed, the ground-based TIR measurements can still be time-consuming and impractical. Thus, remotely collected TIR imagery has been suggested as an alternative tool that can provide crop status information over large regions in a non-invasive manner [26–29]. In particular, unmanned aerial vehicles (UAV) have become a useful remote sensing tool, having significant advantages in terms of cost, versatility, and high spatial resolution [30]. The CWSI studies using UAV-borne sensors have achieved a high correlation with the plant water status measured using ground-based measurements [14,29,31–33].

However, the UAV-borne TIR sensing for plant water stress suffers from the technical issue of the potential degradation of the canopy temperature information by the pixels of a shaded (or shadow) canopy; this is because the surface temperature of sunlit canopy is known to better represent the plant water stress. Existing methods to remove these shaded pixels from remote-sensing images can be divided into two principal steps: shadow detection followed by a de-shadow process [34]. The first step, shadow detection, can be conducted by either thresholding or modeling [35]. The thresholding process is more common as it is less complicated than modeling, because modeling requires prior information of shadows and mathematical conceptualization; consequently, modeling is applied only to specific cases. The thresholding process involves finding the optimal threshold value of a digital number based on histograms to segregate shadow information from other types of information. Previous studies have used different wavelengths to elucidate thresholds for shadow deletion. For example, NIR (757–853 nm) [36], the ratio between blue (450–520 nm) and NIR (760–900 nm) [37], Infrared (10.4–12.5 μm) [38], and indices [39–41] have been used to separate undesired information. However, the TIR information obtained by the commonly used thermal imaging devices (based on an uncooled microbolometer) does not provide sufficient sensitivity for subtle temperature variation [15]; therefore, this method often fails to distinguish shadow canopy pixels from shadow soil pixels. Considering the issues with the shadow canopy pixels, an important process in thermal image processing is shadow pixel removal to improve the resampling of the sunlit canopy information [42]. Zarco-Tejada et al. [43] and Suárez et al. [44] highlight the importance of resampling sunlit canopy pixels using hyperspectral and multispectral imagery, respectively, to assess the plant water stress. Using UAV-borne thermal imagery, several studies have proposed different methodologies to achieve shadow removal and avoid the shadow effect in the case of thermal images. For example, Zarco-Tejada et al. [45] suggested that only the center portion of the canopy row be sampled to minimize the inclusion of shadow canopy pixels. Gonzalez-Dugo et al. [46] sampled the central 50% of the crown pixels of the canopy. Santesteban et al. [29] detailed the complexity of avoiding shadow information, especially in thermal imageries, and proposed a Digital Elevation Model (DEM) and Otsu [47] combined methodology to filter shadows using height differences presented in the ground.

Despite the proposed methodologies of shadow removal for UAV-borne images and, even if capturing images in overcast conditions can minimize the intensity of shadowing [48], the identification of shadow canopy pixels produced by the canopy over itself as information to be deleted in thermal images is not considered. For a drip-irrigated vineyard, Figure 1 shows an example of a thermal image in which the shadow canopy pixels cannot be identified on comparing with the visible imagery VIS (490 nm).

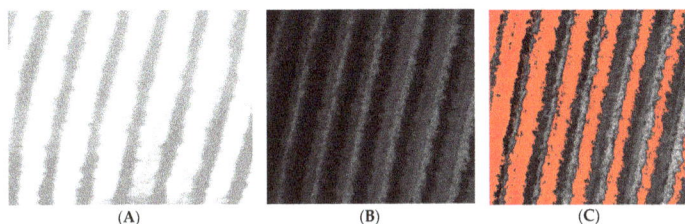

(A)   (B)   (C)

Figure 1. Comparison between the thermal and visible (490 nm) canopy shadow information for a drip-irrigated vineyard: (**A**) thermal image; (**B**) VIS (490 nm) image; and (**C**) VIS (490 nm) image without shadow pixels (represented in red).

Considering the effect of shadow on water stress estimation, it is crucial to determine shaded pixels and remove them [49–51]. Möller et al. [23] proposed a methodology to detect grapevine crop water status using thermal and visible images collected using truck-mount sensors at 15 m above the ground to sample the sunlit canopy information; they used Ground Control Points (GCP) made of cross-marked aluminum plates to geo-reference, align, and coregister the images from two different sensors. Leinonen and Jones [42] also proposed a methodology to assess water stress in grapevine and broad bean fields using ground-obtained thermal and visible images; their methodology was based on non-automatic (by expert user) selection of GCP to overlay the images and later warp and resample the images to obtain the sunlit canopy information. Finally, Smith et al. [52] proposed a methodology to detect regions of soil moisture deficit from a spinach plantation using thermal and visible images. Bulanon et al. [53] proposed a methodology for fruit detection using thermal and VIS imagery in which four corners of a ground-marked region of interest were used to coregister VIS and TIR images and perform shadow removal. However, in all of these studies, challenges in coregistering optical and TIR images were reported when the images were combined for shadow removal [54] using non-automatic coregistration. Considering this, our study proposes an automatic scheme based on Scale Invariant Feature Transformation (SIFT) computer vision algorithm and an improved matching pairs point selection to remove shaded pixels in a UAV-borne thermal image to improve the estimation of the CWSI for a drip-irrigated Cabernet Sauvignon vineyard grown under Mediterranean climate conditions.

2. Materials and Methods

2.1. Site Description and Experimental Design

The study site has a predominant typical Mediterranean climate with a summer period from December to March that is usually dry (2.2% of annual rainfall) and hot with an average daily temperature of 21 °C, and spring that is usually wet (16% of annual rainfall). Average annual rainfall in the region is about 500 mm, which falls primarily during April to August.

Flight campaigns and climate measurements were carried out in a drip-irrigated Cabernet Sauvignon vineyard located in the Pencahue Valley, Maule Region, Chile (35°20′ L.S; 71°46′ L.W). The three-year-old wine grapes were trained on a vertical shoot positioned (VSP) system. The vineyard fractional cover, which represents the dimensionless parameter of ground covered vegetation over uncovered ground [55], was 19%. In addition, the vineyard with east–west oriented rows (at 1 m × 2 m) was irrigated daily using 2 L·h^{-1} drippers spaced at intervals of 1 m. The soil is las doscientas type with a compact arsenic soil texture with high levels of Fe and Mn.

The experimental design consisted of two completely randomized treatments (well-watered and deficit-irrigated vines) with four replications (six vines per replication). The SWP for well-watered vines showed values that ranged between −0.6 and −0.8 MPa and the deficit-irrigated vines showed values that ranged between −0.9 and −1.25 MPa. The SWP was measured at the time of UAV overflight [56] using a pressure chamber (PMS 600, PMS Instrument Company, Corvallis, OR, USA) from the middle vines for each repetition. A total of 32 leaves from the middle zone of the canopy were measured corresponding to two mature and healthy sun-exposed leaves that were previously covered with plastic bags and coated with aluminum foil for at least 1 h before measurements [6].

2.2. Cameras and Image Processing Description

A multispectral camera was used to collect VIS-NIR images for shadow identification. The images were obtained from a Micro MCA-6 camera (Tetracam's Micro Camera Array), which has an array of sensors with band-path filters whose center-wavelengths are 490, 550, 680, 720, 800, and 900 nm with a resolution of 1280 (H) × 1024 (V). For thermal infrared imaging, the FLIR TAU2 640 (FLIR Systems, Inc., Wilsonville, OR, USA) was used. This camera consists of an uncooled microbolometer of 640 (H) × 512 (V) with a pixel pitch of 17 µm and spectral band ranging between 7.5 and 13.5 µm. The thermal calibration was conducted using the methodology proposed by Ribeiro-Gomes et al. [57], in which an artificial neural network is used with the sensor temperature and the digital response of the sensor as input and a Wallis filter to improve the photogrammetry process. Further, the multispectral calibration was performed using the methodology proposed by Poblete et al. [20] in which normalization of the reflectance was performed using a "white reference" Spectralon panel (Labsphere Inc., Sutton, NH, USA) and compared comparison was made with that obtained using a spectroradiometer (SVC HR-1024, Spectra Vista Cooperation, Poughkeepsie, NY, USA) to account for any relative spectral response of each band of the camera as proposed by Laliberte et al. [58].

All images from both sensors were processed using a photogrammetric software PhotoScan (Agisoft LLC, Saint Petersburg, Russia) to stitch the images together to increase the Field of View (FOV) while maintaining the intrinsic characteristics of both cameras [59]; the same software parameters proposed by Ribeiro-Gomes et al. [57] for the same type of sensor were used for stitching.

Finally, the meteorological conditions and flight description on the day of SWP are detailed in Table 1.

Table 1. Day of the year (DOY), Air temperature (Ta), relative humidity (RH), wind speed (u), Radiation (Rn) and phenological stage (PS) at the time of the UAV overpass during the 2016–2017 growing season; Flight and UAV description.

DOY	Flight Time (hh:mm)	Meteorological	Conditions			
		Ta (°C)	RH (%)	u (Km/h)	Rn (W/m^2)	PS
6	15:00	30.81	20.2	11.3	986.7	Berry development
19	14:45	31.71	19.19	9.13	969.6	Berry development
Camera	Wavelenght	Flight	Description			
		Resolution (pixels)	Altitude (m)	Flight Speed (m/s)	Overlapping (%)	Sidelapping (%)
µMCA-6	490, 550, 670, 720, 800, 900 nm	1280 × 1024	30	2	90	75
Tau-2	7.5–13.5 µm	640 × 512	30	2		
	Model	UAV description				
		Navigation controller	Motors model	Number of propellers	Propellers dimension	
	Mikrokopter Okto XL	FlightNav 2.1	MK3638	8	12" × 3.8"	

2.3. CWSI Calculation

The calculation of the CWSI was first proposed by Jones [60] and was described as follows:

$$CWSI = \frac{T_{canopy} - T_{wet}}{T_{dry} - T_{wet}} \tag{1}$$

where T_{canopy} represents the canopy temperature obtained using the UAV-borne TIR. T_{wet} represents the temperature of a fully transpiring canopy and T_{dry} represents the temperature of a non-transpiring fully stressed canopy. As proposed by King et al. [21] and Grant et al. [51], these values do not necessarily need to be an absolute canopy temperature limit value, but serve rather as indicator temperature to scale measured canopy temperatures to the environment for calculating relative water stress. The process for obtaining the values of T_{dry} and T_{wet} was based on the methodology proposed by Park et al. [31]. The process involved using an adaptive approximation based on the TIR histograms derived from the images, and then, identifying the T_{dry} and T_{wet} values after the shadow filtering process by considering the highest and the lowest parts of the histograms, respectively.

2.4. Scale Invariant Feature Transformation (SIFT) and Random Sample Consensus (RANSAC)

This algorithm was originally proposed by Lowe [61] to extract characteristic features from images in a robust manner, which is independent of variations in scaling, rotation, translations, and illumination. The algorithm workflow was summarized and explained in detail by Ghosh and Kaabouch [59]. Based on the study by Ghosh and Kaabouch [59], the five primary steps involved in this algorithm are discussed briefly in the following lines. The scale-space construction step is based on applying several Gaussian filters to the image to compute the differences between the adjacent resulting images. Then, in the scale-space extrema detection, a selection of the highest and smallest values between each point and the 26 consecutive neighbors is conducted. Further, in the keypoint localization step, low contrast and edge response points are discarded. For the resulting keypoints, the orientation assignments based on the gradient directions are computed. To define the keypoint descriptors, histograms over each keypoint orientation is calculated considering the highest peak and values under 80% as predominant directions of the local gradients.

After these five steps are performed, the nearest neighbor of a keypoint in the first image is identified from the keypoints of the second image. To remove the outliers and filtering the incorrectly matched points, the RANSAC algorithm is applied. The RANSAC algorithm was first proposed by Fischler and Bolles [62] as a resampling technique for estimating the parameters of a model, using data that may be contaminated by outliers [63]. As suggested by Derpanis [64], the RANSAC algorithm can be summarized in five principal steps: (1) randomly selectf the minimum number of points required to determine the model parameters; (2) solve for the parameters of the model; (3) determine the number of points under a tolerance value; (4) if the ratio of points resulting from the previous step over the total number of points exceeds a predefined threshold, estimate the model with a new set of points; and (5) otherwise, repeat Steps 1–4 (with a maximum of n iterations). Because the value selected for n is high to avoid mismatching, the RANSAC algorithm is time consuming [63] and has a high computational complexity when coupled with the SIFT algorithm [65]. In addition, as RANSAC is a non-deterministic algorithm [66,67], it does not guarantee the return of an optimal solution [68], resulting in different results for different runs [69]. Furthermore, when computed with few SIFT-derived keypoints, it can be sensitive to initial conditions [70]. Considering these issues, and because thermal and visible images have different characteristics with, for most cases, different spatial resolutions, their coregistering process is complex and the assumption of global statistical dependence is not completely satisfied [71]. The RANSAC algorithm between both images leads to different pairing points, which affects the overall performance and consistency in results. This statement is consistent with Turner, et al. [72], who, using RANSAC algorithm, concluded that thermal mosaics showed lower accuracy when coregistered with multispectral images, compared with visible mosaics. To address this issue, we propose an alternative filtering of matching points based on statistical parameters between previous matched pairs. Image analysis and processing were performed using the MATLAB 2017a (Mathworks Inc., Matick, MA, USA) based on the methodology proposed by Vedaldi and Fulkerson [73].

2.5. Slope Filtering of Matching Points

As discussed above, in our method, a statistical filtering method was applied to filter previously mismatched points and our results were compared with those obtained using the RANSAC method. Our process involved both images (thermal and multispectral) as a continuous image joined by the resulting matching pair point (Figure 2).

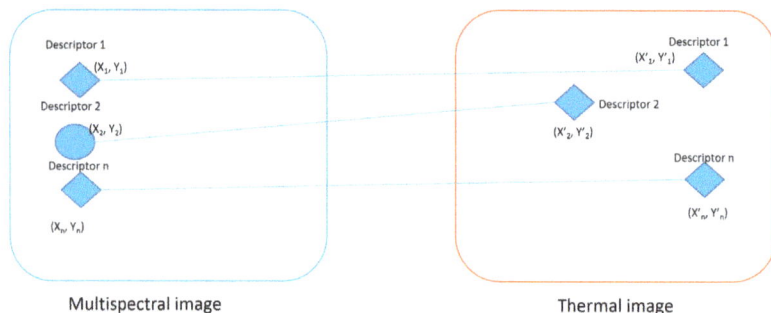

Multispectral image   Thermal image

Figure 2. Slope calculation for previously matched descriptors points as an output of the scale invariant feature transformation (SIFT) algorithm.

The slope of previous matched points was calculated using the Euclidean formula as follows:

$$m = \frac{(y'_1 - y_1)}{(x'_1 - x_1)} \tag{2}$$

where (x'_1, y'_1) corresponds to the thermal image descriptor 1 and (x_1, y_1) corresponds to the multispectral image descriptor 1.

Then, the statistical parameters were calculated for each matched feature and the filtering was conducted based on the mode of the slopes. As an example, in Figure 2, the previous matched descriptor 2 should not be considered because it was identified as a correctly matched feature, but the slope of both descriptors is different from the mode of all the slopes.

2.6. Shadow Filtering

As proposed by Shahtahmasseb et al. [34], histogram-based thresholding methods are commonly employed for shadow detection. In this study, with the aim of identifying the optimal wavelength for shadow detection, histograms for 112 UAV-borne images obtained from the vineyard were analyzed. The K-means clustering algorithm with k-means++ was used to optimize the thresholding [74]; this process was applied for shadow detection to six multispectral bands (490, 550, 680, 720, 800 and 900 nm) and their relative performance was compared. Five clusters and 200 iterations were selected for the classification. After performing the previous steps, the classified clusters for shadows from the six multispectral bands were used to build a mask that was applied to their RGB composition to evaluate the accuracy of the classification. Using the abovementioned process and the previously described SIFT algorithm, the resulting mask was coregistered with the thermal images to delete canopy shadow pixels.

2.7. Statistical Analysis

For assessing the impact of shadow canopy pixels on the linear correlation between CWSI and SWPl the coefficient of determination (R^2) was calculated. In addition, the root mean square error (RMSE) and standard error (SE) parameters were calculated for the comparison.

3. Results

3.1. SIFT and Comparison between RANSAC and Slope Filtering for Filtering Matched Features

The comparison between the abovementioned filtering processes was conducted using a complete orthomosaic obtained from the vineyard built using 112 images. The RANSAC algorithm outputs and its fluctuation on the filtered points is shown in Figure 3. Figure 3a shows the initial matched points, while Figure 3b,c shows randomly selected examples after the application of the RANSAC filter for matched points.

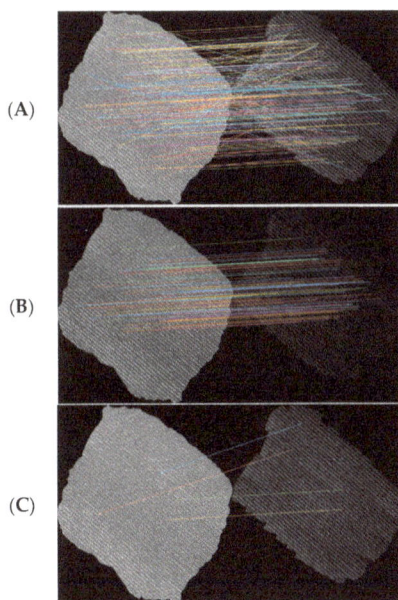

Figure 3. The RANSAC filtered points obtained at different times for a drip-irrigated vineyard: (**A**) initial matched points; (**B**) first execution; and (**C**) second execution.

In addition, the statistical parameters of the matching features slope are listed in Table 2. The selected statistical filter was based on the mode of the slope and its result of filtering is shown in Figure 4.

Table 2. Statistical parameters of slope matched points using the SIFT.

Statistical Parameter	Value
Mode	−0.3066
Mean	0.0605
Standard Deviation	0.1690
Max	1.5890
Min	−0.9278
Median	−0.0514

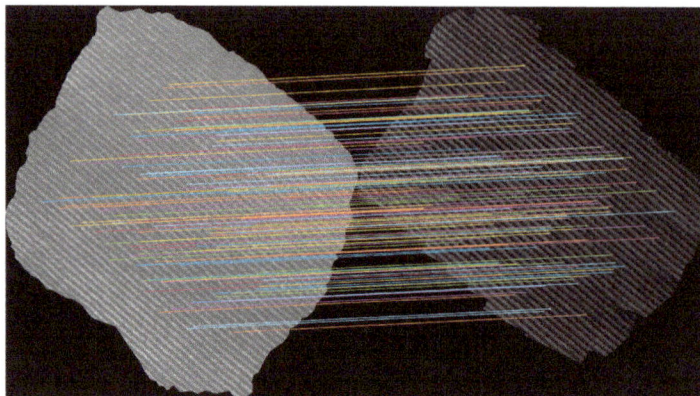

Figure 4. Filtered previous matched points considering the mode of the slope as a filter for a drip-irrigated vineyard.

3.2. Shadow Filtering

Multispectral Band Selection for Shadow Detection

For shadow identification, the histogram distribution was calculated to detect peaks related to shadow information. An example of one image per band and its distribution is shown in Figure 5 for a drip-irrigated vineyard.

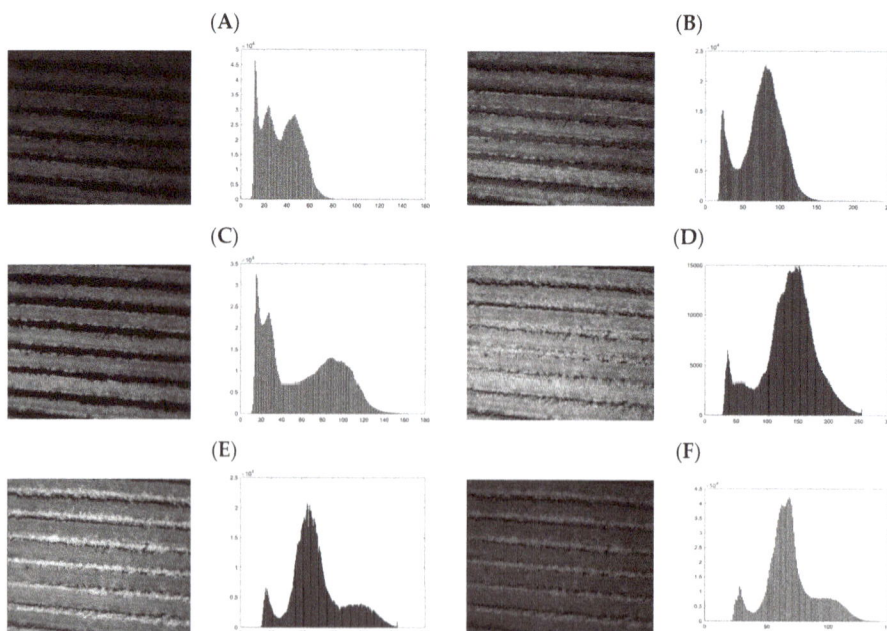

Figure 5. Six spectral band images and its distribution for a drip-irrigated vineyard: (**A**) 490 nm; (**B**) 550 nm; (**C**) 680 nm; (**D**) 720 nm; (**E**) 800 nm; and (**F**) 900 nm.

As mentioned previously, five clusters were selected and 200 iterations were conducted for the classification of all the images. The K-means++ methodology [74] was used to set the thresholds in which shadows should be identified for the six bands. After that, each generated mask was applied to an RGB image composition to identify which better represented the shadow. Figure 6 shows the filtered images and the five identified clusters per band. As is clear in Figure 6, for each band, the clustering process allowed the identification of different types of information.

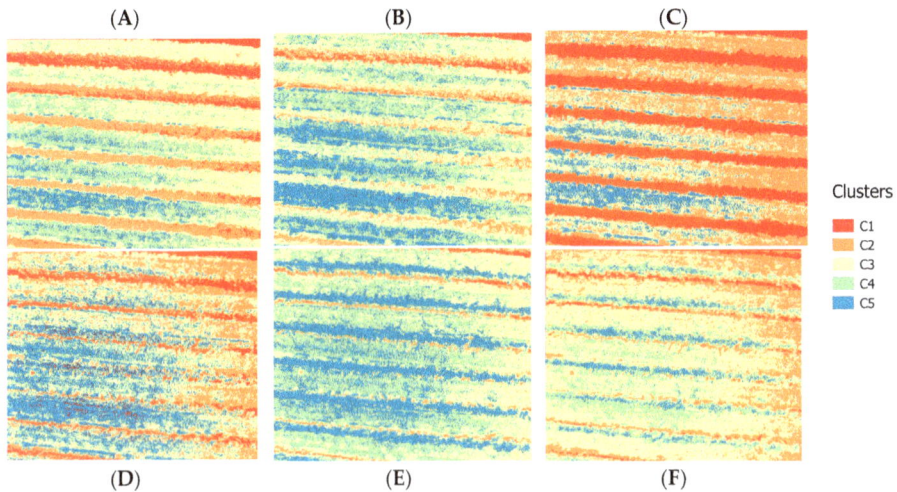

Figure 6. Six spectral clustered images using K-means++ algorithm for a drip-irrigated vineyard: (**A**) 490 nm; (**B**) 550 nm; (**C**) 680 nm; (**D**) 720 nm; (**E**) 800 nm; and (**F**) 900 nm.

For the 490-nm and 550-nm group of images, cluster 1 (C1) tends to identify both soil and internal shadows, while cluster 2 (C2) tends to classify vegetation information. On the other hand, for the 680-nm image, shadow is misclassified, nevertheless C1 allows directly identifying vegetation information. Finally, the 720 nm, 800 nm, and 900 nm images seem to misclassify shadow, mixing classified information in both cases with grassy soil and bare soil. To validate our method, a mask was built from the C1—680 nm image to select just vine canopy which included internal shaded canopy pixels. The resulting mask was applied over the images and K-means++ algorithm was carried out to classify vegetation and internal shaded canopy pixels (Figure 7A).

To assess and validate the accuracy of shadow identification, confusion matrices were calculated for the randomly selected marked winegrapes, as shown in Figure 7B, for the six bands to assess the percentage of correct shadow classification. The percentages of well classified shadow for 490, 550, 680, 720, 800 and 900 nm were 90%, 68%, 89%, 77%, 66% and 58%, respectively (Table 3). Cohen's kappa coefficient value, which is used to assess the chance-corrected agreement between two classifications [75], for each band was 0.77, 0.56, 0.76, 0.71, 0.54, and 0.41, respectively.

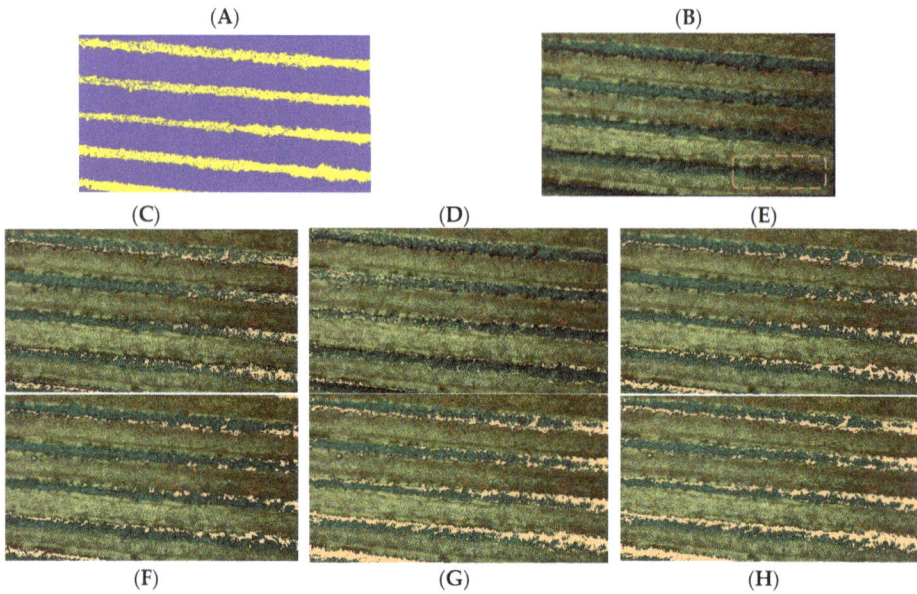

Figure 7. Shadow masks applied to an RGB composition for a drip-irrigated vineyard: (**A**) canopy and internal shadow mask; (**B**) RGB composition; (**C**) 490 nm; (**D**) 550 nm; (**E**) 680 nm; (**F**) 720 nm; (**G**) 800 nm; and (**H**) 900 nm.

Table 3. Confusion matrix for the predicted and observed shadow information. **C1**: Shadow; **C2**: No shadow; %: Percentage of correctly classified shadow pixels; **Ck**: Cohen's Kappa Coefficient.

		Predicted				
		C1	C2	%	Ck	
	C1	8220	1600	90	0.77	
	C2	910	11,630			**B1**
	C1	9090	730	68	0.56	
	C2	4280	8260			**B2**
Observed	C1	8290	1530	89	0.76	
	C2	1030	11,510			**B3**
	C1	9470	350	77	0.71	
	C2	2900	9640			**B4**
	C1	9630	190	66	0.54	
	C2	5060	7480			**B5**
	C1	9820	0	58	0.41	
	C2	7000	5540			**B6**

Based on this information, the 490-nm image, which showed the highest percentage of accuracy and Cohen's kappa coefficient value, was selected to be coregistered with the thermal image and for thermal shadow deletion.

3.3. Effect of Shadow Removal on the Relationship between CWSI and SWP

To assess the impact of shadow removal on the prediction of the SWP using CWSI, UAV-borne TIR images with and without removal of shadow canopy pixels were compared.

Figure 8 shows the thermal image after automatic coregistration and shadow canopy removal. The colored regions correspond to the filtered temperature information, while the background image represents the initial vineyard information without filtered canopy shadow pixels. The mean values of canopy temperature for the cases with and without shadow canopy were 28.84 ± 1.8 °C and 29.95 ± 2.05 °C, respectively. In addition, the relationship between CWSI and SWP is shown in Figure 9. The mean values of CWSI for the non-filtered information were 0.45 ± 0.14, while those for the filtered information were 0.52 ± 0.17. Finally, the results indicated that the relationship between the CWSI and SWP improved after using the automatic coregistration algorithm. In particular, the coefficient of determination (R^2) increased from 0.64 to 0.77. In addition, the values of RMSE and SE decreased from 0.2 to 0.1 MPa and 0.24 to 0.16 MPa, respectively.

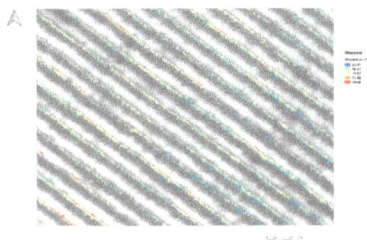

Figure 8. Final resulting thermal image of the drip-irrigated vineyard after automatic coregistration with the 490-nm image and filtered using the proposed shadow removal algorithm.

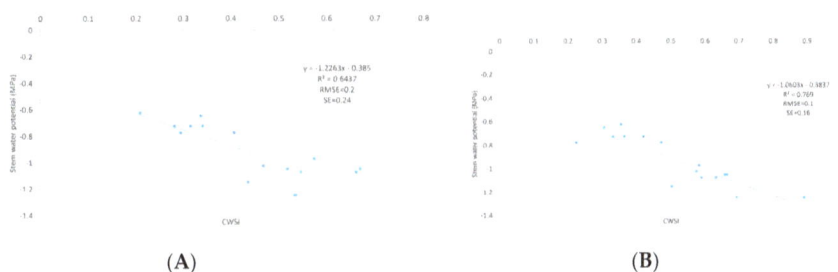

(A)   (B)

Figure 9. Relationships between CWSI and SWP for the vineyard: (**A**) center of the row temperature; and (**B**) temperature after coregistration with the 490-nm image and application of the proposed shadow removal algorithm.

4. Discussion

The selection of B1 (490 nm) as the better multispectral band for classifying shadow canopy pixels was consistent with the previous study by Ünsalan et al. [76], who used the k-means and blue information derived from the RGB spectrum to segment information avoiding shadow pixels to extract street networks and detect houses. This band selection was also proposed by Sirmacek et al. [77], who used the blue wavelength spectrum to detect shadows for building detection, suggesting that this region was dominantly better even compared with green and red for shadow pixels identification [78,79]. The selection of 490 nm image was also preferred when compared with upper wavelengths, in which blue spectrum showed better results for shadow detection increasing the performance for near infrared and shortwave infrared [80]. This validates the previous assumption that internal canopy shadow cannot be identified by TIR imagery. Considering this, the importance of coregister thermal and visible images for detecting shadow pixels was also highlighted by Leinonen et al. [42] who using ground cameras with a non-automatic methodology concluded that one of the principal steps is to correct overlapping VIS and TIR images to assess vine water status.

In the present study, the SWP values between the stressed and well-watered vines [81] can be easily identified. The relationship between the CWSI and SWP improved when the shadow pixels were removed from the vine canopy using the suggested automatic algorithms. For the vineyard, the fractional cover was 19%, while the percentage of canopy shaded pixels was 43%. This indicates that only 8.2% of final vegetation pixels were used to develop the relationship between CWSI and SWP. Although the relationship between CWSI and SWP improved, the impact of the shadow was significant in those vines with more water stress [51]. In contrast, because no reduction of the transpiration rate occurred in well-watered vines [82], the difference between leaf temperature and air temperature was not representative [83]. These results are consistent with those of Van Zyl [49], who suggested that the impact of shadow for SWP relationships in stressed vines was considerably higher compared with sunlit leaves. In addition, Pou et al. [50] suggested that shaded canopy information negatively affects the relationship of the vine water status and CSWI because the leaf temperature decreases. Furthermore, Jones et al. [82] suggested that a greater sensitivity with respect to leaf temperature with water status measurements might be better when sunlit canopy information is considered. The effect of shadow deletion on the relationship between CWSI and SWP for stressed and well-watered vines is shown in Figure 10. Considering the results of Figure 9, in stressed vines, the shadow deletion process significantly improved the CWSI-SWP relationship with values of R^2, increasing from 0.05 to 0.35. However, no differences were observed for well-watered vines.

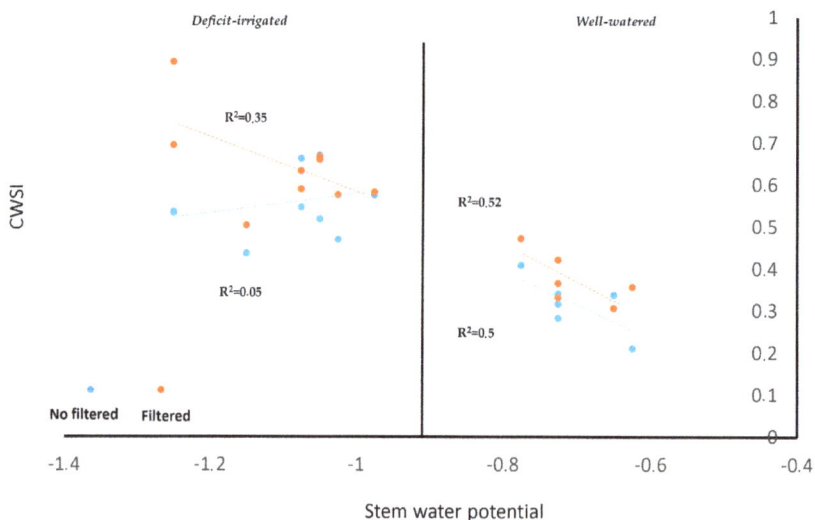

Figure 10. Comparison of the effect of shadow deletion on the CWSI-SWP relationship for a Cabernet Sauvignon vineyard.

5. Conclusions

Using a modified SIFT computer vision algorithm and Kmeans++ clustering, we performed automatic coregister UAV-TIR and UAV-VIS imagery to detect canopy shadow pixels information in thermal images. The deletion of the canopy shadow information in TIR images positively affects the relationship between the CWSI and SWP, showing an increment in R^2 from 0.64 to 0.77. In addition, the relationship showed a decrease in RMSE from 0.2 to 0.1 MPa and in SE from 0.24 to 0.16 MPa. As future work, our methodology should be applied for validation in different cultivars, seasons, and field conditions. In addition, the impact of automatic removal of shadow canopy pixels should be assessed for evapotranspiration modeling using UAV-TIR images of vineyards.

Acknowledgments: This study was supported by the Chilean government through the projects CONICYT-PFCHA (No. 2014-21140229) and FONDECYT (No. 1160997) and by the Universidad de Talca through the research program Adaptation of Agriculture to Climate Change (A2C2). Finally, the authors also express their gratitude to Luis Ahumada, Fernando Fuentes, and Camilo Riveros for their active collaboration on the data collection and field expertise during the flight campaigns.

Author Contributions: All authors conceived and designed this research and data analysis. They prepared the manuscript and consecutive revisions with contribution from all of them. Tomas Poblete contributed in the image processing and algorithms procedures implementation. Dongryeol Ryu also contributed in the analysis and discussion of the remote sensed data.

Conflicts of Interest: The authors declare no conflict of interest.

References

1. Bates, B.; Kundzewicz, Z.W.; Wu, S.; Palutikof, J. *Climate Change and Water: Technical Paper Vi*; Intergovernmental Panel on Climate Change (IPCC): Geneva, Switzerland, 2008.
2. Chaves, M.M.; Santos, T.P.; Souza, C.R.D.; Ortuño, M.; Rodrigues, M.; Lopes, C.; Maroco, J.; Pereira, J.S. Deficit irrigation in grapevine improves water-use efficiency while controlling vigour and production quality. *Ann. Appl. Biol.* **2007**, *150*, 237–252. [CrossRef]
3. Chapman, D.M.; Roby, G.; Ebeler, S.E.; Guinard, J.X.; Matthews, M.A. Sensory attributes of cabernet sauvignon wines made from vines with different water status. *Aust. J. Grape Wine Res.* **2005**, *11*, 339–347. [CrossRef]
4. Berger, T.; Birner, R.; Mccarthy, N.; DíAz, J.; Wittmer, H. Capturing the complexity of water uses and water users within a multi-agent framework. *Water Resour. Manag.* **2007**, *21*, 129–148. [CrossRef]
5. Granier, C.; Aguirrezabal, L.; Chenu, K.; Cookson, S.J.; Dauzat, M.; Hamard, P.; Thioux, J.J.; Rolland, G.; Bouchier-Combaud, S.; Lebaudy, A. Phenopsis, an automated platform for reproducible phenotyping of plant responses to soil water deficit in arabidopsis thaliana permitted the identification of an accession with low sensitivity to soil water deficit. *New Phytol.* **2006**, *169*, 623–635. [CrossRef] [PubMed]
6. Choné, X.; Van Leeuwen, C.; Dubourdieu, D.; Gaudillère, J.P. Stem water potential is a sensitive indicator of grapevine water status. *Ann. Bot.* **2001**, *87*, 477–483. [CrossRef]
7. Romero, P.; García, J.G.; Fernández-Fernández, J.I.; Muñoz, R.G.; del Amor Saavedra, F.; Martínez-Cutillas, A. Improving berry and wine quality attributes and vineyard economic efficiency by long-term deficit irrigation practices under semiarid conditions. *Sci. Hortic.* **2016**, *203*, 69–85. [CrossRef]
8. Balint, G.; Reynolds, A.G. Irrigation level and time of imposition impact vine physiology, yield components, fruit composition and wine quality of ontario chardonnay. *Sci. Hortic.* **2017**, *214*, 252–272. [CrossRef]
9. Nortes, P.; Pérez-Pastor, A.; Egea, G.; Conejero, W.; Domingo, R. Comparison of changes in stem diameter and water potential values for detecting water stress in young almond trees. *Agric. Water Manag.* **2005**, *77*, 296–307. [CrossRef]
10. Espadafor, M.; Orgaz, F.; Testi, L.; Lorite, I.J.; González-Dugo, V.; Fereres, E. Responses of transpiration and transpiration efficiency of almond trees to moderate water deficits. *Sci. Hortic.* **2017**, *225*, 6–14. [CrossRef]
11. Moriana, A.; Pérez-López, D.; Prieto, M.; Ramírez-Santa-Pau, M.; Pérez-Rodriguez, J. Midday stem water potential as a useful tool for estimating irrigation requirements in olive trees. *Agric. Water Manag.* **2012**, *112*, 43–54. [CrossRef]
12. Ahumada-Orellana, L.E.; Ortega-Farías, S.; Searles, P.S.; Retamales, J.B. Yield and water productivity responses to irrigation cut-off strategies after fruit set using stem water potential thresholds in a super-high density olive orchard. *Front. Plant Sci.* **2017**, *8*, 1280. [CrossRef] [PubMed]
13. Acevedo-Opazo, C.; Tisseyre, B.; Guillaume, S.; Ojeda, H. The potential of high spatial resolution information to define within-vineyard zones related to vine water status. *Precis. Agric.* **2008**, *9*, 285–302. [CrossRef]
14. Baluja, J.; Diago, M.P.; Balda, P.; Zorer, R.; Meggio, F.; Morales, F.; Tardaguila, J. Assessment of vineyard water status variability by thermal and multispectral imagery using an unmanned aerial vehicle (uav). *Irrig. Sci.* **2012**, *30*, 511–522. [CrossRef]
15. Vadivambal, R.; Jayas, D.S. Applications of thermal imaging in agriculture and food industry—A review. *Food Bioprocess Technol.* **2011**, *4*, 186–199. [CrossRef]

16. Rapaport, T.; Hochberg, U.; Shoshany, M.; Karnieli, A.; Rachmilevitch, S. Combining leaf physiology, hyperspectral imaging and partial least squares-regression (pls-r) for grapevine water status assessment. *ISPRS J. Photogramm. Remote Sens.* **2015**, *109*, 88–97. [CrossRef]

17. Rallo, G.; Minacapilli, M.; Ciraolo, G.; Provenzano, G. Detecting crop water status in mature olive groves using vegetation spectral measurements. *Biosyst. Eng.* **2014**, *128*, 52–68. [CrossRef]

18. Pôças, I.; Rodrigues, A.; Gonçalves, S.; Costa, P.M.; Gonçalves, I.; Pereira, L.S.; Cunha, M. Predicting grapevine water status based on hyperspectral reflectance vegetation indices. *Remote Sens.* **2015**, *7*, 16460–16479. [CrossRef]

19. Pôças, I.; Gonçalves, J.; Costa, P.M.; Gonçalves, I.; Pereira, L.S.; Cunha, M. Hyperspectral-based predictive modelling of grapevine water status in the portuguese douro wine region. *Int. J. Appl. Earth Observ. Geoinf.* **2017**, *58*, 177–190. [CrossRef]

20. Poblete, T.; Ortega-Farías, S.; Moreno, M.A.; Bardeen, M. Artificial neural network to predict vine water status spatial variability using multispectral information obtained from an unmanned aerial vehicle (uav). *Sensors* **2017**, *17*, 2488. [CrossRef] [PubMed]

21. King, B.; Shellie, K. Evaluation of neural network modeling to predict non-water-stressed leaf temperature in wine grape for calculation of crop water stress index. *Agric. Water Manag.* **2016**, *167*, 38–52. [CrossRef]

22. Gade, R.; Moeslund, T.B. Thermal cameras and applications: A survey. *Mach. Vis. Appl.* **2014**, *25*, 245–262. [CrossRef]

23. Möller, M.; Alchanatis, V.; Cohen, Y.; Meron, M.; Tsipris, J.; Naor, A.; Ostrovsky, V.; Sprintsin, M.; Cohen, S. Use of thermal and visible imagery for estimating crop water status of irrigated grapevine. *J. Exp. Bot.* **2006**, *58*, 827–838. [CrossRef] [PubMed]

24. DeJonge, K.C.; Taghvaeian, S.; Trout, T.J.; Comas, L.H. Comparison of canopy temperature-based water stress indices for maize. *Agric. Water Manag.* **2015**, *156*, 51–62. [CrossRef]

25. Sepúlveda-Reyes, D.; Ingram, B.; Bardeen, M.; Zúñiga, M.; Ortega-Farías, S.; Poblete-Echeverría, C. Selecting canopy zones and thresholding approaches to assess grapevine water status by using aerial and ground-based thermal imaging. *Remote Sens.* **2016**, *8*, 822. [CrossRef]

26. Zhang, C.; Kovacs, J.M. The application of small unmanned aerial systems for precision agriculture: A review. *Precis. Agric.* **2012**, *13*, 693–712. [CrossRef]

27. Ortega-Farías, S.; Ortega-Salazar, S.; Poblete, T.; Kilic, A.; Allen, R.; Poblete-Echeverría, C.; Ahumada-Orellana, L.; Zuñiga, M.; Sepúlveda, D. Estimation of energy balance components over a drip-irrigated olive orchard using thermal and multispectral cameras placed on a helicopter-based unmanned aerial vehicle (uav). *Remote Sens.* **2016**, *8*, 638. [CrossRef]

28. López-Granados, F.; Torres-Sánchez, J.; Serrano-Pérez, A.; de Castro, A.I.; Mesas-Carrascosa, F.-J.; Peña, J.-M. Early season weed mapping in sunflower using uav technology: Variability of herbicide treatment maps against weed thresholds. *Precis. Agric.* **2016**, *17*, 183–199. [CrossRef]

29. Colomina, I.; Molina, P. Unmanned aerial systems for photogrammetry and remote sensing: A review. *ISPRS J. Photogramm. Remote Sens.* **2014**, *92*, 79–97. [CrossRef]

30. Park, S.; Ryu, D.; Fuentes, S.; Chung, H.; Hernández-Montes, E.; O'Connell, M. Adaptive estimation of crop water stress in nectarine and peach orchards using high-resolution imagery from an unmanned aerial vehicle (uav). *Remote Sens.* **2017**, *9*, 828. [CrossRef]

31. Santesteban, L.; Di Gennaro, S.; Herrero-Langreo, A.; Miranda, C.; Royo, J.; Matese, A. High-resolution uav-based thermal imaging to estimate the instantaneous and seasonal variability of plant water status within a vineyard. *Agric. Water Manag.* **2017**, *183*, 49–59. [CrossRef]

32. Bellvert, J.; Zarco-Tejada, P.J.; Girona, J.; Fereres, E. Mapping crop water stress index in a 'pinot-noir' vineyard: Comparing ground measurements with thermal remote sensing imagery from an unmanned aerial vehicle. *Precis. Agric.* **2014**, *15*, 361–376. [CrossRef]

33. Bellvert, J.; Marsal, J.; Girona, J.; Zarco-Tejada, P.J. Seasonal evolution of crop water stress index in grapevine varieties determined with high-resolution remote sensing thermal imagery. *Irrig. Sci.* **2015**, *33*, 81–93. [CrossRef]

34. Shahtahmassebi, A.; Yang, N.; Wang, K.; Moore, N.; Shen, Z. Review of shadow detection and de-shadowing methods in remote sensing. *Chin. Geogr. Sci.* **2013**, *23*, 403–420. [CrossRef]

35. Liu, W.; Yamazaki, F. Object-based shadow extraction and correction of high-resolution optical satellite images. *IEEE J. Sel. Top. Appl. Earth Observ. Remote Sens.* **2012**, *5*, 1296–1302. [CrossRef]

36. Miura, H.; Midorikawa, S.; Fujimoto, K. Automated building detection from high-resolution satellite image for updating gis building inventory data. In Proceedings of the 13th World Conference on Earthquake Engineering, Vancouver, BC, Canada, 1–6 August 2004.

37. Song, M.; Civco, D.L. A Knowledge-Based Approach for Reducing Cloud and Shadow. In Proceedings of the 2002 ASPRS-ACSM Annual Conference and FIG XXII Congress, Washington, DC, USA, 19–26 April 2002; pp. 22–26.

38. Heiskanen, J.; Kajuutti, K.; Jackson, M.; Elvehøy, H.; Pellikka, P. Assessment of glaciological parameters using landsat sat-ellite data in svartisen, northern norway. In Proceedings of the EARSeL-LISSIG-Workshop Observing Our Cryosphere from Space, Bern, Switzerland, 11–13 March 2002.

39. Hendriks, J.; Pellikka, P. Estimation of reflectance from a glacier surface by comparing spectrometer measurements with satellite-derived reflectances. *J. Glaciol.* **2004**, *38*, 139–154.

40. Cai, D.; Li, M.; Bao, Z.; Chen, Z.; Wei, W.; Zhang, H. In Study on shadow detection method on high resolution remote sensing image based on his space transformation and ndvi index. In Proceedings of the 18th International Conference on Geoinformatics, Beijing, China, 18–20 June 2010; pp. 1–4.

41. Sotomayor, A.I.T. *A Spatial Analysis of Different Forest Cover Types Using Gis and Remote Sensing Techniques*; Innovation and Technology Commission (ITC): Geneva, Switzerland, 2002.

42. Leinonen, I.; Jones, H.G. Combining thermal and visible imagery for estimating canopy temperature and identifying plant stress. *J. Exp. Bot.* **2004**, *55*, 1423–1431. [CrossRef] [PubMed]

43. Zarco-Tejada, P.J.; González-Dugo, V.; Berni, J.A. Fluorescence, temperature and narrow-band indices acquired from a uav platform for water stress detection using a micro-hyperspectral imager and a thermal camera. *Remote Sens. Environ.* **2012**, *117*, 322–337. [CrossRef]

44. Suárez, L.; Zarco-Tejada, P.J.; González-Dugo, V.; Berni, J.; Sagardoy, R.; Morales, F.; Fereres, E. Detecting water stress effects on fruit quality in orchards with time-series pri airborne imagery. *Remote Sens. Environ.* **2010**, *114*, 286–298. [CrossRef]

45. Zarco-Tejada, P.J.; González-Dugo, V.; Williams, L.; Suárez, L.; Berni, J.A.; Goldhamer, D.; Fereres, E. A pri-based water stress index combining structural and chlorophyll effects: Assessment using diurnal narrow-band airborne imagery and the cwsi thermal index. *Remote Sens. Environ.* **2013**, *138*, 38–50. [CrossRef]

46. Gonzalez-Dugo, V.; Zarco-Tejada, P.; Nicolás, E.; Nortes, P.; Alarcón, J.; Intrigliolo, D.; Fereres, E. Using high resolution uav thermal imagery to assess the variability in the water status of five fruit tree species within a commercial orchard. *Precis. Agric.* **2013**, *14*, 660–678. [CrossRef]

47. Otsu, N. A threshold selection method from gray-level histograms. *Automatica* **1975**, *11*, 23–27. [CrossRef]

48. Fraser, R.H.; Olthof, I.; Lantz, T.C.; Schmitt, C. Uav photogrammetry for mapping vegetation in the low-arctic. *Arct. Sci.* **2016**, *2*, 79–102. [CrossRef]

49. Van Zyl, J. Diurnal variation in grapevine water stress as a function of changing soil water status and meteorological conditions. *S. Afr. J. Enol. Vitic.* **2017**, *8*, 45–52. [CrossRef]

50. Pou, A.; Diago, M.P.; Medrano, H.; Baluja, J.; Tardaguila, J. Validation of thermal indices for water status identification in grapevine. *Agric. Water Manag.* **2014**, *134*, 60–72. [CrossRef]

51. Grant, O.M.; Chaves, M.M.; Jones, H.G. Optimizing thermal imaging as a technique for detecting stomatal closure induced by drought stress under greenhouse conditions. *Physiol. Plant.* **2006**, *127*, 507–518. [CrossRef]

52. Smith, H.K.; Clarkson, G.J.; Taylor, G.; Thompson, A.J.; Clarkson, J.; Rajpoot, N.M. Automatic detection of regions in spinach canopies responding to soil moisture deficit using combined visible and thermal imagery. *PLoS ONE* **2014**, *9*, e97612.

53. Bulanon, D.; Burks, T.; Alchanatis, V. Image fusion of visible and thermal images for fruit detection. *Biosyst. Eng.* **2009**, *103*, 12–22. [CrossRef]

54. Li, S.; Kang, X.; Fang, L.; Hu, J.; Yin, H. Pixel-level image fusion: A survey of the state of the art. *Infor. Fusion* **2017**, *33*, 100–112. [CrossRef]

55. Morsdorf, F.; Kötz, B.; Meier, E.; Itten, K.; Allgöwer, B. Estimation of lai and fractional cover from small footprint airborne laser scanning data based on gap fraction. *Remote Sens. Environ.* **2006**, *104*, 50–61. [CrossRef]

56. Moriana, A.; Fereres, E. Plant indicators for scheduling irrigation of young olive trees. *Irrig. Sci.* **2002**, *21*, 83–90.

57. Ribeiro-Gomes, K.; Hernández-López, D.; Ortega, J.F.; Ballesteros, R.; Poblete, T.; Moreno, M.A. Uncooled thermal camera calibration and optimization of the photogrammetry process for uav applications in agriculture. *Sensors* **2017**, *17*, 2173. [CrossRef] [PubMed]
58. Laliberte, A.S.; Rango, A. Texture and scale in object-based analysis of subdecimeter resolution unmanned aerial vehicle (uav) imagery. *IEEE Trans. Geosci. Remote Sens.* **2009**, *47*, 761–770. [CrossRef]
59. Ghosh, D.; Kaabouch, N. A survey on image mosaicing techniques. *J. Vis. Commun. Image Represent.* **2016**, *34*, 1–11. [CrossRef]
60. Jones, H.G. *Plants and Microclimate: A Quantitative Approach to Environmental Plant Physiology*; Cambridge University Press: Cambridge, UK, 2013.
61. Lowe, D.G. Object recognition from local scale-invariant features. In Proceedings of the Seventh IEEE International Conference on Computer Vision, Kerkyra, Greece, 20–27 September 1999; pp. 1150–1157.
62. Fischler, M.A.; Bolles, R.C. Random sample consensus: A paradigm for model fitting with applications to image analysis and automated cartography. *Commun. ACM* **1981**, *24*, 381–395. [CrossRef]
63. Raguram, R.; Frahm, J.-M.; Pollefeys, M. A comparative analysis of ransac techniques leading to adaptive real-time random sample consensus. In *Computer Vision–ECCV 2008*; Springer: Berlin/Heidelberg, Germany, 2008; pp. 500–513.
64. Derpanis, K.G. Overview of the ransac algorithm. *Image Rochester N. Y.* **2010**, *4*, 2–3.
65. Vourvoulakis, J.; Kalomiros, J.; Lygouras, J. Fpga-based architecture of a real-time sift matcher and ransac algorithm for robotic vision applications. *Multimed. Tools Appl.* **2017**, 1–23. [CrossRef]
66. Michaelsen, E.; von Hansen, W.; Kirchhof, M.; Meidow, J.; Stilla, U. Estimating the essential matrix: Goodsac versus ransac. In Proceedings of the ISPRS Symposium on Photogrammetric Computer Vision, Bonn, Germany, 20–22 September 2006.
67. Meler, A.; Decrouez, M.; Crowley, J.L. Betasac: A new conditional sampling for ransac. In Proceedings of the British Machine Vision Conference, Aberystwyth, UK, 31 August–3 September 2010.
68. Bush, F.N.; Esposito, J.M. Vision-based lane detection for an autonomous ground vehicle: A comparative field test. In Proceedings of the 2010 42nd Southeastern Symposium on System Theory (SSST), Tyler, TX, USA, 7–9 March 2010; pp. 35–39.
69. Bazin, J.-C.; Seo, Y.; Pollefeys, M. Globally optimal consensus set maximization through rotation search. In *Asian Conference on Computer Vision*; Springer: Berlin/Heidelberg, Germany, 2012; pp. 539–551.
70. Ramos, F.; Kadous, M.W.; Fox, D. In Learning to associate image features with crf-matching. In *Experimental Robotics*; Springer: Berlin/Heidelberg, Germany, 2009; pp. 505–514.
71. Kong, S.G.; Heo, J.; Boughorbel, F.; Zheng, Y.; Abidi, B.R.; Koschan, A.; Yi, M.; Abidi, M.A. Multiscale fusion of visible and thermal ir images for illumination-invariant face recognition. *Int. J. Comput. Vis.* **2007**, *71*, 215–233. [CrossRef]
72. Turner, D.; Lucieer, A.; Malenovský, Z.; King, D.H.; Robinson, S.A. Spatial co-registration of ultra-high resolution visible, multispectral and thermal images acquired with a micro-uav over antarctic moss beds. *Remote Sens.* **2014**, *6*, 4003–4024. [CrossRef]
73. Vedaldi, A.; Fulkerson, B. Vlfeat: An open and portable library of computer vision algorithms. In Proceedings of the 18th ACM International Conference on Multimedia, Firenze, Italy, 25–29 October 2010; ACM: New York, NY, USA, 2010; pp. 1469–1472.
74. Arthur, D.; Vassilvitskii, S. K-means++: The advantages of careful seeding. In Proceedings of the Eighteenth Annual ACM-SIAM Symposium on Discrete Algorithms, New Orleans, LA, USA, 7–9 January 2007; Society for Industrial and Applied Mathematics: Philadelphia, PA, USA, 2007; pp. 1027–1035.
75. Byrt, T.; Bishop, J.; Carlin, J.B. Bias, prevalence and kappa. *J. Clin. Epidemiol.* **1993**, *46*, 423–429. [CrossRef]
76. Ünsalan, C.; Boyer, K.L. A system to detect houses and residential street networks in multispectral satellite images. *Comput. Vis. Image Underst.* **2005**, *98*, 423–461. [CrossRef]
77. Sirmacek, B.; Unsalan, C. Building detection from aerial images using invariant color features and shadow information. In Proceedings of the 23rd International Symposium on Computer and Information Sciences, Istanbul, Turkey, 27–29 October 2008; pp. 1–5.
78. Teke, M.; Başeski, E.; Ok, A.; Yüksel, B.; Şenaras, Ç. Multi-spectral false color shadow detection. In *Photogrammetric Image Analysis, Proceedings of the ISPRS Conference, PIA 2011 Munich, Germany, 5–7 October 2011*; Springer: Berlin/Heidelberg, Germany, 2011; pp. 109–119.

Sensors **2018**, *18*, 397

79. Zhu, Z.; Woodcock, C.E. Object-based cloud and cloud shadow detection in landsat imagery. *Remote Sens. Environ.* **2012**, *118*, 83–94. [CrossRef]
80. Luo, Y.; Trishchenko, A.P.; Khlopenkov, K.V. Developing clear-sky, cloud and cloud shadow mask for producing clear-sky composites at 250-meter spatial resolution for the seven modis land bands over canada and north america. *Remote Sens. Environ.* **2008**, *112*, 4167–4185. [CrossRef]
81. Acevedo-Opazo, C.; Tisseyre, B.; Ojeda, H.; Ortega-Farias, S.; Guillaume, S. Is it possible to assess the spatial variability of vine water status? *OENO ONE* **2008**, *42*, 203–219. [CrossRef]
82. Jones, H.G.; Stoll, M.; Santos, T.; Sousa, C.d.; Chaves, M.M.; Grant, O.M. Use of infrared thermography for monitoring stomatal closure in the field: Application to grapevine. *J. Exp. Bot.* **2002**, *53*, 2249–2260. [CrossRef] [PubMed]
83. Idso, S.B. Non-water-stressed baselines: A key to measuring and interpreting plant water stress. *Agric. Meteorol.* **1982**, *27*, 59–70. [CrossRef]

sensors

MDPI

Article

Rapid 3D Reconstruction for Image Sequence Acquired from UAV Camera

Yufu Qu *, Jianyu Huang and Xuan Zhang

Department of Measurement Technology & Instrument, School of Instrumentation Science & Optoelectronics
Engineering, Beihang University, Beijing 100191, China; Hjy448@buaa.edu.cn (J.H.);
zhangxuanaj@buaa.edu.cn (X.Z.)
* Correspondence: qyf@buaa.eud.cn; Tel.: +86-010-8231-7336

Received: 23 November 2017; Accepted: 11 January 2018; Published: 14 January 2018

Abstract: In order to reconstruct three-dimensional (3D) structures from an image sequence captured by unmanned aerial vehicles' camera (UAVs) and improve the processing speed, we propose a rapid 3D reconstruction method that is based on an image queue, considering the continuity and relevance of UAV camera images. The proposed approach first compresses the feature points of each image into three principal component points by using the principal component analysis method. In order to select the key images suitable for 3D reconstruction, the principal component points are used to estimate the interrelationships between images. Second, these key images are inserted into a fixed-length image queue. The positions and orientations of the images are calculated, and the 3D coordinates of the feature points are estimated using weighted bundle adjustment. With this structural information, the depth maps of these images can be calculated. Next, we update the image queue by deleting some of the old images and inserting some new images into the queue, and a structural calculation of all the images can be performed by repeating the previous steps. Finally, a dense 3D point cloud can be obtained using the depth–map fusion method. The experimental results indicate that when the texture of the images is complex and the number of images exceeds 100, the proposed method can improve the calculation speed by more than a factor of four with almost no loss of precision. Furthermore, as the number of images increases, the improvement in the calculation speed will become more noticeable.

Keywords: UAV camera; multi-view stereo; structure from motion; 3D reconstruction; point cloud

---

1. Introduction

Because of the rapid development of the unmanned aerial vehicle (UAV) industry in recent years, civil UAVs have been used in agriculture, energy, environment, public safety, infrastructure, and other fields. By carrying a digital camera on a UAV, two-dimensional (2D) images can be obtained. However, as the requirements have grown and matured, 2D images have not been able to meet the requirements of many applications such as three-dimensional (3D) terrain and scene understanding. Thus, there is an urgent need to reconstruct 3D structures from the 2D images collected from UAV camera. The study of the methods in which 3D structures are generated by 2D images is an important branch of computer vision. In this field, many researchers have proposed several methods and theories [1–17]. Among these theories and methods, the three most important categories are the simultaneous localization and mapping (SLAM) [1–3], structure from motion (SfM) [4–14] and multiple view stereo (MVS) algorithms [15–17], which have been implemented in many practical applications. As the number of images and their resolution increase, the computational times of the algorithms will increase significantly, limiting them in some high-speed reconstruction applications.

Two major contributions in this paper are methods of selecting key images selection and SfM calculation of sequence images. Key images selection is very important to the success of 3D reconstruction. In this paper, a fully automatic approach to key frames extraction without initial pose information is proposed. Principal Component Analysis (PCA) is used to analyze the correlation of features over frames to automate the key frame selection. Considering the continuity of the images taken by UAV camera, this paper proposes a 3D reconstruction method based on an image queue. To ensure the smooth of two consecutive point cloud, an improved bundle-adjustment named weighted bundle-adjustment is used in this paper. After using a fixed-size image queue, the global structure calculation is divided into several local structure calculations, thus improving the speed of the algorithm with almost no loss of accuracy.

2. Literature Review

The general 3D reconstruction algorithm without a priori positions and orientation information can be roughly divided into two steps. The first step involves recovering the 3D structure of the scene and the camera motion from the images. The problem addressed in this step is generally referred to as the SfM problem. The second step involves obtaining the 3D topography of the scene captured by the images. This step is usually completed by generating a dense point data cloud or mesh data cloud from multiple images. The problem addressed in this step is generally referred to as the MVS problem. In addition, the research into Real-time simultaneous localization and mapping (SLAM) and 3D reconstruction of the environment have become popular over the past few years. Positions and orientations of monocular camera and sparse point map can be obtained from the images by using SLAM algorithm.

2.1. SfM

The SfM algorithm is used to obtain the structure of the 3D scene and the camera motion from the images of stationary objects. There are many similarities between SLAM and SfM. They both estimate the localizations and orientations of camera and sparse features. Nonlinear optimization is widely used in SLAM and SfM algorithms. Researchers have proposed improved algorithms for different situations based on early SfM algorithms [4–6]. A variety of SfM strategies have emerged, including incremental [7,8], hierarchical [9], and global [10–12] approaches. Among these methods, a very typical one was proposed by Snavely [13], who used it in the 3D reconstruction of real-world objects. With the help of feature point matching, bundle adjustment, and other technologies, Snavely completed the 3D reconstruction of objects by using images of famous landmarks and cities. The SfM algorithm is limited in many applications because of the time-consuming calculation. With the continuous development of computer hardware, multicore technologies, and GPU technologies, the SfM algorithm can now be used in several areas. In many applications, the SfM algorithm has higher requirements for the computing speed and accuracy. There are several improved SfM methods such as the method proposed by Wu [8,14]. These methods can improve the speed of the structure calculation without loss of accuracy. Among the incremental SfM, hierarchical SfM, and global SfM, the incremental SfM is the most popular strategy for the reconstruction of unordered images. Two important steps in incremental SfM are the feature point matching between images, and bundle adjustment. As the resolution and number of images increase, the number of matching points and parameters optimized by bundle adjustment will increase dramatically. This results in a significant increase in the computational complexity of the algorithm and will make it difficult to use it in many applications.

2.2. MVS

When the positions and orientations of the cameras are known, the MVS algorithm can reconstruct the 3D structure of a scene by using multiple-view images. One of the most representative methods was proposed by Furukawa [15]. This method estimates the 3D coordinates of the initial points by matching the difference of Gaussians and Harris corner points between different images, followed by patch expansion, point filtering, and other processing. The patch-based matching method is used to match other pixels between images. After that, a dense point data cloud and mesh data cloud can be obtained. Inspired by Furukawa's method, some researchers have proposed several 3D reconstruction algorithms [16–18] based on depth-map fusion. These algorithms can obtain reconstruction results with an even higher density and accuracy. The method proposed by Shen [16] is one of the most representative approaches. The important difference between this method and Furukawa's method is that it uses the position and orientation information of the cameras as well as the coordinates of the sparse feature points generated from the structure calculation. The estimated depth maps are obtained from the mesh data generated by the sparse feature points. Then, after depth–map refinement and depth–map fusion, a dense 3D point data cloud can be obtained. An implementation of this method can be found in the open-source software openMVS [16].

Furukawa's approach relies heavily on the texture of the images. When processing weakly textured images, it is difficult for this method to generate a dense point cloud. In addition, the algorithm must repeat the patch expansion and point cloud filtering several times, resulting in a significant increase in the calculation time. Compared to Furukawa's approach, Shen's method directly generates a dense point cloud using depth-map fusion. This method can easily and rapidly obtain a dense point cloud. Considering the characteristics of the problems that must be addressed in this study, we use a method similar to Shen's approach to generating a dense point data cloud.

2.3. SLAM

SLAM mainly consists in the simultaneous estimation of the localization of the robot and the map of the environment. The map obtained by SLAM is often required to support other tasks. The popularity of SLAM is connected with the need for indoor applications of mobile robotics. As the UAV industry rises, SLAM algorithms are widely used in UAV applications. Early SLAM approaches are based on Extended Kalman Filters, Rao-Blackwellised Particle Filters, and maximum likelihood estimation. Without priors, MAP estimation reduces to maximum-likelihood estimation. Most SLAM algorithms are based on iterative nonlinear optimization [1,2]. The biggest problem of SLAM is that some algorithms are easily converging to a local minimum. It usually returns a completely wrong estimate. Convex relaxation is proposed by some authors to avoid convergence to local minima. These contributions include the work of Liu et al. [3]. Kinds of improved SLAM algorithms have been proposed to adapt to different applications. Some of them are used for vision-based navigation and mapping.

3. Method

3.1. Algorithm Principles

The first step of our method involves building a fixed-length image queue, selecting the key images from the video image sequence, and inserting them into the image queue until full. A structural calculation is then performed for the images of the queue. Next, the image queue is updated, several images are deleted from the front of the queue, and the same number of images is placed at the end of the queue. The structural calculation of the images in the queue is then repeated until all images are processed. On an independent thread, the depth maps of the images are calculated and saved in the depth-map set. Finally, all depth maps are fused to generate dense 3D point cloud data. Without the use of ground control points, the result of our method lost the accurate scale of the model. The algorithm flowchart is outlined in Figure 1.

Start   image sequence

Build image queue   Select key images

‹–––––––––Add m images into queue–––––––––

image queue
SFM

Computed
depth map

Discard k images
at front of the
queue

       Add k
Update image queue images at
Add depth map into   the end of
       the queue

Depth
map
set

Depth-map
fusion

dense point cloud

end

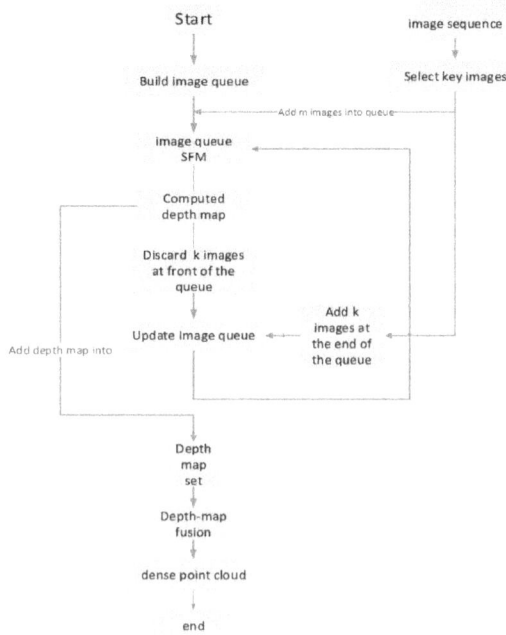

Figure 1. Algorithm flowchart.

3.2. Selecting Key Images

In order to complete the dense reconstruction of the point cloud and improve the computational speed, the key images (which are suitable for the structural calculation) must first be selected from a large number of UAV video images captured by a camera. The selected key images should have a good overlap of area for the captured scenes. For two consecutive key images, they must meet the key image constraint (denoted as $R\,(I_1, I_2)$) if they have a sufficient overlap area. In this study, we propose a method for directly selecting key images for reconstructing the UAV camera's images (the GPS equipped on the UAV can only reach an accuracy on the order of meters; by using GPS information as a reference for the selection of key images, discontinuous images will form). The overlap area between images can be estimated by the correspondence between the feature points of the images. In order to reduce the computational complexity of feature point matching, we propose a method of compressing the feature points based on principal component analysis (PCA). It is assumed that the images used for reconstruction are rich in texture. Three principal component points (PCPs) can be generated from PCA, each reflecting the distribution of the feature points in different images. If the two images are captured almost at the same position, the PCPs of them almost coincide in the same place. Otherwise, the PCPs will move and be located in different positions on the image. The process steps are as follows. First, we use the scale-invariant feature transform (SIFT) [19] feature detection algorithm to detect the feature points of each image (Figure 2a). There must be at least four feature points, and the centroid of these feature points can then be calculated as follows:

$$\bar{p} = \frac{1}{n}\sum_{i=1}^{n} P_i, \; P_i = \begin{pmatrix} x \\ y \end{pmatrix} \tag{1}$$

where P_i is the pixel coordinate of the feature point, and $\bar{p}$ is the centroid. The following matrix is formed by the image coordinates of the feature points:

$$A = \begin{bmatrix} \left(P_1 - \overline{P}\right)^T \\ \vdots \\ \left(P_n - \overline{P}\right)^T \end{bmatrix} \tag{2}$$

Then, the singular value decomposition (SVD) of matrix A yields two principal component vectors. The principal component points (PCPs) are obtained from these vectors (Equations (3) and (4)). To compress a large number of feature points into three PCPs (Figure 2b),

$$U \sum V^* = svd(A) \tag{3}$$

$$p_{m1} = \overline{P}, p_{m2} = \left(V_1 + \overline{P}\right), p_{m3} = \left(V_2 + \overline{P}\right) \tag{4}$$

where p_{m1}, p_{m2}, and p_{m3} are the three PCPs, and V_1 and V_2 are the two vectors of V^*. The PCPs can reflect the distribution of the feature points in the image. After that, by calculating the positional relationship of the corresponding PCPs between two consecutive images, we can estimate the overlap area between images. The average displacement (d_p) between PCPs, as expressed in Equation (5), can be calculated as follows: d_p reflects the relative displacement of feature points; when $d_p < D_l$, it is likely that the two images are almost captured at the same position; and when $d_p > D_h$, the overlap area of two images becomes too small. In this paper, we use 1/100 of the resolution as the value of D_l and 1/10 of the resolution as the value of D_h. When d_p is within the certain range given in Equation (6), the two images will meet the key image constraint $R(I_1, I_2)$:

$$d_p = \frac{1}{3} \sum_{i=1}^3 \left[\left(p_{1i} - p_{2i}\right)^T \times \left(p_{1i} - p_{2i}\right)\right]^{0.5} \tag{5}$$

$$R(I_1, I_2) : D_l < d_p < D_h \tag{6}$$

where p_{1i} is the *i*th PCP of the first image (I_1), and p_{2i} is that of the second image (I_2). The result is presented in Figure 2c. This is a method for estimating the overlap areas between images, and it is not necessary to calculate the actual correlation between the two images when selecting key images. Moreover, the algorithm is not time-consuming for either the calculation of the PCPs or the estimation of the distance between PCPs. Therefore, this method is suitable for quickly selecting key images from a UAV camera's video image sequence.

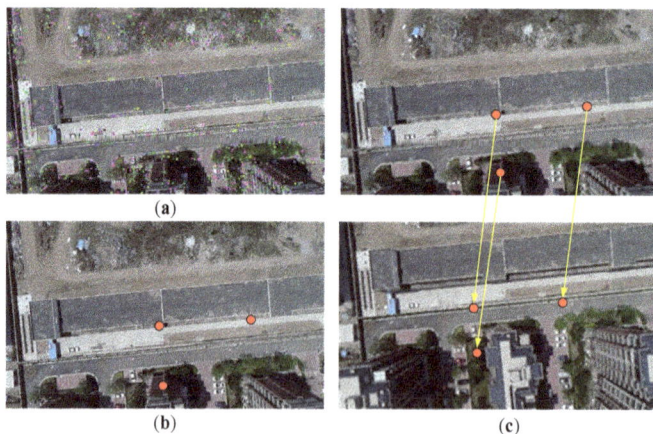

Figure 2. Feature point compression. (**a**) Detecting the feature points of an image; (**b**) calculating the principal component points (PCPs) of the feature points; and (**c**) matching the PCPs.

3.3. Image Queue SfM

This study focuses on the 3D reconstruction of UAV camera's images. Considering the continuity of UAV camera's images, we propose a SfM calculation method based on an image queue. This method constructs a fixed-size image queue and places key images into the queue until full. Then, the structure of the images in the queue is computed, and the queue is updated with new images. Eventually, we will complete the structural calculation of all images by repeating the structural computation and queue update. The image queue SfM includes two steps. The first involves the SfM calculation of the images in the queue. The second involves updating the images in the image queue.

3.3.1. SfM Calculation for the Images in the Queue

We propose the use of the incremental SfM algorithm. The process is illustrated in Figure 3. The collection of all images used for the reconstruction is first recorded as set C. The total number of images in C is assumed to be N. The size of the initial fixed queue is m (it is preferred that any two images in the queue have overlapping areas, and m can be modified according to the requirements of the calculation speed. When m is chosen as a smaller number, the speed increases, but the precision decreases correspondingly). In order to keep the stability of the algorithm, the value of m is generally taken greater than 5, and k is less than half of m. Then, m key images are inserted into the image queue. All of the images in the image queue are recorded as C_q, and the structure of all of the images in C_q is calculated.

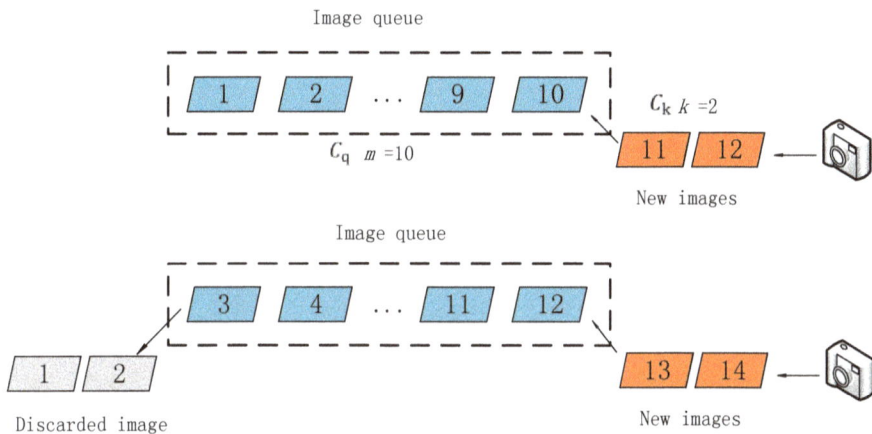

Figure 3. Structure from motion (SfM) calculation of the images in the queue.

Considering the accuracy and speed of the algorithm, the SfM approach used in this study uses an incremental SfM algorithm [7]. The steps of the algorithm are summarized below.

1. The SIFT [19] feature detection algorithm is used to detect the feature points on all images in the queue, and the correspondence of the feature points are then obtained by the feature point matching [20] between every two images in the queue.

2. Two images are selected from the queue as the initial image pair using the method proposed in [21]. The fundamental matrix of the two images is obtained by the random sample consensus (RANSAC) method [22], and the essential matrix between the two images is then calculated when the intrinsic matrix (obtained by the calibration method proposed in [23]) is known. The first two terms of radial and tangential distortion parameters are also obtained and used for image rectification. After remapping the pixels onto new locations on the image based on distortion model, the image

distortion caused by lens could be eliminated. Then, the positions and orientations of the images can be obtained by decomposing the essential matrix according to [24].

3. According to the correspondence of the feature points in different images, the 3D coordinates of the feature points are obtained by triangulation (the feature points are denoted as P_i ($i = 1,\ldots,t$)).

4. The parameters calculated in the previous steps are passed into the bundle adjustment [25] for nonlinear optimization [26].

5. The structure of the initial image pair is calculated, and one of the coordinate systems of the cameras taking the image pair is set as the global coordinate system. The image of the queue that has completed the structure calculation is placed into the set C_{SFM} ($C_{SFM} \subset C_q$).

6. The new image (I_{new}) is placed into the set (C_{SFM}), and the structural calculation is performed. The new image must meet the following two conditions. First, there should be at least one image in C_{SFM} that has common feature points with I_{new}. Second, at least six of these common feature points must be in P_i ($i = 1,\ldots,t$) (in order to improve the stability of the algorithm, this study requires at least 15 common feature points). Finally, all of the parameters from the structure calculation are optimized by bundle adjustment.

7. Repeat step 6 until the structure of all of the images inside the queue is calculated ($C_{SFM} = C_q$).

3.3.2. Updating the Image Queue

After the above steps, the structural calculation of all of the images in C_q can be performed. In order to improve the speed of the structural calculation of all of the images in C, this study proposes an improved SfM calculation method; the structural calculation of the images is processed in the form of an image queue. Figure 4 illustrates the process of the algorithm. We delete k images at the front of the queue, save their structural information, and then place k new images at the tail of the queue; these k images are then recorded as a set C_k. The ($m-k$) images left in the queue are recorded as a set $C_r(C_q = C_r \cup C_k)$, so now $C_{SFM} = C_r$. The structure of the images in C_r is known, and the structural information contains the coordinates of the 3D feature points (marked as P_r). The corresponding image pixels of P_r are marked as a set U_r, and the projection relationship is expressed as $P : P_r \rightarrow U_r$. Then, the pixels of the feature points (marked as U_k) of the images in C_k are detected, and the pixels in U_k and U_r are matched. We obtain the correspondence $M : U_{rC} \leftrightarrow U_{kc}$ ($U_{rc} \in U_r, U_{kc} \in U_k$), and U_{rC} and U_{kc} are the image pixels of the same object points (marked as P_c) in different images from C_r and C_k, respectively, expressed as $P : P_c \rightarrow U_{kc}, P_c \rightarrow U_{rc}$, where P_c is the control point. The projection matrix of the images in C_k can be estimated by the projection relationship between P_c and U_{kc}; then, the positions and orientations of the cameras can be calculated. In contrast, P_c can be used in the later weighted bundle adjustment to ensure the continuity of the structure. Then, we repeat step 6 until $C_{SFM} = C_q$. Finally, the structure of all of the images can be calculated by repeating the following two procedures alternately: calculate the SfM of the images in the queue and update the image queue.

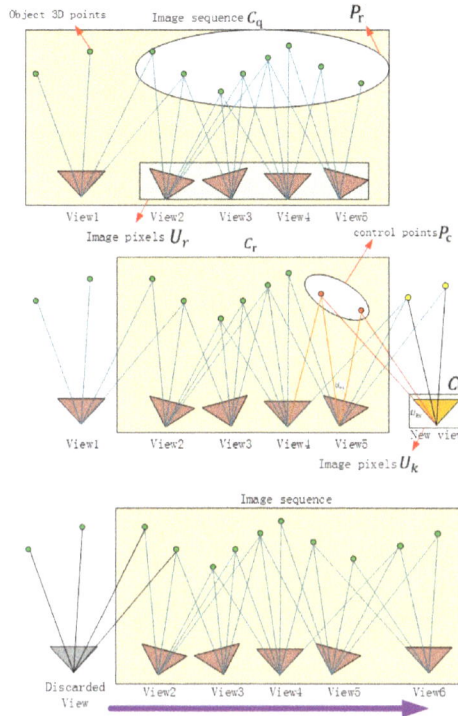

Figure 4. Updating the image queue ($m = 5, k = 1$).

3.3.3. Weighted Bundle Adjustment

An important part of the SfM algorithm is bundle adjustment. Our method divides a large number of images into small groups of images in the form of an image queue. When calculating the structure by the queue, optimization of the bundle adjustment causes the parameters to reach the subregion optimum rather than the global optimum. Small differences in the parameters between the subregions will result in discontinuous structures. This problem can be addressed by using control points, which are the points connecting two sets of adjacent feature points of the image, as shown in Figure 5. When we use bundle adjustment to optimize the parameters, we must keep the control points unchanged or with as little change as possible. This is achieved by weighting the error term of the control points. After the first update of the image queue, the formula for the projection error of the bundle adjustment used in step 6 will be altered.

For a single image, Equation (7) is the projection formula of the 3D point to the image pixel, and Equation (8) is the reprojection error formula:

$$\begin{pmatrix} v_i^f \\ u_i^f \end{pmatrix} = K[R,t] \begin{pmatrix} p_i \\ 1 \end{pmatrix} = f(R,T,P_i) \tag{7}$$

$$e_{projrct} = \sum_{i=1}^{n} \left\{ \left(\begin{pmatrix} v_i \\ u_i \end{pmatrix} - \begin{pmatrix} v_i^f \\ u_i^f \end{pmatrix} \right)^{T} \times \left(\begin{pmatrix} v_i \\ u_i \end{pmatrix} - \begin{pmatrix} v_i^f \\ u_i^f \end{pmatrix} \right) \right\} \tag{8}$$

$$e_{projrct} = \sum_{i=1}^{n} \left\{ \left(\begin{pmatrix} v_i \\ u_i \end{pmatrix} - \begin{pmatrix} v_i^f \\ u_i^f \end{pmatrix} \right)^{T} \times \left(\begin{pmatrix} v_i \\ u_i \end{pmatrix} - \begin{pmatrix} v_i^f \\ u_i^f \end{pmatrix} \right) \right\} + w_j \sum_{j=1}^{c} \left\{ \left(\begin{pmatrix} v_i \\ u_i \end{pmatrix} - \begin{pmatrix} v_i^f \\ u_i^f \end{pmatrix} \right)^{T} \times \left(\begin{pmatrix} v_i \\ u_i \end{pmatrix} - \begin{pmatrix} v_i^f \\ u_i^f \end{pmatrix} \right) \right\} \tag{9}$$

where K is the internal matrix of the camera, R and T are the external parameters, P_i is the 3D feature point, $\begin{pmatrix} v_i \\ u_i \end{pmatrix}$ is the actual pixel coordinate of the feature point, and $\begin{pmatrix} v_i^f \\ u_i^f \end{pmatrix}$ is the pixel coordinate calculated from the structural parameters. The number of control points is k. The calculation of the bundle adjustment is a nonlinear least-squares problem. The structural parameters $\left(R, T, P_{i(i=1,...,n)} \right)$ can be optimized by minimizing $e_{projrct}$ after changing the value of the parameters.

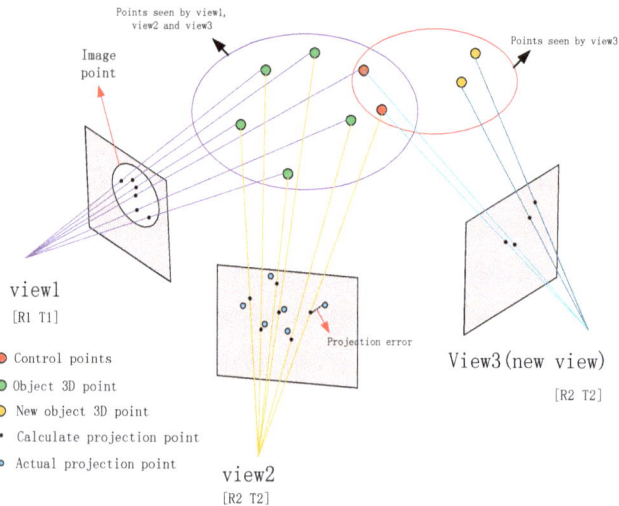

Figure 5. Weighted bundle adjustment.

The difference between the weighted bundle adjustment and the bundle adjustment is the weight of the control points' projection error. The weight is w_j (after an experimental comparison, a value of 20 is suitable for w_j). Equation (9) is the reprojection error formula of the weighted bundle adjustment.

3.3.4. MVS

For the dense reconstruction of the object, considering the characteristics of the problem addressed in this study, we use the method based on depth-map fusion to obtain the dense point cloud. The method is similar to that proposed in [16]. The algorithm first obtains the feature points in the structure calculated by the SfM. By using Delaunay triangulation, we can obtain the mesh data from the 3D feature points. Then, the mesh is used as an outline of the object, which is projected onto the plane of the images to obtain the estimated depth maps. The depth maps are optimized and corrected using the pixel matching algorithm based on the patch. Finally, dense point cloud data can be obtained by fusing these depth maps.

4. Experiments

4.1. Data Sets

In order to test the accuracy and speed of the algorithm proposed in this study, real outdoor photographic images taken from a camera fixed on a UAV and standard images together with standard point cloud provided by roboimagedata [27] are used to reconstruct various dense 3D point clouds. The object models and images provide by roboimagedata are scanned with a high precision structured light setup consisting of two Point Grey Research GS3-U3-91S6C-C industrial cameras with resolution of 9.1 Mp and a LG-PF80G DLP projector with a resolution of 1140 × 912 pixels mounted on a rigid aluminum frame. In addition, a high precision New-mark Systems RT-5 turntable is used to provide automatic rotation of the object). Figure 6a–e present some of the outdoor images (different resolution images taken with the same camera) taken from a camera carried by the DJI Phantom 4 Pro UAV (camera hardware: 1/2.3 inch CMOS, Effective 12.4 million pixels. Lens: FOV 94° 20 mm (35 mm format equivalent) f/2.8 Focal point at infinity). Figure 6f presents some images of an academic building taken by a normal digital camera which moves around the building (the camera's depth of field is near infinity). Figure 6d,e present some of the standard images [28] taken by a camera fixed to a robotic arm (with known positions and orientations) which is provided by roboimagedata. Table 1 lists all of the information for the experimental image data and the parameters used in the algorithm. We used a computer running Windows 7 64-bit with 8 GB of RAM and a quad-core 2.80-GHz Intel (R) Xeon (r) CPU.

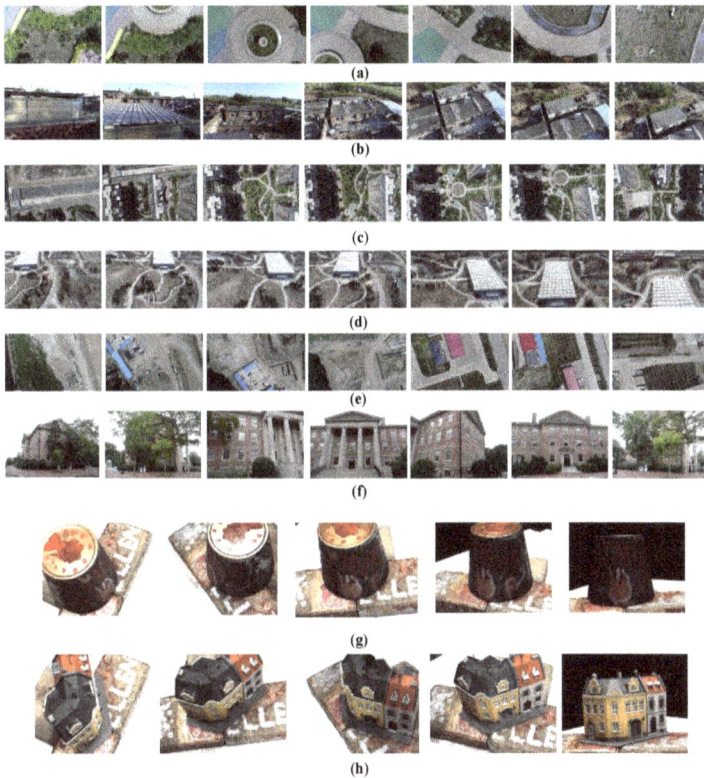

Figure 6. Images for experiment (**a**) Garden; (**b**) Village; (**c**) Building; (**d**) Botanical Garden; (**e**) Factory land; (**f**) Academic building; (**g**) Pot; and (**h**) House.

Table 1. Information for the Experimental Image Data.

Name	Image	Resolution	(m, k)	D_l	D_h
Garden	126	1920 × 1080	(15, 6)(20, 7)(40, 15)	25	150
Village	145	1920 × 1080	(15, 6)(20, 7)(40, 15)	25	150
Building	149	1280 × 720	(15, 6)(20, 7)(40, 15)	20	150
Botanical Garden	42	1920 × 1080	(15, 6)(20, 7)(40, 15)	25	150
Factory Land	170	1280 × 720	(15, 6)(20, 7)(40, 15)	20	150
Academic Building	128	1920 × 1080	(15, 6)(20, 7)(40, 15)	25	150
Pot	49	1600 × 1200	(8, 3)(10, 4)(15, 6)	20	200
House	49	1600 × 1200	(8, 3)(10, 4)(15, 6)	20	200

4.2. Precision Evaluation

In order to test the accuracy of the 3D point cloud data obtained by the algorithm proposed in this study, we compared the point cloud generated by our algorithm (PC) with the standard point cloud PC_{STL} which is captured by structured light scans (The RMS error of all ground truth poses is within 0.15 mm) provided by roboimagedata [27]. The accuracy of the algorithm is determined by calculating the nearest neighbor distance of the two point clouds [28]. First, the position of the point cloud is registered by the iterative nearest point method. For the common part of PC and PC_{STL}, each point p_1 in the PC, PC_{STL} is searched for the nearest point p_1', and the Euclidean distance between p_1 and p_1' is calculated. The distance point cloud is obtained after the distance calculation of each point and marked with different color. We compare the results of our method to those of openMVG [7], openMVS [16] and MicMac [29–31] (three open-source software packages). The main concern of openMVG is SfM calculation, while the main concern of openMVS is dense reconstruction. MicMac is a free open-source photogrammetric suite that can be used in a variety of 3D reconstruction scenarios. They both achieved state-of-the-art results. An open source software named Cloud Compare [32] is used for the test. The results are presented in Figures 7–12.

In the first experiment. As shown in Figure 7, point cloud shown in Figure 7a is generated by our method from 49 images ($m = 15$, $k = 6$). The number of points in the point cloud is 2,076,165. Figure 7b the number of points in point cloud of openMVG + openMVS is 2,586,511. Figure 7c the number of points in point cloud generated by MicMac is 270,802. And Figure 7d is standard point cloud provided by roboimagedata. The number of points is 2,880,879.

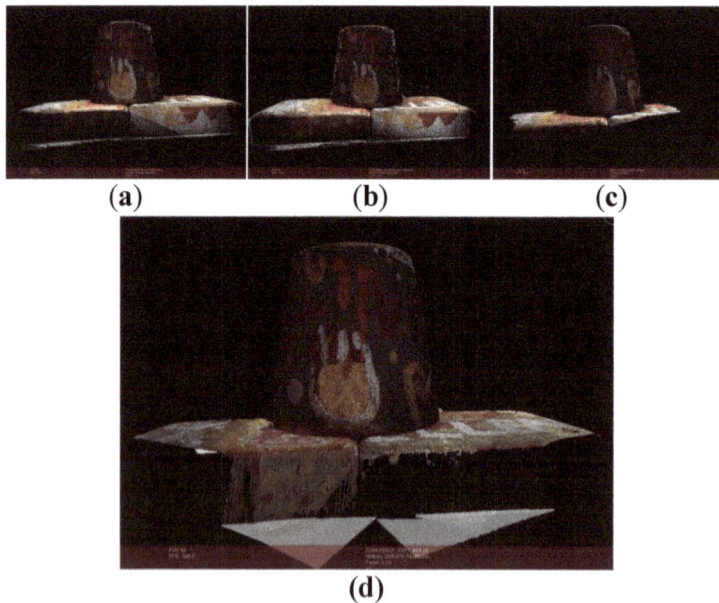

(a) (b) (c)

(d)

Figure 7. Point cloud comparison. (**a**) Point cloud of our method ($m = 15$, $k = 6$); (**b**) Point cloud of openMVG + openMVS; (**c**) Point cloud of MicMac; (**d**) Standard point cloud.

The distance point clouds are shown in Figure 8a–c. The calculation of distance is performed only on the common part of the two point clouds. Different color means different value of distance.

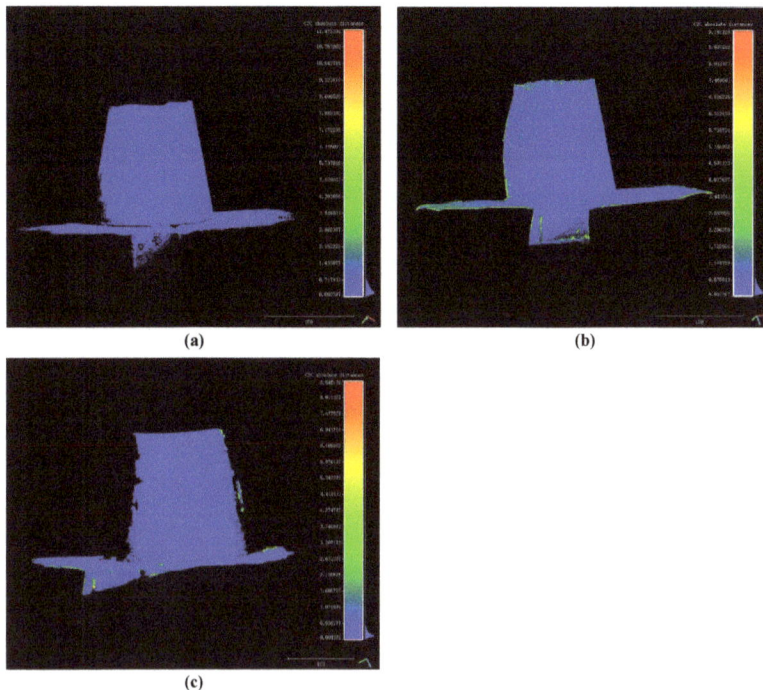

Figure 8. (**a**) Distance point cloud between the proposed method's result and the standard point cloud; (**b**) Distance point cloud between openMVG + openMVS's result and the standard point cloud; (**c**) Distance point cloud of MicMac and the standard point cloud.

Distance histograms in Figure 9a–c is statistics results of distance point cloud in Figure 8a–c. For the pot experiment, most distances are less than 1.5 cm when the pot is higher than 200 cm (the relative error is less than 1%).

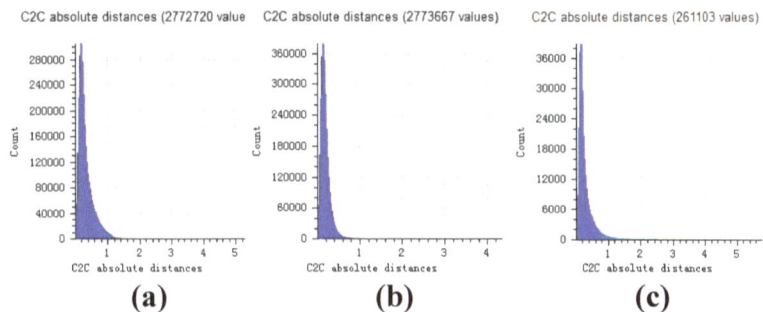

Figure 9. (**a**) Distance histogram of our result; (**b**) Distance histogram of openMVG + openMVS's result; (**c**) Distance histogram of MicMac's result.

In the second experiment. As shown in Figure 10, Figure 10a point cloud is generated by our method from 49 images ($m = 10$, $k = 5$). The number of points in the point cloud is 2,618,918. Figure 10b the number of points in Point cloud of openMVG + openMVS is 2,695,354. Figure 10c the number of points in point cloud generated by MicMac is 321,435. And (d) is standard point cloud provided by roboimagedata. The number of points is 3,279,989.

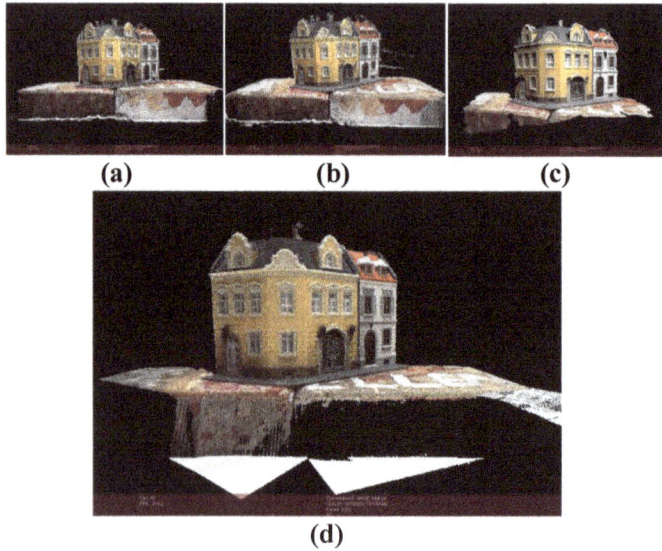

Figure 10. Point cloud comparison. (**a**) Point cloud of our method ($m = 15, k = 6$); (**b**) Point cloud of openMVG + openMVS; (**c**) Point cloud of MicMac; (**d**) Standard point cloud.

Distance histograms in Figure 12a–c are statistics results of distance point clouds in Figure 11a–c. For the house experiment, most distances are less than 1cm when the house is higher than 150 cm (the relative error is less than 1%).

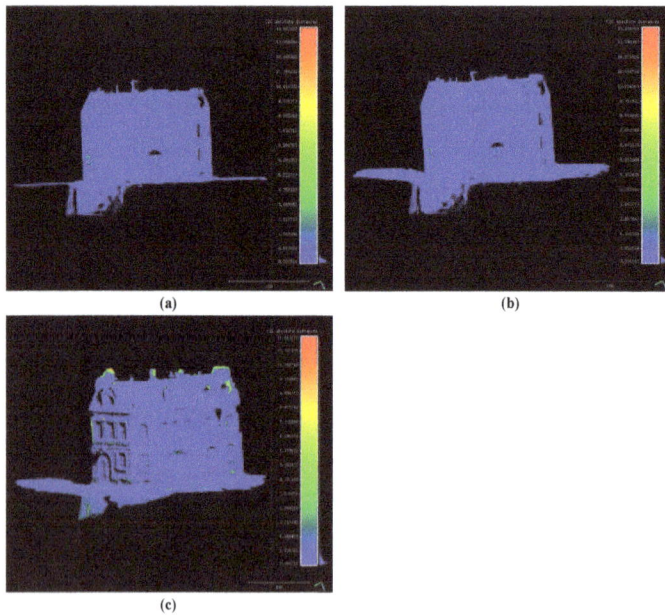

Figure 11. (**a**) Distance point cloud between the proposed method's result and the standard point cloud; (**b**) Distance point cloud between openMVG + openMVS's result and the standard point cloud; (**c**) Distance point cloud between MicMac's result and the standard point cloud.

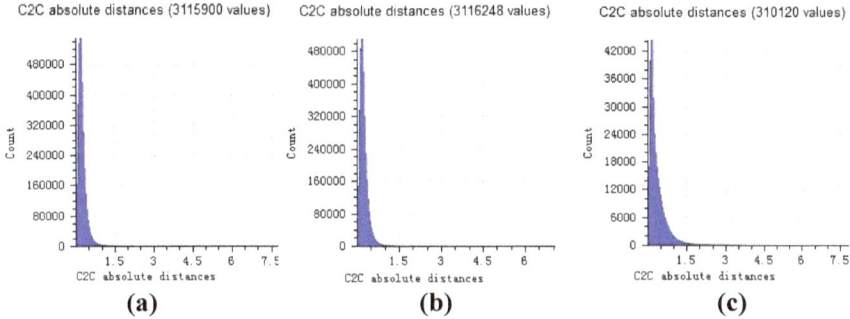

Figure 12. (**a**) Distance histogram of our result; (**b**) Distance histogram of openMVG + openMVS; (**c**) Distance histogram of MicMac.

The number of points of the point clouds generated by our algorithm are almost the same as openMVG + openMVS's results, and much more than those of MicMac. MicMac's result is smoother but less dense. The accuracy of our method is almost the same as openMVG + openMVS and MicMac (state-of-the-art methods), but the speed is much faster than them.

4.3. Speed Evaluation

In order to test the speed of the proposed algorithm, we compared the time consumed by our method with those consumed by openMVG and MicMac. Different m and k values for the algorithm are selected, and the same image data are used to run the program under the same hardware conditions. The running times of the algorithm are recorded in Table 2, and the precision is 1 s.

Table 2. Running Time Comparison.

Name	Images	Resolution	Our Method Time (s)			OpenMVG Time (s)	MicMac Time(s)
			$m = 15, k = 6$	$m = 20, k = 7$	$m = 40, k = 15$		
Garden	126	1920 × 1080	284.0	291.0	336.0	1140.0	3072.0
Village	145	1920 × 1080	169.0	209.0	319.0	857.0	2545.0
Building	149	1280 × 720	171.0	164.0	268.0	651.0	2198.0
Botanical Garden	42	1920 × 1080	77.0	82.0	99.0	93.0	243.0
Factory Land	170	1280 × 720	170.0	207.0	343.0	1019.0	3524.0
Academic building	128	1920 × 1080	124.0	182.0	277.0	551.0	4597.0
			$m = 15, k = 6$	$m = 10, k = 4$	$m = 8, k = 3$		
Pot	49	1600 × 1200	35.0	39.0	47.0	56.0	351.0
House	49	1600 × 1200	59.0	53.0	54.0	74.0	467.0

The accuracy of our result is almost the same as result of openMVG and MicMac, but the speed of our algorithm is faster than them. As is shown in Table 2.

There are two aspects that affect the speed of the algorithm. For most feature point matching algorithms, all images must match each other; thus, the time complexity of matching is $O(N^2)$. After using the methods proposed in this study, the time complexity becomes $\frac{n}{k}O(m \times k)$ because the matching calculation occurs only for the images inside the image queue. Although m and k are fixed and their values are generally much smaller than N, the speed of the matching is greatly improved. Second, for the SfM calculations, most of the time is spent on bundle adjustment. Bundle adjustment itself is a nonlinear least-squares problem that optimizes the camera and structural parameters; the calculation time will increase because of the increase in the number of parameters. The proposed method divides the global bundle adjustment, which optimizes a large number of parameters, into several local bundle adjustments so that the number of the parameters remains small and the calculation speed of the algorithm improves greatly.

4.4. Results

The result is shown in Figure 13 ($m = 15$, $k = 6$).The scene in this case is captured by an UAV camera in a garden of YanJiao. The flight height is about 15 m from the ground and is kept unchanged. The flight distance is around 50 m. The images' resolution is 1920 × 1080. And the number of points in point cloud is 4,607,112.

The result is shown in Figure 14 ($m = 15$, $k = 6$). The scene in this case is captured by a UAV camera in a village. The UAV is launched from the ground and flies over the house. The maximum flight height is around 6 m. The flight distance is around 20 m. The images' resolution is 1920 × 1080. And the number of points in the point cloud is 3,040,551.

The result is shown in Figure 15 ($m = 15$, $k = 6$). The scene in this case is captured by a UAV camera in a village. The UAV flight over the top of the buildings. The flight height is around 80 m and is kept unchanged. The flight distance is around 150 m. The images' resolution is 1280 × 720 and the number of points in point cloud is 2,114,474.

The result is shown in Figure 16 ($m = 10$, $k = 3$). In this case, the UAV flight is over a botanical garden. The flight blocks are integrated for many parallel strips. The flight height is around 40 m and kept unchanged. The flight distance is around 50 m. The images' resolution is 1920 × 1080 and the number of points in point cloud is 2,531,337.

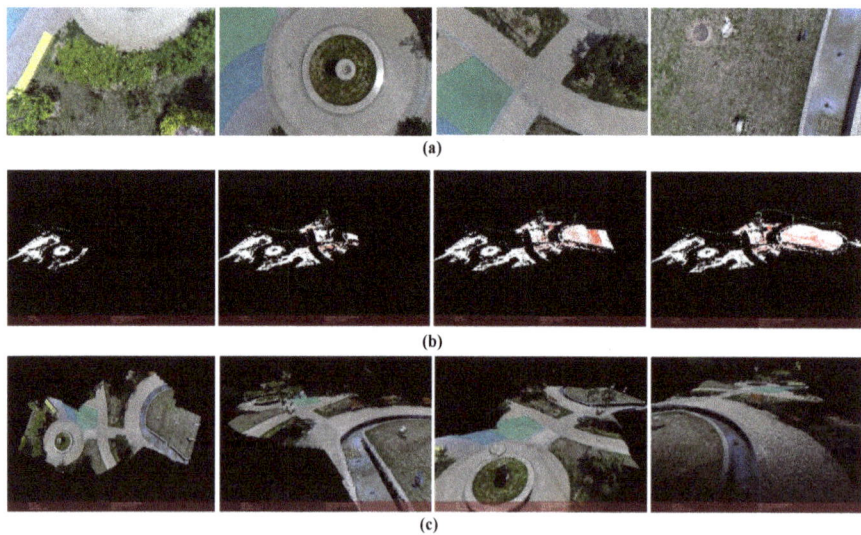

Figure 13. Reconstruction result of a garden. (**a**) Part of the images used for reconstruction; (**b**) Structure calculation of image queue SfM (green points represent the positions of the camera); (**c**) Dense point cloud of the scene.

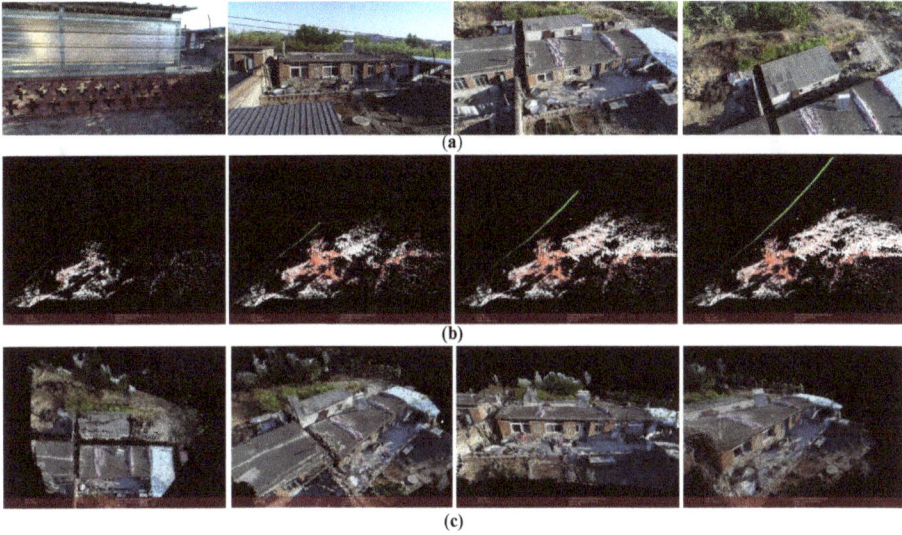

Figure 14. Reconstruction result of a village. (**a**) Part of the images used for reconstruction; (**b**) Structure calculation of image queue SfM (green points represent the positions of the camera); (**c**) Dense point cloud of the scene.

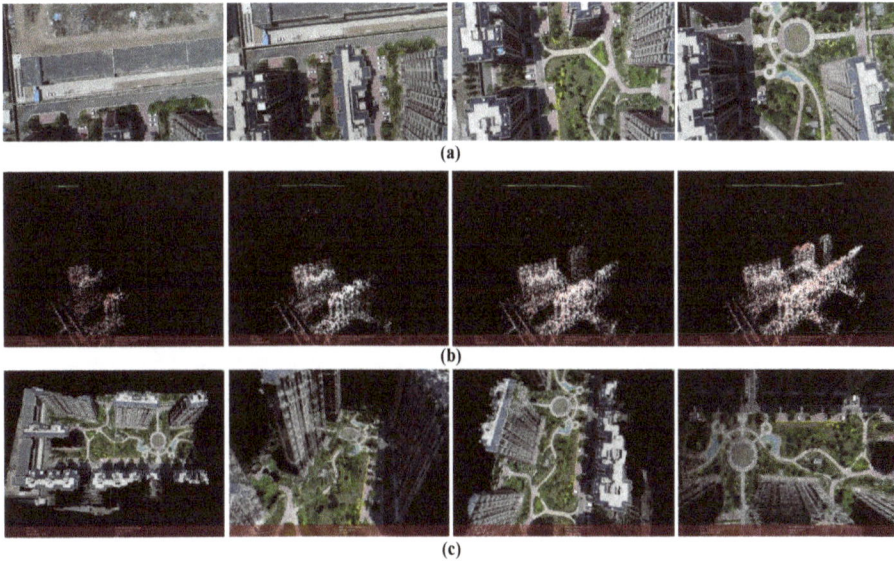

Figure 15. Reconstruction result of buildings. (**a**) Part of the images used for reconstruction; (**b**) Structure calculation of image queue SfM (green points represent the positions of the camera); (**c**) Dense point cloud of the scene.

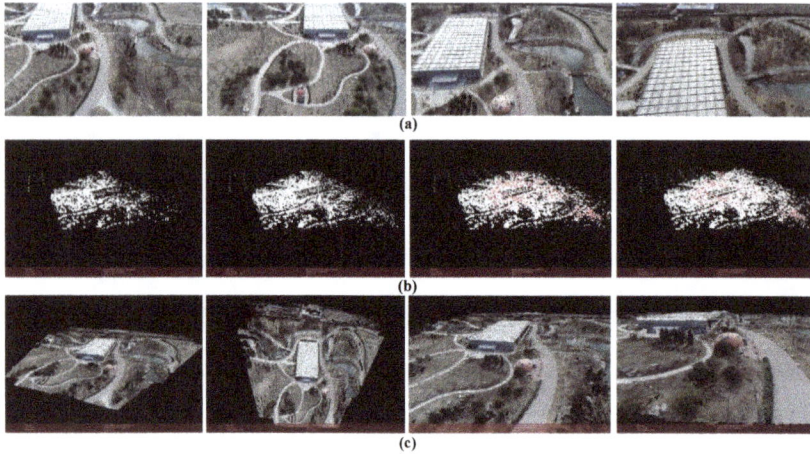

Figure 16. Reconstruction result of botanical garden. (**a**) Part of the images used for reconstruction; (**b**) Structure calculation of image queue SfM (green points represent the positions of camera); (**c**) Dense point cloud of the scene.

The result is shown in Figure 17 ($m = 20$, $k = 5$). In this case, the UAV flight is over a factory land. The flight height is around 90 m and is kept unchanged. The flight distance is around 300 m. The images' resolution is 1280×720 and the number of points in point cloud is 9,021,836.

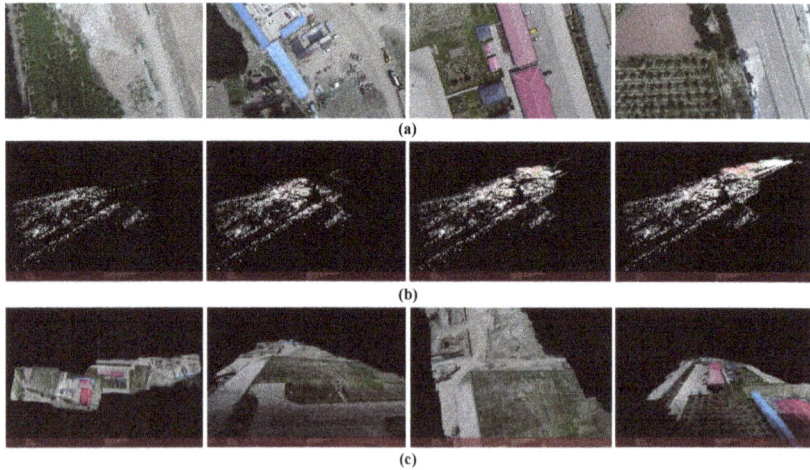

Figure 17. Reconstruction result of botanical garden. (**a**) Part of the images used for reconstruction; (**b**) Structure calculation of image queue SfM (green points represent the positions of camera); (**c**) Dense point cloud of the scene.

The result is shown in Figure 18 ($m = 25$, $k = 8$). In this case, a ground-based camera instead of UAV camera is used to move around the academic building and taken images. The images' resolution is 1920×1080 and the number of points in point cloud is 23,900,173. The result shows that our algorithm can be used in reconstruction from normal digital camera images as long as the images are taken continuously.

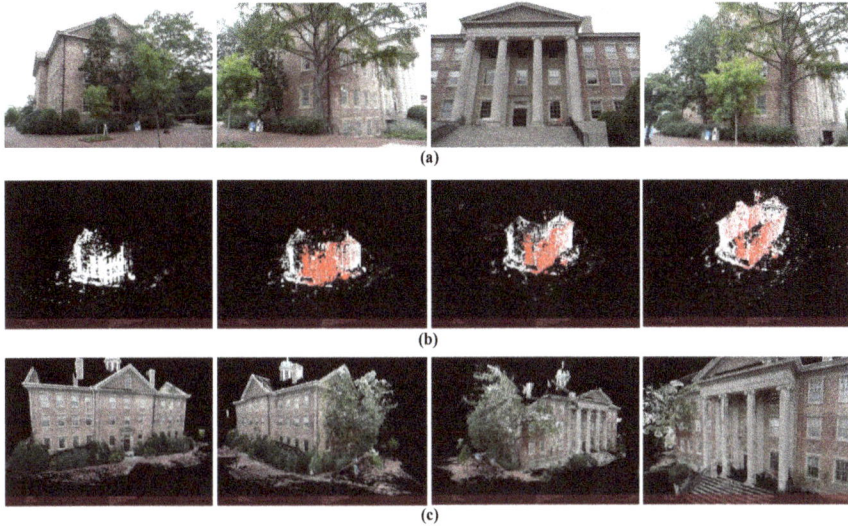

Figure 18. Reconstruction result of botanical garden. (**a**) Part of the images used for reconstruction; (**b**) Structure calculation of image queue SfM (green points represent the positions of camera); (**c**) Dense point cloud of the scene.

The results of experiment images used in this paper are present in Figures 13–18. For each example, Figure 18a shows some of the images used for 3D reconstruction. In the Figure 18b four most representative views of SfM, calculation results are selected to present the process of image queue SfM. Green points represent the positions of camera, and red points are control points, white points are structural feature points. Positions and orientations of cameras together with object feature points are derived in the order of camera movement. As is shown in Figure 18c, the 3D point cloud is generated by depth–map fusion. Accurate result can be obtained by using our method as long as the images are captured continuously. The final results accurately reproduce the appearance of the scenes.

5. Conclusions

In order to reconstruct the 3D structure of scenes using image sequences, we propose a rapid and accurate 3D reconstruction method based on an image queue. First, a principal component analysis method of the feature points is used to select the key images suitable for 3D reconstruction, which ensures that the algorithm improves the calculation speed with almost no loss of accuracy. Then, considering the continuity and relevance of the UAV camera's images, we propose a method based on an image queue. Our method divides a global bundle adjustment calculation into several local bundle adjustment calculations, greatly improving the calculation speed of the algorithm and making the structures continuous. Finally, dense 3D point cloud data of the scene are obtained by using depth–map fusion. The experiments demonstrate that when the texture of the images is complex and the number of images exceeds 100, the proposed method can improve the calculation speed by more than a factor of four with almost no loss of calculation accuracy. Furthermore, when the number of images increases, the improvement in the calculation speed will become more noticeable.

When the scene is too long, such as the flight distance is more than 300 m. The structure of the reconstruction will be distorted due to accumulated errors. This problem is solved in global SfM [7] by using loop closure constraint. Our future work will be aimed at cumulative errors elimination and will obtain higher accuracy. With the rise of artificial intelligence research, the parameters of m and k can be selected automatically by using deep learning and machine learning. Improving the performance of the algorithm in parameter selection is also part of our future work.

Acknowledgments: This work was financially supported by Natural National Science Foundation of China (NSFC) (51675033).

Author Contributions: Yufu Qu analyzed the weak aspects of existing methods and set up the theoretical framework. Jianyu Huang designed the method of selecting key images from image sequence and SfM calculation for the UAV camera's images, then programmed to achieve the methods, performed the experiment. Xuan Zhang collected the experimental image data and helped improving the performance of the algorithm and analyzed the result. Jianyu Huang wrote the paper and Yufu Qu made the modification.

Conflicts of Interest: The authors declare no conflict of interest.

References

1. Polok, L.; Ila, V.; Solony, M.; Smrz, P.; Zemcik, P. Incremental Block Cholesky Factorization for Nonlinear Least Squares in Robotics. *Robot. Sci. Syst.* **2013**, *46*, 172–178.
2. Kaess, M.; Johannsson, H.; Roberts, R.; Ila, V.; Leonard, J.J.; Dellaert, F. iSAM2: Incremental smoothing and mapping using the Bayes tree. *Int. J. Robot. Res.* **2012**, *31*, 216–235. [CrossRef]
3. Liu, M.; Huang, S.; Dissanayake, G.; Wang, H. A convex optimization based approach for pose SLAM problems. In Proceedings of the IEEE/RSJ International Conference on Intelligent Robots and Systems, Vilamoura-Algarve, Portugal, 7–12 October 2012; pp. 1898–1903.
4. Beardsley, P.A.; Torr, P.H.S.; Zisserman, A. 3D model acquisition from extended image sequences. In Proceedings of the European Conference on Computer Vision, Cambridge, UK, 14–18 April 1996; pp. 683–695.
5. Mohr, R.; Veillon, F.; Quan, L. Relative 3-D reconstruction using multiple uncalibrated images. *Int. J. Robot. Res.* **1995**, *14*, 619–632. [CrossRef]
6. Dellaert, F.; Seitz, S.M.; Thorpe, C.E.; Thrun, S. Structure from motion without correspondence. In Proceedings of the IEEE Conference on Computer Vision and Pattern Recognition, Hilton Head Island, SC, USA, 15 June 2000; Volume 552, pp. 557–564.
7. Moulon, P.; Monasse, P.; Marlet, R. Adaptive structure from motion with a contrario model estimation. In Proceedings of the Asian Conference on Computer Vision, Daejeon, Korea, 5–9 November 2012; pp. 257–270.
8. Wu, C. Towards linear-time incremental structure from motion. In Proceedings of the International Conference on 3DTV-Conference, Aberdeen, UK, 29 June–1 July 2013; pp. 127–134.
9. Gherardi, R.; Farenzena, M.; Fusiello, A. Improving the efficiency of hierarchical structure-and-motion. In Proceedings of the Computer Vision and Pattern Recognition, San Francisco, CA, USA, 13–18 June 2010; pp. 1594–1600.
10. Moulon, P.; Monasse, P.; Marlet, R. Global fusion of relative motions for robust, accurate and scalable structure from motion. In Proceedings of the IEEE International Conference on Computer Vision, Portland, OR, USA, 23–28 June 2013; pp. 3248–3255.
11. Crandall, D.J.; Owens, A.; Snavely, N.; Huttenlocher, D.P. SfM with MRFs: Discrete-continuous optimization for large-scale structure from motion. *IEEE Trans. Pattern Anal. Mach. Intell.* **2013**, *35*, 2841–2853. [CrossRef] [PubMed]
12. Sweeney, C.; Sattler, T.; Höllerer, T.; Turk, M. Optimizing the viewing graph for structure-from-motion. In Proceedings of the IEEE International Conference on Computer Vision, Los Alamitos, CA, USA, 7–13 December 2015; pp. 801–809.
13. Snavely, N.; Simon, I.; Goesele, M.; Szeliski, R.; Seitz, S.M. Scene reconstruction and visualization from community photo collections. *Proc. IEEE* **2010**, *98*, 1370–1390. [CrossRef]
14. Wu, C.; Agarwal, S.; Curless, B.; Seitz, S.M. Multicore bundle adjustment. In Proceedings of the Computer Vision and Pattern Recognition, Colorado Springs, CO, USA, 20–25 June 2011; pp. 3057–3064.
15. Furukawa, Y.; Ponce, J. Accurate, dense, and robust multiview stereopsis. *IEEE Trans. Pattern Anal. Mach. Intell.* **2010**, *32*, 1362–1376. [CrossRef] [PubMed]
16. Shen, S. Accurate multiple view 3D reconstruction using patch-based stereo for large-scale scenes. *IEEE Trans. Image Process.* **2013**, *22*, 1901–1914. [CrossRef] [PubMed]

17. Li, J.; Li, E.; Chen, Y.; Xu, L. Bundled depth-map merging for multi-view stereo. In Proceedings of the Computer Vision and Pattern Recognition, San Francisco, CA, USA, 13–18 June 2010; pp. 2769–2776.

18. Schönberger, J.L.; Zheng, E.; Frahm, J.M.; Pollefeys, M. *Pixelwise View Selection for Unstructured Multi-View Stereo*; Springer International Publishing: New York, NY, USA, 2016.

19. Lowe, D.G. Distinctive image features from scale-invariant keypoints. *Int. J. Comput. Vis.* **2004**, *60*, 91–110. [CrossRef]

20. Moulon, P.; Monasse, P. Unordered feature tracking made fast and easy. In Proceedings of the European Conference on Visual Media Production, London, UK, 5–6 December 2012.

21. Moisan, L.; Moulon, P.; Monasse, P. Automatic homographic registration of a pair of images, with a contrario elimination of outliers. *Image Process. Line* **2012**, *2*, 329–352. [CrossRef]

22. Fischler, M.A.; Bolles, R.C. Random sample consensus: A paradigm for model fitting with applications to image analysis and automated cartography. *Read. Comput. Vis.* **1987**, *24*, 726–740.

23. Zhang, Z. A flexible new technique for camera calibration. *IEEE Trans. Pattern Anal. Mach. Intell.* **2000**, *22*, 1330–1334. [CrossRef]

24. Hartley, R.; Zisserman, A. *Multiple View Geometry in Computer Vision*, 2nd ed.; Cambridge University Press: Cambridge, UK, 2003.

25. Triggs, B.; Mclauchlan, P.F.; Hartley, R.I.; Fitzgibbon, A.W. Bundle adjustment—A modern synthesis. In Proceedings of the International Workshop on Vision Algorithms: Theory and Practice, Corfu, Greece, 21–22 September 1999; pp. 298–372.

26. Ceres Solver. Available online: http://ceres-solver.org (accessed on 14 January 2018).

27. Sølund, T.; Buch, A.G.; Krüger, N.; Aanæs, H. A large-scale 3D object recognition dataset. In Proceedings of the Fourth International Conference on 3D Vision, Stanford, CA, USA, 25–28 October 2016; pp. 73–82. Available online: http://roboimagedata.compute.dtu.dk (accessed on 14 January 2018).

28. Jensen, R.; Dahl, A.; Vogiatzis, G.; Tola, E. Large scale multi-view stereopsis evaluation. In Proceedings of the Computer Vision and Pattern Recognition, Columbus, OH, USA, 23–28 June 2014; pp. 406–413.

29. Pierrot Deseilligny, M.; Clery, I. Apero, an Open Source Bundle Adjustment Software for Automatic Calibration and Orientation of Set of Images. In Proceedings of the ISPRS—International Archives of the Photogrammetry, Remote Sensing and Spatial Information Sciences XXXVIII-5/W16, Trento, Italy, 2–4 March 2012; pp. 269–276.

30. Galland, O.; Bertelsen, H.S.; Guldstrand, F.; Girod, L.; Johannessen, R.F.; Bjugger, F.; Burchardt, S.; Mair, K. Application of open-source photogrammetric software MicMac for monitoring surface deformation in laboratory models. *J. Geophys. Res. Solid Earth* **2016**, *121*, 2852–2872. [CrossRef]

31. Rupnik, E.; Daakir, M.; Deseilligny, M.P. MicMac—A free, open-source solution for photogrammetry. *Open Geosp. Data Softw. Stand.* **2017**, *2*, 14. [CrossRef]

32. Cloud Compare. Available online: http://www.cloudcompare.org (accessed on 14 January 2018).

Article

Uncooled Thermal Camera Calibration and Optimization of the Photogrammetry Process for UAV Applications in Agriculture

Krishna Ribeiro-Gomes [1], David Hernández-López [2], José F. Ortega [1], Rocío Ballesteros [1], Tomás Poblete [3] and Miguel A. Moreno [1,*]

[1] Regional Centre of Water Research, University of Castilla-La Mancha, 02071 Albacete, Spain; krishnaribeiro@yahoo.com.br (K.R.-G.); Jose.Ortega@uclm.es (J.F.O.); Rocio.Ballesteros@uclm.es (R.B.)
[2] Institute of Regional Development, University of Castilla-La Mancha, 02071 Albacete, Spain; David.Hernandez@uclm.es
[3] Centro de Investigación y Transferencia en Riegoy Agroclimatología, Universidad de Talca, Talca 3460000, Chile; totopoblete@gmail.com
* Correspondence: MiguelAngel.Moreno@uclm.es; Tel.: +34-967-599-200

Received: 2 August 2017; Accepted: 19 September 2017; Published: 23 September 2017

Abstract: The acquisition, processing, and interpretation of thermal images from unmanned aerial vehicles (UAVs) is becoming a useful source of information for agronomic applications because of the higher temporal and spatial resolution of these products compared with those obtained from satellites. However, due to the low load capacity of the UAV they need to mount light, uncooled thermal cameras, where the microbolometer is not stabilized to a constant temperature. This makes the camera precision low for many applications. Additionally, the low contrast of the thermal images makes the photogrammetry process inaccurate, which result in large errors in the generation of orthoimages. In this research, we propose the use of new calibration algorithms, based on neural networks, which consider the sensor temperature and the digital response of the microbolometer as input data. In addition, we evaluate the use of the Wallis filter for improving the quality of the photogrammetry process using structure from motion software. With the proposed calibration algorithm, the measurement accuracy increased from 3.55 °C with the original camera configuration to 1.37 °C. The implementation of the Wallis filter increases the number of tie-point from 58,000 to 110,000 and decreases the total positing error from 7.1 m to 1.3 m.

Keywords: uncooled thermal camera calibration; microbolometer; unmanned aerial vehicle; image filtering; structure from motion; irrigation management

1. Introduction

Unmanned aerial vehicles (UAVs) provide new alternatives to traditional satellite-based remote sensing for obtaining high-resolution images in real time for precision agriculture and environmental applications [1,2]. When compared with other remote sensing platforms, UAVs have the advantage of being more flexible, lower cost, more independent of climatic variables [1], and they can provide higher-resolution information [3]. Therefore, these platforms offer appropriate resolution for vegetation observation which was not possible with traditional platforms. Different types of sensors, such as RGB digital cameras [4], thermal cameras [5], multispectral and hyperspectral cameras [6], and other sensors [7], allow for the extraction of agriculturally-useful information [8]. UAVs have been used to predict several crop characteristics, such as water status variability, crop region and tree crown mapping, vegetation index calculation, and species phenotyping, among others [5,9–11]. Estimating crop yields is one of the main challenges UAV-based vegetation analysis face today.

With this aim the successful tomato detection using UAV images carried out by [12] seems promising. Additionally, the use of thermal information obtained from these devices in agricultural applications has been proposed, mostly focused on crop phenotyping under stress conditions [13–17]. However, depending on the thermal sensor used, thermal calibration is a crucial problem to be solved as uncooled sensors' temperature measurements are constantly changing [18]. Research on the correct use of thermal cameras in agricultural applications is becoming more frequent, primarily for the development of studies showing the possibilities of using this equipment in crop monitoring. Rud et al. [19] stated that the use of images generated by thermal cameras is a very effective tool in assessing water availability in the cultivation of potatoes. Möller et al. [20] stated that the use of thermal cameras together with the use of digital cameras provide very good accuracy in determining physiological data for vineyards. DeJonge et al. [21] in studies with maize plants, affirmed that the monitoring and quantification of water stress through evaluating the canopy temperature by using thermometry can be useful in the detection of plant stress.

The use of thermal information is a remote sensing technique that is being developed to interpret the state of crops, the detection of pests and diseases related to the moisture content, and the determination of the energy balance and, therefore, the water needs, through the use of data obtained from thermal and multispectral cameras. Energy balance methods generally demand surface reflectance data detected remotely in the visible and near infrared regions of the spectrum to determine the thermal and infrared band [22–24]. Some of these models are described below:

1. The Surface Energy Balance Index (SEBI), developed by [25], is based on the idea of the Crop Water Stress Index (CWSI) and an essential aspect is the variation of the surface temperature with respect to the air temperature. It is a pioneering and widely-used model.
2. The two-source model (TSM), described in [26], is widely used, emphasizing its use in the case of the vineyard.
3. Clumped (three-source model: transpiration of the cover, evaporation from the soil of the row, evaporation from the ground between rows), generated from the works of [27], has been used in vineyards with good results, although it the accuracy of some parameters need to be improved (characterization of the roof architecture or parameterization of soil moisture) [28].
4. Surface Energy Balance Algorithm for Land (SEBAL), one of the most used models, developed by [23], calculates evapotranspiration as a residue of the energy balance of the surface. Within the most used models, SEBAL is designed to calculate the energy balance components, both locally and regionally, with minimum soil data [29,30].
5. The Simplified Surface Energy Index (S-SEBI) is a method based on a simplification of SEBI [25]. It is based on the contrast between a maximum and minimum surface reflectance temperature (albedo) for dry and wet conditions. Thus, it divides the available energy into sensible and latent heat flows. If the maximum and minimum surface temperatures are clearly available in the image, it does not require additional meteorological data, which becomes an advantage.
6. The Surface Energy Balance System (SEBS) is a SEBI modification to estimate the energy balance on the surface [31]. SEBS estimates the sensible and latent heat fluxes from satellite data and commonly-available meteorological data (air temperature and wind speed).
7. The Mapping Evapotranspiration at High Resolution and with Internalized Calibration (METRIC) is a widely-used model and proposes the modification of some parameters of the SEBAL model [22,32]. It is calibrated internally with the inclusion in the images of two reference surfaces (dry or wet pixels and hot or cold pixels) that permits fixing the boundary conditions in the energy balance and simplifying the need for atmospheric corrections.
8. The Surface Energy Balance to Measure Evapotranspiration (MEBES) is a development of SEBAL performed by [33] for application in a wide area of Spain. MEBES is a version developed for applications in regions where the availability of meteorological data is limited (incomplete data).

MEBES was also validated with a lysimetric measurement at the local level. In addition, local actual evapotranspiration values (ETa) were compared using the Penman-Montieth method.

9 Remote Sensing of Evapotranspiration (ReSET) is a SEB model, proposed by [34] on the same principles as METRIC and SEBAL, but with some improvements, such as being able to integrate data from different meteorological stations.

These models each have their advantages and disadvantages [30], but they allow knowing ETa with different irrigation and cultivation management strategies [35], being able to approximate many parameters that determine the criteria for irrigation scheduling [36]. With the use of these methodologies, and based on thermometry parameters, the water status of the crop and the stress can be obtained. Many works are available for different herbaceous or woody species with different spatial scales but, in general, they are focused on the study of wide territories as support systems to the management of water resources [29,33,37,38].

All these models are based on the fact that the temperature of the canopy is an indicator of the water status of the plant, which is linked in turn to the stomatal conductance. Various water stress indices of crops have been developed based on the temperature of the canopy. The crop water stress index (CWSI) was developed by [39] and is increasingly being used to decide on irrigation management [40]. Other indicators are being implemented and applied to try to improve irrigation management.

To develop a functional methodology that utilizes images acquired with thermal cameras on UAVs, it is imperative that these sensors provide quantitative temperature information and that this temperature is measured with high precision, which demands a proper radiometric calibration. Another important problem related to the use of thermal images is the mosaicking in the photogrammetric process due to the low contrast of this type of image. This fact causes failures of the algorithms used for the automatic detection and pairing for the relative orientation of the images.

Thermal cameras are used in a wide range of different applications and, compared with conventional sensors, they do not depend on any external energy source [41]. These devices can be classified according to the type of image detector that they have, being classified as cooled or uncooled. Cooled thermal infrared cameras have the largest use in remote sensing because this type of camera is very sensitive and accurate [42], but the use of cooled sensors has some problematic because they are large, expensive, and consume a large amount of energy. Due to this fact, cooled thermal cameras are not usually mounted on small UAVs [42,43]. In contrast, the use of uncooled cameras coupled to UAVs is viable because they are lighter [5]. One disadvantage of the use of uncooled sensors is that these microbolometers are not as sensitive and accurate as in cooled systems and the majority of them are not calibrated, being only able to measure relative temperatures of a scene (image). For most remote sensing applications, accurate surface temperatures are required, which demands a calibrated thermal camera from the spectral and geometric point of view [5,43].

In uncooled thermal cameras, which mount thermo-electric cooler (TEC)-less microbolometers, the microbolometer is not stabilized to a constant temperature. This fact makes the sensor temperature fluctuate along with the temperature of the camera, which should be taken into account in any camera calibration model used [44]. Additionally, there are other causes that demand a calibration of TEC-less infrared sensors [44]:

1 Non-uniformity correction, which refers to the different operating points of the individual pixels of a microbolometer. A smoothing process is typically carried out in the current uncooled thermal cameras which attempts to equalize their performance.

2 Defective pixel correction, which refers to pixels that either do not work or whose parameters vary greatly from the mean. This is a characteristic of the sensor, which should be specified by the manufacturer. The correction of these pixels is based on their location and their interpolation based on the data obtained from neighbouring pixels. The main objective of this correction is to have a high-quality visual image rather than a high-quality radiometric value.

3   Shutter correction, which refers to the correction required due to the radiance of the camera interior that also varies with sensor temperature. Current uncooled thermal cameras perform an automatic shutter correction based on the time or change in sensor temperature.

4   Radiometric calibration, which refers to establishing the relationship between the response of the sensor and the temperature of the object. It is possible to approximate the sensor output signal with a Planck curve.

5   Temperature dependence correction, which refers to the effect of the sensor temperature on the response of the sensor. A linear correction that considers the signal from the object and the signal from the camera (dependent on camera temperature) is typically used to perform this type of correction.

In current uncooled thermal cameras, the first three sources of inaccuracies are corrected by the firmware included in the system. The fourth correction in devices is corrected by the digital acquisition system, if included in the sensor. However, the fifth correction is not usually performed, which leads to errors in the temperature measurements that are not acceptable for many applications, such as some agricultural or environmental applications. We found that the digital response of the camera is affected by the camera temperature in a non-linear manner. Thus, we will compare linear and non-linear classical models (polynomial models) to calibrate the thermal camera. Also, we will implement Artificial Neural Networks models to calibrate the camera because of their appropriate performance for solving highly non-linear problems [45].

Other sources of inaccuracies in thermal measurements using TEC-less cameras are [46]: size-of-source effect, distance effect, and environmental effects, among others, which are also present in the cooled cameras.

The procedure of generating the mosaic of orthoimages solves the general method of photogrammetry from the original images, obtaining the effect of obtaining an orthogonal perspective and covering the entire study area, which extends the field of view of the cameras [47] without introducing undesirable lens deformation [48]. Different types of photogrammetry software, such as PhotoScan® (Agisoft, St. Petersburg, Russia), Pix4D® (Pix4D, Lausanne, Switzerland), Apero-MicMac [49], VisualSfM [50], and Bundler [51], are used to obtain geomatic products from UAVs, such as georeferenced orthoimages and digital elevation models (DEM). These software packages require adequate image quality to obtain accurate geomatic products with the photogrammetry process.

Some problems related to obtaining orthoimages from thermal imaging are as follows: (1) the spatial resolution of commercial thermal chambers is still low, with commercial product resolutions varying from 160 × 120 to 1280 × 1024 [41], (2) vanadium oxide microbolometers have a higher noise level compared to other sensors [52], (3) compared to optical images, thermal images have low resolution and weak local contrast [53], and (4) in the particular case of non-refrigerated cameras, the acquisition of a time series of thermal images may fluctuate due to changes in sensor temperature [54]. Another problem of the use of thermal images in the photogrammetric process is that the images present very low contrast due to the low variation of the temperature in the observed objects. The low contrast of the images makes it difficult for computer vision algorithms to detect key points automatically. This problem implies that there are not enough matching points and the relative orientation process, with or without autocalibration, is not able to orient the entire set of images so that the mosaic of images, if it is achieved, has outstanding imperfections. In addition, it can lead to imprecision in the geometric calibration process of the camera.

In order to minimize these problems, filters have been developed for the treatment of images. These filters, in a simplified way, are described as part of mathematical procedures that consist of isolating the components of interest, so as to reinforce or soften the spatial contrasts of the grey level that integrate an image. It is performed by transforming the original grey levels of each pixel in such a way that they increase the difference with their corresponding neighbours. The filters can be classified according to the effect they produce to the images, being able to be of low step, of high step, of median, or directional, among others [55]. The purpose of the application of these filters in the context of the problem under study is to facilitate the photogrammetric procedure, primarily in the finding of tie-points.

Studies related to the pre-processing of images through the use of filters relate their functionality and effects in the images. Kou et al. [56] developed a detail enhancement algorithm to produce a detailed image, whereby the fine details can be amplified by enlarging all the gradients in the source image, except those of the pixels at the edges. From fine detail enhancement algorithms, fine details can be improved while avoiding halo artefacts and gradient inversion artefacts around the edges. Guidi et al. [57] analysed the effects of optical preprocessing with polarizing filters and digital preprocessing with high dynamic range (HDR) images to improve the conduction of 3D automated modelling based on Structure from Motion (SfM) and image matching. However, these authors observed that the metallic object does not preserve the polarizations of light and, consequently, is not affected by such an improvement. HDR-based techniques have also been analysed, revealing a moderate improvement in the ceramic object tested, on the order of 5%, compared to the standard images, but definitely a better result in the metal object (+63%).

Within this context, the objective of this work is to develop a procedure and an algorithm, based on machine learning, for radiometric calibration of uncooled thermal cameras. In addition, thermal image filtering to improve the photogrammetry solution is evaluated.

2. Materials and Methods

To obtain accurate high-resolution thermal products usable in precision agriculture, an integrated methodology is proposed. This methodology focuses on the accuracy of the acquired data, through the implementation of a novel calibration process, together with the appropriate data treatment during the mosaicking process to avoid post-processing artefacts.

2.1. Utilized Equipment

Different types of aircraft have different capabilities, with advantages or disadvantages depending on the application [1]. Compromises must be made between ease of flying, stability against wind, handling flight failures, distance covered, load capacity, and take-off/landing requirements. In this study, a Microdrone md4-1000 (Microdrones, Inc., Kreuztal, Germany) was utilized. It is a vertical take-off and landing (VTOL) quadcopter aircraft (Figure 1). The acronym VTOL denotes the capability of a flight vehicle to take off and land again in the vertical direction without the need of a runway. It employs four rotors or propellers on vertical shafts, mounted on one level of the bodywork. The UAV size is 1.030 m from motor to motor.

Regarding the flight performance, this UAV has a maximum rate of climb of 7.5 m·s^{-1}, a maximum horizontal speed of 12 m·s^{-1}, a maximum take-off weight (MTOW) of 6 kg, and a maximum load of 1.2 kg. The autonomy of the UAV reaches 40 min under optimal climatic conditions thanks to the 6S2P LiPo, 22.2 V, 13,000 mAh battery. With these characteristics, it can fly an area of approximately 80 ha per flight using a sensor that weights 0.3 kg.

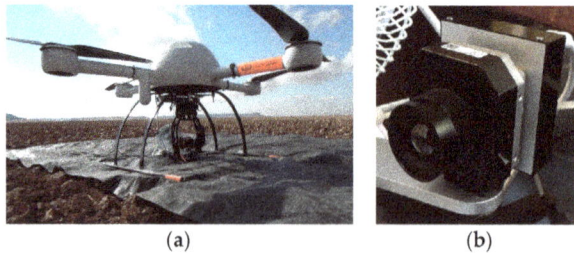

(a) (b)

Figure 1. Unmanned aerial vehicle, Microdrone MD4-1000 (**a**) and uncooled thermal camera (**b**).

The reference targets used to analyse the error of the geomatic products, especially designed for thermal applications as described below, were measured using Leica System 1200 GNSS receivers (Leica Geosystems AG, Wetzlar, Germany), which are dual-frequency systems that receive data from Global Positioning System (GPS) and Global Navigation Satellite System (GLONASS) constellation satellites and allow for measurement in a real-time kinematic (RTK) mode while static observations are recorded at the base receiver. During the subsequent post-processing step, the coordinates of the measured points were obtained in a global system with centimetre-level accuracy. This equipment updated its position with a frequency of 20 Hz (0.05 s) to minimize the possibility of recording false coordinates. The accuracy in RTK mode was 0.02 m in the X and Y axes, and 0.03 mm in the Z axis.

The thermal camera used to obtain the thermal images was the FLIR Tau2 (FLIR Systems, Inc., Wilsonville, OR, USA) (Figure 1), the main features of this camera are a focal of 9 mm (FOV 69° × 56°); an uncooled microbolometer of 640 (H) × 520 (V), a pixel size of 17 µm; a spectral band ranging 7.5–13.5 µm, and a weight of approximately 72 g without a lens.

2.2. Radiometric Calibration Data Acquisition

A radiometric calibration of the thermal cameras was conducted using a blackbody source Hyperion R Model 982 (Isothermal Technology Limited, Pine Grove, Southport, Merseyside, UK), with a large of 50 mm in diameter and 150 mm deep. The temperature range of the blackbody is from −10 to 80 °C. The thermal camera was installed in a fixed position against the black body at a distance of 0.5 m (Figure 2). The temperature of the sensor was measured and recorded for each image acquisition event. Blackbody temperatures used in the calibration process ranged from 5 to 65 °C in steps of 5 °C, which covers the range of temperatures found in agricultural applications. To obtain different values of the temperature of the camera, the experiment was carried out in a cooling room and in a regular room in which the temperature was modified to induce low and high temperatures with the cooling/heater system. Thus, a wide temperature range of the sensor is achieved to generalize the calibration model for any temperature condition. For each temperature of the blackbody images were obtained for sensor temperatures ranging between 5 and 31 °C (Figure 3).

A lower number of measurements were obtained for sensor temperatures lower than 20 °C, which corresponds with those obtained in the cooling room. Some problems related with water condensation in the black body and for reaching constant temperatures in the blackbody (it was set every 10 °C from 5 to 65 °C) made it a difficult task. With these data, it is ensured that a wide range of temperatures during the calibration process covers all the scenarios during agricultural applications. To ease the process of camera calibration a software called TermCal was developed by the authors in the MatLab® (Mathworks, Natick, MA, USA) environment in which the user can select a representative area of the blackbody temperature captured by the thermal camera and determine the digital response of the image, assigning a value of sensor temperature and perform the camera calibration with the different models described in this paper.

Figure 2. Disposal of the blackbody and the camera for calibration.

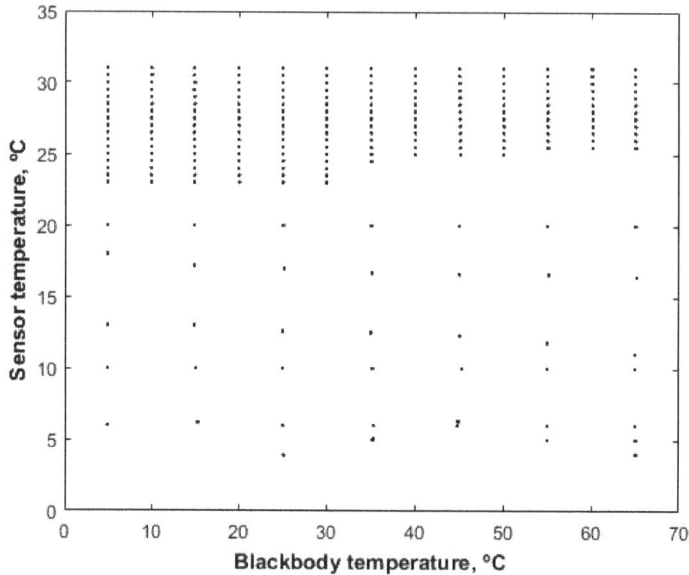

Figure 3. Range of thermal camera temperature and blackbody temperature analysed.

2.3. Analyzed Algorithms for Radiometric Calibration

Three types of models were analysed in this paper to perform the camera calibration: (1) linear models (Equation (1)), (2) polynomial models (Equations (2)–(4)), which are the models that are traditionally applied for the correction for camera temperature variation [44], and (3) an artificial neural network [45], because of their adequate performance for solving highly non-linear regression problems. To determine the best model, 65% of the data were used for calibration and 35% for validation. The calibration and validation subsets were randomly selected ensuring that both data set cover the whole range of measurements:

$$TBB = p00 + p10 \times DL + p01 \times TC \tag{1}$$

$$TBB = p00 + p10 \times DL + p01 \times TC + p20 \times DL2 + p11 \times DL \times TC + p02 \times TC2 \tag{2}$$

$$TBB = p00 + p10 \times DL + p01 \times TC + p20 \times DL2 + p11 \times DL \times TC + p02 \times TC2 + p30 \times DL3 + p21 \times DL2 \times TC + p12 \times DL \times TC2 \tag{3}$$

$$TBB = p00 + p10 \times DL + p01 \times TC + p20 \times DL2 + p11 \times DL \times TC + p02 \times TC2 + p30 \times DL3 + p21 \times DL2 \times TC + p12 \times DL \times TC2 + p03 \times TC3 \tag{4}$$

where TBB is the temperature of the black body; DL is the digital response of the camera; TC is the temperature of the camera; and pij are the regression coefficients.

Artificial neural networks (ANNs) are mathematical models that simulate the functioning of a biological neuron, and these networks have some advantages which make their use possible in different fields of study. The training algorithm type used in the neural network was back propagation. Considering a neuron j in layer i, the sum of the input variables and corresponding weights (S_j) in the input vector may be described according to Equation (5):

$$S_j = w_{0j} + \sum_{i=1}^{n} w_{ij} \cdot x_I \tag{5}$$

where S_j is sum of the input variables with their corresponding weights in the neuron j of layer i; w_{ij} is the weight associated with each of the input neurons with respect to the hidden layer nodes; w_{0j} is the weight associated with the first input neuron with respect to the hidden layer nodes; x_I is the input value stored in each neuron i; and n is the number of input variables.

Equation (6) describes the output variable of neuron j:

$$y_j = f(S_j) \tag{6}$$

where y_j is the output of neuron j; the activation function.

The activation function, for a hyperbolic tangent function responds to Equation (7):

$$y_j = \tanh(S_j) \tag{7}$$

where y_j is the output of neuron j; and tanh is the hyperbolic tangent.

2.4. Analysis of Residuals

In order to analyse the model adjustments, the following statistics were utilized: number of observations (n); representing the amount of data to be evaluated. In this paper $n = 266$ images; the average values of the digital levels and the sensor temperature, which is given in Equation (8); the coefficient of determination (R^2, Equation (9)); the root mean square error (RMSE), given by Equation (10); the relative error (RE, Equation (11)), and the similarity index (SI, Equation (12)).

$$\bar{x} = \frac{\sum_{i=1}^{n} X_i}{n} \tag{8}$$

where $\bar{x}$ is the average of all observed values, X_i, and n is the number of observations:

$$R^2 = \left[\sum_{i=1}^{n} \frac{(O_n - MO)(S_n - MS)}{\sqrt{\sum_{i=1}^{n}(O_n - MO)^2 \sum_{i=1}^{n}(S_n - MS)^2}} \right]^2 \tag{9}$$

where R^2 is the coefficient of determination; O_n are the observed values; S_n simulated values; MO is the average value of the n observed values; MS is the average value of n simulated values; and n is the number of observations:

$$RMSE = \sqrt{\frac{\sum_{i=1}^{n}(S_n - O_n)^2}{n}} \tag{10}$$

where $RMSE$ is the root mean square error ($^\circ$C); n is the number of observations; S_n are the simulated values; and O_n are the observed values:

$$RE = \left(\frac{RMSE}{MO} \right) \cdot 100 \tag{11}$$

where RE is the relative error (%); $RMSE$ is the root mean square error; and MO is the average value of the n observed values:

$$SI = 1 - \left(\frac{\sum_{i=1}^{n}(S_n - O_n)^2}{\sum_{i=1}^{n}((S_n - MO) + (O_n - MO))^2} \right) \tag{12}$$

where SI is the similarity index; n is the number of observations; S_n are the simulated values; O_n are the observed values; and MO is the average value of the n observed values.

In addition to these statistics, error analysis will be performed by: (1) analysis of linear regression between observed and simulated values; (2) adjustment of the residuals to the normal distribution; (3) homoscedasticity; and (4) Cook's distance for outlier detection.

2.5. Photogrammetry Process and Image Filtering

The procedure to acquire georeferenced geomatic products includes: (1) flight planning, which considers photogrammetric data that are provided by flight-planning software; (2) locating ground control points (GCPs) along the observed area; (3) measuring GCPs with a GNSS-RTK; (4) executing flights that follow the uploaded flight plan; (5) visually selecting the best set of images for post-processing by removing any blurred images; (6) entering the images and coordinates of the targets into photogrammetry software, which will self-calibrate the camera with the bundle-adjustment method; and (7) obtaining a georeferenced orthoimage, dense point cloud, and digital terrain model (DTM) [8,58].

As stated in the introduction section, the photogrammetry process using thermal images is a challenging task because of the lack of contrast in the images and the difficulty of locating the GCPs in the images. To solve the low contrast in the image we propose to apply the Wallis filter [59] on thermal imaging, which has been successfully applied in other cases with visible images when the scene presents low contrast [60–62]. This filter applies a contrast enhancement to each zone of the image by adjusting the brightness values in specific areas of the image to make the measurement and the standard deviation match the user default values. This improvement achieves a good local contrast throughout the image (Figure 4a,b), which allows for better detection of key points and corresponding matching, as well as allowing the operator to improve the photointerpretation of the GCPs.

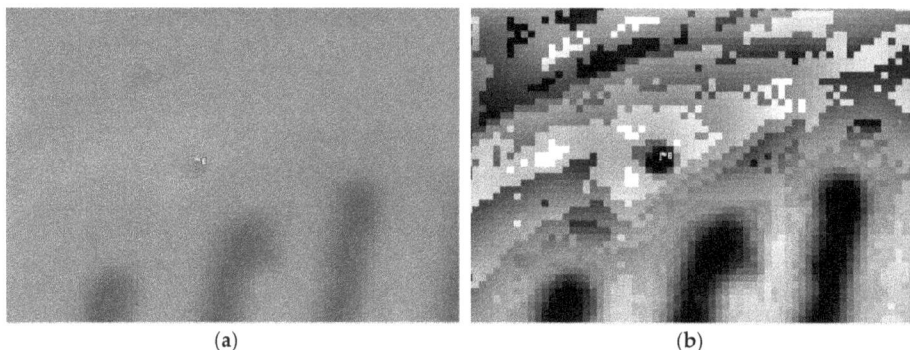

(a)   (b)

Figure 4. Example of photointerpretation in non-filtered images (**a**) and filtered images (**b**).

From the set of thermal images resulting from the planning flight, three sets of images are created: (1) the set of original thermal images, (2) the set of filtered thermal images after applying the Wallis filter to the set of thermal images, and (3) the set of radiometrically-calibrated thermal images after applying the thermal radiometric calibration to the original set of original thermal images. Photogrammetric processing was performed using Agisoft PhotoScan® (Agisoft, St. Petersburg, Russia) software using the parameters described in Table 1.

Table 1. Parameters used in Agisoft PhotoScan®.

Alignment Parameters	
Accuracy	High
Pair preselection	Generic
Key point limit	40.000
Tie point limit	4.000
Adaptive camera model fitting	yes
Optimized parameters	f, b1, b2, cx, cy, k1–k4, p1, p2
Dense point cloud	
Quality	Medium
Depth filtering	Mild
Model	
Surface type	Arbitrary
Source data	Dense cloud
Face count	High
Interpolation	Enabled
Orthomosaic	
Mapping mode	Orthophoto
Blending mode	Mosaic

For the processes of alignment of the images, determination of the dense cloud of points, and creation of the mesh, we used the set of thermal images treated with the Wallis filter. To texture and generate the final orthoimages, these images were replaced by the set of radiometrically-calibrated images. To apply the Wallis filter to the images, the GRAPHOS program was used [61]. In the process of self-calibration of the thermal camera with Agisoft PhotoScan® software the following parameters were determined:

f: which is the focal length measured in pixels.
cx and cy: which are the coordinates of the main point.
b1, b2: which are the biased transformation coefficients.
k1, k2, k3, k4: which are the radial distortion coefficients.
p1, p2: which are the tangential distortion coefficients.

In addition to the results of the self-calibration procedure, errors were calculated in the GCPs. The total error in X, Y, and Z was calculated from Equation (13):

$$\text{Total error} = \sqrt{\left(\sum_{i=1}^{n} \frac{\left[\left(X_{i,est} - X_{i,in} \right) + \left(Y_{i,est} - Y_{i,in} \right) + \left(Z_{i,est} - Z_{i,in} \right) \right]}{n} \right)} \qquad (13)$$

where X_i is the estimated value for the X coordinate for the position of the camera i; $X_{i,in}$ is the input value for the X coordinate for the position of the camera i; $Y_{i,est}$ is the estimated value for the Y coordinate for the position of the camera i; $Y_{i,in}$ is the value input for the Y coordinate for the position of the camera i; $Z_{i,est}$ is the estimated value for the Z coordinate for the position of the camera i; and $Z_{i,in}$ is the input value for the Z coordinate for the position of the camera i.

The photogrammetric processing was performed for three different situations: (1) use of the images as obtained from the camera with the factory settings, without filtering and without radiometric calibration; (2) use of filtered images without radiometric calibration; and (3) use of radiometrically-filtered and corrected images. The analysis of these cases allowed illustrating the improvements provided by the proposed methodology, by comparing the results in the photogrammetric processing of the images filtered and corrected radiometrically with those obtained with the same software when using the set of original thermal images (unfiltered and without radiometric calibration). In addition, the difference in temperature measurement was analysed by the use of radiometrically-corrected and uncorrected images.

The number of tie points obtained with the filtered and unfiltered images and the number of stereoscopic pairs generated were evaluated, which has a significant influence on the accuracy of the aerotriangulation with autocalibration and, consequently, on the quality of the obtained geomatic products.

2.6. Application to a Case Study

The proposed methodology was implemented in a vineyard located in Iniesta (Cuenca, Spain) irrigated with water of the hydrogeological unit (H.U.) 08.29. (Figure 5). This unit is located in the southeast of Spain, on the eastern side of the La Mancha plains, with a total area of 8500 km^2 and with relatively uniform agronomic features. This area is classified as semi-arid according to the aridity index (AI) described by [63] (AI = 0.36). Therefore, proper water management for irrigation is essential for rural development. Air temperature and other meteorological variables were recorded from an agro-meteorological station located 7 km from the plot. For the 2015 season, the precipitation was 268.4 mm year^{-1} and the annual reference evapotranspiration (ETo) was 1321.4 mm year^{-1}.

The total area of the plot is 17.8 ha. The varieties cultivated in this plot were Sauvignon Blanc, Garnacha Tintorera, and Syrah. Plots were irrigated using drip irrigation systems, applying an average of 800 m^3 ha^{-1}, which means a deficit irrigation due to water scarcity and water restriction in the area. The irrigation system is divided into 12 sectors. The cultivation techniques applied were considered typical practices for vineyards.

Figure 5. Location of the case study, and detail of the orthoimage obtained with the UAV. Black squares are ground measurements of validation points (PNOA, 2015).

Black squares are the locations of the ground measurements, which were performed over rain-fed neighbouring vineyards (eastern squares), irrigated vineyards that were irrigated the night before the flight (cantered squares), and irrigated vineyards irrigated seven days before the flight. In addition, soil measurements were obtained in the three placements.

2.7. Flight Planning and UAV Data Acquisition

The objective of flight planning was to generate a navigation file that guides the UAV to automatically capture images, with proper overlapping and sidelapping, according to the requirements of the final product (mainly GSD) and the requirements of the photogrammetry workflow [58]. In this case of study, the GSD employed was 0.20 m.

Overlapping was established in 60% and sidelapping in 40%. To ensure these values, flight planning should consider the errors of GPS, camera angles, etc. Microdrone photogrammetric flight planning software (MFLIP) was developed by the authors in collaboration with the company ICOM 3D (Asturias, Spain) for UAV flight planning, introducing all of the photogrammetry parameters required for each flight. The main result of the flight planning is an ASCII file in which each line includes an order for the UAV. This file is copied into the microSD card on the UAV. In addition, it generates a database with the theoretical footprints of the images, overlapping, among other data, in SHP files that can be opened with QGIS or a similar program. This information is useful in the photogrammetry workflow.

With the dual objective of georeferencing the generated geomatic products and geometrically calibrating the camera lens, a total of eight GCPs were located and measured by GNSS-RTK. In order to allow the measurement of the GCPs in the thermal images, an evaluation was made of combinations of materials suitable for making the thermal targets, with EVA rubber material of dimensions of 0.60 × 0.80 m being chosen for their manufacture, with a polished aluminum plate of 0.34 × 0.29 m (Figure 6) in the centre. The polished aluminum sheet allows reflecting of the temperature of the sky, which is obviously very low compared to the temperature of the ground. Thus, it was easy to detect the targets in the thermal images by means of photointerpretation.

Figure 6. Type of GCPs used in thermal images.

2.8. Ground Temperature Acquisition for Validation

In order to validate the proposed methodology, different temperature measurements were taken in different vines and soils just after the flight. To determine water stress in the plants, thermal measurements were obtained in different sectors. As the irrigation interval was one week, measurements were taken in vines irrigated from one week before the flight (five measurements in sector 3) to the night before (five measurements in sector 9). Additionally, neighboring rain-fed vines (five measurements) and soil (five measurements) were obtained. The measurements were made in the field with a FLIR B660 refrigerated thermal camera (FLIR Systems, Inc., Wilsonville, OR, USA). The emissivity was set to 1 to be able to compare both measurements. A Leica Zeno GNSS-RTK handheld system (Leica Geosystems AG) was used to capture the coordinates of each of the selected temperature sample points immediately after the flight, as well as the plants and the soil, to locate the sampled points in the generated orthoimagery. These temperature values were compared with the temperature values obtained from the orthoimages generated from the photogrammetric procedure using the radiometrically-corrected thermal images.

3. Results

3.1. Error Analysis of the Uncooled Thermal Camera

We obtained a series of data from 266 images of this camera that were analysed as they were captured with the configurations of the manufacturer. These same images were used in the comparison of the adjustments of the lineal and polynomial models and the ANN. Table 2 shows the statistics that determine the measuring accuracy of the camera. It can be observed that the measurement accuracy with the default parameters of the camera is very low, with an RMSE of 3.55 °C and an average RE of 8.47%. When using artificial neural networks, it can be observed that the RMSE decreased up to 1.37 °C and a RE of 8%.

Table 2. Main statistic indices of the manufacturer configuration, linear model, polynomial models (P1–P4), and the artificial neural network model.

	Manufacturer Configuration	Linear	P1	P2	P3	P4	ANN
Data	266	95	95	95	95	95	95
R^2	0.96	0.99	0.99	0.99	0.99	0.99	0.99
RMSE, °C	3.55	1.81	1.81	1.49	1.51	1.51	1.37
Relative Error, %	8.47	5.59	5.57	4.59	4.66	4.66	4.22
Similarity Index	0.99	1.00	1.00	1.00	1.00	1.00	1.00

It can be observed that the RMSE decreases drastically from the manufacturer configuration to the polynomial and even the linear model (from 3.55 °C to 1.49 °C). According to the manufacturer, this camera should have an expected final precision of ±2 °C in absolute temperature measurements. There is an increase of the RMSE for polynomial models 3 and 4, which could be due to overfitting of the model. However, these differences are very slight. Thus, in this case, the best classical model is the polynomial 2 with a RMSE of 1.49 °C. However, when applying the ANN model, a RMSE of 1.37 °C can be reached. The slight difference does not justify the use of complex models based on machine learning over traditional polynomial models. However, a deep analysis of the model's residuals should be performed to select the best model. Figure 7 shows the error analysis for the manufacturer configuration, polynomial 2 and ANN, respectively. It can be seen that the measurement errors are very high and are not randomly distributed, being non-homoscedastic. These errors call into question the usefulness of the camera in its original configuration for agronomic applications and demonstrate the need for an adequate calibration process to increase the accuracy of the measurement.

In agronomic applications, using satellite-derived data and non-local meteorological data, it was observed that a 2 °C error in soil temperature (Ts) corresponds to an error of 0.23 in the latent heat flux (LE), a 1 °C error in canopy temperature (Tc) corresponds to an error of 0.10 °C in LE for satellite data, an error of 1 °C in Ts corresponds to an error of 0.11 in LE, and an error of 0.5 °C in Tc corresponds to an error of 0.06 °C in LE for non-local meteorological data [64]. The RMSE obtained after calibration (1.4 °C) could be acceptable for agronomic applications, while using the manufacturer configuration (3.55 °C) could lead to gross errors in the energy balance model. Therefore, it is essential to perform an adequate calibration of uncooled thermal cameras.

Additionally, to implement the energy balance model from thermal images obtained with UAVs, it is recommended to perform a vicarious calibration with ground temperature measurements. Even if the camera is powered on before the flight for around 30 min and stabilized with climatic conditions, high variation of the sensor temperature could appear during the flight. Thus, the vicarious calibration could become a difficult task, requiring different points for ground measurement along the plot, increasing the cost of this activity. Here, we propose a calibration considering the sensor temperature and to perform a vicarious calibration in accessible, easy to measure points.

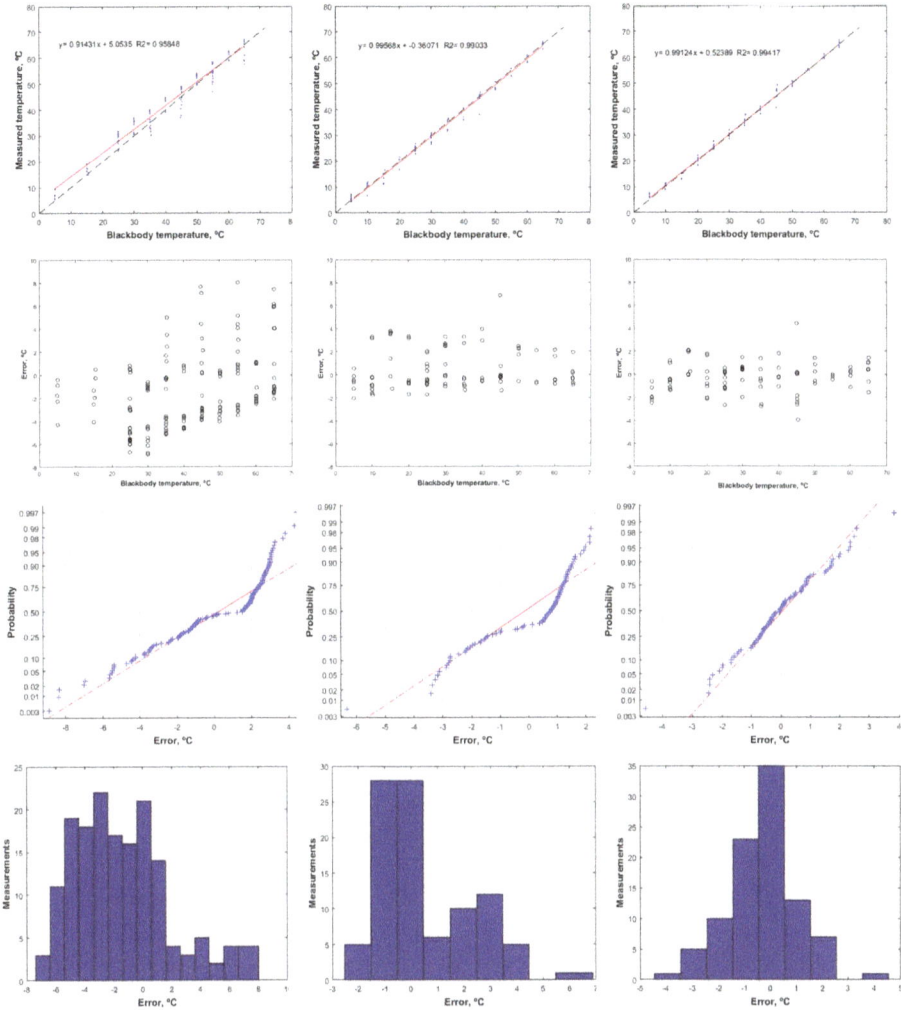

Figure 7. Error analysis for the FLIR Tau2 9 mm camera with the original configuration (column 1), the polynomial 2 model (column 2), and the artificial neural network model (column 3).

3.2. Results of Wallis Filter Application

The flight characteristics obtained from the photogrammetric process in Agisoft PhotoScan® for each of the comparisons made are described in Table 3.

The number of images is the total number of images loaded in the project; the flight height is the average height above the terrain level calculated in the photogrammetric procedure in Agisoft PhotoScan®; ground resolution is the averaged field resolution on all aligned images; the coverage area is the size of the study area; the number of oriented images refers to the images for which the photogrammetric orientation has been corrected; projections is the total number of projections of valid tie-points; and re-projection error is the quadratic mean of the average of re-projection errors on all tie-points in all images. The re-projection error is the distance in the position in the image, in unit pixels, between the position detected by computational vision and the result of applying the collinearity equation from the position in the model system (terrain for relative or absolute orientation).

Table 3. Characteristics of each of the photogrammetric procedures performed for each of the described situations.

	Unfiltered Images	Filtered Images
Number of images	1154	1154
Flight height (m)	81.1	80.4
Ground resolution (cm pix^{-1})	13.8	13.5
Covered area (km^2)	0.366	0.364
Number of images oriented	1.148	1.151
Tie-points	58,193	110,089
Projections	272,078	445,291
Re-projection error (pix)	0.504	0.442

The results presented in Table 3 show that when applying the Wallis filter in the images the number of tie-points increases by 89% for the filtered images compared with unfiltered images. It is also observed that the number of projections increases when the filter is applied, with an increase of 63%. Table 3 also shows that the re-projection error decreases after filtering the images. These results demonstrate that the application of the Wallis filter to the images allows for an improvement in the processing of the images by the greater number of image tie-points recognized in the photogrammetric software used.

The results obtained from the detection of points of interest (key points) and their correspondences (points of connection or matching) on the images with the Wallis filter have greatly improved the results. Figure 8 shows a comparison between the number of tie-points in the alignment process using the original images (X-axis), and the number of coincident points for the filtered images (Y-axis).

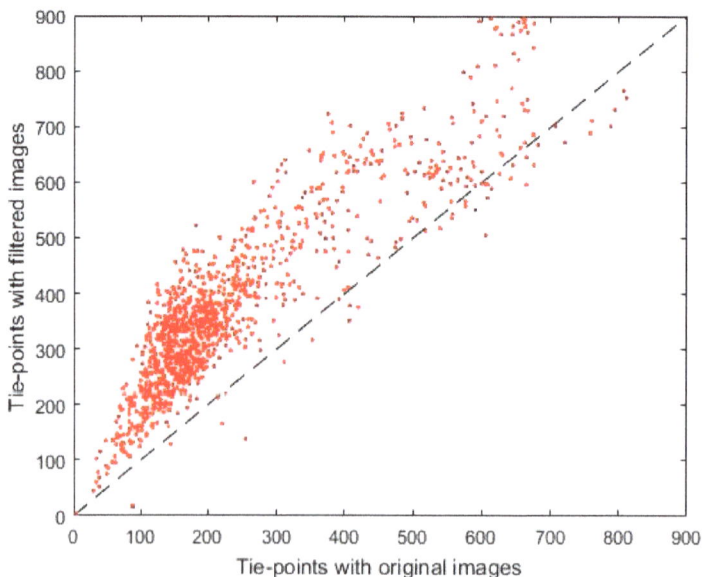

Figure 8. Comparison of the tie-points of the filtered images and the original images.

To evaluate the statistical significance of the difference, a box and whisker diagram was obtained (Figure 9), which resulted in a significant difference.

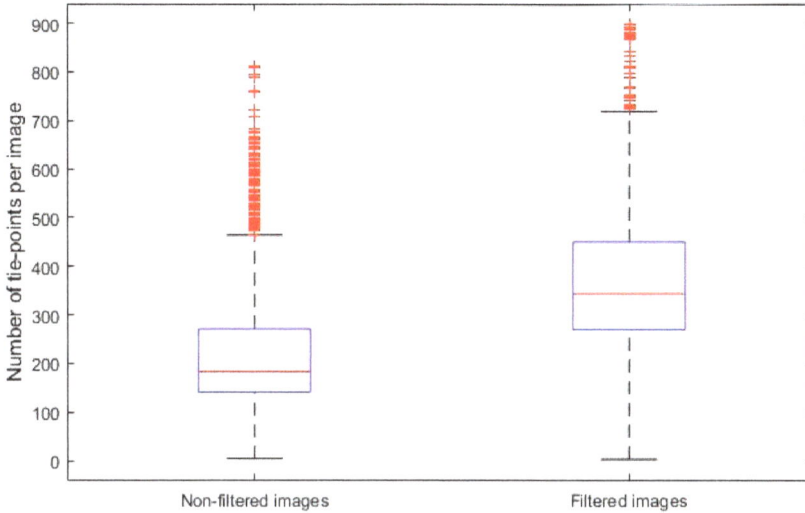

Figure 9. Box and whisker diagram of the number of link points of unfiltered and filtered.

In addition, Figure 10 shows how the percentage of the increase of the tie-points resulting from the filtering process is greater for those images with a low number of points, a key question that allows the orientation of these images and explains that they may not be oriented if this technique is not applied. In this way, the effectiveness of the Wallis filter is confirmed in the automatic detection of tie-points for thermal imaging, which results in greater precision in the geometric calibration of the camera and in the relative and absolute aerotriangulation, which translates, finally, in high-quality geomatic products.

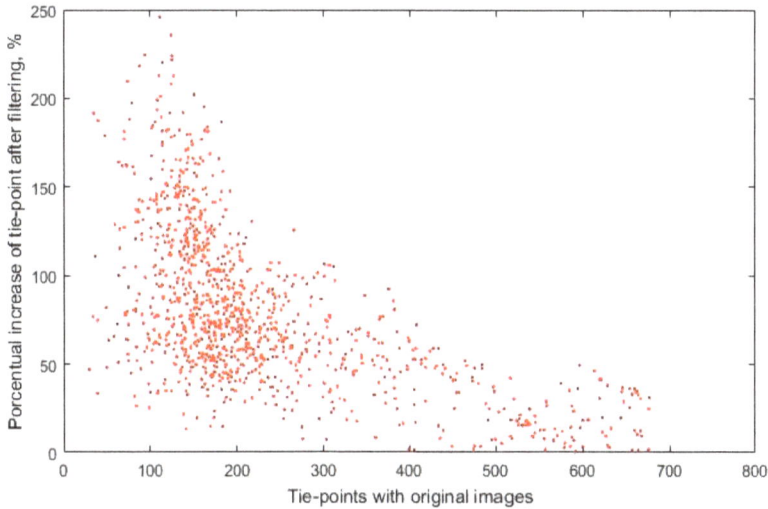

Figure 10. Increase in coincident points after Wallis filter application.

The calculated errors in X, Y, and Z are shown in Tables 4 and 5, showing that the accuracy obtained in the GCPs after the absolute orientation is better for the filtered and geometrically-calibrated images, with reductions of 2.66 to 0.60 m in X, 2.45 to 0.43 m in Y, and from 6.19 to 0.98 m in Z. The 3D error reaches 7.2 m if the filtering process is not implemented and is reduced to 1.2 when filtering. The main error is obtained in Z, which means that the geomatic product obtained is accurate in planimmetry, but innacurate in altimetry. Thus, only the thermal orthoimage is a usable product while the dense cloud cannot be used to obtain the canopy volume. Thus, an interesting solution would be installing together multispectral, RGB, and thermal cameras to obtained different products on the same flights.

Table 4. Control points of images not filtered and not geometrically calibrated.

GCP	Error X (mm)	Error Y (mm)	Error Z (mm)	Total (mm)	Image (pix)
1	1.00	−1.12	0.80	1.70	0.58 (34)
2	−3.81	−0.33	2.78	4.73	0.97 (49)
3	4.82	1.65	−10.87	12.01	0.74 (32)
4	2.69	1.41	5.83	6.57	0.65 (30)
5	0.53	3.56	2.36	4.30	0.36 (32)
7	−2.41	−5.36	−6.14	8.50	0.56 (26)
8	−1.60	0.68	4.89	5.19	0.86 (24)
9	−1.48	0.33	−8.92	9.05	0.53 (31)
Total	2.66	2.45	6.20	7.18	0.70

Table 5. Control points of the filtered and calibrated geometrically images.

GCP	Error X (mm)	Error Y (mm)	Error Z (mm)	Total (mm)	Image (pix)
1	−0.37	0.46	−0.98	1.14	0.57 (34)
2	−0.19	−0.24	−0.04	0.32	1.65 (49)
3	0.02	0.32	0.13	0.35	0.64 (31)
4	0.05	−0.64	1.33	1.48	0.87 (32)
5	1.32	0.61	1.00	1.77	0.38 (32)
7	−0.98	−0.32	−1.78	2.05	0.67 (27)
8	−0.20	0.28	0.90	0.97	0.78 (23)
9	0.08	−0.39	−0.17	0.43	0.51 (31)
Total	0.61	0.43	0.98	1.23	0.92

The geomatic product obtained presents a better quality besides being well georeferenced, as can be observed in Figure 11.

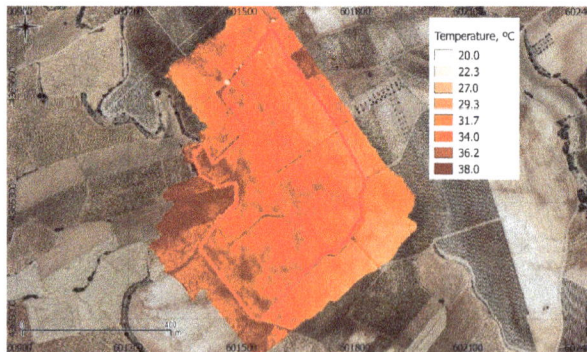

Figure 11. Ortoimage generated using calibrated and filtered images.

In some areas, the temperature difference between the corrected and uncorrected orthoimages reaches −5.06 °C and up to 1.93 °C.

3.3. Results of Temperature Measurements in the Case Study

Table 6 shows the comparison of three sources of information for temperatures in the validation surfaces: the measurements with the FLIR B660 hand-held thermal camera performed after the thermal flight as ground truth, those obtained by measurement in the uncalibrated images, and those obtained in the images resulting from the thermal calibration, both obtained from the generated orthoimages. The measurements, as described in the methodology (Figure 5), were obtained in neighbouring rain-fed vineyards (RV), vineyards irrigated the day before the flight (IV), vineyards irrigated seven days before the flight (7d-IV), and in the soil. Images were obtained for every condition.

Table 6. Results obtained from the points sampled.

	Handheld Camera		Original Configuration		Corrected Data	
	Mean	SD	Mean	SD	Mean	SD
RV 1	30.2	0.3	32.0	0.1	29.7	0.1
RV 2	29.2	0.2	33.8	0.0	31.7	0.1
RV 3	27.7	0.3	33.8	0.2	31.6	0.2
RV 4	29.3	0.3	32.7	0.4	30.5	0.4
RV 5	29.2	0.3	31.4	0.1	29.1	0.2
IV 1	27.4	0.1	28.3	0.0	25.8	0.0
IV 2	26.8	0.3	28.2	0.3	25.6	0.3
IV 3	28.6	0.2	27.2	0.0	24.6	0.0
IV 4	27.4	0.2	28.1	0.5	25.5	0.5
IV 5	26.0	0.2	28.5	0.3	26.0	0.3
7d-IV 1	28.2	0.3	33.4	0.0	31.4	0.0
7d-IV 2	26.7	0.3	32.7	0.3	30.6	0.3
7d-IV 3	28.3	0.1	32.5	0.2	30.4	0.2
7d-IV 4	27.1	0.1	31.4	0.1	29.3	0.1
7d-IV 5	28.3	0.1	30.9	0.0	28.8	0.0
Soil 1	42.8	0.2	42.2	0.1	40.7	0.1
Soil 2	42.3	0.3	41.6	0.1	40.1	0.1
Soil 3	41.7	0.3	40.6	0.1	39.0	0.1
Soil 4	43.2	0.5	40.1	0.1	38.4	0.1
Soil 5	41.3	0.2	39.2	0.0	37.5	0.0

Figure 12 shows the adjustment of the data measured with the hand-held thermal camera and the calibrated camera using the procedure developed and the ortho-mosaic process described. It can be observed that the adjustment between the values is not adequate, with an RMSE of 2.6 °C, a maximum error of 4.7 °C and a SD of 2.7 °C. These errors can be due to: (1) the photogrammetric process of orthoimaging changes the temperature values in each pixel due to the need to apply an operation similar to the resampling method, which implies that the value assigned to each pixel of the final orthoimaging is the result of an average of close values of the different images involved, not allowing the software employed to request the use of the neighbour method closest to the optimal image, which would be ideal according to the principles of remote sensing; (2) the handheld camera is not the best method to perform a vicarious calibration, since its measurement accuracy, although markedly greater than the uncooled sensor on the UAV, is not less than 1 °C. Thus, in future studies we will use a spectroradiometer that measures in the thermal bands. Additionally, a method for selecting the more nadiral image of the set of images that measures the same point will be implemented. Thus, the temperature will be obtained from this image and not from the orthoimage.

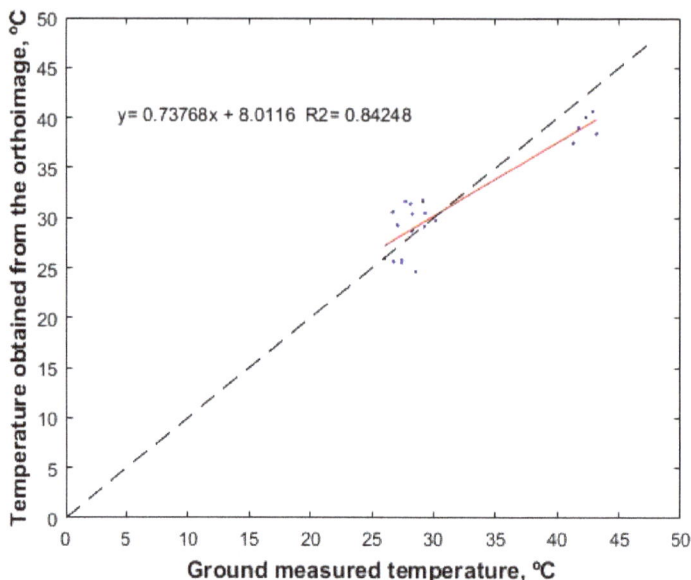

Figure 12. Comparison of the data obtained in the field with the data obtained from the generated geomatic product.

4. Conclusions

To perform accurate measurements of temperature with uncooled thermal cameras it is necessary to perform and adequate camera calibration that considers the effect of the sensor temperature on the measurement. Additionally, a proper photogrammetry process should be implemented to generate high-quality mosaics from low-contrast thermal images. The proposed calibration procedure of uncooled thermal cameras, based on the use of artificial neural networks, which considers as input data the digital level of each pixel and the sensor temperature, noticeably improve accuracy. The temperature of the sensor has a great effect on the measurement accuracy of the thermal sensor. Without calibration errors, close to four degrees can be obtained while calibrating with the proposed methodology the measurement error can be reduced to approximately 1.5 °C.

The application of the Wallis filter to thermal imaging significantly improves the photogrammetric solution, providing a thermal high-quality thermal orthoimage. Similar problems have been found by other researchers that utilized similar cameras [65]. They found temperature differences between −5 °C to 20 °C depending of the time of the day. They performed vicarious calibration but they did not accounted for a laboratory camera calibration as the presented in this manuscript.

Applications based on temperature measurements with uncooled microbolometers should consider accurate camera calibration, proper systems for measuring microbolometer temperature, and a rigorous methodology to perform UAV flights that warranties the measurement precision. Also, images treatment in the photogrammetry workflow should carefully perform to avoid geometric inaccuracies. All these aspects, increase the cost for obtaining a reliable thermal product.

With this research, we detected the need to improve the quality of the vicarious calibration with accurate spectroradiometers and the need to obtain new methodologies to obtain temperature values from the best positioned image for each point. We will focus our future research in these two newly-detected lines of research.

Sensors **2017**, *17*, 2173

Acknowledgments: The authors wish to express their gratitude to the Spanish Ministry of Education and Science (MEC) for funding the project AGL2014-59747-C2-1-R (co-funded by FEDER) and to the IBRASIL Erasmus Mundus for funding the PhD grant of the first author. Part of these results has been obtained thanks to the funding of the BBVA Foundation with the grant *"Beca Leonardo a Investigadores y Creadores Culturales 2017"*.

Author Contributions: All the authors contributed equally to this study.

Conflicts of Interest: The authors declare no conflict of interest.

References

1. Zhang, C.; Kovacs, J.M. The application of small unmanned aerial systems for precision agriculture: A review. *Precis. Agric.* **2012**, *13*, 693–712. [CrossRef]
2. Ballesteros, R.; Ortega, J.F.; Hernández, D.; Moreno, M.A. Applications of georeferenced high-resolution images obtained with unmanned aerial vehicles. Part II: Application to maize and onion crops of a semi-arid region in Spain. *Precis. Agric.* **2014**, *15*, 593–614. [CrossRef]
3. Laliberte, A.S.; Rango, A. Image Processing and Classification Procedures for Analysis of Sub-decimeter Imagery Acquired with an Unmanned Aircraft over Arid Rangelands. *GISci. Remote Sens.* **2011**, *48*, 4–23. [CrossRef]
4. Majidi, B.; Bab-Hadiashar, A. Real time aerial natural image interpretation for autonomous ranger drone navigation. In Proceedings of the Digital Imag Computing Techniques and Application (DICTA 2005), Cairns, QLD, Australia, 6–8 December 2005; pp. 448–453.
5. Berni, J.A.J.; Zarco-Tejada, P.J.; Suárez, L.; Fereres, E. Thermal and narrowband multispectral remote sensing for vegetation monitoring from an unmanned aerial vehicle. *IEEE Trans. Geosci. Remote Sens.* **2009**, *47*, 722–738. [CrossRef]
6. Zarco-Tejada, P.J.; Berjón, A.; López-Lozano, R.; Miller, J.R.; Martín, P.; Cachorro, V.; González, M.R.; De Frutos, A. Assessing vineyard condition with hyperspectral indices: Leaf and canopy reflectance simulation in a row-structured discontinuous canopy. *Remote Sens. Environ.* **2005**, *99*, 271–287. [CrossRef]
7. Kingston, D.B.; Beard, A.W. Real-time Attitude and Position Estimation for Small UAVs using Low-cost Sensors. In Proceedings of the AIAA 3rd Unmanned Unlimited Technical Conference on Workshop and Exhibit, Chicago, IL, USA, 20–23 September 2004.
8. Ribeiro-Gomes, K.; Hernandez-Lopez, D.; Ballesteros, R.; Moreno, M.A. Approximate georeferencing and automatic blurred image detection to reduce the costs of UAV use in environmental and agricultural applications. *Biosyst. Eng.* **2016**, *151*, 308–327. [CrossRef]
9. Baluja, J.; Diago, M.P.; Balda, P.; Zorer, R.; Meggio, F.; Morales, F.; Tardaguila, J. Assessment of vineyard water status variability by thermal and multispectral imagery using an unmanned aerial vehicle (UAV). *Irrig. Sci.* **2012**, *30*, 511–522. [CrossRef]
10. Zaman-Allah, M.; Vergara, O.; Araus, J.L.; Tarekegne, A.; Magorokosho, C.; Zarco-Tejada, P.J.; Hornero, A.; Albà, A.H.; Das, B.; Craufurd, P.; et al. Unmanned aerial platform-based multi-spectral imaging for field phenotyping of maize. *Plant Methods* **2015**, *11*, 35. [CrossRef] [PubMed]
11. Senthilnath, J.; Kandukuri, M.; Dokania, A.; Ramesh, K.N. Application of UAV imaging platform for vegetation analysis based on spectral-spatial methods. *Comput. Electron. Agric.* **2017**, *140*, 8–24. [CrossRef]
12. Senthilnath, J.; Dokania, A.; Kandukuri, M.; Ramesh, K.N.; Anand, G.; Omkar, S.N. Detection of tomatoes using spectral-spatial methods in remotely sensed RGB images captured by UAV. *Biosyst. Eng.* **2016**, *146*, 16–32. [CrossRef]
13. Bellvert, J.; Marsal, J.; Girona, J.; Zarco-Tejada, P.J. Seasonal evolution of crop water stress index in grapevine varieties determined with high-resolution remote sensing thermal imagery. *Irrig. Sci.* **2014**, *33*, 81–93. [CrossRef]
14. Bellvert, J.; Zarco-Tejada, P.J.; Girona, J.; Fereres, E. Mapping crop water stress index in a "Pinot-noir" vineyard: Comparing ground measurements with thermal remote sensing imagery from an unmanned aerial vehicle. *Precis. Agric.* **2014**, *15*, 361–376. [CrossRef]
15. Elvanidi, A.; Katsoulas, N.; Bartzanas, T.; Ferentinos, K.P.; Kittas, C. Crop water status assessment in controlled environment using crop reflectance and temperature measurements. *Precis. Agric.* **2017**, *18*, 332–349. [CrossRef]

16. Ortega-Farías, S.; Ortega-Salazar, S.; Poblete, T.; Kilic, A.; Allen, R.; Poblete-Echeverría, C.; Ahumada-Orellana, L.; Zúñiga, M.; Sepúlveda, D. Estimation of energy balance components over a drip-irrigated olive orchard using thermal and multispectral cameras placed on a helicopter-based unmanned aerial vehicle (UAV). *Remote Sens.* **2016**, *8*, 638.

17. Santesteban, L.G.; Di Gennaro, S.F.; Herrero-Langreo, A.; Miranda, C.; Royo, J.B.; Matese, A. High-resolution UAV-based thermal imaging to estimate the instantaneous and seasonal variability of plant water status within a vineyard. *Agric. Water Manag.* **2017**, *183*, 49–59. [CrossRef]

18. Gómez-Candón, D.; Virlet, N.; Labbé, S.; Jolivot, A.; Regnard, J.L. Field phenotyping of water stress at tree scale by UAV-sensed imagery: New insights for thermal acquisition and calibration. *Precis. Agric.* **2016**, *17*, 786–800. [CrossRef]

19. Rud, R.; Cohen, Y.; Alchanatis, V.; Levi, A.; Brikman, R.; Shenderey, C.; Heuer, B.; Markovitch, T.; Dar, Z.; Rosen, C.; et al. Crop water stress index derived from multi-year ground and aerial thermal images as an indicator of potato water status. *Precis. Agric.* **2014**, *15*, 273–289. [CrossRef]

20. Möller, M.; Alchanatis, V.; Cohen, Y.; Meron, M.; Tsipris, J.; Naor, A.; Ostrovsky, V.; Sprintsin, M.; Cohen, S. Use of thermal and visible imagery for estimating crop water status of irrigated grapevine. *J. Exp. Bot.* **2007**, *58*, 827–838. [CrossRef] [PubMed]

21. DeJonge, K.C.; Taghvaeian, S.; Trout, T.J.; Comas, L.H. Comparison of canopy temperature-based water stress indices for maize. *Agric. Water Manag.* **2015**, *156*, 51–62. [CrossRef]

22. Allen, R.G.; Tasumi, M.; Morse, A.; Trezza, R.; Wright, J.L.; Bastiaanssen, W.; Kramber, W.; Lorite, I.; Robinson, C.W. Satellite-based energy balance for mapping evapotranspiration with internalized calibration (METRIC)—Applications. *J. Irrig. Drain. Eng.* **2007**, *133*, 395–406. [CrossRef]

23. Bastiaanssen, W.G.M.; Menenti, M.; Feddes, R.A.; Holtslag, A.A.M. A remote sensing surface energy balance algorithm for land (SEBAL). 1. Formulation. *J. Hydrol.* **1998**, *212*, 198–212. [CrossRef]

24. Gowda, P.H.; Chavez, J.L.; Colaizzi, P.D.; Evett, S.R.; Howell, T.A.; Tolk, J.A. ET mapping for agricultural water management: Present status and challenges. *Irrig. Sci.* **2008**, *26*, 223–237. [CrossRef]

25. Roerink, G.J.; Su, Z.; Menenti, M. S-SEBI: A simple remote sensing algorithm to estimate the surface energy balance. *Phys. Chem. Earth Part B Hydrol. Ocean. Atmos.* **2000**, *25*, 147–157. [CrossRef]

26. Norman, J.M.; Kustas, W.P.; Humes, K.S. Source approach for estimating soil and vegetation energy fluxes in observations of directional radiometric surface temperature. *Agric. For. Meteorol.* **1995**, *77*, 263–293. [CrossRef]

27. Brenner, A.J.; Incoll, L.D. The effect of clumping and stomatal response on evaporation from sparsely vegetated shrublands. *Agric. For. Meteorol.* **1997**, *84*, 187–205. [CrossRef]

28. Poblete-Echeverría, C.; Ortega-Farias, S. Estimation of actual evapotranspiration for a drip-irrigated merlot vineyard using a three-source model. *Irrig. Sci.* **2009**, *28*, 65–78. [CrossRef]

29. Allen, R.; Irmak, A.; Trezza, R.; Hendrickx, J.M.H.; Bastiaanssen, W.; Kjaersgaard, J. Satellite-based ET estimation in agriculture using SEBAL and METRIC. *Hydrol. Process.* **2011**, *25*, 4011–4027. [CrossRef]

30. Liou, Y.A.; Kar, S.K. Evapotranspiration estimation with remote sensing and various surface energy balance algorithms—A review. *Energies* **2014**, *7*, 2821–2849. [CrossRef]

31. Su, Z. The Surface Energy Balance System (SEBS) for estimation of turbulent heat fluxes. *Hydrol. Earth Syst. Sci.* **2002**, *6*, 85–100. [CrossRef]

32. Allen, R.G.; Tasumi, M.; Morse, A. Satellite-based evapotranspiration by METRIC and Landsat for western states water management. *US Bur. Reclam. Evapotranspiration Workshop* 2005, 1–19. [CrossRef]

33. Ramos, J.G.; Cratchley, C.R.; Kay, J.A.; Casterad, M.A.; Martínez-Cob, A.; Domínguez, R. Evaluation of satellite evapotranspiration estimates using ground-meteorological data available for the Flumen District into the Ebro Valley of N.E. Spain. *Agric. Water Manag.* **2009**, *96*, 638–652. [CrossRef]

34. Elhaddad, A.; Garcia, L.A.; Chavez, J.L. Using a surface energy balance model to calculate spatially distributed actual evapotranspiration. *J. Irrig. Drain. Eng.* **2011**, *137*, 17–26. [CrossRef]

35. Poblete-Echeverría, C.; Ortega-Farias, S. Calibration and validation of a remote sensing algorithm to estimate energy balance components and daily actual evapotranspiration over a drip-irrigated Merlot vineyard. *Irrig. Sci.* **2012**, *30*, 537–553. [CrossRef]

36. Allen, R.G.; Pereira, L.S.; Raes, D.; Smith, M. Crop Evapotranspiration: Guidelines for Computing Crop Requirements FAO Irrigation and Drainage Paper No. 56. *FAO Rome* **1998**, *300*, D05109.

37. Awan, U.K.; Anwar, A.; Ahmad, W.; Hafeez, M. A methodology to estimate equity of canal water and groundwater use at different spatial and temporal scales: A geo-informatics approach. *Environ. Earth Sci.* **2016**, *75*, 1–13. [CrossRef]

38. Mahmoud, S.H.; Alazba, A.A. A coupled remote sensing and the Surface Energy Balance based algorithms to estimate actual evapotranspiration over the western and southern regions of Saudi Arabia. *J. Asian Earth Sci.* **2016**, *124*, 269–283. [CrossRef]

39. Idso, S.B.; Jackson, R.D.; Pinter, P.J.; Reginato, R.J.; Hatfield, J.L. Normalizing the stress-degree-day parameter for environmental variability. *Agric. Meteorol.* **1981**, *24*, 45–55. [CrossRef]

40. King, B.A.; Shellie, K.C. Evaluation of neural network modeling to predict non-water-stressed leaf temperature in wine grape for calculation of crop water stress index. *Agric. Water Manag.* **2016**, *167*, 38–52. [CrossRef]

41. Gade, R.; Moeslund, T.B. Thermal cameras and applications: A survey. *Mach. Vis. Appl.* **2014**, *25*, 245–262. [CrossRef]

42. Sheng, H.; Chao, H.; Coopmans, C.; Han, J.; McKee, M.; Chen, Y. Low-cost UAV-based thermal infrared remote sensing: Platform, calibration and applications. In Proceedings of the 2010 IEEE/ASME International Conference on Mechatronic Embedded Systems and Applications (MESA), Qingdao, China, 15–17 July 2010; pp. 38–43.

43. Jensen, A.M.; McKee, M.; Chen, Y. Procedures for processing thermal images using low-cost microbolometer cameras for small unmanned aerial systems. In Proceedings of the 2014 IEEE International Geoscience and Remote Sensing Symposium, Québec City, QC, Canada, 13–18 July 2014; pp. 2629–2632.

44. Budzier, H.; Gerlach, G. Calibration of uncooled thermal infrared cameras. *J. Sens. Sens. Syst.* **2015**, *4*, 187–197. [CrossRef]

45. Bishop, C.M. Neural networks for pattern recognition. *J. Am. Stat. Assoc.* **1995**, *92*, 482.

46. Mcevoy, H.; Simpson, R.; Machin, G. Quantitative InfraRed Thermography Review of current thermal imaging temperature calibration and evaluation facilities, practices and procedures, across EURAMET. In Proceedings of the 11th International Conference on Quantitative InfraRed Thermography (QIRT 2012), Naples, Italy, 11–14 June 2012.

47. Botterill, T.; Mills, S.; Green, R. Real-time aerial image mosaicing. In Proceedings of the International Conference Image and Vision Computing New Zealand, Queenstown, New Zealand, 8–9 November 2010; pp. 1–8.

48. Ghosh, D.; Kaabouch, N. A survey on image mosaicing techniques. *J. Vis. Commun. Image Represent.* **2016**, *34*, 1–11. [CrossRef]

49. Pierrot-Deseilligny, M.; Clery, I. Apero, an Open Source Bundle Adjusment Software for Automatic Calibration and Orientation of Set of Images. In Proceedings of the ISPRS Symposium, 3DARCH11, Trento, Italy, 2–4 March 2011.

50. Wu, C. VisualSFM: A Visual Structure from Motion System. Available online: http://ccwu.me/vsfm/doc.html (accessed on 21 September 2017).

51. Snavely, N.; Seitz, S.M.; Szeliski, R. Photo tourism: Exploring Photo Collections in 3D. *ACM Trans. Graph.* **2006**, *25*, 835–846. [CrossRef]

52. Kastek, M.; Dulski, R.; Trzaskawka, P.; Bieszczad, G. Sniper detection using infrared camera: Technical possibilities and limitations. In *SPIE Defense, Security, and Sensing*; International Society for Optics and Photonics: Orlando, FL, USA, 2010.

53. Hartmann, W.; Tilch, S.; Eisenbeiss, H.; Schindler, K. Determination of the Uav Position By Automatic Processing of Thermal Images. *ISPRS-Int. Arch. Photogramm. Remote Sens. Spat. Inf. Sci.* **2012**, *XXXIX-B6*, 111–116. [CrossRef]

54. Smoorenburg, M.; Volze, N.; Tilch, S.; Hartmann, W.; Naef, F.; Kinzelbach, W. Evaluation of Thermal Infrared Imagery Acquired with an Unmanned Aerial Vehicle for Studying Hydrological Processes. In Proceedings of the EGU General Assembly Conference Abstracts, Orlando, FL, USA, 5 May 2010.

55. Pérez Álvarez, J.A. Apuntes de Fotogrametría II. *Cent. Univ. Mérida* **2001**, *53*, 1689–1699.

56. Kou, F.; Chen, W.; Li, Z.; Wen, C. Content adaptive image detail enhancement. *IEEE Signal Process. Lett.* **2015**, *22*, 211–215. [CrossRef]

57. Guidi, G.; Gonizzi, S.; Micoli, L.L. Image pre-processing for optimizing automated photogrammetry performances. *ISPRS Ann. Photogramm. Remote Sens. Spat. Inf. Sci.* **2014**, *II-5*, 145–152. [CrossRef]

58. Ballesteros, R.; Ortega, J.F.; Hernández, D.; Moreno, M.A. Applications of georeferenced high-resolution images obtained with unmanned aerial vehicles. Part I: Description of image acquisition and processing. *Precis. Agric.* **2014**, *15*, 593–614. [CrossRef]

59. Wallis, K.F. Seasonal adjustment and relations between variables. *J. Am. Stat. Assoc.* **1974**, *69*, 18–31. [CrossRef]

60. Gaiani, M.; Remondino, F.; Apollonio, F.I.; Ballabeni, A. An advanced pre-processing pipeline to improve automated photogrammetric reconstructions of architectural scenes. *Remote Sens.* **2016**, *8*, 178. [CrossRef]

61. González-Aguilera, D.; López-Fernández, L.; Rodriguez-Gonzalvez, P.; Guerrero, D.; Hernandez-Lopez, D.; Remondino, F.; Menna, F.; Nocerino, E.; Toschi, I.; Ballabeni, A.; et al. Development of an all-purpose free photogrammetric tool. *ISPRS Int. Arch. Photogramm. Remote Sens. Spat. Inf. Sci.* **2016**, *41*, 31–38.

62. Jazayeri, I.; Fraser, C. Interest operators in close-range object reconstruction. *Int. Arch. Photogramm. Remote Sens. Spat. Inf. Sci. Beijing* **2008**, *37*, 69–74.

63. *UNEP World Atlas of Desertification*, 2nd ed.; Publications Office of the European Union: Luxembourg, 1997.

64. Sánchez, J.M.; Kustas, W.P.; Caselles, V.; Anderson, M.C. Modelling surface energy fluxes over maize using a two-source patch model and radiometric soil and canopy temperature observations. *Remote Sens. Environ.* **2008**, *112*, 1130–1143. [CrossRef]

65. Torres-Rua, A. Vicarious calibration of sUAS microbolometer temperature imagery for estimation of radiometric land surface temperature. *Sensors* **2017**, *17*, 1499. [CrossRef] [PubMed]

sensors

MDPI

Article

Accuracy Analysis of a Dam Model from Drone Surveys

Elena Ridolfi *, Giulia Buffi, Sara Venturi and Piergiorgio Manciola

DICA Department of Civil and Environmental Engineering, University of Perugia, 06125 Perugia, Italy;
giulia.buffi@unipg.it (G.B.); sarvent@gmail.com (S.V.); piergiorgio.manciola@unipg.it (P.M.)
* Correspondence: elena.ridolfi@unipg.it; Tel.: +39-075-585-3619

Received: 1 June 2017; Accepted: 31 July 2017; Published: 3 August 2017

Abstract: This paper investigates the accuracy of models obtained by drone surveys. To this end, this work analyzes how the placement of ground control points (GCPs) used to georeference the dense point cloud of a dam affects the resulting three-dimensional (3D) model. Images of a double arch masonry dam upstream face are acquired from drone survey and used to build the 3D model of the dam for vulnerability analysis purposes. However, there still remained the issue of understanding the real impact of a correct GCPs location choice to properly georeference the images and thus, the model. To this end, a high number of GCPs configurations were investigated, building a series of dense point clouds. The accuracy of these resulting dense clouds was estimated comparing the coordinates of check points extracted from the model and their true coordinates measured via traditional topography. The paper aims at providing information about the optimal choice of GCPs placement not only for dams but also for all surveys of high-rise structures. The knowledge a priori of the effect of the GCPs number and location on the model accuracy can increase survey reliability and accuracy and speed up the survey set-up operations.

Keywords: dam survey; monitoring; UAV; ground control point; marker optimization; dense point cloud; vulnerability analysis; accuracy

---

1. Introduction

In recent years, dam safety has acquired increasing attention because of the high number of accidents and failures that have occurred. For instance, in the United States from 2005 to 2013 the State Dam Safety Program reported 173 dam failures and 587 occurrences that, without intervention, would likely have resulted in dam failure [1]. The typical causes of failures are foundation deterioration, including uneven settlement and earthquakes [2]; overtopping as a result of inadequate spillway crests [3], debris blockage of spillways and settlement of the dam crest [4]; piping, including high pore pressure and embankment slips [5] and others, such as improper construction, defective materials and also acts of warfare [6]. The need for sharing experiences related to this issue led to the establishment of the International Commission on Large Dams (ICOLD) whose mission is the dissemination of information to increase designers' and managers' awareness of the events which can lead to a disaster [7,8].

In this framework, monitoring operations and vulnerability assessment of dams plays a crucial role in preventing catastrophes and thus safeguarding human lives. Because of the large dimensions and the low accessibility of dams, a unique opportunity for monitoring is offered by Unmanned Aerial Vehicle (UAV) systems. Indeed, UAVs are extremely advantageous for visual investigation of large-scale structures such as dams and retention walls [9]. The use of UAVs has acquired a key role because of their fast and low cost operations and of the possibility of reaching places that are otherwise difficult or impossible to access directly [10–12]. UAV surveys find applications in several fields, for instance UAVs are used for vegetation monitoring [13], precision agriculture enhancement [14], and information acquisition concerning damages caused by natural disasters such as earthquakes [15,16]. In hydrology, drones are used to measure open channel surface velocities [17,18].

As structures are subject to deterioration due to increasing loads, weather conditions and ageing processes, drones offer a unique opportunity. As a matter of fact, conventional inspections are based on visual investigations which are time consuming, require technical experts and are therefore expensive. In this framework, drones provide an important contribution to strategies for monitoring of structures. For instance, Hallermann et al. [19] presented a methodology for inspecting ageing structures combining conventional inspection measurements with modern photogrammetric computer vision methods for geo-referenced structures, 3D-modeling and automatic post-flight damage detection. Achille et al. [20] made use of drones to survey vertical structures after an earthquake occurred in Mantua (IT). Grenzdörffer et al. [21] coupled terrestrial laser scanner (TLS) measurements with drone photogrammetry to detail a cultural monument in Germany. Hallermann and Morgenthal [22] and Hallermann et al. [23] used a drone to inspect bridges and viaducts.

In this framework, the use of the UAV technique for inspection and geometry recreation of dams is highly appropriate [24] and it can substantially improve monitoring and survey operations. Data acquired during the survey can be reliably considered as the basis for the construction of a 3D model for vulnerability studies. Indeed, a dam is a multi-hazard vulnerable structure that can be severely affected by earthquakes, flooding and terrain stability-related issues. The 3D model of the dam can be utilized to build a finite element (FE) model and perform static and dynamic analyses of the structure. Moreover, many authors have dealt with the simulation of dam breaks in a 3D hydraulic framework to better represent the vertical acceleration influencing the flow [25–30]. Thus, the availability of a 3D model of the dam is of great interest for providing a proper representation of the breaking event. Among the other applications of a 3D dam model, we can also list the 3D model of spillways to investigate the dynamics of the flow that spills over [31,32].

The 3D model of an object is built up from the images acquired during the UAV survey. Indeed, thanks to the Structure from Motion (SfM) technique, it is possible to build a dense point cloud (i.e., a 3D model) of an object from the 2D images acquired during a drone survey [33].

The SfM technique revolutionized three-dimensional topographic surveys in many fields, such as physical geography, by opening data collection and processing to a wider public [34]. Indeed, the SfM is a cheap technique that does not need specialized supervision. These two characteristics played the main role in the dissemination of SfM as highlighted by Micheletti et al. [35] who illustrated the potential of SfM applications in geomorphological research. Micheletti et al. [36] investigated the potential of freely available SfM tools to process high-resolution topographic and terrain data acquired through a smartphone. Among the other applications, SfM is used to investigate wave run-up [37], to determine soil erosion [38,39], to map landslides and to assess glacier movement [40,41]. Many authors have coupled the SfM technique with drones for surveying and modeling structures. Indeed, thanks to drones it is possible to acquire images of entire buildings including the roof and other parts inaccessible to the scanner [42]. For instance, Bolognesi et al. [43] detailed the "Delizia del Verginese" Castle in central Italy; Koutsoudis et al. [44] reconstructed an Ottoman monument located in Greece, estimating the accuracy of the data produced by a multi-image 3D reconstruction technique in terms of surface deviation and distance measurement accuracy. Among the other benefits, 3D modeling of cultural heritage sites has led to scientific and cost-effective improvements in documenting and archiving operations [45]. Westoby et al. [46] applied the SfM technique to a breached moraine-dam; Hallermann et al. [19] obtained a high-resolution orthophoto of a dam located in Germany to inspect its surface and to identify damages. Reagan et al. [47] combined autonomous flight with 3D digital image correlation to inspect bridges.

To georeference the dense point cloud and thus obtain an in-scale model of a specific object, the coordinates of a specific number of points (i.e., ground control points; GCPs) deployed on the object are used. The coordinates of GCPs are estimated by traditional survey techniques acquiring the coordinates of objects named markers. Markers are usually square shaped objects deployed on the dam surface before the survey. Subsequently, to determine the accuracy of the 3D model, it is necessary to compare the coordinates of check points (i.e., CPs) lying on the dense point cloud with their actual coordinates, acquired by traditional topographic methods. Therefore, the accuracy of the 3D model is strongly influenced by the location and number of GCPs used to georeference the model itself.

In literature, some authors have dealt with the effects of several features on the dense point cloud accuracy, such as the flight altitude, the inclination of the camera's optimal axis, the image network geometry, image matching performance, surface texture and lighting conditions, GCPs number, positioning and accuracy [48]. For instance, Bolognesi et al. [43] investigated the accuracy of the dense point cloud of a historic castle by varying the flight altitude, the camera optical axis inclination and the number and location of GCPs. Barry and Coakley [49] estimated the accuracy achievable using a UAV on a 2-hectar site varying the number of GCPs. Tahar [50] evaluated the effect of different number of GCPs on the photogrammetric survey on a hilly area in Malaysia, while Tahar et al. [51] assessed the effect of position and number of GCPs on DEM generation. Mesas-Carrascosa et al. [52,53] analyzed the effect of three flight altitudes, flight modes (stop and cruising modes) and ground control point settings on ortho-mosaicked images. According to literature guidelines, GCPs should be widely distributed across the target area [54] and at the edge or outside [55] to enclose the area of interest [56–59]. According to Harwin and Lucieer [60], GCPs distribution needs to be adapted to the surveyed object and to the distance of the UAV. GCPs should be placed between 1/5 and 1/10 of the distance from the UAV to the surveyed object. Work investigating the effect of marker configurations on the dense point cloud of a high-rise building and, more specifically, of a dam is still to be done.

The placement of control points, especially in giant structures such as dams, is a time and money-consuming task. Moreover, some dam portions can be reached during specific seasons only because of the variable hydrostatic level in the reservoir. In this framework, the knowledge a priori of the effect of GCPs location choice on the accuracy of the dense cloud is relevant information that could not only increase the 3D model reliability but also speed up the experimental set up and thus, the dam survey itself.

This work focuses on the specific issue of marker deployment, while the other features are kept constant. This allows us to investigate this issue in detail. Given that no study on the effects of GCPs deployment on a masonry dam yet exists, the original contribution of this paper is the investigation of the accuracy affecting the 3D model of a dam obtained by a drone survey exploring different GCPs layouts. Results obtained by this study can find application for the improvement of large building surveys.

As the connection between the dam and the terrain plays a key role in ensuring the stability of the dam itself, the survey needs to be as accurate as possible to correctly represent and monitor the interconnection between dam and terrain. Thus, a second original contribution is represented by the investigation of the GCPs location effect on the dam's boundaries (i.e., the abutments). Moreover, the 3D model accuracy is highly affected by the presence of singularities, such as the spillways. Indeed, a proper location of GCPs needs to take into account the presence of openings in order to obtain an accurate dense point cloud.

This paper investigates the case study of the Ridracoli dam (IT); an arch dam located in central Italy constructed principally for drinkable water supply purposes. An accurate assessment of the vulnerability of this structure is of utmost importance to ensure the continuous availability of the water resource from its multipurpose retention basin [61]. It is a masonry arch-shaped dam with a height of 103.5 m and crowning length equal to 32 m. The dam widths at the foundations and at the crowning are equal to 30 m and 7 m, respectively. For this type of structure, the mean vertical error can be assumed acceptable if equal to or lower than the 0.1% of the dam height. In fact, this value is consistent with finite element analysis [62] and hydraulic modeling [63]. The analysis is performed on the upstream face of the dam generating a unique dense point cloud of this portion. The upstream face was chosen for three reasons: (1) markers have a higher concentration and are more uniformly spaced on the upstream face than on the rest of the structure; (2) it offers the possibility of better investigating the spillway openings and the interconnection terrain-structure; (3) it highlights the similarity between the upstream face and other high-rise buildings. In fact, this paper aims to provide useful guidelines on marker location that could be valid also for other high-rise constructions. It is worth underlining that the downstream face of the dam is inaccessible, therefore it was not possible to deploy markers on it.

The paper is organized as follows: first, the case study area is presented and details about the experimental set up are provided. Several configurations of markers are presented and the survey technique to estimate marker coordinates is presented. Then, the drone survey is reported. Secondly, we detail the dense point cloud construction by varying GCPs layout. The errors between CPs coordinates on the dam model and on the actual dam are estimated and analyzed for each layout. In the results and discussion section we summarize our findings, focusing on the effects of error on dam boundaries.

2. Case Study and Experimental Set-Up Description

The markers layout assessment is performed using the Ridracoli dam as a case study. The dam was named after the homonymous lake and it was built on the Bidente River in the province of Forlì –Cesena (central Italy), generating a basin of about 33 million m^3 of water (Figure 1). The dam supplies water to 48 municipalities in provinces throughout central Italy and, since 1989, to the Republic of San Marino [64]. The Ridracoli dam has a double-curvature arch-gravity structure with a maximum height of 103.5 m and a crest length of 432 m at 561 m a.s.l. The structure is divided into 27 independent parts called ashlars that avoid cracks that can result from hydrostatic and thermal loads. Vertical contraction joints ensure the continuity of the structure. Eight free spillway gates, located in the centre of the dam crown, allow the overflow of the dam (Figure 2).

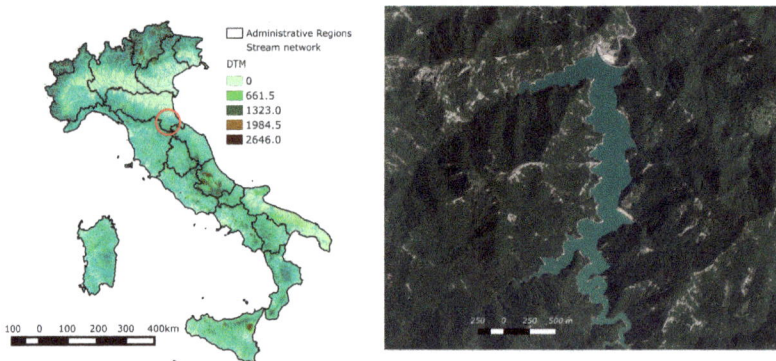

Figure 1. DTM of Italy (left panel), the red circle indicates the case study area, while the dam case study i.e., the Ridracoli dam and the Ridracoli lake are on the right panel.

Figure 2. Ridracoli dam while overflowing. Small panel: HIGHONE 4HSEPRO quadrotor mounting a gimbal system and a SONY Alpha 7R while at rest.

In the following sub-sections the drone survey and the topographical technique to acquire marker coordinates will be detailed. The materialization of targets to be used either as Ground Control Points (GCPs) for georeferencing or as independent Check Points (CPs) during tests is presented.

Experimental Set-Up

The aerial platform is a HighOne 4HSE Pro quadrotor (Italdrone, Ravenna, Italy) mounting a gimbal system and an Alpha 7R, 36.4 Mpix Full frame camera (Sony) oriented with its axis along the perpendicular (Figure 2). The gimbal compensates drone vibrations due to the wind and to the flight operations. The lens is a 35 mm f/9 and the Sony Alpha 7 camera has the AF—Autofocus—Lock feature that manages the framing dimensions in relation to the subject characteristics and also allows an auto-focus procedure. The camera focus, after being tested on a portion of the structure, was set to infinity during the survey. The dam, the ancillary structures and the surrounding land were inspected by drone. Around 200 images of the upstream face were shot. To georeference the frames, before the drone survey, a set of markers were deployed on the dam structure to be used as Ground Control Points. Each marker has a squared shape and is 0.40 m high (Figure 3).

Figure 3. Upstream face of the Ridracoli dam. The red circles indicate marker positions. The small panel shows a marker applied on the structure.

In two different moments of the year, sixty regularly-spaced markers were placed on three rows on the upstream face (Figure 3). In August 2015, two rows of markers were applied on the crest railing and on the upstream surface at the hydrostatic level of the time (i.e., 543.28 m a.s.l.) using a boat. While in October 2015 a third row of markers was deployed at the new hydrostatic level (i.e., 533.65 m a.s.l.). The UAV survey of the upstream face was performed after this last marker deployment to ensure uniform shooting conditions of the frames [65]. The image size is 7360 × 4912 pixels with a resolution of 350 dpi. The images were shot at a distance of about 15 m from the dam; the Ground Sample Distance (GSD) equals 2.1 mm. The images are shot every 1.87 s and the images overlap by more than 70%. An example is provided in Figure 4.

Figure 4. Images overlap by more than 70%.

The camera positions are represented in Figure 5; because of the great extent of the structure an enlargement of the two boundaries is provided.

Figure 5. Camera positions (upper panel) and two enlargements for the left and right abutments (lower left and right panels, respectively).

We recall that this paper aims at evaluating the influence of both GCPs layout and number on the accuracy of a 3D dense point cloud of the structure. Moreover, we want to determine which is the effect of the GCPs location choice on the dense point cloud at the boundaries of the structures, i.e., at both the right and the left abutments. To this end, a high number of markers (i.e., 60) were placed on the dam upstream face. To analyze the effects of marker location, 25 different GCPs layouts were chosen. In each layout, markers are used either as GCPs to georeference the frames or as CPs to test the resulting model accuracy. The former are represented as black circles, the latter as empty circles (Table 1). Marker configurations are grouped according to the GPCs density on a 9 × 9 square marker formation. This representation is useful for analyzing the effects of GCPs location, the density being equal. For instance, to evaluate the effects of different marker configurations, the pattern being equal, the following three groups are investigated: *b*, *g* and *j*; *a*, *d* and *r*; *f*, *n* and *y*. To evaluate the effect of deploying GCPs along one row at different elevations, the layouts *a*, *d*, *e*, *l*, *f*, *p*, *r* and *y* are analyzed. For instance, in layouts *a*, *d* and *r*, GCPs are placed only on the dam crest (i.e., the first row) with consistently different spacing. In a similar way, in layouts f and y GCPs are placed on the upper and lower rows with variable density.

To evaluate how the GPCs location effects the dense point clouds at the boundaries, we chose configurations that are similar but with different GCPs numbers and positions at the abutments. This is the case of the two groups of configurations c and e and i and n. Each couple has the same GCPs density, however, only one of the two supplies markers at the dam boundaries. It is worth underlining that also configurations with a very low number of GCPs were chosen to investigate the widest range possible of GCPs configurations.

Table 1. Marker layouts with IDs ranging from a to z. The black circles are the markers used as GCPs, the real coordinates of which are used to georeference the dense point cloud, while the empty circles are the markers used as control points (CPs). The coordinates of CPs are extracted from the model and compared with their actual coordinates to evaluate the accuracy of the model.

Density		Layout					
1/9	a		b				
1/6	c		d		e		
2/9	f		g				
1/3	h		i		j		
1/3	l		m		n		
1/3	o		p				
1/3	q		r				
1/2	s		t		u		
5/9	v						
2/3	w		x		y		
5/6	z						

The coordinates of the markers were acquired through a traditional survey technique using a Total Station TS30 (Leica-Geosystems). To this end, a pre-existing geodetic network was used for monitoring purposes. The existing network consists of four vertices materialized by little pillars (i.e., BS, SS, DS and BD, Figure 6). Besides the existing network, a new one was setup with 7 vertices materialized by topographic nails fixed on the ground (Figure 6). The new network was connected to the first one through topographic triangulation measurements.

To improve the accuracy of the measured coordinates, a hyper-deterministic scheme was adopted and the acquired observations were treated rigorously, executing least-squares compensation of topographic measurements. The standard deviation of each point is lower than 1 cm along the three directions. The mean value of all standard deviations equals 1.0 cm, 1.0 cm and 0.8 cm along the three directions, respectively. Due to the low error values, the points are suitable for either georeferencing the images or for validating the model. For more details about the traditional topographic survey, the reader can refer to the work of Buffi et al. [65].

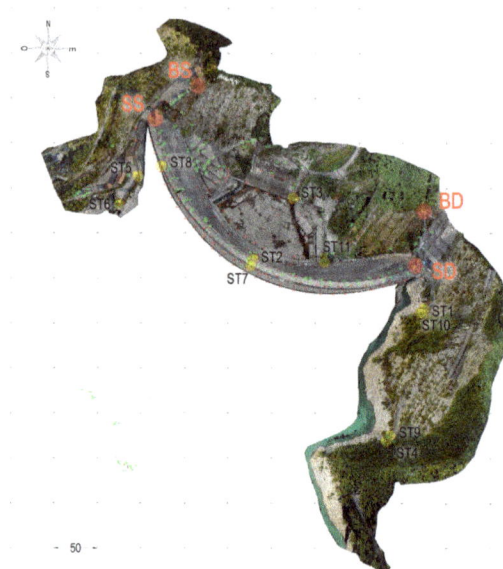

Figure 6. Orthophoto of the dam and the materialization vertices of the two geodetic networks used for the topographic survey. The vertices of the existing network materialized by little pillars are red, the vertices of a new network materialized by topographic nails fixed on the ground are yellow.

3. Image Error Analysis

The frames acquired by UAV are used to build a 3D dense point cloud model of the upstream face through the Structure from Motion (SfM) technique (Figure 7). The SfM technique allows the construction of a three-dimensional dense point cloud model of an object starting from the automatic collimation of frames [33]. At first, the procedure consists in performing a feature point detection and matching, using automatic algorithms. Then, incrementally, this procedure adds images, triangulates matching features and refines the scene using bundle adjustment, allowing for the construction of the 3D point cloud. The Agisoft Photoscan® (vers. 1.2.4) software was employed to build the 3D dense point cloud. To derive the dense point cloud, the procedure is presented in the study by Jaud et al. [66].

Figure 7. UAV dense 3D point cloud of a portion of the upstream face of the Ridracoli dam.

First, the frames are uploaded. On photographs, common tie points are found and matched. The external camera orientation parameters are detected for each image. The Brown's distortion model is used to simulate the lens distortion, and camera calibration parameters are found. This model allows for correcting both radial and tangential distortion. The coordinates of the markers used as GCPs are associated to the corresponding marker centre. This procedure has the purpose of georeferencing the frames on which the markers are used to build the most accurate possible dense point cloud. Secondly, the dense point cloud is built using the camera positions and the images and the CPs coordinates are extracted from the dense point cloud. The camera model has been optimized after the photos aligning. The parameters of the camera model are shown in Table 2.

Table 2. Parameters of the camera model.

Focal lenght (pix)	fx	7424.32
	fy	7428.37
Principal point offset (pix)	cx	3654.48
	cy	2434.65
Skew coefficient (pix)	skew	−5.89
Radial distortion coefficient	k1	0.05
	k2	−0.25
	k3	0.04
Tangential distortion coefficent	p1	0.00
	p2	0.00

The model is built using the Photoscan setting values reported in Table 3. Concerning image coordinates accuracy, the tie point accuracy equals 1 pixel, marker accuracy equals 0.1 pixel. Agisoft Photoscan® (ver. 1.2.4) provides an average RMS error for tie points equal to 0.798 pixel and an RMS average projection error equal to 1.216 pixel (Table 3).

Table 3. Workflow and parameters used as Photoscan® input to build the dense point cloud.

Workflow	
Align Photo	
Accuracy	Medium
Pair pre-selection	Disabled
Point Limit	40,000
Build Preliminary Mesh	
Surface type	Arbitrary
Source data	Sparse
Interpolation	Enabled
Polygon count	Custom
Point classes	All
Import GCPs (GCPs Settings)	
Camera accuracy (m)	10
Marker accuracy (m)	0.005
Tie point accuracy (pix)	1
Build Dense Cloud	
Quality	Medium
Depth filtering	Aggressive

It is possible to assess the errors between the coordinates of the CPs on the dense point cloud and compare their actual coordinates measured by total station. The accuracy of the dense point cloud model depends on the number and configuration of the set of GCPs used to georeference the frames. The CPs coordinates extracted from the dense point cloud are hereafter referred to as 'extracted coordinates'. The accuracy analysis is performed in accordance with geospatial positioning accuracy standards [67]. For each layout, the errors along the elevation (i.e., ε_z), the North and East directions (i.e., ε_Y and ε_X respectively) are evaluated for each CP, comparing measured and extracted coordinate values:

$$\varepsilon_z = Z_{obs} - Z_{estimated} \tag{1}$$

$$\varepsilon_Y = Y_{obs} - Y_{estimated} \tag{2}$$

$$\varepsilon_X = X_{obs} - X_{estimated} \tag{3}$$

Moreover, the error vector lying on the North-East plane (i.e., ε_{XY}) is evaluated for each CP:

$$\varepsilon_{XY} = \sqrt{\varepsilon_X^2 + \varepsilon_Y^2} \tag{4}$$

To estimate the overall quality of each GCPs layout, the Mean Absolute Error (MAE) is evaluated as follows:

$$MAE_j = \frac{1}{N}\sum_{i=1}^{N}|\varepsilon_{j,i}| \tag{5}$$

where $\varepsilon_{j,i}$ is the error along the j-th direction (i.e., North, East, elevation and on the North-East plane, therefore $j = 1, \ldots, 4$) of the i-th check point measured by the topographic survey and M is the number of check points for each specific layout. The Mean Absolute Error measures the overall match between observed and simulated coordinate values. A perfect GCPs layout would result in an MAE equal to zero. This metric estimation does not provide any information about under- or over-estimation, but it determines all deviations from the observed values regardless of the sign.

4. Results and Discussion

To analyze the effect of GCPs layout on the dense point cloud accuracy, the errors between measured and extracted coordinates of CPs are estimated for each layout. As expected, both in the altitude direction and on the North-East plane the MAE value decreases when the number of GCPs increases (Figure 8).

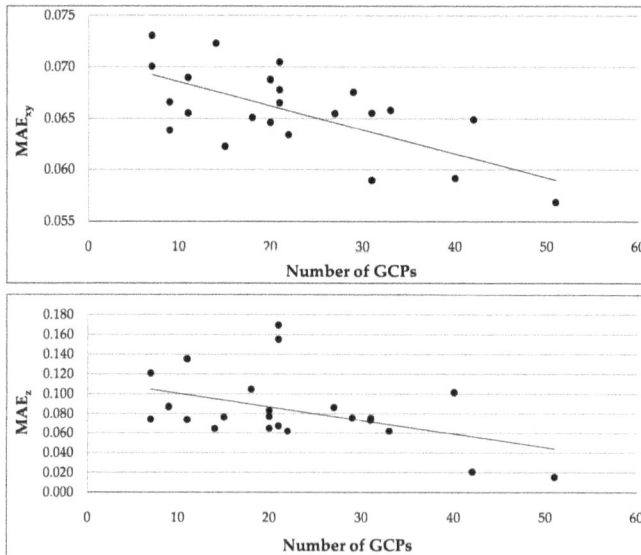

Figure 8. Mean Absolute Error (MAE) on the North-East plane and along the elevation (i.e., z direction) against the number of Ground Control Points (GCPs).

However, it is possible to observe that configurations with the same number of GCPs can have a consistently different MAE value. For instance, this is the case of the configurations *q*, *s* and *t*. They are all characterized by 31 GCPs, but at the same time by a consistently different MAE in the altitude direction and on the North-East plane, as also shown in Table 4. It is interesting to note that they have the same number of GCPs but different density as their pattern is different. Configurations *j*, *l* and *r* consist of 21 GCPs, but are characterized by different error values in all directions. Layouts with a similar pattern (e.g., configurations *f* and *y*) can be characterized by different error values along the elevation. This is due to the fact that it is the combination of both GCPs density and pattern that defines the accuracy of the dense point cloud. Both density and layout are useful as they can consistently speed up the operations for the experimental setting.

Table 4. Ground Control Points (GCPs) ID for each layout grouped by density, corresponding number of GCPs for each layout, Mean Absolute Error values (MAE) for the North, East, elevation (z) directions and the MAE evaluated on the North-East plane for each marker layout.

Density	Combination ID	Number of GCP	MAE_x	MAE_y	MAE_xy	MAE_z
		-	m	m	m	m
1/9	a	7	0.058	0.032	0.070	0.121
	b	7	0.059	0.032	0.073	0.074
1/6	c	9	0.053	0.033	0.067	0.087
	d	11	0.054	0.031	0.066	0.135
	e	9	0.052	0.031	0.064	0.087
2/9	f	14	0.061	0.030	0.072	0.065
	g	11	0.056	0.032	0.069	0.074
1/3	h	20	0.053	0.030	0.065	0.077
	i	22	0.053	0.029	0.063	0.062
	j	21	0.054	0.029	0.066	0.066
	l	21	0.053	0.037	0.068	0.155
	m	20	0.055	0.032	0.069	0.083
	n	20	0.056	0.030	0.069	0.065
	o	15	0.051	0.029	0.062	0.076
	p	18	0.052	0.033	0.065	0.105
	q	31	0.049	0.027	0.059	0.073
	r	21	0.056	0.030	0.067	0.170
1/2	s	31	0.054	0.031	0.066	0.075
	t	31	0.049	0.027	0.059	0.073
	u	29	0.055	0.030	0.068	0.076
5/9	v	33	0.053	0.031	0.066	0.062
2/3	w	40	0.048	0.029	0.059	0.102
	x	27	0.054	0.029	0.065	0.086
	y	42	0.055	0.027	0.065	0.021
5/6	z	51	0.050	0.023	0.057	0.015

The boxplots of errors in the three directions North, East and Elevation and on the North-East plane are reported in Figure 9. The boxplots visually represent the median, the second and third quartiles of errors for each GCPs combination. Moreover, the boxplots highlight the presence of outliers, which are those CPs with higher error value than any other CPs. Along the elevation, the marker layout with the smallest MAE is the *z*, followed by *y* as also shown in Table 4. The former layout is characterized by the highest number of GCPs (i.e., 51), and GCPs are deployed in a way that ensures the almost complete coverage of the dam upstream face. Indeed, in all rows except in the intermediate one, the GCPs are placed in every position.

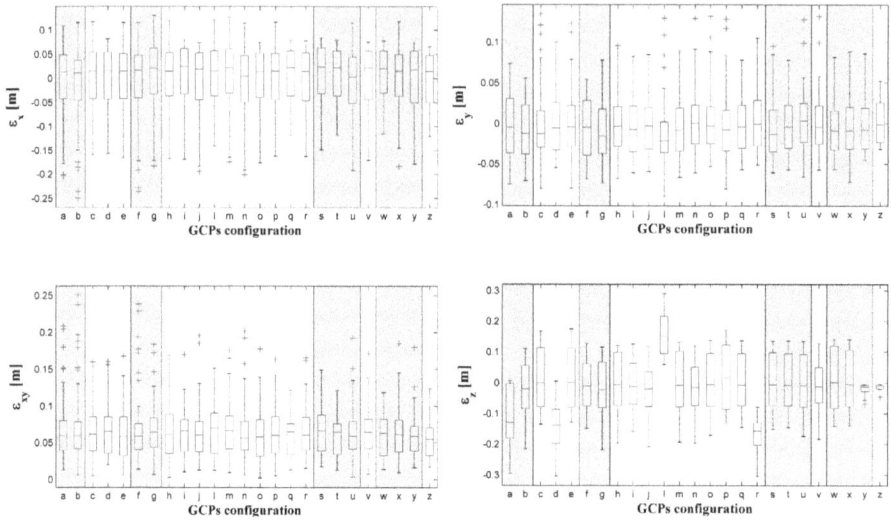

Figure 9. Boxplots of errors (ε) along North, East, elevation (y, x and z) directions and on the North-East plane for each GCPs layout. The background colour groups layout with the same density as in Table 1.

It is worth underlining that the configuration *y* has the second lowest MAE value, even though the second row in not secured with any GCP. This is due to the fact that the first and the lowest rows are responsible for the accuracy of the model if not provided by markers as they are the most difficult to capture during the UAV survey, for two different reasons. Indeed, the last row is close to the lake level, therefore, the UAV cannot move close to the markers and cannot shoot images from every angle, while the low accuracy of the first row is due to the fact that the first row of markers is applied to the balustrade on the dam crest. The frames shot of the balustrade capture the empty spaces in between the balustrade sticks. It results in overexposed frames because of the automatic brightness adjustment of the camera shutter and are then characterized by a low quality. The RMS error of the images that constitute the dense point cloud close to the spillway and to the dam crowning is higher than the other images because it includes the sky and/or water. An example is provided by Figure 10. The images show the same part of the dam. From left to right, the greater the quantity of water in the images, the higher the RMS error of the image. In fact, the RMS errors are, respectively, 0.975, 1.011 and 1.368 pixel. To improve the accuracy of the georeferenced dense point cloud, it is important to deploy a high number of equally-spaced GCPs on the first row. Configurations with a low number of GCPs on the balustrade (i.e., the first row on the upstream face) have high error values.

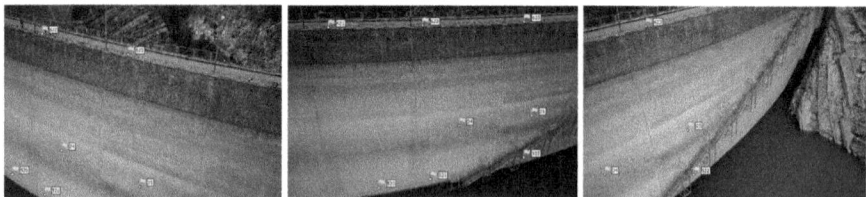

Figure 10. An example of the influence of the percentage of water on the RMS image value: from left to right, the RMS error of the image increases with the percentage of water. In the images, the green flags represent the markers.

The error values along the altitude prove the influence of GCPs pattern (Figure 11). If the GCPs are deployed on the crest, as in combinations *d* and *r*, the crest is characterized by the lowest error values, while the highest error values are concentrated on the lowest row, at the bottom of the upstream face. Markers on the lowest row are difficult to survey because of the proximity to the water. In this case, the MAE value is more conditioned by the GCPs pattern than by their number. A similar behavior can be observed in combinations *l* and *p*, where the GCPs are deployed either on the middle row or on the lowest row, respectively.

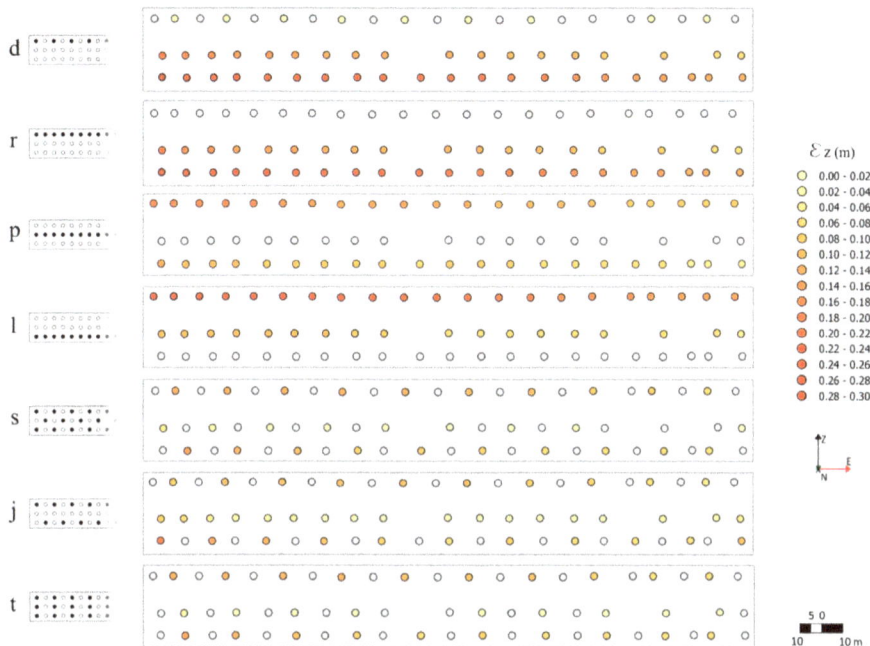

Figure 11. Magnitude of error along the elevation (ε_z) for each marker, considering seven different marker configurations.

In both cases, the highest error values are concentrated on the first row as the balustrade effects the frames quality, as mentioned before. It is interesting to note that when the markers are placed on both the upper and lower rows, the errors decrease consistently, as for combinations *s*, *j*, *t*. The combination pattern and number of GCPs defines the accuracy of the resulting 3D model. As a matter of fact, configurations *r*, *l* and *j* are characterized by the same number of GCPs (i.e., 21), however, their different layouts result in very different error values. It is worth noting that the error along the elevation in percentage is very small when compared to the dam height. The minimum error value is 0.015% of the dam height (i.e., 103.5 m). Indeed, the error along the elevation of configuration *z* is 1.5 cm. Configurations *z*, *y*, *i*, *v*, *f*, *n* and *j* are all characterized by an MAE_z (see Table 4) equal to 0.015 m, 0.021 m, 0.062 m, 0.062 m, 0.065 m, 0.065 m and 0.066 m, respectively, and thus lower than or equal to 0.067% of the dam height. Thus, these 3D models seem to be characterized by error values that are compatible with a Finite Element analysis. Indeed, for a FE analysis, an error of 0.1% of the dam height is acceptable, in this case equal to 0.1 m [62]. It is important to point out that the model of the spillways must be accurate. Indeed, an inaccurate representation of the spillway geometry can cause an inaccurate representation of the overflowing water. Taking into account an error on the evaluation of the overflow equal to the aforementioned 0.067% of the dam height, the consequent error on the peak discharge would be around 4.5% of the design peak. This error is compatible with the uncertainties in the flood peak estimation [63].

The MAE values on the North-East plane lead to the same conclusions (Figure 9, lower left panel). Indeed, the configuration z has the lowest value in this direction, as it does along the elevation. It is followed by the configuration t, in which the GCPs are placed on the three rows in every other position, ensuring a good coverage of all the upstream face.

It is interesting to highlight that the best performing configurations are in agreement with the recommendations of literature. Indeed, GCPs are widely distributed across the target area [54] and at the edge of the upstream face [55]. Results suggest that GCPs distribution needs to be adapted to the object surveyed and to the distance of the UAV from the observed object, as in Harwin and Lucieer [60]. In our work, we found that the "threshold distance" between the GCPs (distance beyond which results do not improve) is 13 m. This distance is greater than that recommended by Harwin and Lucieer [60]. A minimum spacing of 13 m has shown a good performance both for a drone survey and for a classic laser scanning survey, as confirmed by Buffi et al. [68].

The Dense Point Cloud Accuracy at the Dam Boundaries

The boundaries of the dam are among the most important parts of the dam since they define the connections between the structure and the surrounding terrain. However, because of the shape of both right and left abutments, dam boundaries are not easily accessible by drone survey (Figure 12a,b).

Figure 12. (**a**) Left rock abutment and (**b**) right rock abutment and the stairway element on the upstream face.

Thus, images shot of the boundaries are characterized by perspective distortion. Moreover, the presence of the stairways on the right side made it difficult to survey that portion of the dam. To increase the accuracy of the resulting dense point cloud it was necessary to place markers in the proximity of the rocks and under the stairways on the right side of the structure. The high number of frames and the presence of GCPs made it possible to increase the accuracy of the dense point cloud as is evident from the performance of configuration z. To deeply investigate the effects of the GCPs pattern at the boundaries, the two configuration couples c, e and i, n are analyzed. Configurations c and e both have the same GCPs density and the same pattern. However, GCPs are placed in the proximity of the boundaries only in configuration c. This results in high error values at the boundaries, as testified by the number of outliers along the north direction (Figure 9). Layout c leads to high error values at the left and right boundaries because of the absence of the GCPs (Figure 13). The error is higher at the right side where the presence of the stairways makes it difficult to capture the dam face. Layout e differs from c as markers are placed at both sides of the upstream face. However, both configurations are characterized by high error values on the dam crest. It is worth noting that the spillway gates are characterized by a high error value as the openings imply an uncertain dense point cloud when no markers are placed in the proximity. Similarly, configuration b is effected by the absence of GCPs in the proximity of the right abutment where the presence of a stairway and of rocks makes it difficult to capture the dam surface on the North-East plane (Figure 13). The effect on the boundaries is more evident comparing layout *i* with layouts *n* and *b* (Figure 13). GCPs are placed at both sides of the upstream face only in configuration *i*. This leads to a high error value at the boundaries for configuration *n* and *b*, while the error is lower for configuration *i* at the same location.

Moreover, it is worth noting that all configurations except *b* in Figure 13 present a systematic error in areas characterized by a low GCPs density. A more uniform distribution of the GCPs across the dam upstream face would substantially mitigate the systematic errors. This finding is in accordance with James et al. [69] who noticed this behavior while estimating the accuracy of DEMs georeferenced with different sets of GCPs.

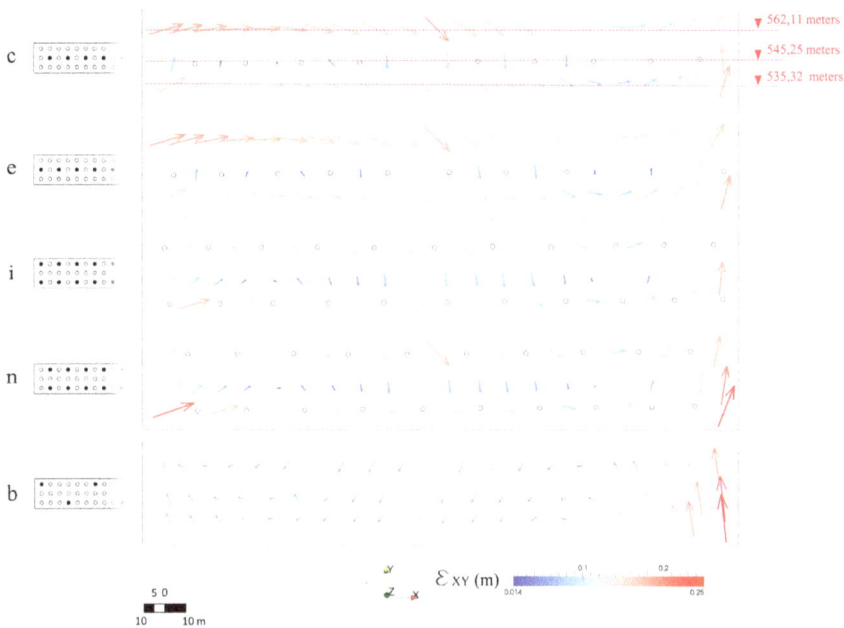

Figure 13. Error along the North-East plane (ε_{XY}) for each CP for GCPs configurations *c, e, i, n, b*. The placement of the arrows (indicating the error magnitude and direction) and of the grey circle is in accordance with the location of the GCPs—positioned at different levels on the dam upstream face—to facilitate the reading of the data. In particular, the figure on the top shows the altitude (meters above sea level) of the location of the GCPs.

5. Conclusions

This work explores the effects of ground control point position and number on the accuracy of a dam dense point cloud, obtained by a drone survey. As GCP deployment is a time and money-consuming task, especially on large structures, this paper aims to provide principles for supporting the GCP deployment on high-rise buildings in order to speed up operations on site.

As expected, the results show that the model performs better when the density of markers is high. However, it is the combination of both GCP pattern and GCP density that determines the gain in accuracy. Results highlight that the error values show a higher variability along the elevation than along any other direction because of the high-rise characteristics of the dam. Therefore, GCPs should be placed at different elevations to increase the accuracy of the resulting dense point cloud. Moreover, where the structure is characterized by discontinuities such as spillway gates, it is necessary to place

GCPs in the proximity of the openings to gain in accuracy. In addition, the presence of a balustrade, water, sky and uniform texture increases the RMS error of the images. In order to increase the accuracy of the georeferenced model special attention should be paid to the marker placement, in particular near the spillways, balustrade and hydrostatic level.

This paper also draws attention to the dam boundaries, where the presence of rocks reduces the accessibility to the dam and reduces the image quality. Since modeling the connection between the structure and the surrounding terrain is a relevant issue, it is necessary to deploy GCPs in the proximity of the abutments, avoiding patterns without GCPs at the boundaries. Moreover, we noticed that it is important to shoot a high number of images from either side of a singularity. For instance, this is the case of the right abutment, which was captured by a high number of shots from left to right and by a small number from right to left. In this case, the accuracy can be substantially lower, especially if a small number of markers are placed at the boundary.

These findings can be extended to other high-rise structures. For instance, if the façade of a structure is characterized by openings, as in the case of the spillways, it is necessary to deploy GCPs in the direct proximity of these openings to increase the accuracy of the photogrammetric survey.

In accordance with literature guidelines, GCPs should be widely distributed across the target area and at the edge of the upstream face. To reduce the occurrence of systematic errors, GCPs should be better distributed across the dam upstream face. Future developments could include an analysis of any changes in the accuracy of the model obtained by varying the overlapping of the frames or removing poor quality images.

Acknowledgments: The main author is supported by the Cassa di Risparmio di Perugia Foundation which is gratefully acknowledged. Giulia Buffi's participation in this work is supported by Romagna Acque Società delle Fonti S.p.A. This work was supported by the Italian Ministry of Education, University and Research under PRIN grant No. 20154EHYW9 "Combined numerical and experimental methodology for fluid structure interaction in free surface flows under impulsive loading". The authors would like to express their gratitude to Romagna Acque Società delle Fonti S.p.A. CEO Eng. Andrea Gambi, to the chief engineers Eng. Giuseppe Montanari and Eng. Franco Farina, to the technical manager Fabrizio Cortezzi for the availability of the Ridracoli dam. The authors are also grateful to Eng. Barberini and Italdrone, who supported the acquisition of the drone images. Moreover, the authors acknowledge Eng. Silvia Grassi, who directed the acquisition of the topographic data. Three anonymous reviewers are acknowledged for their insightful comments and suggestions that consistently improved the work.

Author Contributions: Elena Ridolfi, Giulia Buffi, Sara Venturi and Piergiorgio Manciola conceived and designed the analysis; Elena Ridolfi, Giulia Buffi and Sara Venturi built the model and made the experiments with different configurations of markers; Elena Ridolfi, Giulia Buffi, Sara Venturi and Piergiorgio Manciola analyzed results; Elena Ridolfi, Giulia Buffi and Sara Venturi wrote the paper."

Conflicts of Interest: The authors declare no conflict of interest.

References

1. Association of State Dam Safety Officials. Available online: http://www.damsafety.org/news/?p=412f29c8-3fd8-4529-b5c9-8d47364c1f3e (accessed on 27 March 2016).
2. Xu, F.; Yang, X.; Zhou, J. Dam-break flood risk assessment and mitigation measures for the Hongshiyan landslide-dammed lake triggered by the 2014 Ludian earthquake. *Geomat. Nat. Hazards Risk* **2016**, 1–19. [CrossRef]
3. Biscarini, C.; di Francesco, S.; Ridolfi, E.; Manciola, P. On the simulation of floods in a narrow bending valley: The malpasset dam break case study. *Water* **2016**, *8*, 545. [CrossRef]
4. Kuo, J.; Yen, B.; Hsu, Y.; Lin, H. Risk Analysis for Dam Overtopping—Feitsui Reservoir as a Case Study. *J. Hydraul. Eng.* **2007**, *133*, 955–963. [CrossRef]
5. Zhang, L.M.; Xu, Y.; Jia, J.S. Analysis of Earth Dam Failures: A Database Approach. *Georisk Assess. Manag. Risk Eng. Syst. Geohazards* **2009**, *3*, 184–189. [CrossRef]
6. Costa, J.E. *Floods from Dam Failures*; United States Geological Survey: Denver, CO, USA, 1985; pp. 1–59.
7. International Commission on Large Dams (ICOLD). *Lessons from Dam Incidents*; ICOLD: Paris, France, 1974.
8. International Commission on Large Dams (ICOLD). *Dam Failures Statistical Analysis, Bulletin 99*; ICOLD: Paris, France, 1995.

9. Hallermann, N.; Morgenthal, G.; Rodehorst, V. Vision-based deformation monitoring of large scale structures using Unmanned Aerial Systems. *IABSE Symp. Rep.* **2014**, *102*, 2852–2859. [CrossRef]
10. Silvagni, M.; Tonoli, A.; Zenerino, E.; Chiaberge, M. Multipurpose UAV for search and rescue operations in mountain avalanche events. *Geomat. Nat. Hazards Risk* **2016**, 1–16. [CrossRef]
11. Trasviña-Moreno, C.; Blasco, R.; Marco, Á.; Casas, R.; Trasviña-Castro, A. Unmanned Aerial Vehicle Based Wireless Sensor Network for Marine-Coastal Environment Monitoring. *Sensors* **2017**, *17*, 460. [CrossRef] [PubMed]
12. Rashed, S.; Soyturk, M. Analyzing the Effects of UAV Mobility Patterns on Data Collection in Wireless Sensor Networks. *Sensors* **2017**, *17*, 413. [CrossRef] [PubMed]
13. Venturi, S.; di Francesco, S.; Materazzi, F.; Manciola, P. Unmanned aerial vehicles and Geographical Information System integrated analysis of vegetation in Trasimeno Lake, Italy. *Lakes Reserv. Res. Manag.* **2016**, *21*, 5–19. [CrossRef]
14. Honkavaara, E.; Saari, H.; Kaivosoja, J.; Pölönen, I.; Hakala, T.; Litkey, P.; Mäkynen, J.; Pesonen, L. Processing and Assessment of Spectrometric, Stereoscopic Imagery Collected Using a Lightweight UAV Spectral Camera for Precision Agriculture. *Remote Sens.* **2013**, *5*, 5006–5039. [CrossRef]
15. Cao, C.; Liu, D.; Singh, R.P.; Zheng, S.; Tian, R.; Tian, H. Integrated detection and analysis of earthquake disaster information using airborne data. *Geomat. Nat. Hazards Risk* **2016**, *7*, 1099–1128. [CrossRef]
16. Dominici, D.; Alicandro, M.; Massimi, V. UAV photogrammetry in the post-earthquake scenario: Case studies in L'Aquila. *Geomat. Nat. Hazards Risk* **2016**, 1–17. [CrossRef]
17. Bolognesi, M.; Farina, G.; Alvisi, S.; Franchini, M.; Pellegrinelli, A.; Russo, P. Measurement of surface velocity in open channels using a lightweight remotely piloted aircraft system. *Geomat. Nat. Hazards Risk* **2016**, 1–14. [CrossRef]
18. Tauro, F.; Porfiri, M.; Grimaldi, S. Surface flow measurements from drones. *J. Hydrol.* **2016**, *540*, 240–245. [CrossRef]
19. Hallermann, N.; Morgenthal, G.; Rodehorst, V. Unmanned Aerial Systems (UAS)—Case Studies of Vision Based Monitoring of Ageing Structures. In Proceedings of the International Symposium Non-Destructive Testing In Civil Engineering, Berlin, Germany, 15–17 September 2015.
20. Achille, C.; Adami, A.; Chiarini, S.; Cremonesi, S.; Fassi, F.; Fregonese, L.; Taffurelli, L. UAV-based photogrammetry and integrated technologies for architectural applications—Methodological strategies for the after-quake survey of vertical structures in Mantua (Italy). *Sensors* **2015**, *15*, 15520–15539. [CrossRef] [PubMed]
21. Grenzdörffer, G.J.; Naumann, M.; Niemeyer, F.; Frank, A. Symbiosis of UAS photogrammetry and TLS for surveying and 3D modeling of cultural heritage monuments-a case study about the cathedral of St. Nicholas in the city of Greifswald. In Proceedings of the International Archives of the Photogrammetry, Remote Sensing and Spatial Information Sciences International Conference on Unmanned Aerial Vehicles in Geomatics, Toronto, ON, Canada, 30 August–2 September 2015; Volume 40, pp. 91–96.
22. Hallermann, N.; Morgenthal, G. Visual inspection strategies for large bridges using Unmanned Aerial Vehicles (UAV). In Proceedings of the 7th International Conference on Bridge Maintenance, Safety, Management and Life Extension, Shanghai, China, 7–11 July 2014; pp. 661–667.
23. Hallermann, N.; Morgenthal, G.; Rodehorst, V. *Unmanned Aerial Systems (UAS)—Survey and Monitoring Based on High-Quality Airborne*; International Association for Bridge and Structural Engineering: Geneva, Switzerland, 2015; pp. 1–8. [CrossRef]
24. *Federal Energy Regulatory Commission Division of Dam Safety and Inspections*; Arch Dams: Washington, DC, USA, 1999.
25. Zech, Y.; Soares-Frazão, S. Dam-break flow experiments and real-case data. A database from the European IMPACT research. *J. Hydraul. Res.* **2007**, *45*, 5–7. [CrossRef]
26. Soares-Frazão, S. Experiments of dam-break wave over a triangular bottom sill. *J. Hydraul. Res.* **2007**, *45*, 19–26. [CrossRef]
27. Biscarini, C.; di Francesco, S.; Manciola, P. CFD modeling approach for dam break flow studies. *Hydrol. Earth Syst. Sci.* **2010**, *14*, 705–718. [CrossRef]
28. Di Francesco, S.; Biscarini, C.; Manciola, P. Numerical simulation of water free-surface flows through a front-tracking lattice Boltzmann approach. *J. Hydroinform.* **2015**, *17*, 1–6. [CrossRef]
29. Di Francesco, S.; Falcucci, G.; Biscarini, C.; Manciola, P. LBM method for roughness effect in open channel flows. *AIP Conf. Proc.* **2012**, *1947*, 1777–1779. [CrossRef]

30. Di Francesco, S.; Zarghami, A.; Biscarini, C.; Manciola, P. Wall roughness effect in the lattice Boltzmann method. *AIP Conf. Proc.* **2013**, *1558*, 1677–1680.
31. Dargahi, B. Experimental Study and 3D Numerical Simulations for a Free-Overflow Spillway. *J. Hydraul. Eng.* **2006**, *132*, 899–907. [CrossRef]
32. Tabbara, M.; Chatila, J.; Awwad, R. Computational simulation of flow over stepped spillways. *Comput. Struct.* **2005**, *83*, 2215–2224. [CrossRef]
33. Ullman, S. The Interpretation of Structure from Motion. *Proc. R. Soc. Lond. Ser. B Biol. Sci.* **1979**, *203*, 405–426. [CrossRef]
34. Smith, M.W.; Carrivick, J.L.; Quincey, D.J. Structure from motion photogrammetry in physical geography. *Prog. Phys. Geogr.* **2016**, *40*, 247–275. [CrossRef]
35. Micheletti, N.; Chandler, J.H.; Lane, S.N. *Structure from Motion (SFM) Photogrammetry*; Clarke, L.E., Nield, J.M., Eds.; British Society for Geomorphology: London, UK, 2015.
36. Micheletti, N.; Chandler, J.H.; Lane, S.N. Investigating the geomorphological potential of freely available and accessible structure-from-motion photogrammetry using a smartphone. *Earth Surf. Process. Landf.* **2015**, *40*, 473–486. [CrossRef]
37. Casella, E.; Rovere, A.; Pedroncini, A.; Mucerino, L.; Casella, M.; Cusati, L.; Vacchi, M.; Ferrari, M.; Firpo, M. Study of wave runup using numerical models and low-altitude aerial photogrammetry: A tool for coastal management. *Estuar. Coast. Shelf Sci.* **2014**, *149*, 160–167. [CrossRef]
38. Castillo, C.; Pérez, R.; James, M.; Quinton, N.; Taguas, E.; Gómez, J. Comparing the accuracy of several field methods for measuring gully erosion. *Soil Sci. Soc. Am. J.* **2012**, *76*, 1319–1332. [CrossRef]
39. Gomez-Gutierrez, A.; Schnabel, S.; Berenguer-Sempere, F.; Contador, F.L.; Rubio-Delgado, J. Using 3D photo-reconstruction methods to estimate gully headcut erosion. *Catena* **2014**, *120*, 91–101. [CrossRef]
40. Lucieer, A.; de Jong, S.; Turner, D. Mapping landslide displace ments using Structure from Motion (SfM) and image correlation of multi-temporal UAV photography. *Prog. Phys. Geogr.* **2014**, *38*, 97–116. [CrossRef]
41. Ryan, J.; Hubbard, A.; Box, J.; Todd, J.; Christoffersen, P.; Carr, J.; Holt, T.; Snooke, N. UAV photogrammetry and structure from motion to assess calving dynamics at Store Glacier, a large outlet draining the Greenland ice sheet. *Cryosphere* **2015**, *9*, 1–11. [CrossRef]
42. Hashim, K.A.; Ahmad, A.; Samad, A.M.; Tahar, K.N.; Udin, W.S. Integration of low altitude aerial &terrestrial photogrammetry data in 3D heritage building modeling. In Proceedings of the 2012 IEEE Control and System Graduate Research Colloquium (ICSGRC), Shah Alam, Malaysia, 16–17 July 2012; pp. 225–230.
43. Bolognesi, M.; Furini, A.; Russo, V.; Pellegrinelli, A.; Russo, P. Accuracy of cultural heritage 3D models by RPAS and terrestrial photogrammetry. *Int. Arch. Photogramm. Remote Sens. Spat. Inf. Sci.* **2014**, *40*, 113–119. [CrossRef]
44. Koutsoudis, A.; Vidmar, B.; Ioannakis, G.; Arnaoutoglou, F.; Pavlidis, G.; Chamzas, C. Multi-image 3D reconstruction data evaluation. *J. Cult. Herit.* **2014**, *15*, 73–79. [CrossRef]
45. De Reu, J.; Plets, G.; Verhoeven, G.; de Smedt, P.; Bats, M.; Cherretté, B.; de Maeyer, W.; Deconynck, J.; Herremans, D.; Laloo, P.; et al. Towards a three-dimensional cost-effective registration of the archaeological heritage. *J. Archaeol. Sci.* **2013**, *40*, 1108–1121. [CrossRef]
46. Westoby, M.J.; Brasington, J.; Glasser, N.F.; Hambrey, M.J.; Reynolds, J.M. "Structure-from-Motion" photogrammetry: A low-cost, effective tool for geoscience applications. *Geomorphology* **2012**, *179*, 300–314. [CrossRef]
47. Reagan, D.; Sabato, A.; Niezrecki, C. Unmanned Aerial Vehicle Acquisition of Three—Dimensional Digital Image Correlation Measurements for Structural Health Monitoring of Bridges. In *SPIE Smart Structures and Materials+ Nondestructive Evaluation and Health Monitoring*; International Society for Optics and Photonics: Portland, OR, USA, 2017; p. 1016909. [CrossRef]
48. Eltner, A.; Kaiser, A.; Castillo, C.; Rock, G.; Neugirg, F.; Abellán, A. Image-based surface reconstruction in geomorphometry-merits, limits and developments. *Earth Surf. Dyn.* **2016**, *4*, 359–389. [CrossRef]
49. Barry, P.; Coakley, R. Field Accuracy Test of RPAS Photogrammetry. *Arch. Photogramm. Remote Sens. Spat. Inf. Sci.* **2013**, *40*, 27–31. [CrossRef]
50. Tahar, K.N. An evaluation on different number of ground control points in unmanned aerial vehicle photogrammetric block. *Int. Arch. Photogramm. Remote Sens. Spat. Inf. Sci* **2013**, *XL-2/W2*, 93–98.
51. Tahar, K.N.; Ahmad, A.; Abdul, W.; Wan, A.; Akib, M.; Mohd, W.; Wan, N. Assessment on Ground Control Points in Unmanned Aerial System Image Processing for Slope Mapping Studies. *Int. J. Sci. Eng. Res.* **2012**, *3*, 1–10.

52. Mesas-Carrascosa, F.-J.; Torres-Sánchez, J.; Clavero-Rumbao, I.; García-Ferrer, A.; Peña, J.-M.; Borra-Serrano, I.; López-Granados, F. Assessing Optimal Flight Parameters for Generating Accurate Multispectral Orthomosaicks by UAV to Support Site-Specific Crop Management. *Remote Sens.* **2015**, *7*, 12793–12814. [CrossRef]
53. Mesas-Carrascosa, F.-J.; García, M.N.; de Larriva, J.M.; García-Ferrer, A. An Analysis of the Influence of Flight Parameters in the Generation of Unmanned Aerial Vehicle (UAV) Orthomosaicks to Survey Archaeological Areas. *Sensors* **2016**, *16*, 1838. [CrossRef] [PubMed]
54. Smith, M.W.; Vericat, D. From experimental plots to experimental landscapes: Topography, erosion and deposition in sub-humid badlands from Structure-from-Motion photogrammetry. *Earth Surf. Process. Landf.* **2015**, *40*, 1656–1671. [CrossRef]
55. James, M.R.; Robson, S. Straightforward reconstruction of 3D surfaces and topography with a camera: Accuracy and geoscience application. *J. Geophys. Res. Earth Surf.* **2012**, *117*, F03017. [CrossRef]
56. Smith, M.W.; Carrivick, J.L.; Hooke, J.; Kirkby, M.J. Reconstructing flash flood magnitudes using "Structure-from-Motion": A rapid assessment tool. *J. Hydrol.* **2014**, *519*, 1914–1927. [CrossRef]
57. Javernick, L.; Brasington, J.; Caruso, B. Modeling the topography of shallow braided rivers using Structure-from-Motion photogrammetry. *Geomorphology* **2014**, *213*, 166–182. [CrossRef]
58. Rippin, D.M.; Pomfret, A.; King, N. High resolution mapping of supra-glacial drainage pathways reveals link between micro-channel drainage density, surface roughness and surface reflectance. *Earth Surf. Process. Landf.* **2015**, *40*, 1279–1290. [CrossRef]
59. James, M.R.; Robson, S.; Smith, M.W. 3-D uncertainty-based topographic change detection with structure-from-motion photogrammetry: Precision maps for ground control and directly georeferenced surveys. *Earth Surf. Process. Landf.* **2017**, *19*. [CrossRef]
60. Harwin, S.; Lucieer, A. Assessing the Accuracy of Georeferenced Point Clouds Produced via Multi-View Stereopsis from Unmanned Aerial Vehicle (UAV) Imagery. *Remote Sens.* **2012**, *4*, 1573–1599. [CrossRef]
61. Pierleoni, A.; Bellezza, M.; Casadei, S.; Manciola, P. Multipurpose water use in a system of reservoirs. *IAHS-AISH Publ.* **2007**, *315*, 107–116.
62. Buffi, G.; Manciola, P.; de Lorenzis, L.; Cavalagli, N.; Comodini, F.; Gambi, A.; Gusella, V.; Mezzi, M.; Niemeier, W.; Tamagnini, C. Calibration of finite element models of concrete arch-gravity dams using dynamical measures: The case of Ridracoli. In Proceedings of the X International Conference on Structural Dynamics, EURODYN 2017, Rome, Italy, 10–13 September 2017.
63. Manciola, P.; Di Francesco, S.; Biscarini, C. Flood protection and risk management: The case of Tescio River basin. In Proceedings of the Role of Hydrology inWater Resources Management, Capri, Italy, 13–16 October 2008.
64. Ravenna, Alpina S.p.A. and Consorzio Acque Forlì, Diga di Ridracoli. 1985. Available online: http://www.alpina-spa.it/resources/pdf/pubblicazioni/985-06-aavv-diga-di-ridracoli-alpina-06-1985.pdf (accessed on 2 August 2017).
65. Buffi, G.; Manciola, P.; Grassi, S.; Gambi, A.; Barberini, M. Survey of the Ridracoli Dam: UAV-Based Photogrammetry and Traditional Topographic Techniques in the inspection of Vertical Structures. *Geomat. Nat. Hazards Risk* **2017**. [CrossRef]
66. Jaud, M.; Passot, S.; le Bivic, C.; Delacourt, R.; Grandjean, N.; le Dantec, P. Assessing the Accuracy of High Resolution Digital Surface Models Computed by PhotoScan® and MicMac® in Sub-Optimal Survey Conditions. *Remote Sens.* **2016**, *8*, 465. [CrossRef]
67. U.S. Geological Survey. *Geospatial Positioning Accuracy Standards, Part 3: National Standard for Spatial Data Accuracy*; Federal Geographic Data Committee: Reston, VA, USA, 1998.
68. Buffi, G.; Niemeier, W.; Manciola, P.; Grassi, S. Comparison of 3D Model Derived from UAV and TLS—The Experience at Ridracoli Dam, Italy. In Proceedings of the 154th DVW-Seminar, Fulda, Germany, 28–29 November 2016.
69. James, M.R.; Robson, S.; D'Oleire-Oltmanns, S.; Niethammer, U. Optimising UAV topographic surveys processed with structure-from-motion: Ground control quality, quantity and bundle adjustment. *Geomorphology* **2017**, *280*, 51–66. [CrossRef]

sensors

MDPI

Article

Towards a Transferable UAV-Based Framework for River Hydromorphological Characterization

Mónica Rivas Casado [1,*], Rocío Ballesteros González [2], José Fernando Ortega [2], Paul Leinster [1] and Ros Wright [3]

[1] School of Water, Energy and Environment, Cranfield University, Cranfield, Bedfordshire MK430AL, UK; paul.leinster@cranfield.ac.uk

[2] Regional Centre of Water Research, Universidad de Castilla-La Mancha, Carretera de las Peñas km 3.2, 02071 Albacete, Spain; rocio.ballesteros@uclm.es (R.B.G.); jose.ortega@uclm.es (J.F.O.)

[3] National Fisheries Services, Environment Agency, Threshelfords Business Park, Inworth Road, Feering, Essex CO61UD, UK; ros.wright@environment-agency.gov.uk

* Correspondence: m.rivas-casado@cranfield.ac.uk; Tel.: +44-123-475-0111

Received: 23 June 2017; Accepted: 21 September 2017; Published: 26 September 2017

Abstract: The multiple protocols that have been developed to characterize river hydromorphology, partly in response to legislative drivers such as the European Union Water Framework Directive (EU WFD), make the comparison of results obtained in different countries challenging. Recent studies have analyzed the comparability of existing methods, with remote sensing based approaches being proposed as a potential means of harmonizing hydromorphological characterization protocols. However, the resolution achieved by remote sensing products may not be sufficient to assess some of the key hydromorphological features that are required to allow an accurate characterization. Methodologies based on high resolution aerial photography taken from Unmanned Aerial Vehicles (UAVs) have been proposed by several authors as potential approaches to overcome these limitations. Here, we explore the applicability of an existing UAV based framework for hydromorphological characterization to three different fluvial settings representing some of the distinct ecoregions defined by the WFD geographical intercalibration groups (GIGs). The framework is based on the automated recognition of hydromorphological features via tested and validated Artificial Neural Networks (ANNs). Results show that the framework is transferable to the Central-Baltic and Mediterranean GIGs with accuracies in feature identification above 70%. Accuracies of 50% are achieved when the framework is implemented in the Very Large Rivers GIG. The framework successfully identified vegetation, deep water, shallow water, riffles, side bars and shadows for the majority of the reaches. However, further algorithm development is required to ensure a wider range of features (e.g., chutes, structures and erosion) are accurately identified. This study also highlights the need to develop an objective and fit for purpose hydromorphological characterization framework to be adopted within all EU member states to facilitate comparison of results.

Keywords: hydromorphology; intercalibration; unmanned aerial vehicle; photogrammetry; artificial neural network; water framework directive

1. Introduction

A recognition of the importance of protecting and improving the water environment and the increasing pressures on water resources resulting from population growth, patterns of use and climate change has seen the introduction of major water protection laws in many countries. This includes the development of regulatory frameworks based on an assessment of the ecological quality of freshwater systems such as the European Union Water Framework Directive (EU WFD) [1], which aims to achieve good ecological status or potential of inland and coastal waters.

Each member state has adopted a specific methodology within the overall WFD to classify the ecological status of a water body based on an assessment of the biological, chemical, physico-chemical and supporting elements. The measures derived for individual parameters are based on significantly different methods and are therefore subject to multiple sources of difference. This variation in approach between member states means that it can be difficult to compare the results obtained in the various countries. In the particular case of supporting elements, recent studies have looked at the multiple methods used for hydromorphological characterization. Hydromorphology has a key role in the assessment of hydrology (i.e., the quantity and dynamics of water flow and connection to groundwater bodies), morphology (i.e., reach depth and width variation), structure and substrate of the river, structure of the riparian zone and river continuity [2].

The strengths and limitations in relation to the implementation of the WFD of a total of 139 available (European and non-European) hydromorphological characterization methods were identified under the REFORM project [3]. Overall, the majority of methods relied on the assessment of the physical habitat and did not provide sufficient consideration of physical processes or were pressure-specific. As a result, over the last few years there has been an increasing effort to develop geomorphologically based methods that facilitate the understanding of river functioning and evolution as a basis for assessing current conditions [4]. Examples of such methods are the Morphological Quality Index (MQI), the Morphological Quality Index for Monitoring (MQIm), the Geomorphic Units survey and classification System (GUS) [4–9] and the Hydromorphological Evaluation Tool (HYMET) [10].

The variables used for the characterization of hydromorphology depend upon the method being implemented. For example, the GUS system focuses on the identification of areas containing landforms created by erosion and or deposition inside or outside the river channel [6,9]. These can correspond to sedimentary units or include living or dead vegetation. Examples of geomorphic units are alternate bars, abandoned channels, cut-off channel, glides, island and riffles. The HYMET tool focuses on artificiality and sediment budget assessment [10]. The implementation of the method requires (i) the evaluation of the connectivity of the reach to the sediment production in its catchment; (ii) the analysis of the sediment transfer through the river network to the downstream reach and (iii) the quantification of the river sediment budget and artificiality. Variables such as the number of reservoirs present, the total reservoir storage volume and the quantity of produced sediment with free access to reach are estimated. More traditional methods such as the River Habitat Survey (RHS) [11] are based on descriptors of physical attributes (e.g., predominant bank material, predominant substrate within the channel, flow types and channel modification indicators).

Recent studies have looked at the harmonization of existing methods for hydromorphological characterization. Raven et al. [12] compared qualitatively three survey methods namely, the German Länder-Arbeitsgemeinschaft Wasser-Field Survey (LAWA-FS), the United Kingdom River Habitat Survey (RHS) and the French Système d'Evaluation de la Qualité du Milieu Physique (SEQ-MP). Differences in the quality assessment between the methods arose from the survey strategy and scale, data collection and analysis. Raven et al. [13] developed a methodology to test whether an UK hydromorphological assessment method (RHS) was suitable for rivers outside the UK. The aim was to assess whether the sampling scale (500 m) used was adequate and to suggest benchmarking strategies. Although some modifications were needed, the sampling scale was considered effective for the characterization of small rivers with the designation of hydro-ecological regions being identified as a key step towards a benchmarking program. Further work [14] compared the Ukrainian Field-Survey (UA-FS) and German LAWA-FS approaches to identify similarities and differences to the conformity of the outputs with the WFD requirements. Differences between methods included the assessment and interpretation of lateral erosion, sinuosity, type and depth profile, substrate diversity and special structures whereas similarities where observed for parameters describing land use, current diversity and water depth variation. Benjankar et al. [15] compared three hydromorphological characterization methods (i.e., the German LAWA, a special approach for urban rivers and a hydraulic based method), with results showing that assessment approaches developed for a certain geographic region may not be suitable for rivers in different contexts (i.e., ecoregion) and a transferable characterization approach was required. Langhans et al. [16] proposed a seven step procedure to combine assessment methods and harmonize metric outputs to a common scale from 0 to 1. The method was successfully tested for the integration of three hydromorphological methods developed in the USA (the Rapid Bioasessment Protocol-RBP), Switzerland (the Swiss Modular Concept of stream assessment-SMC) and Germany (LAWA) for four river sections. Similarly, in [2] a total of 121 hydromorphological assessment methods currently in use were compared to identify strengths, limitations and the potential to integrate different approaches. The results highlighted the need for an assessment approach based on integrated analysis, where the morphological and hydrological components drive the hydromorphological characterization. Fernandez et al. [17] reviewed more than 50 methods for river habitat characterization used worldwide. Results indicated that the key differences are the objectives for which the methods were designed, the time required for their implementation and whether the methods measure or evaluate characteristics. Overall, the parameters recorded most often in the various approaches were bank stability, channel substrate, artificial structures, riparian vegetation structure, channel dimensions, flow types, adjacent land uses and bars. The study concluded that methods that can be implemented at multiple spatial scales and provide quantitative information (i.e., indices vs. characterization protocols) were the most effective.

The WFD intercalibration exercise between member states [18] seeks to ensure comparability of ecological status boundaries and national assessment methods across Europe [19–21]. For this purpose, a set of six geographical intercalibration groups (GIGs) comprising waters of similar biogeophysical types (i.e., common intercalibration types) have been defined and are: Alpine, Eastern Continental, Central-Baltic, Mediterranean, Northern and Very Large Rivers (VLR) [18]. These are equivalent to the standard geographical stratification by ecoregion, where the ecoregions are geographical zones influenced by similar geophysical drivers.

The need to find pragmatic and cost-effective assessment approaches that can be adopted as common assessment methods across all countries or even as a baseline benchmark for the intercalibration exercise has been highlighted by some authors [19]. Recent studies have proposed the use of remote sensing data as a way to help achieve the harmonization of hydromorphological sampling protocols [2,22,23] at multiple spatial scales. Several tools have been developed to estimate key hydromorphological variables from remote sensing data. For example, [24] developed a method to derive valley width, active channel width and channel slope automatically from a 5 m spatial resolution DEM and 0.25–0.5 m spatial resolution orthoimage in a geographic information system (GIS) environment. Demarchi et al. [25] developed a semi-automated methodology for mapping hydromorphology indicators of rivers at regional scale using remote sensing data (i.e., high resolution near-infrared and LIDAR topography). The method enables the delineation of the natural fluvial corridor and primary riverscape units (e.g., water channel, unvegetated sediment bars, riparian densely-vegetated units) and in-stream mesohabitats. However, the resolution of commercially available remote sensing products (e.g., Lidar, aerial photography or synthetic aperture radar (SAR) data) is coarser than 10 cm and may not be sufficiently detailed for accurate hydromorphological characterization. Based on the work by [26], resolutions coarser than 2.5 cm provide biased characterizations of hydromorphological features when using artificial neural networks (ANNs) for their identification. The trade-off between high aerial imagery resolution and survey area coverage needs to be taken into account within this context. Aerial imagery at resolutions finer than 2 cm is difficult to obtain for wide-area characterization due to limitations in the current state-of-the art of the technology. Other limitations exist for commercially available remote sensing products: satellite data do not provide on-demand or temporally continuous data sets and provide oblique imagery which may result in partial spatial coverage when structures obstruct the view angle. In addition, both satellite and aircraft based remote sensing products fail to provide information under low cloud conditions.

Methodologies based on high resolution aerial imagery captured from Unmanned Aerial Vehicles (UAVs) have been proposed by several authors as potential approaches to overcome some of these limitations. For example, in [27] an UAV-based methodology for the automated recognition of hydromorphological features using ANNs was developed. Similarly, other authors [28,29] have identified photogrammetric methodologies based on high resolution UAV aerial imagery as the future tool of choice for reliable hydromorphological (i.e., physical river habitat) assessment. These techniques, albeit within their limitations [30], provide a rapid, inexpensive and increasingly accessible alternative to traditional remote sensing methods for physical habitat assessment whilst enabling quantitative microscale characterization [28]. The findings from these studies [26–28,30,31] raise the question as to whether frameworks based on the automated recognition of hydromorphological features from high resolution UAV aerial imagery could provide an objective and unbiased approach that can be adopted as a common assessment method across all EU member states and provide benchmark metrics for the intercalibration exercise.

In [27], we developed an UAV-based framework for the automated recognition of hydromorphological features and demonstrated an 81% average level of accuracy in feature classification for a 1.4 km reach (River Dee, Wales, UK). In [26], the level of resolution required for unbiased identification of hydromorphological features was determined for the same river reach and estimated to be 2.5 cm. Here, the aim is to assess the application of the framework to a range of different fluvial settings. This will be achieved through the following three core objectives:

Objective 1. To test the validity of the framework for a range of fluvial settings as identified by the WFD GIGs.
Objective 2. To compare the accuracy of the framework in hydromorphological feature identification within and between fluvial settings.
Objective 3. To interpret the outputs from (1) and (2) in line with the WFD regulatory framework.

2. Materials and Methods

2.1. Study Sites

Three sites corresponding to different GIGs (i.e., Central-Baltic, Mediterranean and VLR were selected for analysis (Figure 1 and Table 1). The sites were identified to maximize the diversity in hydromorphological characteristics along a ≈ 1 km reach and included the key hydromorphological features expected to be found in the named GIGs. From an initial set of reaches considered for analysis, only those that were representative of the spatial river variability within the waterbody where selected for further analysis. To ensure the selection was based on the current hydromorphological configuration of the waterbody, the full length of the river within each selected waterbody was assessed via multiple field-visits. The total reach length (≈1 km) was selected to exceed that used by the majority of the hydromorphological assessment methodologies used for WFD purposes, specifically those used in UK and Spain [2,3].

The Central-Baltic GIG was represented by a reach within the upper catchment of the River Dee, Wales, UK. The reach was located 30 km from the river source and 11.5 km downstream from the Bala reservoir. The river bed along the reach was characterized by gravel substrate with silt deposition in areas with low to non-perceptible flow. Bank erosion was visible along the reach with some sections presenting eroding cliffs. The banks were primarily covered by grazed grassland with few riparian trees scattered along the reach. The river length within the water body was 27.87 km [32], with the selected reach representing 5% of its totality. The overall designation for the reach was "heavily modified" [32].

The reaches representing both the Mediterranean and the VLR GIGs fell within the Jucar river basin in Spain. The Mediterranean GIG reach was located within the midland section of the river, near Motilleja (Albacete) and was characterized by a calcareous river bed with associated high entrenchment ratio. The reach contained small man-made structures that regulated the flow for irrigation purposes.

The majority of the calcareous bed channel had a layer of fine sediments due to intensive agricultural practices. The river length within the water body was 21.89 km [33,34], with the selected reach accounting for 5.5% of the total length. The overall WFD designation of the water body was "natural" [33,34].

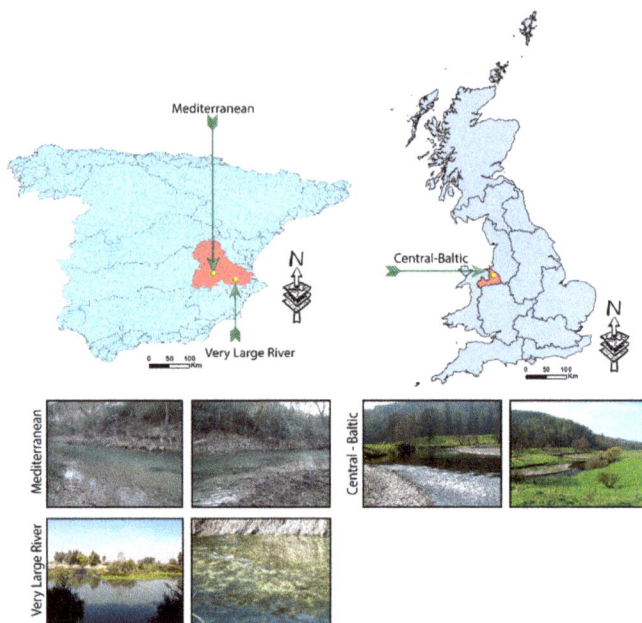

Figure 1. Schematic diagram showing the location of the selected study sites within each Geographical Intercalibration Group (GIG) and detailed imagery of the selected reaches. The maps of Spain and UK show the delineation of the main river basins with those basins containing the study sites highlighted in red.

Table 1. Characteristics of the selected reaches within each of the Water Framework Directive Geographical Intercalibration Groups (GIGs). WFD D, WB ID, WB L, RL, RW and HMWB stand for Water Framework Directive Designation, Water Body Identification code, Water Body Length, Reach Length, mean wetted Reach Width and Heavily Modified Water Body, respectively. VLR stands for Very Large Rivers GIG. Area (m²) refers to the total area within which hydromorphological features were identified. RW was estimated based on a total of 20 width measurements randomly taken along the reach. [1] [35] and [2] [33].

	Geographical Intercalibration Group		
Descriptor	Central-Baltic	Mediterranean	VLR
River	Dee	Jucar	Jucar
Country	UK	Spain	Spain
WFD D	HMWB [1]	Natural [2]	Natural [2]
WB ID	GB111067052240 [1]	ES080MSPF18.12 [2]	ES080MSPF18.28 [2]
WB L (km)	27.73 [1]	21.89 [2]	4.54 [2]
RL (km)	1.4	1.2	0.96
RW (m)	32.62	11.09	18.78
Area (m²)	46,385	30,859	21,784

The VLR GIG reach was located within the lower catchment of the Jucar, 500 m downstream of the Antella sluice (Antella, Valencia). The sluice diverted more than 25% of the river flow to irrigation [33,34]. Gabions were present along both banks for over 500 m at both the normal channel width and bankfull heights. The substrate within the reach was characterized by gravels with fine sediments intermittently present in some areas. The banks were dominated by invasive species (i.e., *Arundo donax*) with both submerged and emergent vegetation frequently present along the reach. The river length within the waterbody was 4.54 km [33,34], with the selected reach accounting for 26.4% of the total length. The WFD designation of the overall water body was "natural" [33,34].

2.2. Selection of Hydromorphological Variables

A set of variables that are major components in the WFD hydromorphological assessment in the UK and Spain were selected for automated identification (Table 2). These were specifically derived from: (i) the River Habitat Survey [11], the key method for the WFD hydromorphological assessment of rivers in UK; (ii) the UK methodology used for the designation of the WFD water body as heavily modified or artificial [36] and (iii) the equivalent Spanish methodologies [37–41]. The selection was based on the potential of these variables to contribute to the implementation of other WFD methods [2,3,5,7,8].

Table 2. Description for each of the hydromorphological features identified within the selected study sties. The features are adapted from the River Habitat Survey [11], the key method for the hydromorphological assessment of rivers in UK and for the designation of the water body [36]. This was further supported by the features used by the Spanish methodologies [37–41]. The water features recorded extended for over 5 m or >1% of the channel length following [11]. These features were recorded even if they were the result of an artificial structure.

	Feature	Description
Substrate	Side Bar	Consolidated river bed material along the margins of a reach which is exposed at low flow.
	Erosion	Predominantly derived from eroding cliffs which are vertical or undercut banks, with a minimum height of 0.5 m and less than 50% vegetation cover.
Water	Riffle	Area within the river channel presenting shallow and fast-flowing water. Generally over gravel, pebble or cobble substrate with disturbed (rippled) water surface.
	Deep Water (Glides and Pools)	Deep glides: deep homogeneous areas with visible flow movement along the surface. Pools: localized deeper parts of the channel created by scouring. Both features present fine substrate, non-turbulent and slow flow.
	Shallow Water	Includes any slow flowing and non-turbulent areas.
	Chute	Low curving fall in contact with substrate.
	Major impacts (pollution)	Indicators of water quality pollution (e.g., accumulation of white/sluggish foam, tipping, litter, sewage, abstraction).
Vegetation	Tree	Trees obscuring the aerial view of the river channel. The distinction between perennial and tree in dormant period was made when possible.
	Vegetated Side Bar	Side bar presenting plant cover in more than 50% of its surface.
	Vegetated Bank	Banks not affected by erosion. When possible the difference was made between grass and shrub cover.
	Submerged Free Floating Vegetation	Plants rooted on the river bed with floating leaves.
	Emergent Free Floating Vegetation	Plants rooted on the river bed with floating leaves on the water surface.
	Grass	Present along the banks and floodplain as a result of intense grazing regime.
	Nuisance plant specie	Invasive species covering a large proportion of the banks or river channel.
	Shadows	Extent of direct, overhead, tree canopy shade. Includes shading of channel and overhanging vegetation.
	Artificial	Any weir, sluices, culverts, bridges, fords, deflectors or equivalent that are not underwater.

The scale at which the automated identification took place was an important consideration for the selection of hydromorphological features. This is because rivers are hierarchically organized systems; at each spatial and temporal scale there are a set of variables that are the most important in determining system behaviors and capacities [7,42]. The scale also determines the remote sensing data required for the characterization of hydromorphology [22]. Within this project, the variables selected focused on reach scale characterization which is the scale at which UAVs show optimal performance. The selected variables were divided into substrate, water, vegetation, shadows and artificial features.

Substrate features (i.e., side bars and erosion) provided information on the contemporary evidence of channel adjustment [7]. From a total of 121 hydromorphological characterization methodologies reviewed in [2], substrate related features were considered by 83 methods. Water features (i.e., riffles, glides, pools, shallow water, chute and major impacts) informed the characteristics and dimensions of the channel type and provide the physical base for habitats [7,9]. The identification of such features (i.e., flow type and change in depth [2]) was carried out within at least 34 of the hydromorphological characterization methods out of the 121 methods reviewed in [2]. Vegetation features (i.e., tree, vegetated side bar, vegetated bank, submerged free floating vegetation, emergent free floating vegetation, grass and nuisance plant species) provided evidence of the vegetation dynamics within the reach and allowed the estimation of derived parameters such as the percentage of riparian corridor under riparian vegetation or the riparian vegetation age structure [7]. They also provided information on the riparian vegetation continuity, species composition, coverage and distribution [2]. Belletti et al. [2] estimated that vegetation features (i.e., in channel vegetation and woody debris) were assessed within at least 54 hydromorphological characterization methods out of the 121 reviewed. Shadows were used to estimate the percentage of the channel with shade present whilst artificial features provided an indication of the constrains forced onto natural channel adjustments [7,9]. The presence of artificial features was recorded in over 80 methods out of the 121 reviewed in [2]. There was no clear indication of how many methods recorded shadows/shade was available.

2.3. Sampling Design and Data Collection

Aerial imagery in the visible spectrum was collected over a 19-month period (21 Apirl 2015–24 November 2016) with a range of platforms and sensors specifically selected for the area to be surveyed (Table 3 and Figures 2 and 3). Ground Control Points (GCPs) were deployed along the river banks to obtain parameters for external orientation. At the Central-Baltic GIG reach, the locations of the GCP centroids were obtained from a Leica GS14 Base and Rover Real Time Kinematic (RTK) GPS (Leica Geosystems AG, Heerbrugg, Switzerland) with a position accuracy of more than 0.02 m in the X, Y and Z dimensions. For all the other reaches, a Leica 1200 GPS linked to a GNSS permanent reference station with a position accuracy of 0.02 m in planimetry and 0.03 m in altimetry was used. Two different GCP designs were used in the study: 1 m × 1 m black squared GCPs with white opposite facing triangles where used for the Central-Baltic reach whereas white circular (0.30 m diameter) GCPs with concentric black rings were used for the Mediterranean and VLR reaches (Figures 2 and 3).

Table 3. Key characteristics of the platforms and sensors used to gather the imagery at each river reach. GIG, VLR, GCP, GSD, Mill. effect. pix. and FLA stand for Geographical Intercalibration Group, Very Large Rivers, Ground Control Point, Ground Sampling Distance, Million Effective Pixels and Focal Length Applied, respectively. The time required for the photogrammetric process (PT) is estimated based on the performance of a computer with an Intel Core i7-5820k 3.30 GHz processor, 32 Gb RAM and 2 graphic cards (Geoforce GTX 980 and Qadro K2200, NVIDIA, Santa Clara, CA, USA). [1] Sony Corporation, Tokio, Japan. [2] Canon[TM], Tokio, Japan.

GIG	Central-Baltic	Mediterranean	VLR
GCPs	60	20	8
GSD	2.5	2.17	2.21
Flight altitude	100	77.6	120
No. Flights	4	2	2
Platform	Falcon 8 Trinity	IRIS9+	md4-1000
Camera	Sony Alpha 6000 [1]	Canon IXUS 115 HS [2]	Sony Alpha ILCE-5100 [1]
Sensor type	CMOS APS-C type Exmor[TM] HD [1]	BCI-CMOS [2]	CMOS APS-C type Exmor[TM] [1]
Mill. effect. pix.	24.3	12.1	24.3
Pixel size (mm)	0.00391	0.02169	0.02214
FLA (mm)	20	5	20
PT (h)	12	12	12

Various vertical-take-off-and-landing UAV platforms (Figure 3) were used for image collection and included a Falcon 8 Trinity (ASCTEC, Krailling, Germany), an IRIS9+ (3DR, Berkeley, CA, USA) and a md4-1000 (Microdrones, Inc., Kreuztal, Germany). The Falcon 8 Trinity platform is a 0.77 m × 0.82 m × 0.12 m octocopter equipped with an uBlox LEA 6S GPS (uBlox, Thalwil, Switzerland). The platform has a vertical-take-off weight of 1.8 kg with the sensor payload (Sony Alpha) only requiring 0.34 kg. Flight endurance under optimal conditions and with fully charged lithium polymer (LiPo) batteries (6250 mAh) is 22 min. The IRIS 9+ is a 0.50 m diameter (from rotor hub to rotor hub) quadcopter equipped with a 3DR u-Blox GPS (uBlox, Thalwil, Switzerland) and a 0.4 kg payload capacity, with the camera requiring only 0.14 kg. Flight endurance under optimal conditions with the payload incorporated and with fully charged LiPo batteries (5100 mAh) is between 16–22 min. The md4-1000 is a 1.00 m diameter (from rotor hub to rotor hub) quadcopter with a 0.224 kg embedded camera, a total payload capacity of 1.2 kg and a maximum take of weight of 6 kg. The md4-1000 is equipped with an mdIMU and GNSS u-Blox6 positioning system. The flight endurance with fully charged 13,000 mAh LiPo batteries is 45 min. The ground sampling distance (GSD) for each reach was calculated to achieve a desired target imagery resolution of 2.5 cm based on the camera focal length and the regulatory airspace constraints within each country (Table 3). All flights were carried out by qualified UAV pilots following the country specific aviation regulation.

The methodology was first used and tested in the Central-Baltic GIG reach [26,27]. The lessons learnt from this deployment were transferred to the other reaches. Data were collected under low flow conditions and with a constant volumetric flow rate (Table 4). The number of flights required was determined by the platform, sensor and reach configuration. Each platform was configured to follow a flight plan that captured images at waypoints that achieved 60% along track and 40% across track image overlap. Longitudinal and cross-sectional multi-passes ensured full spatial coverage.

Table 4. Description of weather and reach characteristics during the flight at each of the Geographical Intercalibration Groups (GIGs) study sites. Weather conditions at the Central-Baltic reach were estimated based on the Shawbury (Shropshire, UK) meteorological aerodrome report. Similar information was obtained from the Agroclimatic Information for Irrigation Service weather stations at Motilleja (Albacete, Spain) and Xátiva (Valencia, Spain) for the Mediterranean and Very Large Rivers (VLR) GIGs reaches, respectively. Q stands for flow. [1] [43], [2] [44], [3] [45].

GIG	Central-Baltic [1]	Mediterranean [2,3]	VLR [2,3]
Date	21 Apirl 2015	28 January 2016	24 November 2016
Discharge (m^3 s^{-1})	4.8	2.5	3.4
Percentile Q (m^3 s^{-1})	Q80	Q80	Q80
Surface wind	1 m s^{-1}–3 m s^{-1}	0.46 m s^{-1}	0.93 m s^{-1}
Wind direction	60–350°	307°	293.5°

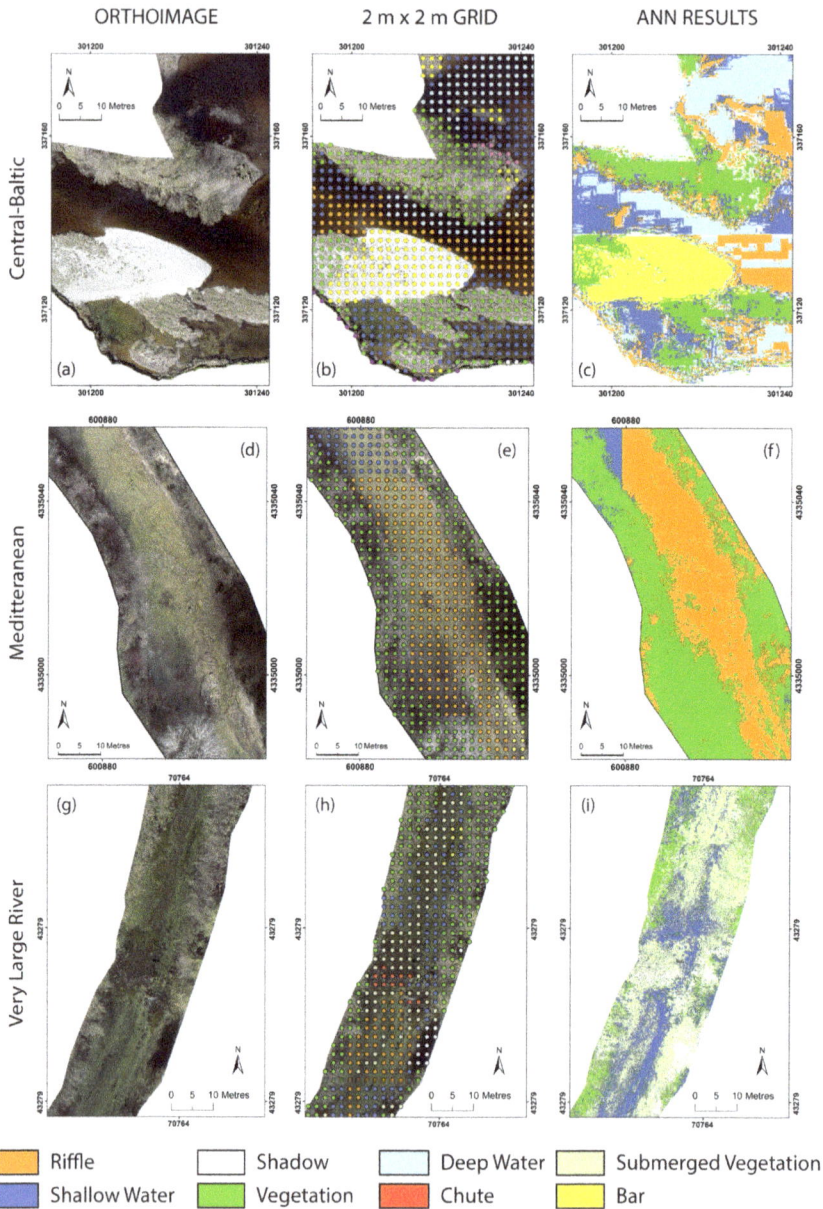

Figure 2. Example of classification outputs obtained for each Geographical Intercalibration Group (GIG). From left to right, orthoimage, 2 m × 2 m ground truth grid and classified outputs from the Artificial Neural Network (ANN); (**a–c**) Outputs for the Central-Baltic GIG reach; (**d–f**) Outputs for the Mediterranean GIG reach; (**g–i**) Outputs for the Very Large Rivers GIG reach.

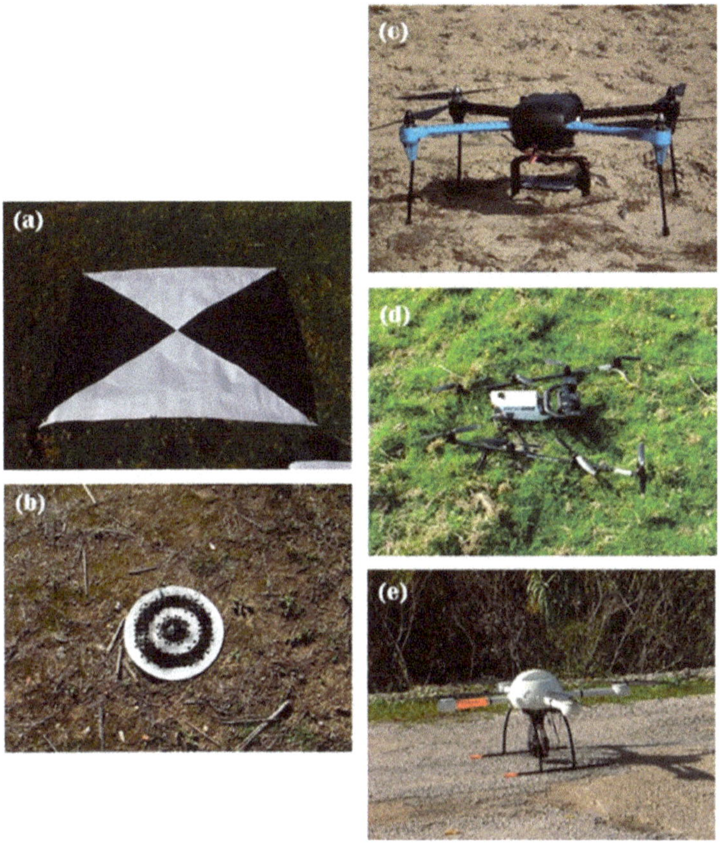

Figure 3. Types of Ground Control Points (GCPs) and Unmanned Aerial Vehicles (UAVs) used to collect the imagery at each reach. (**a**) 1 m × 1 m Squared GCP used in the Central-Baltic reach; (**b**) 0.30 m diameter GCP used in the Mediterranean and Very Large Rivers reaches; (**c**) IRIS9+ UAV (3DR, Berkeley, CA, USA) used at the Mediterranean reach; (**d**) Falcon 8 Trinity (ASCTEC, Krailling, Germany) used at the Central-Baltic reach; (**e**) md4-1000 UAV (Microdrones, Inc., Kreuztal, Germany) used at the Very Large Rivers reach.

An ancillary data set was collected to aid visual identification of hydromorphological features. This consisted of a detailed rapid habitat map of the reach, associated photographic evidence of the records, accurate RTK GPS location of key features and measurements of water depth and velocity. The habitat map was obtained by walking along the reach and drawing polygons indicating the extent of the observed features described in Table 2. The RTK measurements and photographs were collected wherever key features were present to indicate their position and extent (subject to access). Both water depth and velocity readings were obtained by use of a radio controlled boat with integrated ADCP or a hand held ADCP and a standard surveying metric rod. All the information for a single reach was collected within a maximum of a five day period to ensure spatio-temporal collocation of multiple-sensor measurements.

2.4. Photogrammetry

The imagery collected at each reach was assessed based on quality and spatial coverage for inclusion in the photogrammetric process (Figure 4, step 3). Blurred and distorted frames were excluded [46] for the generation of the standard geomatic products (i.e., orthoimage, digital terrain model and point cloud) via Photoscan Pro version 1.1.6 (Agisoft LLC, St. Petersburg, Russia). The photogrammetric process requires all the frames to be georeferenced (i.e., scale, translate and rotate) into a target Geodetic System (i.e., the World Geodetic System, WGS84) using the GCPs coordinates to minimize geometric distortion. For this purpose, the centroid of each GCP was manually identified in all frames and assigned the corresponding field RTK GPS coordinates. The processing time required for the photogrammetric process for each reach based on the performance of a computer with an Intel Core i7-5820k 3.30 GHz processor, 32 Gb RAM and two graphic cards (Geoforce GTX 980 and Qadro K2200, NVIDIA, Santaclara, CA, USA) is summarized in Table 3. The coregistration errors in X, Y and Z were automatically derived by Photoscan Agisoft the difference between the positions of the GCP centroids measured through RTK GPS and the coordinates derived from the imagery.

2.5. Automatic Classification

The automated identification of hydromorphological features was carried out via the Leaf Area Index Calculation (LAIC) software [47] (Figure 4, step 4), originally developed to discriminate green canopy cover from ground, stones and shadow background from high resolution aerial imagery [47,48]. LAIC bases the classification on an ANN approach (i.e., supervised classification technique) that segments the spectral domain of the RGB imagery into areas that directly relate to the features of interest. Key to the overall approach is the training process by which representative samples of target hydromorphological features are selected using a k-means clustering algorithm and used as the reference signature for the classification of the entire orthoimage.

The classification process was described in more detail in [27] but in brief, prior to the ANN training process, the RGB image was transformed into CIE $L \times a \times b$ (CIELAB) space to perform a cluster analysis to identify the different river features. From the three parameters describing the CIELAB space (i.e., lightness (L), green to red scale (a) and blue to yellow scale (b)), only a and b were taken into account by the clustering algorithm. The number of clusters (k) depended on the feature being identified and was determined following an iterative process that increased k by one up to a maximum of ten clusters until visually satisfactory results were obtained. Within this context, visually satisfactory results required the image outputs to show that the feature of interest had been adequately identified. This supervised method was used as a basis to calibrate a Multilayer Perceptron (MP) ANN, which was applied to the remaining image. The ANN was based on three consecutive layers composed of inter-connected nodes. Within the context of this study, the outputs from the software were a classified map of the reach and estimates of the total area allocated to each of the hydromorphological features identified.

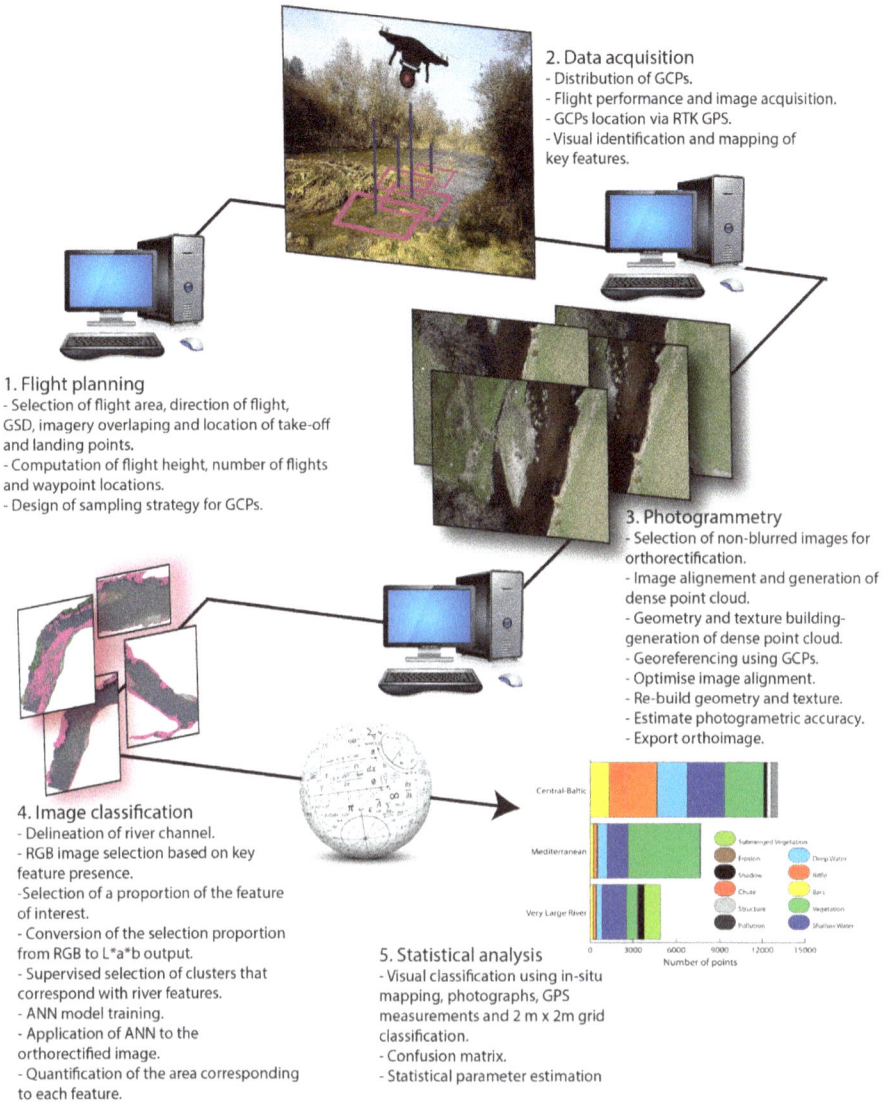

2. Data acquisition
- Distribution of GCPs.
- Flight performance and image acquisition.
- GCPs location via RTK GPS.
- Visual identification and mapping of key features.

1. Flight planning
- Selection of flight area, direction of flight, GSD, imagery overlaping and location of take-off and landing points.
- Computation of flight height, number of flights and waypoint locations.
- Design of sampling strategy for GCPs.

3. Photogrammetry
- Selection of non-blurred images for orthorectification.
- Image alignement and generation of dense point cloud.
- Geometry and texture building-generation of dense point cloud.
- Georeferencing using GCPs.
- Optimise image alignment.
- Re-build geometry and texture.
- Estimate photogrametric accuracy.
- Export orthoimage.

4. Image classification
- Delineation of river channel.
- RGB image selection based on key feature presence.
- Selection of a proportion of the feature of interest.
- Conversion of the selection proportion from RGB to L*a*b output.
- Supervised selection of clusters that correspond with river features.
- ANN model training.
- Application of ANN to the orthorectified image.
- Quantification of the area corresponding to each feature.

5. Statistical analysis
- Visual classification using in-situ mapping, photographs, GPS measurements and 2 m x 2m grid classification.
- Confusion matrix.
- Statistical parameter estimation

Figure 4. Workflow followed from imagery collection to multiple comparison analysis. ANN, GCP and GSD stand for artificial neural network, ground control point and ground sampling distance, respectively.

2.6. Statistical Analysis

The performance of the automated classification technique was assessed independently for each reach. The hydromorphological features were identified at each point of a 2 m × 2 m regular grid super imposed on each of the sampled reaches. The identification was based on the high resolution aerial imagery, the ancillary data sets encompassing the in-situ maps, the RTK measurements of key hydromorphological features and the photographic evidence of their distribution. The resulting output was considered to be the ground truth data set.

The performance of the ANN was assessed following [26,27] through confusion matrices and the estimation of derived standard metrics (Equations (1)–(5)) known as: overall accuracy (*AC*), true positive ratio (*TPR*), true negative ratio (*TNR*), false positive ratio (*FPR*) and false negative ratio (*FNR*). These were calculated as follows:

$$AC = \sum_{i=1}^{I} \left(\frac{TN_i + TP_i}{TN_i + TP_i + FN_i + FP_i} \right) \tag{1}$$

$$TPR_i = \frac{TP_i}{FN_i + TP_i} \tag{2}$$

$$TNR_i = \frac{TN_i}{TN_i + FP_i} \tag{3}$$

$$FNR_i = \frac{FN_i}{FN_i + TP_i} \tag{4}$$

$$FPR_i = \frac{FP_i}{TN_i + FP_i} \tag{5}$$

where TP_i (True Positives) is the number of points correctly identified as class i, FN_i (False Negatives) is the number of points incorrectly rejected as class i, TN_i (True Negatives) is the number of points correctly rejected as class i, FP_i (False Positives) is the number of points incorrectly identified as class i and I is the total number of classes identified.

TPR_i, TNR_i FNR_i and FPR_i were estimated for each of the features of interest whereas AC was a single value of overall classification performance. AC, as well as all the other ratios, ranged from 0 to 1. TPR_i and TNR_i quantifed the power of LAIC at classifying features correctly when compared to the ground truth whereas FNR_i and FPR_i showed misclassification rates.

3. Results

The image coregistration errors in X, Y and Z were below 2.7 cm for all reaches, with the VLR reach presenting the largest errors in X and Y ($\approx$2.7 cm) and the smallest errors in Z ($\approx$1 cm) (Table 5). The smallest coregistration errors in X and Y were found for the Mediterranean ($\approx$1.06 cm) and the Central-Baltic ($\approx$1 cm) reaches, respectively. The dimensions of the sampled reach (i.e., width and length) (Table 1) determined the number of points defined by the 2 m × 2 m regular grid and considered for analysis (Table 5). These varied between 4915 points (VLR) and 13,085 points (Central-Baltic).

Table 5. Parameters describing the coregistration errors and the overall performance in hydromorphological feature identification for each of the Geographical Intercalibration Group (GIG) sites. N and AC stand for the number of points in the 2 m × 2 m grid and the accuracy in feature classification. GCP and VLR stand for Ground Control Point and Very Large Rivers, respectively.

GIG	Central-Baltic	Mediterranean	VLR
Total GCP error in X (cm)	1.1	1.06	2.65
Total GCP error in Y (cm)	1.0	1.49	2.52
Total GCP error in Z (cm)	1.6	1.42	1.01
N	13,085	7716	4915
AC (%)	81	71	50

Based on the visual classification of the 2 m × 2 m grid, the type and extent of hydromorphological features varied considerably between reaches (Figure 5). The configuration within the Central-Baltic was predominantly dominated by riffles (26%), vegetation (25%), shallow waters (20%), deep water (16%) and bars (11%) with each of the other classes identified accounting for less than 3% of the total number of points. The dominant features within the Mediterranean reach include vegetation (65%), shallow water (19%) and deep water (10%) with all the other classes not accounting for more than 5% of the points each. These results were consistent with the pattern observed for the VLR reach, where vegetation, shallow water and deep water accounted for 39%, 35% and 8% of the points, respectively. Each of the remaining classes identified accounted for less than 8% of the classified points. The presence of shadows was significant along the Central-Baltic ($\approx$2%) and VLR (8%) reaches with no shadows being identified along the Mediterranean reach.

The overall accuracy of the automated classification was above 50% for all the reaches, with the Central-Baltic reach presenting the best performance (81%), followed by the Mediterranean (71%) and the VLR (50%) reaches. The patterns of performance, as shown by the *TPR*, *TNR*, *FNR* and *FPR* values (Table 6), differed between reaches. Overall, the ANN was able to identify vegetation successfully with *TPR* values above 74% for all sites. Misclassification of vegetation features primarily occurred with riffles (Central-Baltic) and shallow waters (Central-Baltic, Mediterranean and VLR) (Tables 6 and 7). Vegetation features were also classified as shadow (Central-Baltic and VLR) and deep waters (Mediterranean). These results were consistent with the performance of the ANN for the detection of riffles and side bars (*TPR* > 73% and *FNR* < 27% and *FPR* $\leq$ 6%) for the Central-Baltic and Mediterranean reaches. However, the performance in riffle and side bar identification decreased down to 0 for the VLR reach (*FNR* = 1). Overall, riffles were misclassified as shallow water (Central-Baltic and VLR), shadow (Central-Baltic), vegetation (Central-Baltic, Mediterranean and VLR) and deep water (VLR), whereas side bars were mainly misclassified as vegetation (Central-Baltic and VLR), riffles (Mediterranean) and shallow waters (Mediterranean and VLR) (Tables 6 and 7).

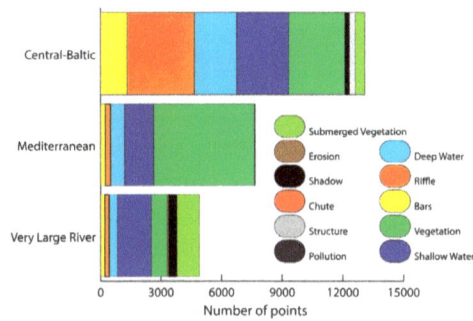

Figure 5. Number of points of the 2 m × 2 m ground truth grid allocated to each feature for each of the Geographical Intercalibration Groups (GIGs).

Table 6. Summary of the ANN performance in hydromorphological feature identification per Geographical Intercalibration Group (GIG) and feature. *TPR*, *TNR*, *FPR* and *FNR* stand for true positive ratio, true negative ratio, false positive ratio and false negative ratio, respectively.

Feature	TPR	TNR	FPR	FNR
Central-Baltic				
Side bar	0.822	0.765	0.000	0.178
Erosion	0.077	0.786	0.001	0.923
Riffle	0.814	0.756	0.060	0.074
Deep water	0.926	0.741	0.008	0.074
Shallow water	0.588	0.815	0.051	0.412
Shadow	0.818	0.770	0.073	0.182
Vegetation	0.810	0.758	0.081	0.192
Mediterranean				
Side bar	0.758	0.706	0.000	0.241
Riffle	0.736	0.707	0.014	0.263
Deep water	0.550	0.724	0.044	0.449
Shallow water	0.515	0.753	0.093	0.484
Vegetation	0.785	0.565	0.299	0.214
Pollution	0.500	0.708	0.001	0.500
Structure	0.000	0.709	0.000	1.000
Chute	0.000	0.708	0.000	1.000

Table 6. *Cont.*

Feature	TPR	TNR	FPR	FNR
		Very Large Rivers		
Side bar	0.000	0.524	0.002	1.000
Riffle	0.000	0.527	0.000	1.000
Deep water	0.665	0.488	0.034	0.334
Shallow water	0.555	0.475	0.136	0.444
Shadow	0.531	0.500	0.073	0.468
Vegetation	0.743	0.481	0.364	0.256
Structure	0.000	0.503	0.000	1.000
Chute	0.283	0.505	0.003	0.716

Table 7. Confusion matrix obtained per Geographical Intercalibration Group (GIG) site and feature considered. The features identified are described in Table 4 and are as follows: Side Bars (SB), Riffles (RI), Erosion (ER), Deep Water (DW), Shallow Water (SW), Chute (CH), Shadow (SH), Vegetation (VG), Pollution (PL), Structure (ST) and Georeferencing Error (GE). ANN refers to the features identified with the Artificial Neural Network algorithms. Visual refers to the features identified through visual observation and considered to be the ground truth data set.

ANN / Visual	SB	RI	ER	DW	SW	CH	SH	VG	PL	ST	GE	Total
					Central-Baltic							
SB	1097	8	-	-	2	-	10	214	-	-	3	1334
RI	-	2717	1	-	318	-	219	76	-	-	8	3339
ER	-	13	22	1	3	-	10	238	-	-	-	287
DW	-	60	-	1927	54	-	8	29	-	-	4	2082
SW	-	262	-	80	1514	-	493	217	-	-	7	2573
CH	-	-	-	-	-	-	-	-	-	-	-	-
SH	-	-	4	-	5	-	180	31	-	-	-	220
VG	-	245	11	7	156	-	197	2129	-	-	9	3250
PL	-	-	-	-	-	-	-	-	-	-	-	-
ST	-	-	-	-	-	-	-	-	-	-	-	-
Total	1097	3305	38	2015	2052	-	1117	3430	-	-	31	13,085
					Meditteranean							
SB	176	19	-	-	16	-	-	-	-	-	21	232
RI	-	198	-	-	1	-	-	65	-	-	5	269
ER	-	-	-	-	-	-	-	-	-	-	-	-
DW	-	-	-	385	77	-	-	188	-	-	47	697
SW	-	25	-	59	752	-	-	541	-	-	71	1448
CH	-	-	-	-	2	-	-	1	-	-	-	3
SH	-	-	-	-	-	-	-	-	-	-	-	-
VG	-	61	-	248	495	-	-	3919	14	-	253	4990
PL	-	-	-	-	-	-	-	5	8	-	3	16
ST	-	-	-	-	-	-	-	1	1	-	8	10
Total	176	303	0	692	1343	-	-	4720	23	-	408	7665
					Very Large River							
SB	-	-	-	-	49	-	5	150	-	-	22	226
RI	-	-	-	104	75	1	1	44	-	-	5	230
ER	-	-	-	-	-	-	-	-	-	-	-	-
DW	-	-	-	268	4	-	12	119	-	-	25	428
SW	7	-	-	3	942	9	137	597	-	-	116	1811
CH	-	-	-	12	-	15	4	22	-	-	50	103
SH	-	-	-	1	36	-	214	152	-	-	402	805
VG	6	-	-	35	276	5	171	1431	-	-	551	2475
PL	-	-	-	-	-	-	-	-	-	-	*	*
ST	-	-	-	-	-	-	3	5	-	-	1	9
Total	13	-	-	423	1382	30	547	2520	-	-	1172	6087

For the specific case of deep waters, the ANN performance (*TPR*) was above 55% in all instances, with the best performances being observed in the Central-Baltic (*TPR* > 92%, *FNR* < 8%) and the VLR reaches (*TPR* > 66%, *FNR* < 34%). Misclassification occurred with riffles (Central-Baltic), shallow water (Central-Baltic and Mediterranean) and vegetation (Central-Baltic, Mediterranean and VLR) (Tables 6 and 7). The performance in shallow water identification was consistent across the three reaches (*TPR* > 51%, *FNR* < 49% and *FPR* < 14%). Here, misclassification was present in shadows (Central-Baltic and VLR), vegetation (Central-Baltic, Mediterranean and VLR), riffles (Central-Baltic) and deep-water (Mediterranean) (Tables 6 and 7). Shadows, when present, were successfully identified with *TPR*, *FNR* and *FPR* values >53%, <47% and <8%, respectively with misclassification mainly occurring with vegetation (VLR). Poor results were obtained across the three reaches when the ANN focused on the identification of chutes ($TPR < 29\%$), erosion ($TPR < 8\%$), structures ($TPR = 0\%$) and major impacts (Table 2) such as pollution ($TPR = 50\%$). The number of validation points available for these features was below 20 for the Mediterranean GIG site and 55 for the VLR GIG site (Table 7).

4. Discussion

The work reported in this paper aimed at assessing the transferability of an already tested framework for hydromorphological feature characterization [26,27] to a range of different fluvial settings. The three objectives addressed within this context were: (i) to test the validity of the framework to a range of fluvial settings as identified by the GIGs; (ii) to compare the accuracy of the framework in hydromorphological feature identification within and between fluvial settings and (iii) to interpret the outputs from (i) and (ii) in line with the WFD regulatory framework. The following sections discuss the outcomes within the context of these objectives.

4.1. UAV Framework Performance at the GIGs Sites

The proposed framework enables the identification of hydromorphological features with accuracies above 50% for all the GIGs considered. Overall, the ANN works robustly for the identification of vegetation, deep water, shallow water, side bars, riffles and shadows. For the particular case of the VLR GIGs, the ANN appears to underperform in the detection of both riffles and side bars. Riffles were systematically classified as deep water, shallow water or vegetation (submerged) whereas side bars were classified as vegetation or shallow waters. The underperformance detected for the VLR site may be the result of the configuration of the reach—i.e., transitional (from deep to shallow and vice versa) flowing water along a wide gravel bed channel that generated smooth rippled surfaces undetectable by the ANN. Within this setting, the ANN seem to be unable to detect the difference between riffles and shallow/deep waters. This is because all the riffles were present within the extent of the transitional water and could be classified as both feature categories simultaneously (e.g., riffles and shallow waters); any validation point falling under a riffle and being classified as shallow water (or vice versa) cannot be automatically considered a misclassification error.

The calcareous nature of the VLR reach may also affect the characteristics of the RGB aerial imagery captured; the water presented a low turbidity and a turquoise-like color that made feature identification (i.e., identification of differences in depth, velocity and substrate, amongst other parameters) challenging. This could explain why side bars (and riffles) were confused with vegetation as the color of both was similar under the survey conditions and why the distinction between riffles, deep and shallow waters cannot be accurately identified by the ANN. In [27], we explored in detail where the confusion between shallow and deep waters occurred along the Central-Baltic GIG site; the transition zone from deep to shallow areas were the key sources of feature misclassification. This lack of ability to identify the transition zones applies to all the reaches analyzed in the present study. The transition zone occupied larger proportions of the surveyed area in the Mediterranean and VLR reaches.

The ANN performs poorly on the classification of chutes, artificial structures, erosion and pollution. These features appeared occasionally along some of the reaches and were not representative enough for the ANN to register them as independent clusters (i.e., they accounted for less than 8% of the classified points). In the particular case of pollution, the ANN provides encouraging results ($TPR > 53\%$). However, the limited number of validation points (i.e., 16 points along the Mediterranean reach) available for that particular feature class does not allow for a conclusive statement on the performance of the ANN for pollution identification. Note that pollution appeared in the Mediterranean GIG reach only and along a backwater that collected white foam and plastics derived from upstream activity.

4.2. Potential Technical Improvements

For the proposed framework to be used for large scale (>1 km) hydromorphological characterization, it is necessary to increase its cost-effectiveness via the reduction of the number of GCP to be deployed. Recent work by several authors (e.g., [48,49]) has already contributed to address this gap in knowledge. This, coupled with increased UAV battery performance, data capture and less CPU demanding software makes UAV based frameworks a plausible option for wide-area (>1 km reach) robust and accurate hydromorphological assessment. Further work should compare the performance of the proposed framework with existing classification techniques for geomorphological environments (e.g., [50,51]).

The work presented here focuses on the characterization of hydromorphological variables at reach level. It is not yet certain whether the framework can be applied to larger spatial scales (e.g., catchment [7]) and achieve the same efficiency as current remote sensing methodologies [22]. Woodget et al. [30] identify the spatial coverage as a key challenge for the routine operational use of UAVs and digital photogrammetry for river habitat mapping. The strong trade-off between resolution and spatial coverage [30] suggests that some aspects of the hydromorphological characterization for WFD purposes may be achieved from UAV imagery but that complementary remote sensing methods may be required to address the remaining aspects [22]. For example, the strong reliance on image texture required for the implementation of Structure-from-Motion (SfM) may also compromise the wide-area implementation of the framework. If the texture of the imagery is compromised, SfM will not detect matching features between overlapping images and fail to produce an orthoimage of the surveyed area [30]. Further research is therefore required to address these points before the framework can be adopted for wide-area monitoring. In addition, further consideration needs to be given to the type of sensors to use on the UAV platform. The RGB imagery collected for this study enables the qualitative assessment of hydromorphological features (e.g., deep/shallow waters) but does not facilitate the quantitative estimation of depth and velocity within the channel reach [30]. It has been highlighted by some authors that the RGB imagery presents limited radiometric resolution which obstructs the restitution of topography in darker parts (e.g., shadow and deep water) [30]. Sensors able to provide multispectral and RAW format imagery may be beneficial and should be considered in further developments of the framework. The variability in the quality of the geomatic products generated from high resolution UAV aerial imagery has also been raised as a general concern by some authors [52]. For example, the optimal selection of GCP can improve the variability of the Digital Elevation Model from 37 mm to 16 mm. The magnitude of these changes in the DEM may be relevant when changes in hydromorphological characteristics over time are required [52] but not relevant where regulatory assessments are the primary consideration.

4.3. Framework Output Interpretation in Line with the WFD

The results obtained indicate that the methodology can be successfully transferred within and between river GIG types and countries but that water turbidity, color and underlying river bed substrate may affect classification performance. The hydromorphological features successfully identified (i.e., vegetation, deep water, shallow water, side bars, riffles and shadows) are common to the majority of hydromorphological characterization methods reviewed in [2] and are therefore of great relevance for the transferability of this framework. They will be able to contribute to the implementation of the WFD in UK as they align with the descriptors included in the RHS methodology [11] and the designation of WFD artificial and heavily modified water bodies [36]. Similarly, they will aid the implementation of the Spanish hydromorphological assessments [37–41] and contribute to the implementation of WFD methodologies in other countries [2,3,5,7,8].

The framework outlined in this paper is a step forward towards the accurate identification of in-channel and bank features and could be integrated with complementary frameworks. Recent work by other authors have shown that other riverine features can be identified with similar UAV based methodologies. In [53], supervised machine learning approaches were used to identify different types of macrophytes. In [54], object oriented classification was used to map dead wood presence, whereas in [29], geomatic products derived from UAV aerial imagery were used to characterize channel substrate. Further research should focus on integrating these scientific advances into a single methodological framework that enables the objective and automated characterization of key WFD water elements. This will address the need identified in previous works [2] for the development of an integrated hydromorphological analysis that includes both morphological and hydrological components to evaluate and classify hydromorphological state and quality. In turn, this will contribute to addressing some of the current limitations in the WFD intercalibration exercise.

At present it is not possible to identify fully the consequences of the EU member states using different methodological approaches to assess the ecological status of water bodies. The intercalibration exercises have not provided a clear answer yet [14,21] and as a result, there is still insufficient information about how the different methods should be implemented and integrated to achieve readily comparable assessments [20]. It is possible that the consequences will be both financial and legal in terms of the measures and efforts undertaken to achieve good ecological status or potential under the WFD [12,14]. An objective and fit for purpose hydromorphological characterization framework that is widely adopted within all member states will facilitate comparison of the results obtained across Europe and will provide a level playing field. The UAV based framework presented here would contribute for example, to a more consistent designation and overall hydromorphological assessment by addressing some of the current limitations, namely: inadequate characterization of pressure gradients within the GIGs, difficulty comparing results obtained from methods with different metrics, lack of comparable data sets and implementation of subjective methods [14] through the provision of a standardized approach to understanding the hydromorphological quality in rivers.

An interesting application of the framework described in this paper would be its use for the designation of the waterbody as "natural", "heavily modified" or "artificial". Currently such designations can be influenced to a significant extent by management and external influence. For the case study areas, the rivers studied representing the Mediterranean and VLR GIGs had been designated by the Spanish authorities as having "natural" hydromorphological conditions [33]. In the UK, the reach representing the Central-Baltic GIGs was less modified than the Spanish reaches but had been designated by the authorities as "highly modified" [35]. The differences in the approaches taken for the designation of the waterbodies translates into different requirements in terms of the measures to be implemented and the associated level of investment required to improve the ecological status of the water bodies. This mismatch in the implementation of the WFD could be addressed if a common framework for hydromorphological characterization were to be adopted. The techniques developed could also be useful in a number of situations where hydromorphological characterization is required aside from the WFD implementations such as river restoration appraisal [55].

5. Conclusions

This paper reports the transferability of an already validated ANN based framework for hydromorphological characterization of three different fluvial settings as defined by the WFD GIGs. The framework provides overall accuracies greater than 50% for all the reaches but shows some limitations when applied to calcareous rivers falling under the VLR GIG. Deep waters, shallow waters, vegetation, side bars, riffles and shadows are successfully identified within all reaches, with the ANN underperforming on the identification of side bars and riffles for the VLR GIG. Further work is required to develop more reliable algorithms that can be incorporated into the existing ANN for the accurate detection of a wider range of features, including chutes, erosion, structures and major impacts (i.e., pollution). This work also highlights the need to develop objective and reliable hydromorphological assessment frameworks that are widely adopted by all EU member states. The framework presented here is a step forward towards that goal.

Acknowledgments: We would like to thank the Environment Agency and EPSRC for funding part of this project under the EPSRC Industrial Case Studentship voucher number 08002930. The underlying data are confidential and cannot be shared. Special thanks go to Natural Resources Wales for their help and support with data collection. The authors acknowledge financial support from the Castilla-La Mancha Regional Government (Spain) under the Postdoctoral Scholarship Programa Operative 2007–2013 de Castilla-La Mancha. Special thanks go to the journal reviewers and editor; they made this manuscript stronger with their positive criticism and valuable comments.

Author Contributions: M.R.C. is the principal investigator and corresponding author. She led and supervised the overall research and field data collection. M.R.C. structured and wrote the paper in collaboration with R.B.G. R.B.G. helped with the flight planning and overall data collection in UK and Spain. R.B.G. processed the UAV imagery and helped with the interpretation of the results. J.F.O. enabled equipment provision and data collection in Spain. J.F.O. coordinated the financial aspects of data collection in the Spanish reaches. He also provided advice and guidance on the regulatory implication of the WFD in Spain. P.L. reviewed the robustness of the approach. P.L. provided input on the policy and regulatory implications of the research outcomes and coordinated communication with key project partners. R.W. coordinated the financial aspects of the research, enabled equipment provision and facilitated fieldwork arrangements in UK. She highlighted research priorities and provided operational advice as and when required. All authors were key for the interpretation of the results.

Conflicts of Interest: The authors declare no conflict of interest.

References and Notes

1. European Commission. Directive 2000/60/EC of the European Parliament and of the Council of 23 October 2000 establishing a framework for Community action in the field of water policy. *Off. J. Eur. Union* **2000**, *L327*, 1–72.

2. Belletti, B.; Rinaldi, M.; Buijse, A.D.; Gurnell, A.M.; Mosselman, E. A review of assessment methods for river hydromorphology. *Environ. Earth Sci.* **2014**, *73*, 2079–2100. [CrossRef]

3. Rinaldi, M.; Belleti, B.; Van de Bund, W.; Bertoldi, W.; Gurnell, A.; Buijse, T.M.E.; Mosselman, E. Review on Eco-Hydromorphological Methods. Available online: http://www.reformrivers.eu/d-11-review-eco-hydromorphological-methods (accessed on 26 September 2017).

4. Rinaldi, M.; Belletti, B.; Bussettini, M.; Comiti, F.; Golfieri, B.; Lastoria, B.; Marchese, E.; Nardi, L.; Surian, N. New tools for the hydromorphological assessment and monitoring of European streams. *J. Environ. Manag.* **2016**, *202*, 363–378. [CrossRef] [PubMed]

5. Gurnell, M.; Bussettini, M.; Camenen, B.; González Del Tánago, M.; Grabowski, R.; Hendriks, D.; Henshaw, A.; Latapie, A.; Rinaldi, M. Multi-Scale Framework and Indicators of Hydromorphological Processes and Forms I. Main Report. Available online: http://www.reformrivers.eu/multi-scale-framework-and-indicators-hydromorphological-processes-and-forms-i-main-report (accessed on 26 September 2017).

6. Belletti, B.; Rinaldi, M.; Bussettini, M.; Comiti, F.; Gurnell, A.M.; Mao, L.; Nardi, L.; Vezza, P. Characterising physical habitats and fluvial hydromorphology: A new system for the survey and classification of river geomorphic units. *Geomorphology* **2017**, *283*, 143–147. [CrossRef]

7. Gurnell, A.M.; Rinaldi, M.; Belletti, B.; Bizzi, S.; Blamauer, B.; Braca, G.; Buijse, A.D.; Bussettini, M.; Camenen, B.; Comiti, F.; et al. A multi-scale hierarchical framework for developing understanding of river behaviour to support river management. *Aquat. Sci.* **2016**, *78*, 1–16. [CrossRef]

8. Rinaldi, M.; Gurnell, A.M.; Belletti, B.; Berga Cano, M.I.; Bizzi, S.; Bussettini, M.; González del Tánago, M.; Grabowski, R.; Habersak, H.; Klösch, M.; et al. Final Report on Methods, Models, Tools to Asses the Hydromorphology of Rivers. Available online: http://www.reformrivers.eu/deliverables/d62-final-report-methods-models-tools-assess-hydromorphology-rivers (accessed on 26 September 2017).

9. Rinaldi, M.; Belletti, B.; Comiti, F.; Nardi, L.; Bussettini, M.; Mao, L.; Gurnell, A.M. The Geomorphic Unit Survey and Classification System (GUS). Available online: http://www.reformrivers.eu/geomorphic-units-survey-and-classification-system-gus (accessed on 26 September 2017).

10. Klösch, M.; Habersack, H. The Hydromorphological Evaluation Tool (HYMET). *Geomorphology* **2017**, *291*, 143–158. [CrossRef]

11. Environment Agency. *River Habitat Survey in Britain and Ireland*; Environment Agency: Bristol, UK, 2003.

12. Raven, P.J.; Holmes, N.T.H.; Charrier, P.; Dawson, F.H.; Naura, M.; Boon, P.J. Towards a harmonized approach for hydromorphological assessment of rivers in Europe: A qualitative comparison of three survey methods. *Aquat. Conserv. Mar. Freshw. Ecosyst.* **2002**, *12*, 405–424. [CrossRef]

13. Raven, P.J.; Holmes, N.T.H.; Vaughan, I.P.; Dawson, F.H.; Scarlett, P. Benchmarking habitat quality: Observations using River Habitat Survey on near-natural streams and rivers in northern and western Europe. *Aquat. Conserv. Mar. Freshw. Ecosyst.* **2010**, *20*, S13–S30. [CrossRef]

14. Scheifhacken, N.; Haase, U.; Gram-Radu, L.; Kozovyi, R.; Berendonk, T.U. How to assess hydromorphology? A comparison of Ukrainian and German approaches. *Environ. Earth Sci.* **2012**, *65*, 1483–1499. [CrossRef]

15. Benjankar, R.; Koenig, F.; Tonina, D. Comparison of hydromorphological assessment methods: Application to the Boise River, USA. *J. Hydrol.* **2013**, *492*, 128–138. [CrossRef]

16. Langhans, S.D.; Lienert, J.; Schuwirth, N.; Reichert, P. How to make river assessments comparable: A demonstration for hydromorphology. *Ecol. Indic.* **2013**, *32*, 264–275. [CrossRef]

17. Fernández, D.; Barquín, J.; Raven, P.J. A review of river habitat characterisation methods: Indices vs. characterisation protocols. *Limnetica* **2011**, *30*, 217–234.

18. European Commission. Commission Decision of 30 October 2008 establishing, pursuant to Directive 2000/60/EC of the European Parliament and the Council, the values of the Member State monitoring system classifications as a result of the intercalibration exercise 2008/915/EC. *Off. J. Eur. Union* **2008**, *L332*, 20–44.

19. Poikane, S.; Zampoukas, N.; Borja, A.; Davies, S.P.; van de Bund, W.; Birk, S. Intercalibration of aquatic ecological assessment methods in the European Union: Lessons learned and way forward. *Environ. Sci. Policy* **2014**, *44*, 237–246. [CrossRef]

20. Poikane, S.; Birk, S.; Böhmer, J.; Carvalho, L.; de Hoyos, C.; Gassner, H.; Hellsten, S.; Kelly, M.; Lyche Solheim, A.; Olin, M.; et al. A hitchhiker's guide to European lake ecological assessment and intercalibration. *Ecol. Indic.* **2015**, *52*, 533–544. [CrossRef]

21. Poikane, S.; Johnson, R.K.; Sandin, L.; Schartau, A.K.; Solimini, A.G.; Urbanič, G.; Arbačiauskas, K.; Aroviita, J.; Gabriels, W.; Miler, O.; et al. Benthic macroinvertebrates in lake ecological assessment: A review of methods, intercalibration and practical recommendations. *Sci. Total Environ.* **2016**, *543*, 123–134. [CrossRef] [PubMed]

22. Bizzi, S.; Demarchi, L.; Grabowski, R.C.; Weissteiner, C.J.; Van de Bund, W. The use of remote sensing to characterise hydromorphological properties of European rivers. *Aquat. Sci.* **2016**, *78*, 57–70. [CrossRef]

23. Gurnell, A.M.; Rinaldi, M.; Buijse, A.D.; Brierley, G.; Piégay, H. Hydromorphological frameworks: Emerging trajectories. *Aquat. Sci.* **2016**, *78*, 135–138. [CrossRef]

24. Martínez-Fernández, V.; Solana-Gutiérrez, J.; González del Tánago, M.; García de Jalón, D. Automatic procedures for river reach delineation: Univariate and multivariate approaches in a fluvial context. *Geomorphology* **2016**, *253*, 38–47. [CrossRef]

25. Demarchi, L.; Bizzi, S.; Piégay, H. Hierarchical Object-Based Mapping of Riverscape Units and in-Stream Mesohabitats Using LiDAR and VHR Imagery. *Remote Sens.* **2016**, *8*, 97. [CrossRef]

26. Rivas Casado, M.; Ballesteros González, R.; Wright, R.; Bellamy, P. Quantifying the effect of aerial imagery resolution in automated hydromorphological river characterisation. *Remote Sens.* **2016**, *8*, 650. [CrossRef]

27. Rivas Casado, M.; Ballesteros González, R.; Kriechbaumer, T.; Veal, A. Automated identification of river hydromorphological features using UAV high resolution aerial imagery. *Sensors* **2015**, *15*, 27969–27989. [CrossRef] [PubMed]

28. Woodget, A.S.; Visser, F.; Maddock, I.P.; Carbonneau, P.E. The accuracy and reliability of traditional surface flow type mapping: Is it time for a new method of characterizing physical river habitat? *River Res. Appl.* **2016**, *39*, 1902–1914. [CrossRef]

29. Woodget, A.S.; Carbonneau, P.E.; Visser, F.; Maddock, I.P. Quantifying submerged fluvial topography using hyperspatial resolution UAS imagery and structure from motion photogrammetry. *Earth Surf. Process. Landf.* **2015**, *40*, 47–64. [CrossRef]

30. Woodget, A.; Austrums, R.; Maddock, I.P.; Habit, E. Drones and digital photogrammetry: From classifications to continuums for monitoring river habitat and hydromorphology. *Wiley Interdiscip. Rev. Water* **2017**, *4*, e1222. [CrossRef]

31. DeBell, L.; Anderson, K.; Brazier, R.E.; King, N.; Jones, L. Water resource management at catchment scales using lightweight UAVs: Current capabilities and future perspectives. *J. Unmanned Veh. Syst.* **2016**, *4*, 7–30. [CrossRef]

32. Environment Agency. Catchment Data Explorer. Available online: http://environment.data.gov.uk/catchment-planning/ManagementCatchment/3024 (accessed on 30 December 2016).

33. Ministerio de Agricultura Alimentación y Medio Ambiente. Confederación Hidrográfica del Júcar. Plan Hidrológico de la Demarcación Hidrográfica del Júcar. Available online: http://www.chj.es/es-es/medioambiente/planificacionhidrologica/Paginas/PHC-2015-2021-Plan-Hidrologico-cuenca.aspx (accessed on 5 March 2017).

34. Ministerio de Agricultura Alimentación y Medio Ambiente. Real Decreto 1/2016, de 8 de enero, por el que se aprueba la revisión de los Planes Hidrológicos de las demarcaciones hidrográficas del Cantábrico Occidental, Guadalquivir, Ceuta, Melilla, Segura y Júcar, y de la parte española de las demarcaciones hidrográficas del Cantábrico Oriental, Miño-Sil, Duero, Tajo, Guadiana y Ebro. *Boletín Of. Estado* **2016**, *16*, 2972–4301.

35. Water Watch Wales Map Gallery. WFD Cycle 2 Rivers and Water-Bodies in Wales. Available online: https: //nrw.maps.arcgis.com/apps/webappviewer/index.html?id=2176397a06d64731af8b21fd69a143f6 (accessed on 8 March 2017).

36. Environment Agency. Technical Assessment Method (Rivers): Morphological Alteration/Identification of pHMWBs and pAWBs (Drainage Channels).

37. Ministerio de Agricultura Alimentación y Medio Rural y Marino. Orden ARM/2656/2008, de 10 de septiembre, por la que se aprueba la instruccion y planificacion hidrologica. *Boletín Of. Estado* **2008**, *229*, 38472–38582.

38. Ministerio de Agricultura Alimentación y Medio Ambiente Real Decreto 817/2015, de 11 de septiembre, por el que se establecen los criterios de seguimiento y evaluacion del estado de las aguas superficiales y las normas de calidad ambiental. *Boletín Of. Estado* **2015**, *219*, 80582–80677.

39. Ministerio de Alimentación Agricultura y Medio Ambiente. Confederación Hidrográfica del Júcar. Plan hidrológico de la demarcación hidrográfica del Júcar. Memoria-Anejo 1. Designación de Masas de Agua Artificiales y muy Modificadas. Available online: http://www.chj.es/descargas/ProyectosOPH/Consulta% 20publica/Anejos/PHJ_Anejo01_MAMM.pdf (accessed on 5 March 2017).

40. Ministerio de Alimentación Agricultura y Medio Ambiente. Confederación Hidrográfica del Júcar. Plan hidrológico de la demarcación hidrográfica del Júcar. Memoria-Anejo 12. Evaluación del estado de las masas de agua superficiales y subterránea. Available online: http://www.chj.es/es-es/medioambiente/planificacionhidrologica/Paginas/PHC-2015-2021-Plan-Hidrologico-cuenca.aspx (accessed on 5 March 2017).

41. Ministerio de Alimentación Agricultura y Medio Ambiente; Confederación Hidrográfica del Ebro. Protocolos de muestreo y análisis para indicadores hidromorfológicos. In *Metodología para establecimiento del estado ecológico según la Directiva Marco del Agua*; Ministerio de Alimentación Agricultura y Medio Ambiente: Madrid, Spain, 2013.

42. Frissell, C.A.; Liss, W.J.; Warren, C.E.; Hurley, M.D. A Hierarchical Framework for Stream Habitat Classification: Viewing Streams in a Watershed Context. *Environ. Manag.* **1986**, *10*, 199–214. [CrossRef]

43. The National Registry for Adoption (NRFA). Station Mean Flow Data. Available online: http://nrfa.ceh.ac. uk/data/station/meanflow/67001 (accessed on 31 December 2016).

44. Ministerio de Agricultural y Pesca Alimentación y Medio Ambiente. Sistema de información agroclimática para el regadío. Available online: http://eportal.mapama.gob.es/websiar/SeleccionParametrosMap.aspx? dst=1 (accessed on 1 May 2017).

45. Ministerio de Alimentación Agricultura y Medio Ambiente. Confederación Hidrográfica del Júcar. Plan hidrológico de la demarcación hidrográfica del Júcar. Memoria-Anejo 5. Régimen de caudales ecológicos. Available online: http://www.chj.es/descargas/ProyectosOPH/Consulta%20publica/Anejos/ PHJ_Anejo05 QEco.pdf (accessed on 5 March 2017).

46. Ribeiro-Gomes, K.; Hernández-Lopez, D.; Ballesteros González, R.; Moreno, M.A. Approximate georeferencing and automatic blurred image detection to reduce the costs of UAV use in environmental and agricultural applications. *Biosyst. Eng.* **2016**, *151*, 308–327. [CrossRef]

47. Córcoles, J.I.; Ortega, J.F.; Hernández, D.; Moreno, M.A. Estimation of leaf area index in onion (Allium cepa L.) using an unmanned aerial vehicle. *Biosyst. Eng.* **2013**, *115*, 31–42. [CrossRef]

48. Ballesteros, R.; Ortega, J.F.; Hernández, D.; Moreno, M.A. Applications of georeferenced high-resolution images obtained with unmanned aerial vehicles. Part I: Description of image acquisition and processing. *Precis. Agric.* **2014**, *15*, 579–592. [CrossRef]

49. Carbonneau, P.E.; Dietrich, J.T. Cost-Effective non-metric photogrammetry from consumer-grade sUAS: Implications for direct georeferencing of structure from motion photogrammetry. *Earth Surf. Process. Landf.* **2017**, *42*, 473–486. [CrossRef]

50. Brodu, N.; Lague, D. 3D terrestrial lidar data classification of complex natural scenes using a multi-scale dimensionality criterion: Applications in geomorphology. *ISPRS J. Photogramm. Remote Sens.* **2012**, *68*, 121–134. [CrossRef]

51. Lague, D.; Brodu, N.; Leroux, J. Accurate 3D comparison of complex topography with terrestrial laser scanner: Application to the Rangitikei canyon (N-Z). *ISPRS J. Photogramm. Remote Sens.* **2013**, *82*, 10–26. [CrossRef]

52. James, M.R.; Robson, S.; d'Oleire-Oltmanns, S.; Niethammer, U. Optimising UAV topographic surveys processed with structure-from-motion: Ground control quality, quantity and bundle adjustment. *Geomorphology* **2017**, *280*, 51–66. [CrossRef]

53. Göktoğan, A.H.; Sukkarieh, S.; Bryson, M.; Randle, J.; Lupton, T.; Hung, C. A rotary-wing unmanned air vehicle for aquatic weed surveillance and management. *J. Intell. Robot. Syst. Theory Appl.* **2010**, *57*, 467. [CrossRef]

54. Dunford, R.; Michel, K.; Gagnage, M.; Piégay, H.; Trémelo, M.L. Potential and constraints of Unmanned Aerial Vehicle technology for the characterization of Mediterranean riparian forest. *Int. J. Remote Sens.* **2009**, *30*, 4915–4935. [CrossRef]

55. Poppe, M.; Kail, J.; Aroviita, J.; Stelmaszczyk, M.; Giełczewski, M.; Muhar, S. Assessing restoration effects on hydromorphology in European mid-sized rivers by key hydromorphological parameters. *Hydrobiologia* **2016**, *769*, 21–40. [CrossRef]

Article

Towards the Automatic Detection of Pre-Existing Termite Mounds through UAS and Hyperspectral Imagery

Juan Sandino *,†, Adam Wooler † and Felipe Gonzalez †

Robotics and autonomous systems, Queensland University of Technology (QUT), 2 George Street, Brisbane City QLD 4000, Australia; adam_wooler@yahoo.com.au (A.W.); felipe.gonzalez@qut.edu.au (F.G.)
* Correspondence: juan.sandinomora@qut.edu.au; Tel.: +61-7-3138-1363
† These authors contributed equally to this work.

Received: 7 August 2017; Accepted: 21 September 2017; Published: 24 September 2017

Abstract: The increased technological developments in Unmanned Aerial Vehicles (UAVs) combined with artificial intelligence and Machine Learning (ML) approaches have opened the possibility of remote sensing of extensive areas of arid lands. In this paper, a novel approach towards the detection of termite mounds with the use of a UAV, hyperspectral imagery, ML and digital image processing is intended. A new pipeline process is proposed to detect termite mounds automatically and to reduce, consequently, detection times. For the classification stage, several ML classification algorithms' outcomes were studied, selecting support vector machines as the best approach for their role in image classification of pre-existing termite mounds. Various test conditions were applied to the proposed algorithm, obtaining an overall accuracy of 68%. Images with satisfactory mound detection proved that the method is "resolution-dependent". These mounds were detected regardless of their rotation and position in the aerial image. However, image distortion reduced the number of detected mounds due to the inclusion of a shape analysis method in the object detection phase, and image resolution is still determinant to obtain accurate results. Hyperspectral imagery demonstrated better capabilities to classify a huge set of materials than implementing traditional segmentation methods on RGB images only.

Keywords: pre-existing termite mounds; UAV; hyperspectral camera; machine learning; image segmentation; support vector machines

---

1. Introduction

The contribution of termites to Australia's environment is significant. Not only their colonies have positively contributed to enhancing the Australian landscape in savannah regions, but also they have been essential in decomposition processes [1–3]. Their diet allows the rapid disintegration of dead wood and plant debris, incrementing, therefore, nutrients' quantification at the surroundings of their mounds [4,5]. Termites also play a significant role in creating dense nutrient patches through the erosion of soil from the aboveground mounds to the ground surrounding the termite mounds, as shown in Figure 1 [6,7]. Their nests provide the necessary alteration of the structure of many trees for the habitats of various animals including birds, reptiles and some mammals. They have also been vital in Aboriginal communities, whereby termites have been used for medicines, dietary supplements and mounds for campfire [8].

Figure 1. Pre-existing termite mound at Cape Range National Park, WA.

Termite distribution across Australia varies according to the average weather temperature in each one of its states. As an illustration, there is a very high termite density distribution on coastal areas of the mainland; high density towards the centre regions including Coastal New South Wales and Coastal South Australia; moderate in both deserts and cold regions such as Alice Springs and Canberra respectively; low and very low density in very cold regions including Melbourne and Tasmania [9,10].

Regardless of the positive or negative influence of termites in Australia, an adequate detection system would provide an accurate and efficient outcome for their study, monitoring and control. Amongst the various methods to accomplish this task, remote sensing is useful due to the easy management of large spatial distributions. For example, the contributions of Bunting et al. [11] demonstrated that through the use of image texture, vast improvements could be made in the classification of plant species to derive texture-based information. He et al. [12] reported what hyperspectral remote sensing could offer for invasion ecologists and review recent progress made in plant invasion research using hyperspectral remote sensing. Wallis et al. [13] demonstrated how remote sensing improved the performance for modelling biodiversity across space and time and for developing effective ecological indicators. As stated by Colomina and Molina [14], the use of UAS in remote sensing has led to a broad development of low-cost diverse agriculture applications, with plenty of facilities to build the products according to the needs of the offered services. Moreover, computer vision and open source philosophy has improved the manageability of UAVs and data processing. Therefore, the increased literature and technological developments in Unmanned Aircraft Systems (UAS) design and path planning, combined with hyperspectral imagery and machine learning approaches have opened the possibility of improving image quality [15] and increasing research outcomes in biosecurity [16], air quality, precision agriculture and environmental sensing [17–20].

Over the last few decades, there has been little historical research on the detection of mounds through remote sensing. What is remarkable, however, the work presented by Vogt [21] showing an approach of photointerpretation from satellite imagery with overall accuracy rates of nearly 50%, and similarly, the one by Vogt and Wallet [22] analysing the feasibility of using shape-based and object-based models with satellite imagery and how unique models must be collected at new studied areas to ensure accurate results. Therefore, in this paper, a first and novel approach towards the detection of pre-existing termite mounds with the use of a UAS, hyperspectral imagery, machine learning and digital image processing is intended. A new pipeline process is proposed to detect termite mounds automatically and reduce, consequently, detection times in a studied zone.

2. Methods and Materials

The primary method is contained in a pipeline composed of four main stages, as depicted in Figure 2. Firstly, there is an image acquisition phase, in which raw hyperspectral images are obtained by using a UAS. Later, a pre-processing stage is run to get a filtered hyper-cube, spatially cropped in regions with pre-existing termite mounds. Afterwards, a classification process is executed to extract valuable layers amongst the cropped image. Finally, a segmentation stage is carried out to filter termite mounds from other objects with similar material characteristics yet different shapes. Hence, a new algorithm is proposed to perform the classification and segmentation processes automatically.

Figure 2. Primary proposed detection pipeline.

2.1. Site

As shown in Figure 3, four sites with pre-existing termite mounds at Cape Range National Park, Western Australia, Australia (−22.190138, 113.865371), were chosen for this study. Each area comprises bush lands, buffel grass, eroded soil regions and remains of decomposed wood branches. The field trip was performed on 10 July of 2016, and a couple of mission routes were conducted between 1:00 p.m. and 2:00 p.m. Meteorological conditions included open sky, 46% average relative humidity, SE winds between 17 and 26 km/h, 0.0 mm rainfall and 21.2 °C maximum temperature [23].

Figure 3. Primary location of the studied sites at Cape Range National Park, WA.

2.2. Image Sensors

A Headwall Nano Hyperspec hyperspectral camera was utilised to generate information regarding the spectrum for each of the pixels in the studied region. This camera supports up to 274 spectral bands, wavelengths ranging from 385 nm–1000 nm (VNIR), 2.2 nm dispersion per pixel, 300 Hz maximum frame rate, 640 spatial bands and 480 GB of storage capacity. Additionally, a Canon EOS 5DS R digital camera was also used in order to capture high-resolution images from the same mission route and identify by photointerpretation the present pre-existing termite mounds in the studied area to save their GPS coordinates into a register file. The camera specifications include 50.6 MP resolution, 28 mm focal length, ISO-400 speed, full frame CMOS sensor of 36 mm by 24 mm, 625 μs exposure time and a GPS sensor.

2.3. UAV and Sample Acquisition

A Hexa-Rotor DJI S800 UAV, featuring a customised integrated gimbal system, high-performance brushless motors, dimensions of 1180 mm × 1000 mm × 500 mm and a total weight of 3.9 kg cameras included, flew on the mentioned site following a mission route in DJI Ground Station 4.0 software. The mission followed an altitude of 66.9 ± 4.6 m, overlap of 80%, side lap of 50% and 4.5 m/s mean speed for a path length of 6.6 km. During the sampling period, the vertical and horizontal Ground Sample Distances (GSD) of the RGB and hyperspectral images were approximately

1.0152 cm/pixel and 4.7 cm/pixel, respectively. By performing manual interpretation on the acquired samples from both cameras, a total of 23 pre-existing termite mounds were identified and targeted for this study. The mounds feature a mean rectangular area of 1.951 m^2 and oval-based and diamond-based shapes in their centre, where vegetation density was virtually null and, conversely, a large grass concentration at their outer boundaries as illustrated in Figure 4.

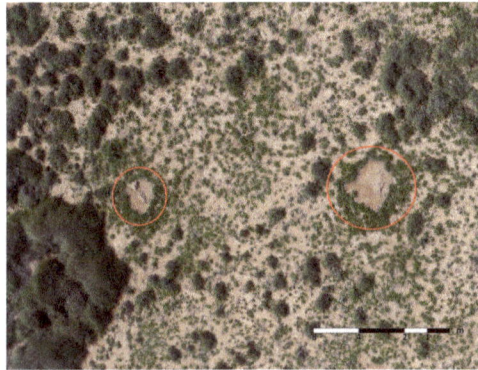

Figure 4. Highlighted pre-existing termite mounds in the planned mission route at -22.190228 latitude, 113.865197 longitude and 68.3 m altitude.

2.4. Software

Raw hyperspectral data were firstly ortho-rectified. Afterwards, various regions of interest (ROI) were extracted by checking their coordinates with the recorded digital image dataset. Next, each ROI was processed with Scyven Software [24]. Based on the freeware Scyllarus^{TM} toolbox provided by Data61 I CSIRO, this application is capable of reading native HDR files and executing many hyperspectral image processing techniques. Amongst them, spatial and spectral filtering, reflectance recovery, material recovery and material classification through Support Vector Machines (SVM), Linear Spectral Unmixing (LSU) and Principal Component Analysis (PCA) methods were utilised. Scyven software allowed a complete analysis of various material reflectance spectra and identification of termite mound components. These spectra curves were accordingly applied at the classification stage as described in Section 2.5. The digital image processing segmentation methods were run with the implementation of OpenCV library [25]. A detailed description of the applied methods is described in the following section.

2.5. Algorithm

An algorithm was designed and tested to identify pre-existing termite mounds in given regions of interest (ROI). As shown in Algorithm 1, the process consists of a preprocessing, classification and a segmentation stage, mentioned beforehand in the primary pipeline method.

Algorithm 1 Detection of pre-exiting termite mounds with hyperspectral imaging.

Input: raw hyperspectral image file
Pre-processing
1: Calculate hyper-cube radiance
2: Perform Orthorectification process
3: Load GPS coordinates from register file, and obtain a ROI
4: Apply closing operator once $\qquad$ ▷ structuring element: 3×3 rectangle
   Material Classification
5: Load illuminant spectrum from white reference
6: Calculate reflectance
7: Load material reflectance library
8: Run SVM classifier
9: Filter segmented material layers: "Eroded Soil" and "Light Grass"
   Object Detection
10: Apply Smooth-Median Filter on soil layer $\qquad$ ▷ kernel size: 3
11: Apply closing operator on soil layer once $\qquad$ ▷ structuring element: 3×3 rectangle
12: Create a temporal Image $T1$
13: $T1 \leftarrow$ double dilation operator on soil layer $\qquad$ ▷ structuring element: 3×3 ellipse
14: Apply Smooth-Median Filter on grass layer $\qquad$ ▷ kernel size: 3
15: Create a temporal Image $T2$
16: $T2 \leftarrow$ AND operator between grass layer and $T1$
17: **if** Mean($T2$) $= 0.0$ **then return** null
18: **end if**
19: Find contours from the soil layer
20: **for** $i \leftarrow 0, n$ **do** $\qquad$ ▷ $n =$ number of detected soil contours
21: $\quad$ Select $SC(i)$ $\qquad$ ▷ $SC(i) =$ soil contour at i index
22: $\quad$ Run *MatchShapes* method $\qquad$ ▷ output: *ratio*
23: $\quad$ $Psc \leftarrow$ perimeter of SC(i)
24: $\quad$ **if** *ratio* ≤ 0.15 && $Psc \geq 10$ **then**
25: $\quad\quad$ Discard $SC(i)$
26: $\quad$ **else**
27: $\quad\quad$ Find contours from the grass layer
28: $\quad\quad$ **for** $j \leftarrow 0, m$ **do** $\qquad$ ▷ $m =$ number of detected grass contours
29: $\quad\quad\quad$ Select $GC(j)$ $\qquad$ ▷ $GC(j) =$ grass contour at j index
30: $\quad\quad\quad$ $Pgc \leftarrow$ perimeter of $GC(j)$
31: $\quad\quad\quad$ $d \leftarrow$ distance between $GC(j)$ and $SC(i)$ centroids
32: $\quad\quad\quad$ $rad \leftarrow$ radius of $SC(i)$
33: $\quad\quad\quad$ $Prop \leftarrow Pgc/Psc$
34: $\quad\quad\quad$ If $d <= 1.2 * rad$ && ($Pgc \geq Psc$ || $Prop > 0.8$) **then**
35: $\quad\quad\quad\quad$ $ILD \leftarrow$ Minimum intersection line distance between $SC(i)$ and $GC(j)$
36: $\quad\quad\quad\quad$ **if** $ILD <= 2.5$ **then**
37: $\quad\quad\quad\quad\quad$ Draw contours at ROI image
38: $\quad\quad\quad\quad$ **end if**
39: $\quad\quad\quad$ **end if**
40: $\quad\quad$ **end for**
41: $\quad$ **end if**
42: **end for**
43: **return** ROI Image

2.5.1. Pre-Processing

Based on the fact that raw imagery obtained by Headwall Nano Hyperspec camera contains intensity values in radiance per pixel in each of the 274 available wavelength bands and dimensional data of 640 by 7000 pixels, digital processing calculations were heavy due to significantly large data file sizes (≃3 GB). Several preprocessing methods were subsequently implemented to retrieve cropped orthorectified hyperspectral cubes, gathering smaller files that could be rapidly processed in Scyven and OpenCV.

With raw hyperspectral cubes as the starting point, radiance correction and orthorectification techniques were implemented by the inbuilt methods from Headwall Hyperspec® III and SpectralView™ software. Then, the mentioned GPS coordinates database in Section 2.2 was read so as to crop and produce several ROIs with pre-existing termite mounds. After performing some random manual inspections, considerable noise and blur were discovered in many of the sample bands from each cropped hyper-cube. Consequently, a single iteration of the closing mathematical morphology operator was run firstly, using a rectangular 3 × 3 structuring element, as depicted in Figure 5.

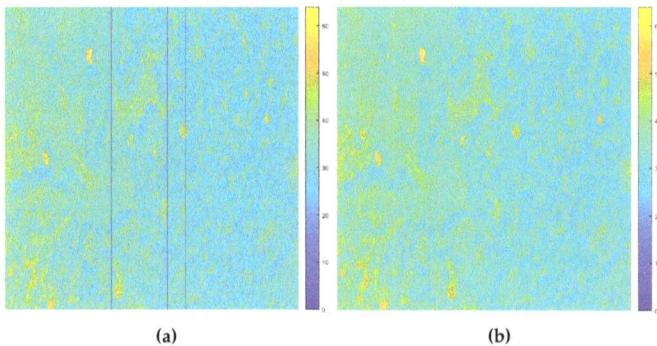

(a)   (b)

Figure 5. False colour representation of an ROI image in radiance during the pre-processing stage. (a) Raw image at 415 nm wavelength band. (b) Resultant filtered image.

2.5.2. Material Classification

In the first step of the classification process, the filtered hyperspectral cube in radiance was processed in order to obtain a new pixel-wise reflectance hyper-cube. Pixel reflectance was, therefore, obtained by importing the white reference spectrum into Scyven, acquired from the testing phase of the hyperspectral camera prior to the execution of the mission route, as shown in Figure 6. The use of a constant illuminant spectrum enhances the validation of the algorithm results instead of estimating dynamically various illuminant spectra [26] and, subsequently, different reflectance intensities for each loaded ROI into the software.

In order to run a machine learning classifier into the hyper-cube, it is required to load a set of material signatures that are wanted to be classified (targets). Prior and only once during this study, the spectra library was retrieved and clustered by selecting a demonstrative scene with an abundant amount of objects by using a deterministic annealing method [27]. Some of the input parameters for deterministic annealing are the maximum number of clusters, which was set to 10, a maximum temperature of 0.02, a minimum temperature of 0.00025, a cooling factor of 0.8, a split threshold of 0.8 and a maximum of 5 materials calculated per pixel. Figure 7 illustrates the spectral reflectance from the clustered materials.

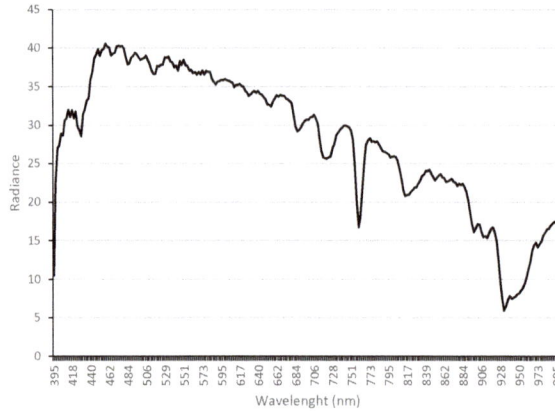

Figure 6. White reference spectrum acquired by the Headwall Nano Hyperspec camera.

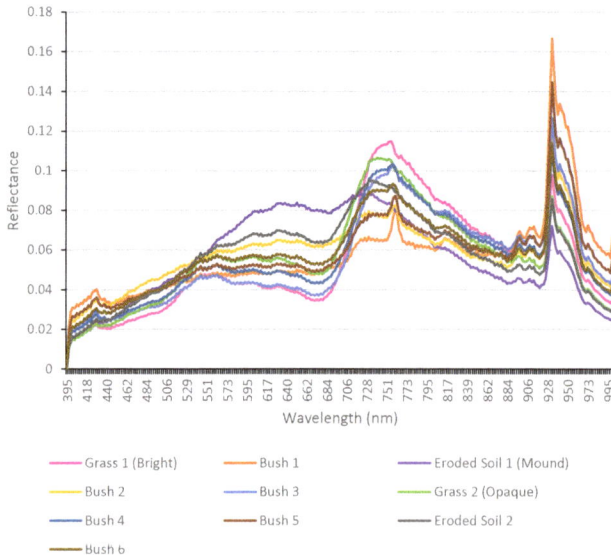

Figure 7. Clustered reflectance spectra of 10 materials from the demonstrative scene.

Based on the materials library as mentioned above, Support Vector Machines (SVM), Linear Spectral Unmixing (LSU) and Principal Component Analysis (PCA) algorithms were tested prior the inclusion of only one of them in the proposed method. The spectra library mentioned above was loaded into the classifiers as targets to retrieve the outcomes that are shown, for example, in Figure 8. The SVM result (Figure 8b) depicts unique colour layers for each targeted material. In contrast, albeit similar, the LSU outcome (Figure 8c) presents certain dark areas, which represent pixels where none of the loaded materials had a significant correlation with the real signature at that location, resulting in colour layers with considerable noise. Finally, The first component of the PCA outcome illustrated in false colour (Figure 8d) is able to highlight alterations at the surroundings of each one of the mounds. Nevertheless, a segmentation process to identify mounds automatically based on PCA represents a challenging task due to the ambiguousness of the given results. In spite of the representation of high variations at the ROI, the available data were not conclusive in comparison to the classifiers as mentioned earlier, so that the primary test results were carried out by using the SVM classifier only.

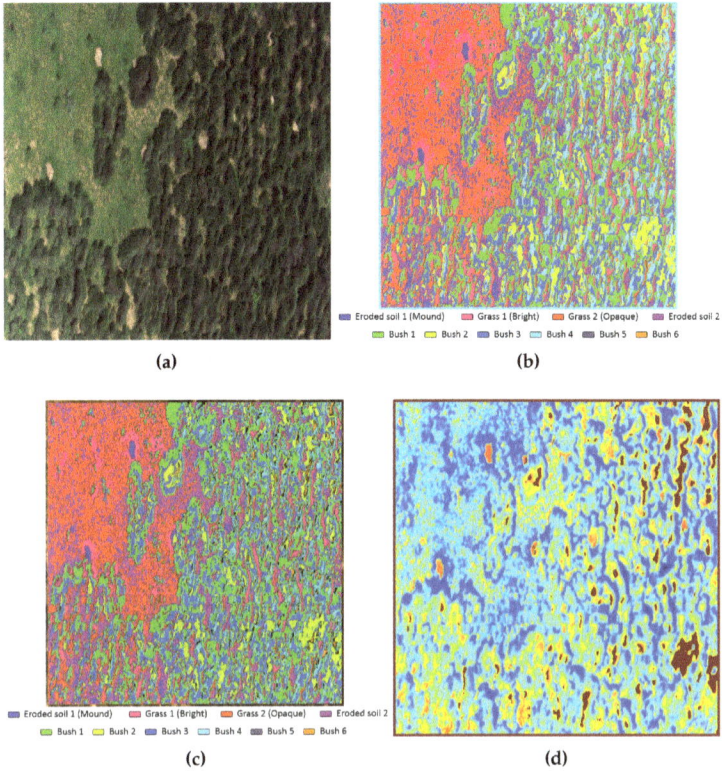

Figure 8. Classification algorithm outputs (**a**) Demonstrative ROI in RGB colour. (**b**) SVM output. (**c**) Linear Spectral Unmixing (LSU) output (**d**) PCA output in false colour.

From the output image of the SVM classifier, two out of the ten layer materials were filtered in order to distinguish authentic termite mounds from other components that shared similar material properties. As illustrated in Figure 9, it is possible to separate by photointerpretation the areas that belong to termite mounds from others regarding bushes and other vegetation objects. Therefore, an additional image segmentation process is required to segment those areas with pre-existing termite mounds.

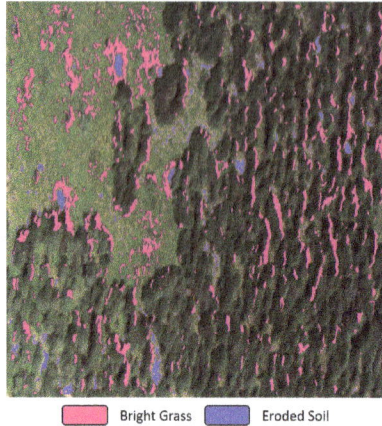

Figure 9. ROI with clustered termite mounds materials.

2.5.3. Object Detection

Steps 10–43 from Algorithm 1 permit the detection of pre-defined termite mounds based on a contour and shape analysis from the generated colour layers on previous steps. This phase allows one to count and display the location of the pre-existing termite mounds at each ROI, as depicted in Figure 10.

The starting point is the manipulation of the binary colour layers from Figure 9. Filtration and morphological techniques were applied in order to attenuate their noise, as shown from Steps 10–14 (Algorithm 1) and depicted in Figure 10a,b. Following this, it was found that termite mounds could be identified from other objects by calculating the distance among their two layers. That gap is considerably small compared with other objects at the ROI. Figure 11 depicts, for instance, this finding using a cropped region with a mound at its centre.

Given a small frame of a termite mound at Figure 11a, the SVM classifier will generate a blue colour layer for the eroded soil signature and the pink one for the bright grass area. After detecting the contour from the blue layer at Figure 11c, all the mean distances between that contour and the exposed pink regions are calculated. If a distance threshold is satisfied (yellow contour from Figure 11d), the detected contour in Figure 11c will be identified as a pre-existing termite mound.

Accordingly, the pink colour layer was set to be slightly dilated in order to determine a good mound centre (Step 14). A temporary image is, therefore, computed by performing a logic AND operation between both layers with the aim to check whether the ROI indeed includes any region with mounds by calculating, thus, its mean intensity defined in Equation (1).

$$Mean = \frac{\sum_{I=0}^{n} Src(I)}{n} \tag{1}$$

where *Src* belongs to the pixel value from the input image at the position *I* and *n* to the total number of pixels. If the mean threshold parameter from Step 17 of Algorithm 1 is not satisfied, the studied image will be discarded with zero found termite mounds as the final result. Figure 10c illustrates, for instance, the relevance of this operation.

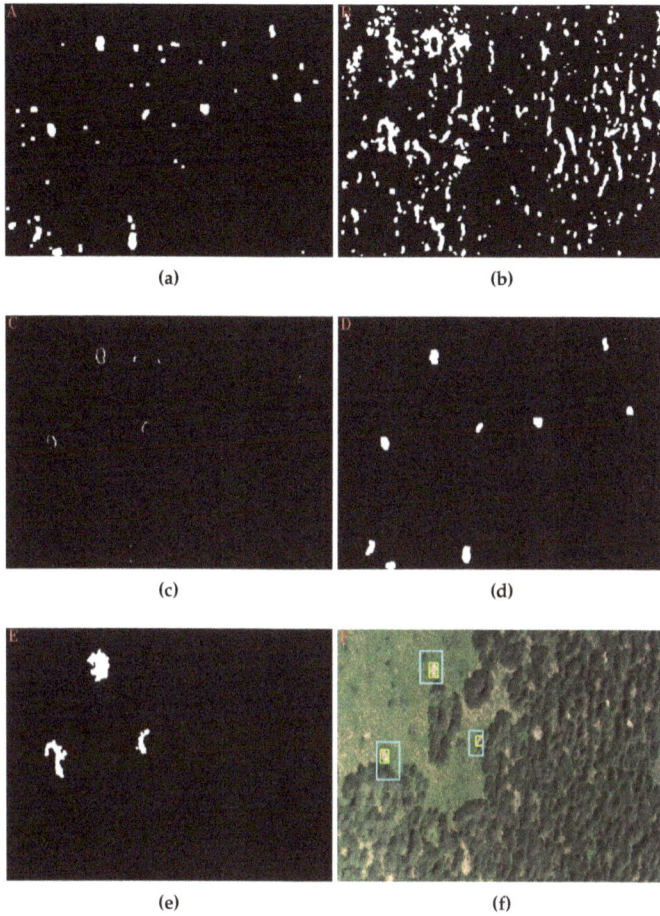

Figure 10. Main stages of the segmentation process. (**a**) Binary soil layer. (**b**) Binary grass layer. (**c**) AND logic operator between (**a**) and (**b**). (**d**) Filtered soil contours. (**e**) Filtered grass contours. (**f**) Identified contours in the ROI.

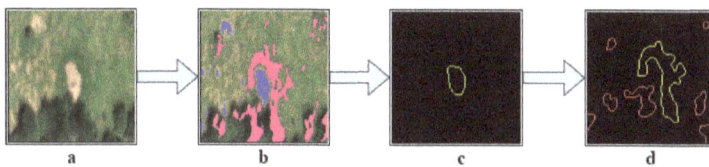

Figure 11. Contour analysis. (**a**) Cropped region. (**b**) classified material layers. (**c**) Detected contour at the soil layer. (**d**) Detected contours at the grass layer. Yellow contour: closest region from the contour (**c**).

Once the first threshold condition is satisfied, a contour detection and shape analysis were run on the soil layer. Firstly, a contour detection algorithm from Suzuki and Abe [28] was executed at Step 19.

Then, the detected soil contours were filtered using their shape properties by comparing their seven Hu moments [29] against some predefined pattern shapes. These patterns consist of a selected library of binary image templates that depicts key mound shapes, as illustrated in Figure 12.

Figure 12. Termite mound templates for the shape matching method.

By using a shape matching method in OpenCV, various contours with similar shapes from the loaded patterns can be detected regardless of their rotation, translation and scale properties. This method, mentioned in Step 22, returns a ratio value, calculated by following Equations (2)–(4).

$$m_i^S = sign(h_i^S) \times log(h_i^S) \tag{2}$$

$$m_i^T = sign(h_i^T) \times log(h_i^T) \tag{3}$$

$$ratio = \sum_{i=1}^{7} |\frac{1}{m_i^S} - \frac{1}{m_i^T}| \tag{4}$$

where h_i^S and h_i^T relate to the Hu moments of the soil and template contour, respectively. The lower the ratio value is, the closer the contour is to the analysed pattern. Finally, some thresholds are established in Step 24 to discard undesirable soil contours. Given that a set of points defines a contour, the surface contour area was calculated using Green's theorem. Figure 10d depicts, for instance, contours that surpassed the conditions mentioned above.

A similar contour detection process was applied in Step 27 into the grass layer in order to save those regions and analyse the location relationship between these and the soil contours that were filtered previously. Following Step 31 from Algorithm 1, the distance d between grass and soil contour centroids was estimated with Equation (5).

$$d = \sqrt{(P_x^S - P_x^G)^2 + (P_y^S - P_y^G)^2} \tag{5}$$

where P_x^S and P_y^S and P_x^G and P_y^G are the spatial x and y coordinates of the soil and grass contour centroids respectively. Additionally, the minimum intersection linear distances between these contours were calculated to check their proximity, as well. Following Equation (5), the distance between each one of the sets of points that defines those contours was estimated. Ultimately, thresholds were set concerning the calculated variables in Step 34 (Figure 10e), depicting only the ones that satisfied the conditions in the ROI image in RGB colour (Figure 10f).

3. Results

In order to test the proposed algorithm, eight cropped hyperspectral images with different characteristics in their scenery were chosen and analysed, the accuracy results of which are shown in Table 1. Those regions include scenes with pre-existing mounds and bush lands in their surroundings, rotated mounds and regions with the absence of mounds. For each one of the cropped images, mounds were visually identified and denoted as samples; the effectively found mounds were named True Positives (*TP*), undetected mounds as False Negatives (*FN*) and detected mounds by the algorithm yet missed at the real scene as False Positives (*FP*). The algorithm's accuracy was calculated by following Equation (6).

$$Accuracy = \frac{TP}{TP + FN + FP} \tag{6}$$

Table 1. Accuracy results of the proposed algorithm.

ROI	Termite Mounds	True Positives	False Negatives	False Positives	Accuracy (%)
1	6	5	1	0	83.3
2	9	4	5	1	40.0
3	1	1	0	0	100.0
4	2	1	1	0	50.0
5	13	11	2	3	68.8
6	2	2	0	0	100.0
7	9	7	2	0	77.8
8	2	2	0	0	100.0
9	3	3	0	1	75.0
10	2	2	0	0	100.0
11	4	3	1	0	75.0
12	2	1	1	0	50.0
13	7	4	3	1	50.0
14	1	1	0	0	100.0
15	6	4	2	0	66.7
16	6	5	1	0	83.3
17	7	5	2	0	71.4
18	6	3	3	0	50.0
19	1	1	0	0	100.0
20	7	4	3	1	50.0
21	4	3	1	1	60.0
22	7	5	2	1	62.5
23	3	3	0	0	100.0
24	8	6	2	0	75.0
25	1	1	0	0	100.0
Total	119	87	32	9	**68.0%**
Proportion	100%	73.1%	26.9%	7.6%	

4. Discussion

By adding an image processing procedure to filter contours by their shapes and identify termite mounds (the segmentation stage of Algorithm 1), the primary outcomes of the proposed method became highly dependent on the GSD values (spatial resolution) and distortion properties from the studied hyperspectral images. As illustrated in Figure 13a, for example, several targets were misclassified owing to a considerable distortion in their oval-based centres and tiny sizes in the scene. Similarly, Figure 13b denotes some limitations in detecting mounds when the oval-based eroded regions are not clearly depicted in the studied area. Both cases revealed how distortion is presented in random areas from any ROI and not only near cornering regions as primarily expected. Conversely, there were successfully identified pre-existing termite mounds regardless of their rotation and location properties in the scene. Additionally, high distortion and low natural light exposure provoked the outcome of false positives, as shown specially for ROI 5 from Table 1. In summary, the given results from detecting appropriately medium and big sized mounds with little distortion presented in the processed images suggest that the proposed algorithm is capable of offering reliable results when decreasing the GSD values and improving the acquisition standards (reducing distortion). However, the unavailability of obtaining new samples at lower altitudes goes beyond the scope of this investigation, and further research is recommended to validate this affirmation.

(a)   (b)

Figure 13. Algorithm outcomes of two ROIs with the remarked missing mounds in red: (**a**) 40% accurate; (**b**) 67% accurate.

The Orthorectification Process (ORP) tended to distort some objects' properties smoothly. Small termite mounds, like the ones depicted in Figure 13, were unlikely to be identified owing to the generated distortion by that process. Incrementing the number of available shape templates might minimise this issue, but conversely, a large number of hyperspectral images would be required to obtain those templates, augmenting, therefore, project costs. As maintained above, acquiring images at lower altitudes might improve the results of the proposed method.

The material clustering process denoted the low viability to identify a unique element signature (spectrum) that could detect termite mounds at the selected sites in Cape Range National Park. Conversely, a minimum of two primary materials was distinguished as shown previously in Figure 9. The main contribution of Algorithm 1 is, therefore, the adequate integration of hyperspectral image processing methods to classify multiple materials and traditional RGB image processing to segment pre-existing termite mounds from that set of material layers. Additionally, the dataset of recovered materials enhances the importance of incorporating hyperspectral technology in similar research.

Amongst the feasible options to optimise the accuracy of Algorithm 1, Evolutionary Algorithms (EA) have been widely used when considering optimisation of multi-objective systems in remote sensing applications [30–32]. Through the implementation of EA, input parameters that were manually calculated could now be estimated and optimised. For this case, EA outputs will correspond to the location of the mound regions, and inputs will be composed of all threshold parameters, such as the minimum and maximum contour area, the distance between selected soil and grass contours, the match-shaping tolerance rate and other inner constant parameters from the polygonal approximation method. A set of various algorithms will be treated as population individuals in order to get a resulting one whose threshold values might obtain the best segmentation result.

Working only with RGB digital imagery rather than hyperspectral imaging is challenging or not possible for the task of identifying pre-existing mounds. The available data on digital RGB photos, even though of high resolution, mostly rely on the colour representation of the materials in three bands of the visible spectrum where different vegetation species are symbolised by the same RGB colour representation. The recovered materials spectra of the different vegetation and soil highlight substantial differences on their magnitude from 540–700 nm, 720–800 nm and 924–1000 nm wavelength ranges and small differences at the remaining bands.

5. Conclusions

This paper presented an algorithm to detect automatically pre-existing termite mounds by following a pipeline process based on acquired hyperspectral imagery with a UAS, an SVM classifier and shape-based image processing detection methods. Slight differences were found in pixel intensity between many of the green vegetation clustered materials, which meant a challenging task at the classification stage. A combination of "bright grass" and "eroded soil" retrieved material signatures was utilised to classify termite mounds instead of extracting a unique material spectrum for their detection. Results showed an overall accuracy of 68%, in which nearly 27% of the targeted mounds were undetected due to high distortion and low-resolution properties from the hyperspectral dataset. On the other hand, the algorithm misclassified less than 8% of random objects as termite mounds. Due to the execution of shape-based detectors, image resolution is still determinant to obtain accurate results. Nonetheless, judging by regions with similar visible colour representations, hyperspectral imagery demonstrated better capabilities to classify a huge set of materials than implementing traditional segmentation methods on RGB images only. Further research by acquiring samples at lower altitudes, as well as increasing the number of targeted mounds is highly recommended. Additionally, optimisation with EAs will boost the accuracy of the proposed method.

Acknowledgments: The authors would like to acknowledge the Plant Biosecurity Cooperative Research Centre (PBCRC) for the financial support to conduct the airborne data collection campaign at Cape Range National Park, Western Australia, Australia. We would also like to acknowledge the QUT Research Engineering Facility (REF) team for their flight operational campaign.

Author Contributions: Juan Sandino and Adam Wooler performed the experimentation phase. Juan Sandino analysed the data, designed and tested the algorithm and wrote the manuscript. Felipe Gonzalez contributed with materials, analysis tools and important advice. Juan Sandino Adam Wooler and Felipe Gonzalez revised the paper.

Conflicts of Interest: The authors declare no conflict of interest.

References

1. Lee, K.E.; Wood, T.G. *Termites and Soils*; Academic Press: London, UK, 1971; p. 251.
2. Dawes-Gromadzki, T.Z. Abundance and diversity of termites in a savannah woodland reserve in tropical Australia. *Aust. J. Entomol.* **2008**, *47*, 307–314.
3. Dawes, T.Z. Impacts of habitat disturbance on termites and soil water storage in a tropical Australian savannah. *Pedobiologia* **2010**, *53*, 241–246.
4. Holt, J.A. Carbon mineralization in semi-arid northeastern Australia: The role of termites. *J. Trop. Ecol.* **1987**, *3*, 255.
5. Jamali, H.; Livesley, S.J.; Grover, S.P.; Dawes, T.Z.; Hutley, L.B.; Cook, G.D.; Arndt, S.K. The Importance of Termites to the CH4 Balance of a Tropical Savanna Woodland of Northern Australia. *Ecosystems* **2011**, *14*, 698–709.
6. Lobry De Bruyn, L.A.; Conacher, A.J. Soil modification by termites in the central wheatbelt of Western Australia. *Aust. J. Soil Res.* **1995**, *33*, 179–193.
7. Gosling, C.M.; Cromsigt, J.P.; Mpanza, N.; Olff, H. Effects of Erosion from Mounds of Different Termite Genera on Distinct Functional Grassland Types in an African Savannah. *Ecosystems* **2012**, *15*, 128–139.
8. Foti, F.L. The Possible Nutritional/Medicinal Value of Some Termite Mounds Used by Aboriginal Communities of Nauiyu Nambiyu (Daly River) and Elliott of the Northern Territory, with an Emphasis on Mineral Elements. Ph.D. Thesis, Charles Darwin University, Casuarina, Australia, 1994.
9. Milner, R.J.; Staples, J.A. Biological control of termites: Results and experiences within a CSIRO project in Australia. *Biocontrol Sci. Technol.* **1996**, *6*, 3–9.
10. Chouvenc, T.; Su, N.Y.; Kenneth Grace, J. Fifty years of attempted biological control of termites—Analysis of a failure. *Biol. Control* **2011**, *59*, 69–82.
11. Bunting, P.; He, W.; Zwiggelaar, R.; Lucas, R. Combining Texture and Hyperspectral Information for the Classification of Tree Species in Australian Savanna Woodlands. In *Innovations in Remote Sensing and Photogrammetry*; Jones, S., Reinke, K., Eds.; Springer: Berlin/Heidelberg, Germany, 2009; pp. 19–26.
12. He, K.S.; Rocchini, D.; Neteler, M.; Nagendra, H. Benefits of hyperspectral remote sensing for tracking plant invasions. *Divers. Distrib.* **2011**, *17*, 381–392.
13. Wallis, C.I.; Brehm, G.; Donoso, D.A.; Fiedler, K.; Homeier, J.; Paulsch, D.; Süßenbach, D.; Tiede, Y.; Brandl, R.; Farwig, N.; et al. Remote sensing improves prediction of tropical montane species diversity but performance differs among taxa. *Ecol. Indic.* **2017**, doi:10.1016/j.ecolind.2017.01.022.

14. Colomina, I.; Molina, P. Unmanned aerial systems for photogrammetry and remote sensing: A review. *ISPRS J. Photogramm. Remote Sens.* **2014**, *92*, 79–97.

15. Zeng, C.; King, D.J.; Richardson, M.; Shan, B. Fusion of Multispectral Imagery and Spectrometer Data in UAV Remote Sensing. *Remote Sens.* **2017**, *9*, 696.

16. Mahlein, A.K.; Rumpf, T.; Welke, P.; Dehne, H.W.; Plümer, L.; Steiner, U.; Oerke, E.C. Development of spectral indices for detecting and identifying plant diseases. *Remote Sens. Environ.* **2013**, *128*, 21–30.

17. Lee, D.S.; Gonzalez, L.F.; Périaux, J.; Srinivas, K. Evolutionary optimisation methods with uncertainty for modern multidisciplinary design in aeronautical engineering. In *Notes on Numerical Fluid Mechanics and Multidisciplinary Design*; Springer: Berlin, Germany, 2009; Volume 100, pp. 271–284.

18. Lee, D.S.; Gonzalez, L.F.; Whitney, E.J. *Multi-objective Multidisciplinary Multi-fidelity Design Tool: HAPMOEA-User Guide*; Technical report; The University of Sydney (USYD)/Queensland University of Technology (QUT): Brisbane City, Australia, 2007.

19. González, L.; Whitney, E.; Srinivas, K.; Périaux, J. Optimum Multidisciplinary and Multi-Objective Wing Design in CFD Using Evolutionary Techniques. In Proceedings of the Computational Fluid Dynamics 2004: Third International Conference on Computational Fluid Dynamics, ICCFD3, Toronto, ON, Canada, 12–16 July 2004.

20. Alvarado, M.; Gonzalez, F.; Fletcher, A.; Doshi, A. Towards the Development of a Low Cost Airborne Sensing System to Monitor Dust Particles after Blasting at Open-Pit Mine Sites. *Sensors* **2015**, *15*, 19667–19687.

21. Vogt, J.T. Detection of Imported Fire Ant (Hymenoptera: Formicidae) Mounds with Satellite Imagery. *Environ. Entomol.* **2004**, *33*, 1718–1721.

22. Vogt, J.T.; Wallet, B. Feasibility of using template-based and object-based automated detection methods for quantifying black and hybrid imported fire ant (Solenopsis invicta and S. invicta x richery) mounds in aerial digital imagery. *Rangel. J.* **2008**, *30*, 291–295.

23. Bureau of Meteorology. *Learmonth, WA—July 2016—Daily Weather Observations*; Bureau of Meteorology: West Perth, WA, Australia, 2016.

24. Habili, N.; Oorloff, J. Scyllarus: From Research to Commercial Software. In Proceedings of the 24th Australasian Software Engineering Conference, Adelaide, Australia, 28 September–1 October 2015; Volume 2, pp. 1–4.

25. Bradski, G. The OpenCV library. *Dr. Dobb's J. Softw. Tools* **2000**, *25*, 120–123.

26. Robles-Kelly, A.; Huynh, C.P. Illumination Spectrum Recovery. U.S. Patent 8,953,906, 10 February 2011.

27. Rose, K. Deterministic Annealing for Clustering, Compression, Classification, Regression, and Related Optimization Problems. *Proc. IEEE* **1998**, *86*, 2210–2239.

28. Suzuki, S.; Abe, K. Topological structural analysis of digitized binary images by border following. *Comput. Vis. Graph. Image Process.* **1985**, *30*, 32–46.

29. Hu, M.K. Visual pattern recognition by moment invariants. *IRE Trans. Inf. Theory* **1962**, *8*, 179–187.

30. Bandyopadhyay, S.; Maulik, U.; Mukhopadhyay, A. Multiobjective Genetic Clustering for Pixel Classification in Remote Sensing Imagery. *IEEE Trans. Geosci. Remote Sens.* **2007**, *45*, 1506–1511.

31. De la Fraga, L.G.; Coello, C.A. A Review of Applications of Evolutionary Algorithms in Pattern Recognition. In *Pattern Recognition, Machine Intelligence and Biometrics*; Wang, P.S.P., Ed.; Springer: Berlin/Heidelberg, Germany, 2011; pp. 3–28.

32. Ma, A.; Zhong, Y.; Zhang, L. Adaptive Multiobjective Memetic Fuzzy Clustering Algorithm for Remote Sensing Imagery. *IEEE Trans. Geosci. Remote Sens.* **2015**, *53*, 4202–4217.

sensors MDPI

Article

UAVs and Machine Learning Revolutionising Invasive Grass and Vegetation Surveys in Remote Arid Lands

Juan Sandino [1,*,†], Felipe Gonzalez [1,†], Kerrie Mengersen [2] and Kevin J. Gaston [3]

1   Institute for Future Environments; Robotics and Autonomous Systems, Queensland University of Technology (QUT), 2 George St, Brisbane City QLD 4000, Australia; felipe.gonzalez@qut.edu.au
2   School of Mathematical Sciences; ARC Centre of Excellence for Mathematical & Statistical Frontiers (ACEMS), Queensland University of Technology (QUT), 2 George St, Brisbane City QLD 4000, Australia; k.mengersen@qut.edu.au
3   Environment and Sustainability Institute, University of Exeter, Penryn, Cornwall TR10 9FE, UK; K.J.Gaston@exeter.ac.uk
*   Correspondence: j.sandinomora@qut.edu.au; Tel.: +61-7-3138-1363
†   These authors contributed equally to this work.

Received: 30 November 2017; Accepted: 11 February 2018; Published: 16 February 2018

Abstract: The monitoring of invasive grasses and vegetation in remote areas is challenging, costly, and on the ground sometimes dangerous. Satellite and manned aircraft surveys can assist but their use may be limited due to the ground sampling resolution or cloud cover. Straightforward and accurate surveillance methods are needed to quantify rates of grass invasion, offer appropriate vegetation tracking reports, and apply optimal control methods. This paper presents a pipeline process to detect and generate a pixel-wise segmentation of invasive grasses, using buffel grass (*Cenchrus ciliaris*) and spinifex (*Triodia* sp.) as examples. The process integrates unmanned aerial vehicles (UAVs) also commonly known as drones, high-resolution red, green, blue colour model (RGB) cameras, and a data processing approach based on machine learning algorithms. The methods are illustrated with data acquired in Cape Range National Park, Western Australia (WA), Australia, orthorectified in Agisoft Photoscan Pro, and processed in Python programming language, scikit-learn, and eXtreme Gradient Boosting (XGBoost) libraries. In total, 342,626 samples were extracted from the obtained data set and labelled into six classes. Segmentation results provided an individual detection rate of 97% for buffel grass and 96% for spinifex, with a global multiclass pixel-wise detection rate of 97%. Obtained results were robust against illumination changes, object rotation, occlusion, background cluttering, and floral density variation.

Keywords: biosecurity; buffel grass; *Cenchrus ciliaris*; drones; remote surveillance; spinifex; *Triodia* sp.; unmanned aerial vehicles (UAV); vegetation assessments; xgboost

---

1. Introduction

Over recent decades, invasive grasses have resulted in very substantial losses to native ecosystems around the world. Governmental, scientific, and community efforts to monitor and control these and other introduced plant species have been extremely challenging due to restricted and difficult access to remote areas, expensive operational costs, and, in some cases, hazardous data collection campaigns [1]. In Australia, the U.S., South Africa, and other parts of the world, introduced grasses have flourished in arid landscapes due to their tenacity under hot, heavy grazing, and drought conditions [2,3]. Moreover, they have been fostered by farmers because of the economic benefits that they bring through land rehabilitation and livestock production. However, many of these plant species have invaded some of the wetter and more fertile parts of the landscape and affected the survival of native plant and animal populations [4–6]. As a result, these species have now been catalogued as invasive plants or weeds. There is an increasing body of research focused on assessing the biodiversity effects of invasive grasses, which shows that their expansion rates are likely to exceed new high records due to climate change effects [6–11]. Straightforward, efficient, and accurate surveillance methods are required to quantify expansion rates of invasive grasses and apply reliable and efficient control methods.

Among the present efforts to monitor invasive grasses and other vegetation, different investigations have developed diverse solutions, using various image sensors and detection methods to meet a range of needs. Previously, satellite and manned aircraft imagery was used to map invasive grass infestations. Nonetheless, advances in unmanned aerial vehicles (UAV) design and path planning [12,13] have seen an increased application of remote sensing for ecological assessments and biosecurity applications [14–17]. Research from Olsson et al. [18], for instance, demonstrated the importance of using hyperspectral imagery for invasive grass detection as compared with satellite imagery. A feasibility study of sensing technology by Marshall et al. [19], for example, illustrates the potential for regional mapping of buffel grass infestations in arid landscapes using high-resolution aerial photography in red, green, blue (RGB) colour model at a cm/pixel scale over multi- and hyperspectral technologies for overall detection rates.

Weed mapping using different image sensors capable of sensing multiple spectral bands is also an active field of research. Alexandridis et al. [20], for instance, developed an approach by integrating UAVs and multispectral imagery for weed mapping, achieving detection rates of up to 96%. Moreover, Blaschke et al. [21] and Torres-Sánchez et al. [22] showed the use of Geographic Object-Based Image Analysis (GEOBIA) through UAVs and multispectral imagery to obtain detection rates of approximately 90%.

Development of image and data processing techniques for vegetation assessments is also increasing. Amongst the popular methods, the use of spectral indexes for weed detection has gained considerable popularity, as explored by Ashourloo et al. [23], Robinson et al. [24], and Lin et al. [25]. In these cases, both supervised and unsupervised segmentation algorithms were greatly influenced by image quality, spectral bands, and ground sampling distance (GSD), among other complex considerations of the scene. In sum, a universal criterion has not yet been defined for choosing a feasible sensing technology and data processing pipeline that meets every application need [26]. This paper proposes the creation of a global approach for the surveillance of invasive grasses and related biosecurity applications by developing an automatic surveillance solution integrating UAV technology with high-resolution RGB cameras and a machine learning-based classification algorithm to process and segment the data. The presented pipeline process is illustrated with the automatic detection of buffel grass (*Cenchrus ciliaris*) and spinifex (*Triodia* sp.) in arid and semi-arid ecosystems in Australia.

2. Materials and Methods

2.1. Process Pipeline

We developed a pipeline process consisting of four main components: Acquisition, Preprocessing, Training, and Prediction, as illustrated in Figure 1. High-resolution digital images are initially captured from a UAV flight mission. Images are downloaded, orthorectified, and preprocessed in order to extract samples with key features and label them subsequently. Data are then fed into a supervised machine learning classifier to train and optimise its detection capabilities. Finally, the entire orthorectified imagery is processed to predict the location of invasive grasses and vegetation in the studied area.

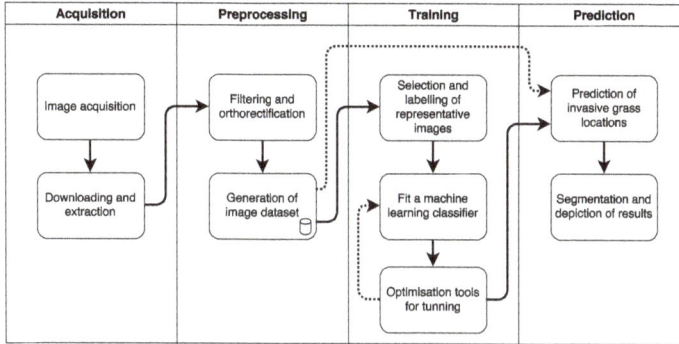

Figure 1. Primary pipeline for mapping of invasive grasses and related vegetation.

2.2. Site

The study site is located in the Cape Range National Park, Western Australia (WA), Australia (−22.190429, 113.865478). The site contains buffel grass, spinifex, remains of dry and decomposed vegetation, bushes, and arid soil. Images were taken in a successive series of four flight campaigns, conducted on the 10 July 2016, from 12:20 p.m. until 2:20 p.m. Meteorological conditions for that day were sunny, with south-easterly winds from 17 to 26 km/h, 46% relative humidity, 21.2 °C mean temperature, and no precipitation [27].

In the site, invasive grass species such as buffel grass and spinifex were found with negligible size variation, viewpoint variation, background clutter, and occlusion. However, they occurred at various densities, as shown in Figure 2.

Figure 2. Main features of the study site. (**a**) Geographical location. (**b**) Area with high density of buffel grass. (**c**) Area with high density of spinifex. (**d**) Area with low density of invasive grasses. (**e**) Buffel grass. (**f**) Spinifex.

2.3. Image Sensors

A Canon EOS 5DsR digital camera (Canon Inc., Tokyo, Japan) was utilised to capture high-resolution images. The camera specifications include 50.6 MP resolution, 28 mm focal length, ISO-400 speed, a full-frame complementary metal–oxide–semiconductor (CMOS) sensor of 36 mm × 24 mm, a 625 μs exposure time, and a global positioning system (GPS) sensor.

2.4. The UAV and Sample Acquisition

A DJI S800 EVO Hexa-rotor UAV (DJI, Guangdong, China) was employed in the study area following a designed mission route with DJI Ground Station 4.0 software. As shown in Figure 3, the UAV featured high-performance brushless motors, a customised dampened gimbal providing active three-axis stabilisation of the sensor payload (levelled out to ensure the sensor was pointing permanently in the direction of the ground), a total weight of 3.9 kg, and dimensions of 1180 mm × 1000 mm × 500 mm. The flight mission was performed at an altitude of 66.9 ± 4.6 m, an overlap of 80%, side lap of 50%, and a route length of 6.6 km at 16.2 km/h. The horizontal and vertical GSD were approximately 1.0152 cm/pixel in both cases.

Figure 3. The DJI S800 EVO (DJI, Guangdong, China) unmanned aerial vehicle (UAV) flying in Cape Range National Park, Western Australia (WA), Australia.

2.5. Software

Various software solutions were used through the development of this research. In order to prepare the data, more than 500 raw images were filtered and orthorectified using Agisoft PhotoScan 1.2. With this software, an orthomosaic image of 44,800 × 17,200 pixels of 2.4 GB was generated. Due to the huge image size and possible random-access memory (RAM) limitations, this image was split into 4816 items of 400 × 400 pixels in Tagged Image File (TIF) and Keyhole Markup Language (KML) formats. A group of representative samples in cropped regions was extracted and subsequently labelled using GNU Image Manipulation Program (GIMP) 2.8.22 to fit the classifier. The generated image set was processed using Python 2.7.14 programming language and several third-party libraries for data manipulation and machine learning, including eXtreme Gradient Boosting (XGBoost) 0.6 [28], Scikit-learn 0.19.1 [29], OpenCV 3.3.0 [30], and Matplotlib 2.1.0 [31].

2.6. Data Labelling

Due to the variety of conditions in which invasive grasses were found in Cape Range National Park, 10 images were selected and analysed using photo interpretation. Invasive grasses (buffel grass and spinifex), as well as common objects in the area, were highlighted using bright distinguishable colours as depicted in Figure 4. Regions were coloured through the "Bucket fill" tool of the GIMP software.

To perform image labelling, a mask for each image sample was generated by assigning integer values for every highlighted pixel. Each bright coloured pixel was filtered from every sample using Equation (1).

$$H_{(x,y)} = \begin{cases} a & \text{if } S_{(x,y)} = F_{(R,G,B)} \\ 0 & \text{otherwise} \end{cases} \tag{1}$$

where H is the mask for each sample S and a is the integer value for every bright colour value $F_{(R,G,B)}$. Values for a were set as follows: 1 = buffel grass; 2 = soil and road; 3 = bushes; 4 = shadow; 5 = dry vegetation (Dry Veg.); 6 = spinifex.

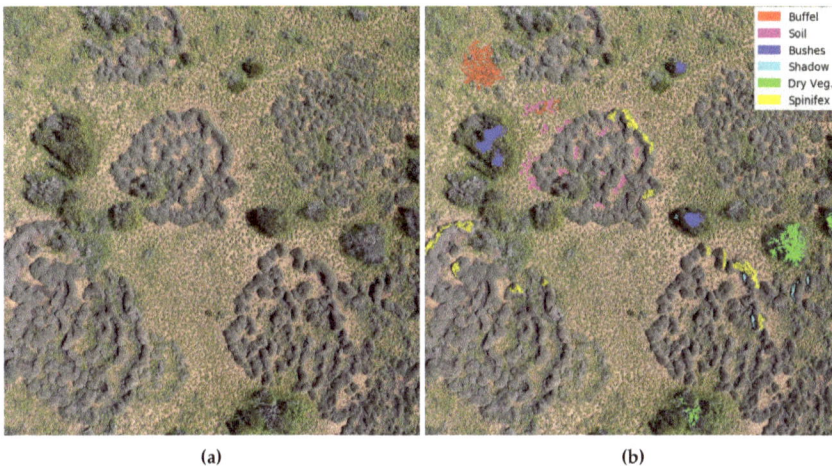

Figure 4. Image labelling. (**a**) Representative sample. (**b**) Highlighted regions using bright colours.

2.7. Classification Algorithm

Algorithm 1 was utilised for the training and prediction stages. It identifies and filters the highlighted regions mentioned in Section 2.6, trains a gradient boosted decision tree classifier, cross-validates the classification rates, predicts unlabelled data, and displays the results.

The training section of Algorithm 1 comprises several steps to load, preprocess the data, and fit an XGBoost classifier. The processing stage transforms the read data into an array of features or attributes, which are consequently processed by the classifier. As described in the algorithm, in order to obtain the feature array D, representative sample images G are firstly converted from their default RGB colour model into the hue, saturation, value (HSV) colour model in Step 3. Then, a set of filters are applied on G and their outputs inserted into D subsequently, as mentioned in Step 5. The two-dimensional (2D) filters calculate the variance into a subset of pixel neighbours contained in a window, following Equations (2) and (3).

Algorithm 1 Detection and segmentation of invasive grasses using high-resolution RGB images.

Required: orthorectified image set I. Representative samples set G. Sample masks set H

Training

1: **for** $i \leftarrow 1, n$ **do**   ▷ $n =$ total images in G (labelled data)
2:     Load G_i and H_i images
3:     Convert colour space of G_i into HSV
4:     Insert each colour channel into a feature array D
5:     Use 2D filters on G_i and insert their outputs into D
6:     From G_i and H_i, filter only the pixels with assigned labelling on D
7: **end for**
8: Split D into training data D_T and testing data D_E
9: Create a XGBoost classifier X and fit it using D_T
10: Use K-fold cross validation with D_E ▷ number of folds = 10
11: Perform grid search to tune X parameters

Prediction

12: **for** $i \leftarrow 1, m$ **do**   ▷ $m =$ total images in I
13:     Load I_i image
14:     Convert colour space of I_i into HSV
15:     Scan every pixel and predict the object using X
16:     $O_i \leftarrow$ Convert the data into a 2D image
17:     Export O_i into TIF format
18: **end for**
19: **return** O_i

$$X = \frac{1}{w^2} \begin{bmatrix} 1 & 1 & \cdots & 1 \\ 1 & 1 & \cdots & 1 \\ \vdots & \vdots & \ddots & \vdots \\ 1 & 1 & \cdots & 1 \end{bmatrix} \tag{2}$$

$$s^2 = E[X^2] - E[X]^2 \tag{3}$$

where X is the kernel of the filter to estimate the mean value of the processed image, w is the window size, and s^2 is the variance defined as the subtraction between the estimation of mean of square and the square of mean. Thus, the array of features D for this case study contains 10 items as follows: hue, saturation, value, variance filters on hue where w equals 3 and 15, variance filters on saturation where w equals 3 and 15, and variance filters of the grayscale image from G_i where w equals 3, 7, and 15. Later, as described in Step 6, pixel locations that were previously labelled are filtered using masks H, following Equation (4).

$$D_j = \begin{cases} [G_i(x,y), \ H_i(x,y)] & \text{if } H_i(x,y) \neq 0 \\ \text{null} & \text{otherwise} \end{cases} \tag{4}$$

where D_j is the 2D output array of the operation, $G_i(x,y)$ is the sample image, and $H_i(x,y)$ is the labelled counterpart of G_i at position (x,y). In total, 342,626 pixel-wise samples were filtered and subsequently split randomly into a training (75%) and testing (25%) data array. In Step 9, data are processed into the XGBoost classifier, which is a state-of-art decision tree and gradient boosting based model created by Chen and Guestrin [28] that is optimised for large tree structures, high execution speed, and excellent performance. Hyper-parameters for this classifier such as the number of estimators, the learning rate, and maximum depth are estimated by running a grid search method in Step 11. This technique evaluates a combination of multiple values for each hyper-parameter, returning the optimal combination of those for the classifier. For this case study, the optimal hyper-parameter values to obtain an accuracy-robustness balance without causing over-fitting are:

$$\text{estimators} = 100, \text{ learning rate} = 0.1, \text{ maximum depth} = 3$$

where "estimators" is the number of trees, "learning rate" is the step size of each boosting step, and "maximum depth" is the maximum depth per tree that defines the complexity of the model. For the prediction stage (Steps 13–17), all the orthorectified images are processed in a loop using the trained classifier and the same data conversion considerations applied at the training stage. Finally, classified pixels for each image are painted in distinguishable colours and exported in TIF format, compatible with geographic information system (GIS) platforms.

3. Results

Segmented images for photo interpretation as well as accuracy indicators were implemented for validation purposes. In total, 85,657 labelled pixels were evaluated from the test data set D_E to assess Algorithm 1. The confusion matrix of the classifier is presented in Table 1.

Table 1. The eXtreme Gradient Boosting (XGBoost) classifier confusion matrix.

Predicted	Buffel	Soil	Bushes	Shadow	Dry vegetation	Spinifex
Buffel	25,256	17	156	0	4	362
Soil	15	25,196	1	0	1	0
Bushes	632	1	3913	2	21	81
Shadow	0	1	0	7729	0	0
Dry vegetation	8	10	6	2	5734	159
Spinifex	508	2	20	0	171	15,649

(Labelled)

From 25,795 instances of pixels labelled as buffel grass, the algorithm predicted correctly 25,256 pixels and reported misclassifications of 362 pixels as spinifex, 156 pixels as bushes, 17 pixels as soil and 4 pixels as dry vegetation. Similarly, 25,196 pixels were successfully predicted as soil, with 17 misclassifications; 3913 pixels as bushes with 737 misclassifications; 7729 pixels as shadow with 1 misclassification; 5734 pixels as dry vegetation with 185 misclassifications; and 15,649 pixels as spinifex with 701 misclassifications. Based on these numbers, a classification report is generated as shown in Table 2.

Table 2. Classification report from confusion matrix of Table 1.

Class	Precision (%)	Recall (%)	F-Score (%)	Support
Buffel	95.60	97.91	96.75	25,795
Soil	99.88	99.93	99.90	25,213
Bushes	95.53	84.15	89.84	4650
Shadow	99.95	99.99	99.97	7730
Dry vegetation	96.68	96.87	96.78	5919
Spinifex	96.30	95.71	96.00	16,350
Mean	97.32	95.76	96.54	$\sum$= 85,657

Here, precision is the ratio between true positives and the sum of true positives and false positives, recall is the ratio between true positives and the sum of true positives and false negatives, f-score is the mean value between precision and recall, and support is the number of tested pixels per class. For this case study, precision errors indicate the output of misleading results by labelling wrong classes, whereas recall errors show the output of incomplete class detection.

Overall, the majority of the classes were successfully classified. For buffel grass and spinifex, most of the misclassified pixels were attributed to their counterpart class; these misclassification rates were small, representing for buffel grass precision and recall errors of 1.92% and 1.40%, respectively, and for spinifex values of 2.23% and 3.11%, respectively. Due to the high variation in greenness values of labelled buffel grass, specific areas of dry grass were classified as spinifex and vice versa. Similarly, misclassification of dry vegetation and spinifex instances (2.88% and 2.69%, 0.98% and 1.05%) occurred owing to many occurrences of this plant in senescence conditions. The classification rates were excellent for the shadow and soil classes, mainly due to the small variation in their visual properties such as their colour intensity, luminosity, and smooth texture. In contrast, the classification of the bushes class was not satisfactory at all, especially its recall rates, as indicated by a greater proportion of pixels classified as buffel grass (3.81% and 13.59%) and to a lesser degree, spinifex (0.49% and 1.74%).

The proposed algorithm is capable of classifying invasive grasses and other vegetation with remarkable global precision rates of 97% and recall rates of 95.76%. The proposed method increases, nevertheless, the likelihood of classifying certain bush regions as buffel grass, with a recall rate of 84.15%. Considering an equal relevance of precision and recall for this investigation, the overall detection rate of the proposed method is 96.54%. The 10-fold cross-validation analysis achieved mean accuracy and standard deviation values of 97.54% and 0.042%, respectively. Furthermore, a feature's relevance analysis was conducted for the XGBoost classifier. The embedded estimation function performs the sum of the instances each feature is split in its decision-tree-based structure. The importance of each feature is depicted in Figure 5.

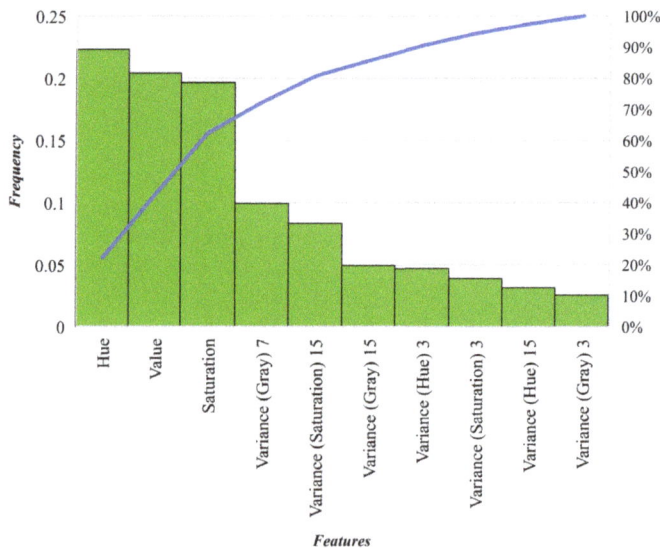

Figure 5. Relevance of each feature for the tuned classifier.

Where each bar item from the *x*-axis represents the relative frequency each feature has in the classifier. Here, hue, value, and saturation scores demonstrate the significant relevance the features had on the model, representing up to 65% of the total instances. These ratings are followed by 2D variance filter images such as the grayscale image with window size of 7 pixels, and the saturation image with a 15-pixel window size. The filters with substantially large window sizes showed clearly the importance of classifying accurately certain objects whose pixels are presented in a set of textures, such as bushes and spinifex. An illustration of the prediction and segmentation outputs is depicted in Figure 6.

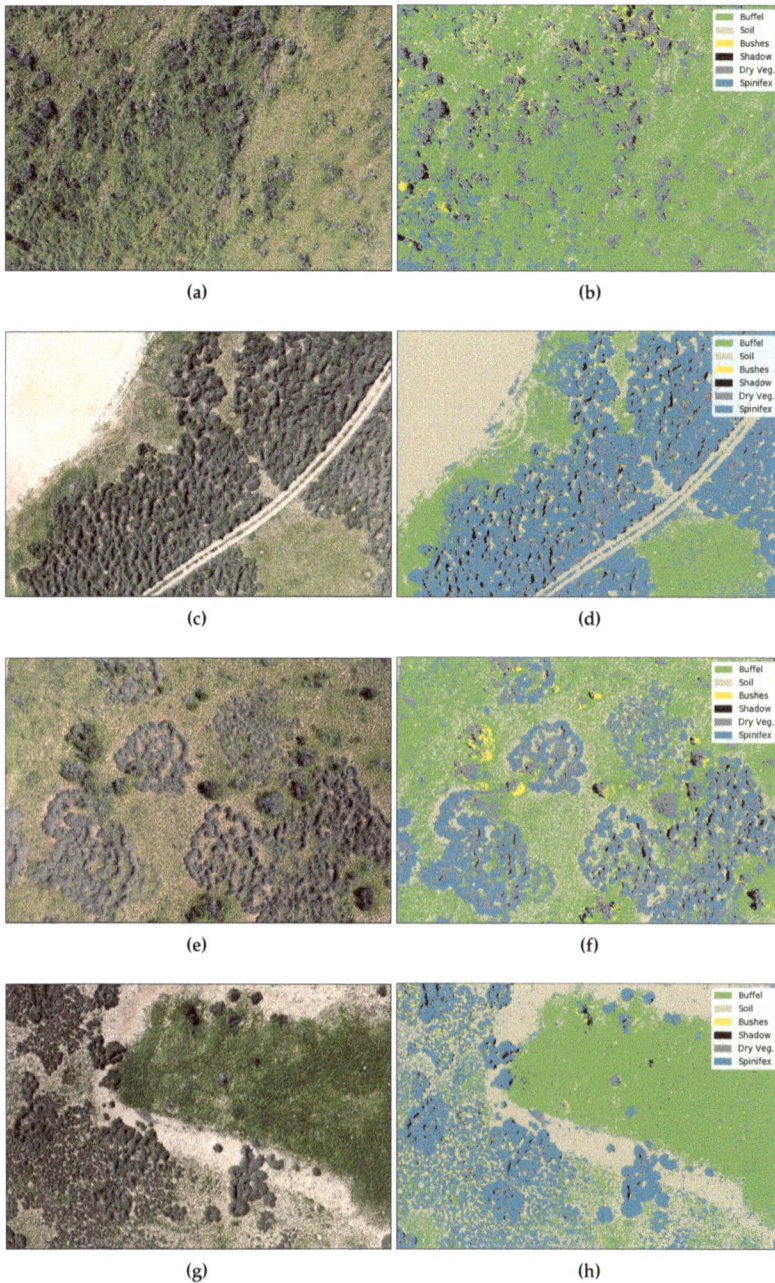

Figure 6. Pixel-wise segmentation from acquired red, green, blue (RGB) colour model images using Algorithm 1. (**a, c, e, g**) Orthorectified RGB images. (**b, d, f, h**) Final segmentation with predicted classes.

Figure 6a,c,e,g depict representative samples where buffel grass, spinifex, bushes, soil, and dry vegetation are displayed at different densities and light conditions, whereas Figure 6b,d,f,h show the segmentation obtained from the proposed algorithm. As seen in the confusion matrix from Table 1, it is possible to obtain highly accurate segmentation results for the buffel grass, spinifex, soil and shadow classes. However, the segmentation results for the "bushes" class is unstable in some images and can be regarded in many cases as image noise. The segmented images can be loaded and displayed in any GIS software, as shown in Figure 7.

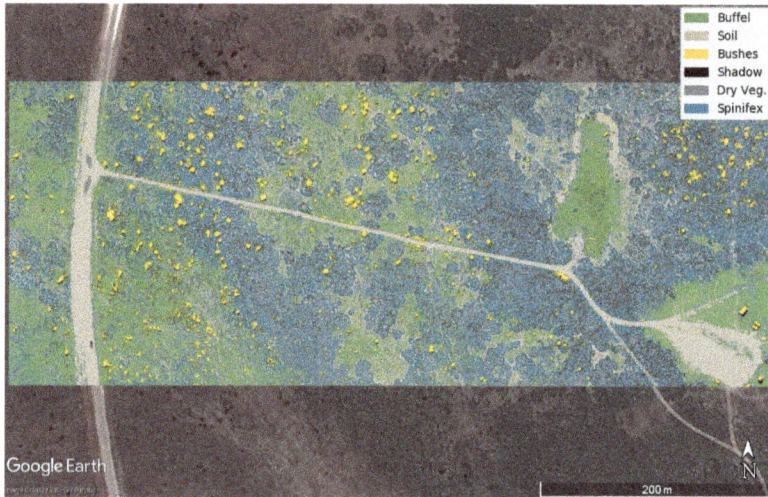

Figure 7. Prediction of invasive grasses in Cape Range National Park and its display in Google Earth.

4. Discussion

Accuracy and segmentation indicators presented in Section 3 validate the proposed pipeline approach to map vegetation and invasive grasses in arid lands. Negligible proportions of observed misclassifications for "buffel" and "spinifex" classes may be attributed to human error during the labelling of sample data. That is strongly evidenced by evaluating the results for the "bushes" class where the number of misclassified pixels is attributed to a challenging image labelling task. These inaccuracies occurred because the visible colour properties of bushes from the RGB sensor showed many similarities with other vegetation. From environmental monitoring and biosecurity perspectives, the proposed method is capable of providing critical information such as the distribution of invasive grasses, density values of invasive species in arid lands, and estimation of their expansion values for the short and mid-term, among others.

The present study represents a competitive approach for the use of UAVs and machine learning-based classification models compared with alternative solutions. It complements the research outcomes on buffel grass of Marshall et al. [19] by confirming a feasible, accurate, lightweight and relatively cheap solution for invasive grass mapping. With regard to invasive grasses in arid lands, this paper has demonstrated that using only high-resolution RGB images and single pixel-wise classification satisfies the need for accurate and efficient detection and segmentation solutions.

It is noteworthy that the invasive grasses in this study had negligible size variation, background clutter, occlusion, and viewpoint variation, constituting, apparently, an advantage. As opposed to senescence conditions, varied levels of grass density and illumination variation did not represent additional challenges. However, acquired data is insufficient for performing further classification tests with changes in illumination in the study area, such as acquisition tasks at different times during the day and under cloudy conditions. These parameters might alter the detection rates of the presented approach, and further research should be conducted under these conditions. The processing of imagery with small GSD values demonstrates how UAV-based remote sensing equipment has improved sensing capabilities compared with satellite and manned aircraft for invasive grass assessments.

Future research should analyse the efficacy of supervised and unsupervised algorithms to label vegetation and specifically invasive grass species accurately, and integrate the best approaches in the proposed pipeline. Additionally, new efforts should be focused on improving the performance of the entire pipeline process as well as the aggregation and evaluation of unsupervised classification algorithms for image labelling tasks using RGB pictures only. Although the amount of previous research in optimising machine learning models is significant, specific areas might be improved for real-time applications, such as orthomosaic-based processes and a better software integration into a single solution.

5. Conclusions

This paper proposed an integrated pipeline methodology for mapping vegetation and invasive grasses in arid lands. The methods were demonstrated by mapping buffel grass and spinifex in remote areas of WA through the use of UAVs, high-resolution RGB imagery, and gradient boosted decision trees. The presented approach illustrates detection rates of 96.75% and 96.00% for single mapping of buffel grass and spinifex, respectively, and a multiclass detection rate of 96.54%. Invasive grasses were accurately detected at different spatial concentrations with a GSD of up to 1.015 cm/pixel, demonstrating how UAV data collection can be useful for invasive grass detection at early stages. This case study demonstrates the implementation of unmanned aerial systems and machine learning for a feasible, accurate, and lightweight assessment of invasive grasses in arid and semi-arid lands. Future work will focus on integrating unsupervised and supervised methods for vegetation data labelling in order to reduce processing times.

Acknowledgments: This work was funded by the Plant Biosecurity Cooperative Research Centre (PBCRC) 2164 project, the Agriculture Victoria Research and the Queensland University of Technology (QUT). The authors would like to acknowledge Derek Sandow and WA Parks and Wildlife Service for the logistic support and permits to access the survey areas at Cape Range National Park. The authors would also like to acknowledge Eduard Puig-Garcia for his contributions in co-planning the experimentation phase. The authors gratefully acknowledge the support of the QUT Research Engineering Facility (REF) Operations Team (Dirk Lessner, Dean Gilligan, Gavin Broadbent and Dmitry Bratanov), who operated the DJI S800 EVO UAV and image sensors, and performed ground referencing. We thank Gavin Broadbent for the design, manufacturing, and tuning of a two-axis gimbal for the camera. We also acknowledge the High-Performance Computing and Research Support Group at QUT, for the computational resources and services used in this work.

Author Contributions: Felipe Gonzalez and Kerrie Mengersen contributed to experimentation and data collection planning. Felipe Gonzalez supervised the ground and airborne surveys, quality of acquired data, and logistics. Juan Sandino designed the proposed pipeline and conducted the data processing phase. Felipe Gonzalez, Kerrie Mengersen, and Kevin J. Gaston provided definitions, assistance and essential advice. Juan Sandino analysed the generated outputs, and validated and optimised the algorithm. All the authors contributed significantly to the composition and revision of the paper.

Conflicts of Interest: The authors declare no conflict of interest. The founding sponsors had no role in the design of the study; in the collection, analyses, or interpretation of data; in the writing of the manuscript, and in the decision to publish the results.

Abbreviations

The following abbreviations are used in this manuscript:

2D   Two-dimensional
CMOS   Complementary metal–oxide–semiconductor
Dry Veg.   Dry vegetation
GEOBIA   Geographic Object-Based Image Analysis
GIMP   GNU Image Manipulation Program
GIS   Geographic information system
GPS   Global positioning system
GSD   Ground sampling distance
HSV   Hue, saturation, value colour model
KML   Keyhole Markup Language
MDPI   Multidisciplinary Digital Publishing Institute
RAM   Random-access memory
RGB   Red, green, blue colour model
TIF   Tagged Image File
UAV   Unmanned Aerial Vehicles
WA   Western Australia
XGBoost   eXtreme Gradient Boosting

References

1. Godfree, R.; Firn, J.; Johnson, S.; Knerr, N.; Stol, J.; Doerr, V. Why non-native grasses pose a critical emerging threat to biodiversity conservation, habitat connectivity and agricultural production in multifunctional rural landscapes. *Landsc. Ecol.* **2017**, *32*, 1219–1242.
2. Schlesinger, C.; White, S.; Muldoon, S. Spatial pattern and severity of fire in areas with and without *buffel grass* (*Cenchrus ciliaris*) and effects on native vegetation in central Australia. *Austral Ecol.* **2013**, *38*, 831–840.
3. Fensham, R.J.; Wang, J.; Kilgour, C. The relative impacts of grazing, fire and invasion by *buffel grass* (*Cenchrus ciliaris*) on the floristic composition of a rangeland savanna ecosystem. *Rangel. J.* **2015**, *37*, 227.
4. Grice, A.C. The impacts of invasive plant species on the biodiversity of Australian rangelands. *Rangel. J.* **2006**, *28*, 27.
5. Marshall, V.; Lewis, M.; Ostendorf, B. *Buffel grass* (*Cenchrus ciliaris*) as an invader and threat to biodiversity in arid environments: A review. *J. Arid Environ.* **2012**, *78*, 1–12.
6. Bonney, S.; Andersen, A.; Schlesinger, C. Biodiversity impacts of an invasive grass: Ant community responses to Cenchrus ciliaris in arid Australia. *Biol. Invasions* **2017**, *19*, 57–72.
7. Jackson, J. Impacts and Management of *Cenchrus ciliaris* (*buffel grass*) as an Invasive Species in Northern Queensland. Ph.D. Thesis, James Cook University, Townsville, Australia, 2004.
8. Jackson, J. Is there a relationship between herbaceous species richness and *buffel grass* (*Cenchrus ciliaris*)? *Austral Ecol.* **2005**, *30*, 505–517.
9. Martin, T.G.; Murphy, H.; Liedloff, A.; Thomas, C.; Chadès, I.; Cook, G.; Fensham, R.; McIvor, J.; van Klinken, R.D. Buffel grass and climate change: A framework for projecting invasive species distributions when data are scarce. *Biol. Invasions* **2015**, *17*, 3197–3210.
10. Miller, G.; Friedel, M.; Adam, P.; Chewings, V. Ecological impacts of *buffel grass* (*Cenchrus ciliaris* L.) invasion in central Australia—Does field evidence support a fire-invasion feedback? *Rangel. J.* **2010**, *32*, 353.
11. Smyth, A.; Friedel, M.; O'Malley, C. The influence of *buffel grass* (*Cenchrus ciliaris*) on biodiversity in an arid Australian landscape. *Rangel. J.* **2009**, *31*, 307.
12. Gonzalez, L.; Whitney, E.; Srinivas, K.; Periaux, J. Multidisciplinary aircraft design and optimisation using a robust evolutionary technique with variable fidelity models. In Proceedings of the 10th AIAA/ISSMO Multidisciplinary Analysis and Optimization Conference, Albany, NY, USA, 30 August–1 September 2004; pp. 3610–3624.
13. Whitney, E.; Gonzalez, L.; Periaux, J.; Sefrioui, M.; Srinivas, K. A robust evolutionary technique for inverse aerodynamic design. In Proceedings of the 4th European Congress on Computational Methods in Applied Sciences and Engineering, Jyväskylä, Finland, 24–28 July 2004.
14. Gonzalez, L.; Montes, G.; Puig, E.; Johnson, S.; Mengersen, K.; Gaston, K. Unmanned Aerial Vehicles (UAVs) and Artificial Intelligence Revolutionizing Wildlife Monitoring and Conservation. *Sensors* **2016**, *16*, 97.
15. Chahl, J. Unmanned Aerial Systems (UAS) Research Opportunities. *Aerospace* **2015**, *2*, 189–202.

16. Allison, R.S.; Johnston, J.M.; Craig, G.; Jennings, S. Airborne Optical and Thermal Remote Sensing for Wildfire Detection and Monitoring. *Sensors* **2016**, *16*, 1310.

17. Thomas, J.E.; Wood, T.A.; Gullino, M.L.; Ortu, G., Diagnostic Tools for Plant Biosecurity. In *Practical Tools for Plant and Food Biosecurity: Results from a European Network of Excellence*; Gullino, M.L., Stack, J.P., Fletcher, J., Mumford, J.D., Eds.; Springer International Publishing: Cham, Switzerland, 2017; pp. 209–226.

18. Olsson, A.D.; van Leeuwen, W.J.; Marsh, S.E. Feasibility of Invasive Grass Detection in a Desertscrub Community Using Hyperspectral Field Measurements and Landsat TM Imagery. *Remote Sens.* **2011**, *3*, 2283–2304.

19. Marshall, V.M.; Lewis, M.M.; Ostendorf, B. Detecting new Buffel grass infestations in Australian arid lands: Evaluation of methods using high-resolution multispectral imagery and aerial photography. *Environ. Monit. Assess.* **2014**, *186*, 1689–1703.

20. Alexandridis, T.; Tamouridou, A.A.; Pantazi, X.E.; Lagopodi, A.; Kashefi, J.; Ovakoglou, G.; Polychronos, V.; Moshou, D. Novelty Detection Classifiers in Weed Mapping: Silybum marianum Detection on UAV Multispectral Images. *Sensors* **2017**, *17*, 2007.

21. Blaschke, T.; Hay, G.J.; Kelly, M.; Lang, S.; Hofmann, P.; Addink, E.; Queiroz Feitosa, R.; van der Meer, F.; van der Werff, H.; van Coillie, F.; Tiede, D. Geographic Object-Based Image Analysis—Towards a new paradigm. *ISPRS J. Photogramm. Remote Sens.* **2014**, *87*, 180–191.

22. Torres-Sánchez, J.; López-Granados, F.; Peña, J. An automatic object-based method for optimal thresholding in UAV images: Application for vegetation detection in herbaceous crops. *Comput. Electron. Agric.* **2015**, *114*, 43–52.

23. Ashourloo, D.; Aghighi, H.; Matkan, A.A.; Mobasheri, M.R.; Rad, A.M. An Investigation Into Machine Learning Regression Techniques for the Leaf Rust Disease Detection Using Hyperspectral Measurement. *IEEE J. Sel. Top. Appl. Earth Obs. Remote Sens.* **2016**, *9*, 4344–4351.

24. Robinson, T.; Wardell-Johnson, G.; Pracilio, G.; Brown, C.; Corner, R.; van Klinken, R. Testing the discrimination and detection limits of WorldView-2 imagery on a challenging invasive plant target. *Int. J. Appl. Earth Obs. Geoinform.* **2016**, *44*, 23–30.

25. Lin, F.; Zhang, D.; Huang, Y.; Wang, X.; Chen, X. Detection of Corn and Weed Species by the Combination of Spectral, Shape and Textural Features. *Sustainability* **2017**, *9*, 1335.

26. Schmittmann, O.; Lammers, P. A True-Color Sensor and Suitable Evaluation Algorithm for Plant Recognition. *Sensors* **2017**, *17*, 1823.

27. Bureau of Meteorology. *Learmonth, WA—July 2016—Daily Weather Observations*; Bureau of Meteorology: Learmonth Airport (station 005007), Australia, 2016.

28. Chen, T.; Guestrin, C. XGBoost: A Scalable Tree Boosting System. In Proceedings of the 22nd ACM SIGKDD International Conference on Knowledge Discovery and Data Mining, San Francisco, CA, USA, 13–17 August 2016; ACM: New York, NY, USA, 2016; pp. 785–794.

29. Pedregosa, F.; Varoquaux, G.; Gramfort, A.; Michel, V.; Thirion, B.; Grisel, O.; Blondel, M.; Prettenhofer, P.; Weiss, R.; Dubourg, V.; et al. Scikit-learn: Machine Learning in Python. *J. Mach. Learn. Res.* **2011**, *12*, 2825–2830.

30. Bradski, G. The OpenCV library. *Dr. Dobb's J. Softw. Tools* **2000**, *25*, 122–125.

31. Hunter, J.D. Matplotlib: A 2D graphics environment. *Comput. Sci. Eng.* **2007**, *9*, 90–95.

Article

Aerial Mapping of Forests Affected by Pathogens Using UAVs, Hyperspectral Sensors, and Artificial Intelligence

Juan Sandino [1,*], Geoff Pegg [2], Felipe Gonzalez [1] and Grant Smith [3]

[1] Insitute for Future Environments; Robotics and Autonomous Systems, Queensland University of Technology (QUT), 2 George St, Brisbane City, QLD 4000, Australia; felipe.gonzalez@qut.edu.au

[2] Horticulture & Forestry Science, Department of Agriculture & Fisheries, Ecosciences Precinct, 41 Boggo Rd Dutton Park, QLD 4102, Australia; geoff.pegg@daf.qld.gov.au

[3] BioProtection Technologies, The New Zealand Institute for Plant & Food Research Limited, Gerald St, Lincoln 7608, New Zealand; grant.smith@plantandfood.co.nz

* Correspondence: j.sandinomora@qut.edu.au; Tel.: +61-7-3138-1363

Received: 22 February 2018; Accepted: 21 March 2018; Published: 22 March 2018

Abstract: The environmental and economic impacts of exotic fungal species on natural and plantation forests have been historically catastrophic. Recorded surveillance and control actions are challenging because they are costly, time-consuming, and hazardous in remote areas. Prolonged periods of testing and observation of site-based tests have limitations in verifying the rapid proliferation of exotic pathogens and deterioration rates in hosts. Recent remote sensing approaches have offered fast, broad-scale, and affordable surveys as well as additional indicators that can complement on-ground tests. This paper proposes a framework that consolidates site-based insights and remote sensing capabilities to detect and segment deteriorations by fungal pathogens in natural and plantation forests. This approach is illustrated with an experimentation case of myrtle rust (*Austropuccinia psidii*) on paperbark tea trees (*Melaleuca quinquenervia*) in New South Wales (NSW), Australia. The method integrates unmanned aerial vehicles (UAVs), hyperspectral image sensors, and data processing algorithms using machine learning. Imagery is acquired using a Headwall Nano-Hyperspec® camera, orthorectified in Headwall SpectralView®, and processed in Python programming language using eXtreme Gradient Boosting (XGBoost), Geospatial Data Abstraction Library (GDAL), and Scikit-learn third-party libraries. In total, 11,385 samples were extracted and labelled into five classes: two classes for deterioration status and three classes for background objects. Insights reveal individual detection rates of 95% for healthy trees, 97% for deteriorated trees, and a global multiclass detection rate of 97%. The methodology is versatile to be applied to additional datasets taken with different image sensors, and the processing of large datasets with freeware tools.

Keywords: *Austropuccinia psidii*; drones; hyperspectral camera; machine learning; *Melaleuca quinquenervia*; myrtle rust; non-invasive assessment; paperbark; unmanned aerial vehicles (UAV); xgboost

1. Introduction

Exotic pathogens have caused irreversible damage to flora and fauna within a range of ecosystems worldwide. Popular outbreaks include the enormous devastations of chestnut blight (*Endothia parasitica*) on American chestnut trees (*Castanea dentata*) in the U.S. [1–3], sudden oak death (*Phytophthora ramorum*) on oak populations (*Quercus agrifolia*) in Europe, California, and Oregon [4–6], dieback (*Phytophthora cinnamomi*) on hundreds of hosts globally [7–9], and myrtle rust (*Austropuccinia psidii*) on Myrtaceae family plants in Australia [10–13]. The effects of the latter case have raised national alerts and response programmes given the extensive host range and the ecological and economic importance of Myrtaceae plants in the Australian environment [14–17]. As a result, various surveillance and eradication programmes have been applied in an attempt to minimise the impacts invasive pathogens cause on local hosts such as dieback in the Western Australia Jarrah forests [18], sudden oak death in the tan oak forests of the U.S. [19], and rapid ohia death (*Ceratocystis fimbriata*) on ohia trees (*Metrosideros polymorpha*) in Hawaii [20].

Modern surveillance methods to map hosts vulnerable to and affected by exotic pathogens can be classified in site-based and remote sensing methods, according to Lawley et al. [21]. Site-based approaches are commonly small regions used to collect exhaustive compositional and structural indicators of vegetation condition with a strong focus on biophysical attributes of single vegetation communities [22,23]. These methods, nonetheless, require deep expertise and time to conduct experimentation, data collection, and validation that, along with their limited area they can cover, represent a challenge while assessing effects on a broad scale [24]. Research has also suggested the design of decision frameworks to monitor and control the most threatened species [17,25,26]. Although these models can determine flora species that require immediate management control, limitations on the amount of tangible, feasible, and broad quantified data of vulnerable host areas [21] have resulted in lack of support from state and federal governments [11].

The role of remote sensing methods to assess and quantify the impacts of invasive pathogens in broad scale has increased exponentially [27,28]. Standard approaches comprise the use of spectral and image sensors through satellite, manned, and unmanned aircraft technology [29]. Concerning sensing technology by itself, applied methods by research communities include the use of non-imaging spectroradiometers, fluorescence, multispectral, hyperspectral, and thermal cameras, and light detection and ranging (LiDAR) technology [30–33]. These equipment are usually employed for the calculation of spectral indexes [34–37] and regression models in the host range [38,39]. Nevertheless, these methods are mainly focused on quantification and distribution, among other physical properties of flora species.

Satellite and manned aircraft surveys have reported limitations concerning resolution, operational costs, and unfavourable climate conditions (e.g., cloudiness and hazard winds) [40]. In contrast, the continuous development of unmanned aerial vehicles (UAVs) designs, navigation systems, portable image sensors and cutting-edge machine learning methods allow unobtrusive, accurate, and versatile surveillance tools in precision agriculture and biosecurity [41–44]. Many studies have positioned UAVs for the collection of aerial imagery in applications such as weed, disease, and pest mapping and wildlife monitoring [21,45–47]. More recently, unmanned aerial systems (UASs) have been deployed in cluttered and global positioning system (GPS)-denied environments [48].

Approaches to the use of UAVs, hyperspectral imagery and artificial intelligence are gaining popularity. For example, Aasen et al. [49] deployed UAS to boost vegetation monitoring efforts using hyperspectral three-dimensional (3D) imagery. The authors of Nasi et al. [50] developed techniques to assess pest damages at canopy levels using spectral indexes and k-nearest neighbour (k-NN) classifiers, achieving global detection rates of 90%. Similar research focused on disease monitoring, however, has been limited. The authors of Calderon et al. [51], for instance, evaluated the early detection and quantification of verticillium wilt in olive plants using support vector machines (SVMs) and linear discriminant analysis (LDA), obtaining mixed accuracy results among the evaluated classes of infection severity (59–75%) [52]. The authors of Albetis et al. [53] presented a system to discriminate asymptomatic and symptomatic red and white vineyard cultivars by *Flavescence doree*, using UAVs, multispectral imagery, and up to 20 data features, collecting contrasting results between the cultivars and maximum accuracy rates of 88%. In sum, the integration of site-based and remote sensing frameworks have boosted the capabilities of these surveillance solutions by combining data from abiotic and biotic factors and spectral responses, respectively [54]. However, this synthesis is still challenging due to the high number of singularities, data correlation, and validation procedures presented in each case study.

Considering the importance of site-based and remote sensing methods to obtain reliable and broader assessments of forest health, specifically, for pest and fungal assessments [55], this paper presents an integrated system that classifies and maps natural and plantation forests exposed and deteriorated by fungal pathogens using UAVs, hyperspectral sensors, and artificial intelligence. The framework is exemplified by a case study of myrtle rust on paperbark tea trees (*Melaleuca quinquenervia*) in a swamp ecosystem of Northeast New South Wales (NSW), Australia.

2. System Framework

A novel framework was designed for the assessment of natural and plantation forests exposed and potentially exposed to pathogens as presented in Figure 1. It comprises four sections linked to each other, denoted as Data Acquisition, Data Preparation, Training, and Prediction. The system interacts directly and indirectly with the surveyed area to acquire information, preprocess and arrange obtained data into features, fit a supervised machine learning classifier, tune the accuracy and performance indicators to process vast amounts of data, and provide prediction reports through segmented images.

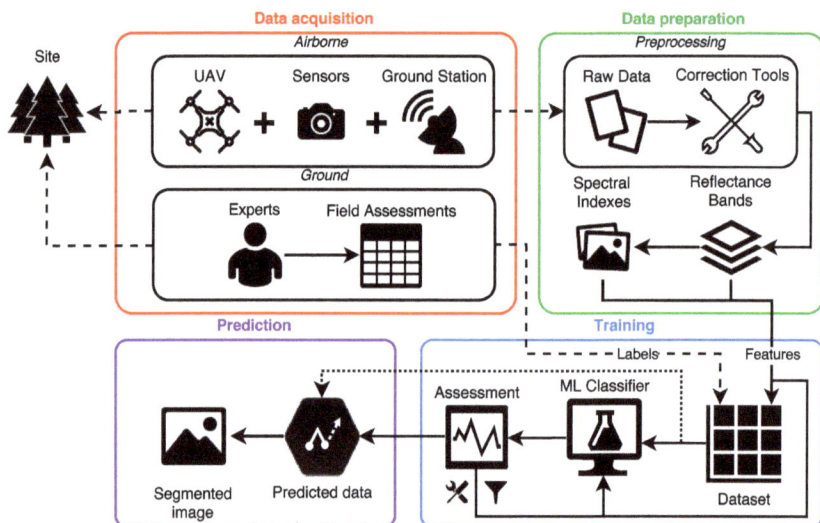

Figure 1. Pipeline process for the detection and mapping of alterations in natural and plantation forests by fungal diseases.

2.1. Data Acquisition

The data acquisition process involves an indirect data collection campaign using an airborne system, and direct ground assessments of the studied pathogen through exclusion trials, which are controlled by biosecurity experts. Airborne data is compiled using a UAV, image sensors, and a ground workstation in the site to acquire data above the tree canopy. Similarly, ground data is collected through field assessment insights by biosecurity experts. Field assessments bring several factors such as growth, reproduction, and regeneration on coppiced trees exposed to any specific pathogen. A database is created by labelling, georeferencing, and correlating relevant insights into every tested plant. Details on the studied area, flight campaigns, and field assessments can be found in Sections 3.1–3.3.

2.1.1. UAV and Ground Station

This methodology incorporates a hexa-rotor DJI S800 EVO (DJI, Guangdong, China) UAV. The drone features high-performance brushless rotors, a total load capacity of 3.9 kg, and dimensions of 118 cm × 100 cm × 50 cm. It follows an automatic mission route using the DJI Ground Station 4.0 software, controlling the route, speed, height above the ground, and overlapping values remotely. It is worth mentioning, however, that other UAVs with similar characteristics can also be used and included into the airborne data collection process of Figure 1.

2.1.2. Sensors

Spectral data is collected using a Headwall Nano-Hyperspec® hyperspectral camera (Headwall Photonics Inc., Bolton, MA, USA). This visible near infrared (VNIR) sensor provides spectral wavelength responses up to 274 bands, a wavelength range from 385 to 1000 nm, a spectral resolution of 2.2 nm, a frame rate of 300 Hz, spatial bands of 640 pixels, and a total storage limit of 480 GB. The camera is mounted in a customised gimbal system that augments data quality by minimising external disturbances such as roll, pitch, and yaw oscillations, as depicted in Figure 2.

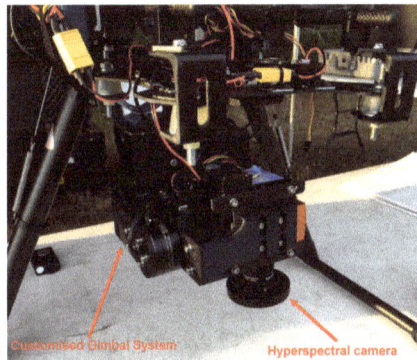

Figure 2. Assembly of the gimbal system and the Headwall Nano-Hyperspec® camera into the DJI S800 EVO unmanned aerial vehicle (UAV).

2.2. Data Preparation

Imagery is downloaded and fed into a range of software solutions to transform raw data into filtered and orthorectified spectral bands in reflectance. Using Headwall SpectralView® software, raw hypercubes from the surveyed area were automatically processed. The preprocessing operations included radiance, orthorectification (through ground control points), and the white reference illumination spectrum (WRIS). The WRIS spectrum comes from the extraction of the radiance signature of a Spectralon target from an acquired image in the surveyed area. Using the orthorectified imagery and the WRIS, reflectance data and scene shading are calculated through CSIRO|Data61 Scyven 1.3.0. software [56,57]. Considering the massive amount of data contained for any orthorectified hyperspectral cube (4–12 GB), each cube is cropped to process regions of interest only (0.5–1 GB) and handle system resources efficiently, as shown in Figure 3.

An additional image that contains the localisation of tested trees in the exclusion trial is created. From a recovered red–green–blue colour model (RGB) image of the cropped hypercube, each tree is graphically labelled using their tracked GPS coordinates in Argis ArcMap 10.4. To handle all the data processing from the proposed system framework, Algorithm 1 was developed. These tasks were conducted with Python 2.7.14 programming language and several third-party libraries for data manipulation and machine learning, including Geospatial Data Abstraction Library (GDAL) 2.2.2 [58], eXtreme Gradient Boosting (XGBoost) 0.6 [59], Scikit-learn 0.19.1 [60], OpenCV 3.3.0 [61], and Matplotlib 2.1.0 [62].

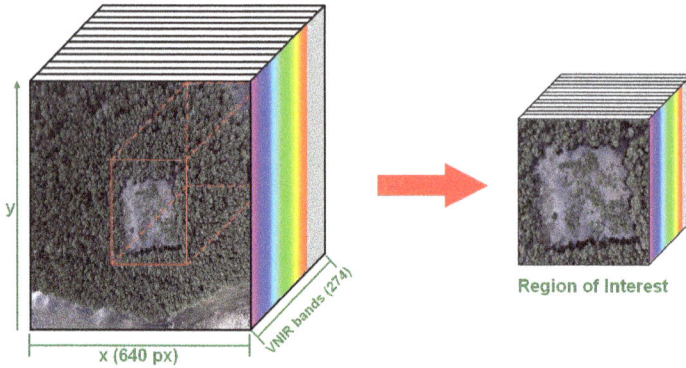

Figure 3. Creation of regions of interest in hyperspectral cubes.

Algorithm 1 Detection and mapping of vegetation alterations using spectral imagery and sets of features.

Required: orthorectified layers (bands) in reflectance I. Labelled regions from field assessments L.

Data Preparation

1: Load I data.
2: $S \leftarrow$ Spectral indexes array from I.
3: $X \leftarrow$ Features array $[I, S]$.

Training

4: $Y \leftarrow$ Labels array from dataset L.
5: $D \leftarrow$ filtered dataset of features X with corresponding labelled pixel from Y.
6: Split D into training data D_T and testing data D_E.
7: Fit an XGBoost classifier C using D_T.
8: $R \leftarrow$ List of unique relevance values of processed features X from C.
9: **for all** values **in** R **do**
10:    $D_{TF} \leftarrow$ Filtered underscored features from D_T.
11:    Fit C using D_{TF}.
12:    Append accuracy values from C into T.
13: **end for**
14: Fit C using the best features threshold from T.
15: Validate C with k-fold cross-validation from D_{TF}.   ▷ number of folds = 10

Prediction

16: $P \leftarrow$ Predicted values for each sample in X.
17: Convert P array into a 2D orthorectified image.
18: $O \leftarrow$ Displayed/overlayed image.
19: **return** O

Reflectance cube bands are loaded into the program to calculate suitable spectral indexes and improve the detection rates as mentioned in Step 2. For this approach, the most traditional indexes, such as the normalised difference vegetation index (NDVI) [63], the green normalised difference vegetation index (GNDVI) [64], the soil-adjusted vegetation index (SAVI) [65], and the second modified adjusted vegetation index (MSAVI2) [66], are calculated. Additionally, two-dimensional (2D) smoothing kernels are applied into such indexes, following Equation (1).

$$K = \frac{1}{w^2} \begin{bmatrix} 1 & 1 & \cdots & 1 \\ 1 & 1 & \cdots & 1 \\ \vdots & \vdots & \ddots & \vdots \\ 1 & 1 & \cdots & 1 \end{bmatrix} \tag{1}$$

where K is the kernel of the filter, and w is the size of the window. Here, up to three kernels for $w = 3$, $w = 7$, and $w = 15$ are calculated per vegetation index. All the hypercube bands in reflectance, as well as calculated vegetation spectral indexes, are denominated data features. In Step 3, an array of features is generated from all the retrieved bands I and the calculated indexes S.

2.3. Training and Prediction

The labelled regions from the ground-based assessments are exported from ArcMap and loaded into an array Y. In Step 5, an array D is created by filtering the features from X with their corresponding labels in Y only. Filtered data is separated into a training (80%) and testing (20%) data array. In Step 7, data is processed into a supervised XGBoost classifier. This model is utilised considering the moderate amount of labelled data (insufficient to run a deep learning model), the amount of information to be processed for a single hypercube, and the nature of the data from the exclusion trial (ground-based test). This classifier is currently a cutting-edge decision tree and gradient boosting model optimised for large tree structures, excellent performance, and fast execution speeds, outperforming detection rates of standard non-linear models such as random forests, k-NN, LDA, and SVM [59]. Moreover, input data features do not require any scaling (normalisation) in comparison with other models. Once the model is fitted, Step 8 retrieves the relevance of each processed feature (reflectance bands, spectral indexes, and transformed images) from the array X in a list R. The list is sorted to discard irrelevant features, increment the detection rates of the algorithm, avoid over-fitting, decrease the complexity of the model, and reduce computer processing loads.

Algorithm 1 executes a loop to evaluate the optimal number of ranked features that can offer the best balance between accuracy and data processing load. At each instance of the loop, specific features from the training set D_T is filtered based on a threshold of relevance scores (Step 10). Later, the XGBoost model is fit using the filtered training set (Step 11) to record all accuracy values per combination. In Step 14, the best combination of accuracy and number of features is retrieved to re-fit the classifier. Finally, the fitted model is validated using k-fold cross-validation (Step 15).

In the prediction stage, unlabelled pixels are processed in the optimised classifier, their values displayed in the same 2D spatial image from the orthorectified hyperspectral cube. Ultimately, classified pixels are depicted using distinguishable colours and exported in tagged image file (TIF) format, a compatible file with georeferencing-based software.

3. Experimentation Setup

3.1. Site

As displayed in Figure 4a, the experimentation occurred in a paperbark tea tree forest located near 71 Boggy Creek Rd, Bungawalbin, NSW 2469, Australia (29°04′42.9″ S 153°15′10.0″ E). Data acquisition was conducted on the 25 August 2016 at 11:40 a.m. Weather Observations for that day stated conditions of a partially cloudy day, with a mean temperature of 18.8 °C, a relative humidity of 46%, west-northwest winds of 20 km/h, a pressure of 1009.5 hPa, and no precipitation [67]. As seen in Figure 4b, the site includes selected paperbark trees that are monitored to assess the effects of myrtle rust on them.

(a)   (b)

Figure 4. Site of the case study. (**a**) Location and covered area of the farm in red and experiment trees in blue; (**b**) Overview of paperbark tea trees under examination.

3.2. Flight Campaign

The acquired data from flight campaign incorporated a single mission route. The UAV was operated with a constant flight height of 20 m above the ground, an overlap of 80%, a side lap of 50%, a mean velocity of 4.52 km/h, and a total distance of 1.43 km. The acquired hyperspectral dataset had a vertical and horizontal ground sample distances (GSD) of 4.7 cm/pixel.

3.3. Field Assessments

In order to evaluate the effects and flora response of myrtle rust on paperbark trees, a biological study in an exclusion trial on the mentioned site was conducted. Several on-ground assessments were conducted in individual trees within a replicated block design using four treatments: trees treated with fungicides (F), insecticides (I), fungicides and insecticides (F + I), and trees without any treatment action. Figure 5 illustrates the treatment methods the studied trees received.

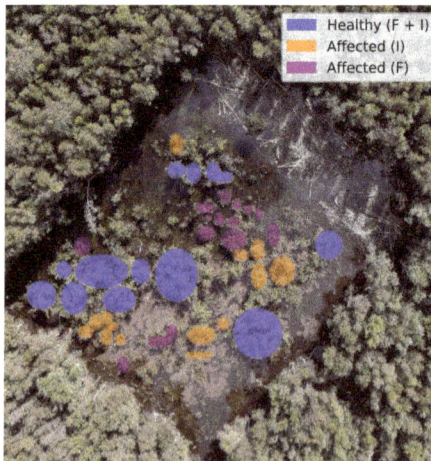

Figure 5. Aerial view of individual trees through on-ground assessments.

From the indicators generated, insect and disease assessments were extracted to label every tree. Overall, the assessment report showed that only the trees that received insecticide and fungicide treatments remained healthy under direct exposure to the rust. In contrast, trees treated with insecticide were affected by rust and those treated with fungicide were affected by insects. Thus, trees treated with fungicides and insecticides were consequently labelled as healthy and the others as affected in the database.

3.4. Preprocessing

As mentioned in Section 2.2, the entire surveyed area included an exclusion trial. As a result, an orthorectified hyperspectral cube in reflectance with spatial dimensions of 2652 × 1882 pixel of a 5.4 GB size was generated. This area was cropped from the original hypercube, reducing computational costs and discarding irrelevant data. Eventually, a 1.7 GB cube of 1200 × 1400 pixel as depicted in Figure 6 was extracted.

Figure 6. Red–green–blue (RGB) colour model representation of the orthorectified cube with its extracted region of interest.

3.5. Training and Prediction

The XGBoost classifier contains several hyper-parameters to be set. Following a grid search technique, in which the classifier performance and detection rates were tracked with a set of possible hyper-parameters, the optimal values found for this case study were

$$\text{estimators} = 100, \text{ learning rate} = 0.1, \text{ maximum depth} = 3$$

where "estimators" is the number of trees, "learning rate" is the step size of each boosting step, and "maximum depth" is the maximum depth per tree that defines the complexity of the model.

4. Results and Discussion

To visualise the benefits of inserting an optimisation scheme in Step 8 of Algorithm 1, detection rates were tracked by training and running the classifier multiple times with only a set of filtered features per instance. The features were ranked with their relevance by the XGBoost classifier and sorted consequently, as illustrated in Figure 7.

The classifier can achieve high accuracy rates exceeding 97% of global accuracy when it processes data using from 10 to 40 features only, with an optimal number of features of 24. On the other hand, the classifier merely improves their registers when the number of processed features is more substantial. With this capability, the proposed approach can process fewer data and reduce the number of calculations to achieve high detection values. Additionally, this boosts the capability of the algorithm of processing large datasets in less time, an ideal scenario for mapping vast rural areas. The most relevant features of this study case are depicted in Table 1 and Figure 8.

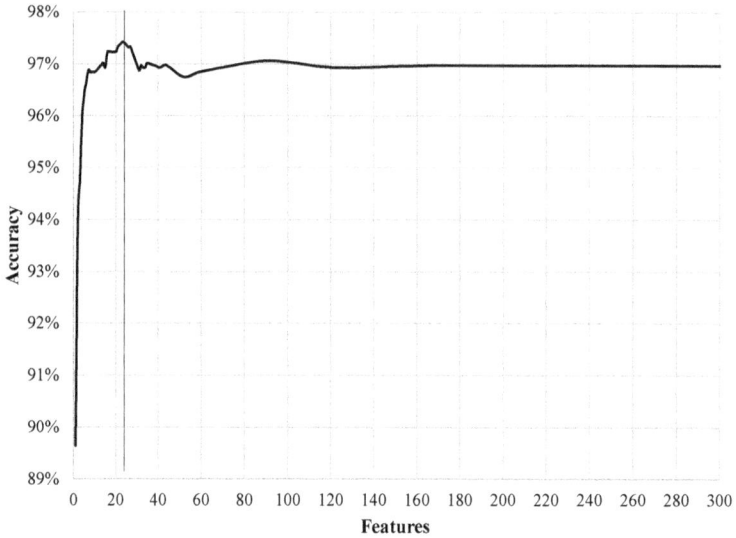

Figure 7. Performance of the classifier using different filtered features. Optimal number of features: 24.

Table 1. Ranking of the most 30 relevant features.

#	Feature	Score	#	Feature	Score	#	Feature	Score
1	NDVI_Mean15	0.0933	11	NDVI_Mean3	0.0247	21	975.3710	0.0119
2	Shading_Mean15	0.0780	12	759.9730	0.0212	22	671.1490	0.0109
3	GNDVI_Mean7	0.0563	13	999_Mean3	0.0212	23	893.2090	0.0109
4	NDVI_Mean7	0.0558	14	999_Mean7	0.0202	24	990.9150	0.0099
5	999_Mean15	0.0504	15	997.5770	0.0188	25	877.6650	0.0094
6	444.6470	0.0494	16	764.4140	0.0148	26	966.4890	0.0094
7	Specularity_Mean15	0.0380	17	444_Mean15	0.0143	27	766.6350	0.0084
8	999.7980	0.0341	18	462.4120	0.0133	28	853.2380	0.0079
9	GNDVI_Mean15	0.0286	19	NDVI	0.0133	29	935.4000	0.0079
10	444_Mean7	0.0267	20	Shading_Mean7	0.0133	30	GNDVI_Mean3	0.0079

It is shown how the first four features for this classification task come from specific vegetation indexes and processed images by 2D kernels—specifically, NDVI, shading, and GNDVI features (Figure 8a–d). Although their illustrations show insights of distinguishable intensities between healthy and affected tree regions from Figure 5, these sets of features are insufficient for segmenting areas of other objects. Thus, specific reflectance wavelengths bands such as 999 and 444 nm (Figure 8e,f) are also determinant. Additionally, features processed with 2D kernels obtained better relevance scores than their unprocessed counterparts. That difference was even greater for processed features using big window kernels considering that high amounts of noise, common in raw hyperspectral imagery, altered the performance of the approach. Nonetheless, these rankings do not suggest that these features can be used as global indicators to detect and map similar case studies (myrtle rust); the feature ranking table showed here is relevant to the fitted XGBoost model only, and results may differ if the same features are processed through other machine learning techniques. It is recommended, therefore, to perform individual analyses for every case study.

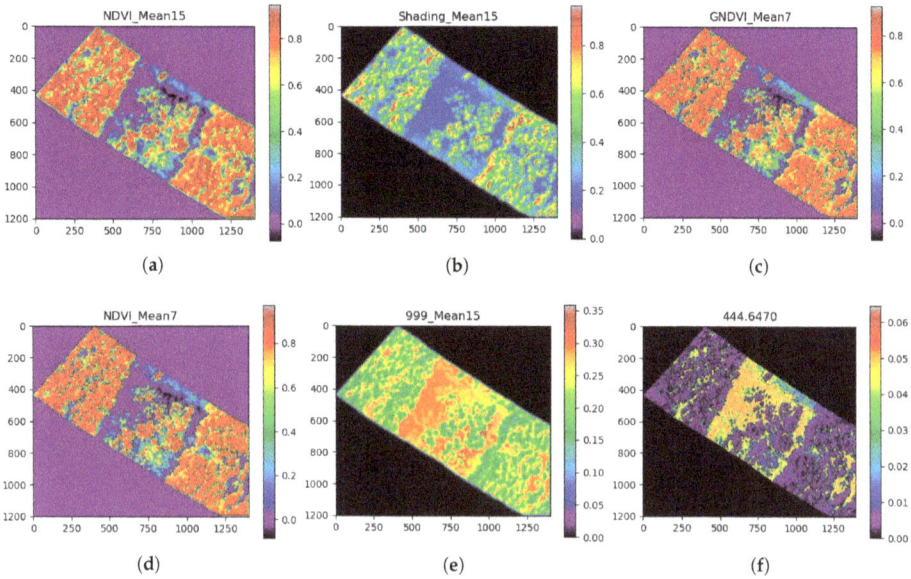

Figure 8. False colour representation of the first six features by relevance. (**a**) Smoothed NDVI with $k = 15$. (**b**) Smoothed Shading with $k = 15$. (**c**) Smoothed GNDVI with $k = 7$. (**d**) Smoothed NDVI with $k = 7$. (**e**) Smoothed 999 nm reflectance band with $k = 15$. (**f**) Raw 444 nm reflectance band.

A total of 11,385 pixel contained in 23 features filtered by their relevance were read again in Step 14 of Algorithm 1. Data was divided into a training array D_E with 9108 pixel and a testing array D_T with 2277 pixel. The generated confusion matrix of the classifier and its performance report is shown in Tables 2 and 3.

Table 2. Confusion matrix of the eXtreme Gradient Boosting (XGBoost) classifier.

	Predicted	Healthy	Affected	Background	Soil	Stems
	Healthy	1049	15	0	0	0
	Affected	45	531	0	0	0
Labelled	Background	0	0	158	0	0
	Soil	0	0	0	321	0
	Stems	0	0	0	1	157

Table 3. Classification report of the confusion matrix of Table 2.

Class	Precision (%)	Recall (%)	F-Score (%)	Support
Healthy	95.89	98.59	97.24	1064
Affected	97.25	92.19	94.72	576
Background	100.00	100.00	100.00	158
Soil	99.69	100.00	99.68	321
Stems	100.00	99.37	99.68	158
Mean	97.32	97.32	97.35	$\Sigma = 2277$

In sum, most of the classes were predicted favourably. The majority of misclassifications between the "Healthy" and "Affected" classes are possibly caused by human errors while labelling the regions manually in the raw imagery. Considering a weighed importance of precision and recall of 1:1, the F-support scores highlight a detection rate of 97.24% for healthy trees, 94.25% for affected trees, and an overall detection rate of 97.35%. Validation through k-fold cross-validation shows that the presented approach has an accuracy of 96.79%, with a standard deviation of 0.567%.

The performance of Algorithm 1 was tested in a computer with the following characteristics: Processor Intel® Core™ i7-4770, 256 GB SSD, 16 GB RAM, Windows 7 64bit, and AMD Radeon™ HD 8490. It contains a report of the elapsed seconds for the application to accomplish the primary data processing, training, and prediction tasks, as illustrated in Table 4.

Table 4. Performance in seconds of the main tasks from Algorithm 1.

Sub-Section	Instance 1	Instance 2	Instance 3	Instance 4	Instance 5	Mean	Std. Dev.
Data preparation							
Loading Hypercube	11.927	10.944	11.954	11.766	11.521	11.622	0.417
Calculating indexes	46.864	51.860	51.901	52.622	47.322	50.114	2.779
Training							
Preprocessing	0.152	0.141	0.149	0.148	0.140	0.146	0.005
Fitting XGBoost	8.948	8.654	8.758	8.679	8.692	8.746	0.119
Features Filtering	53.236	55.433	60.364	57.253	53.446	55.946	2.962
Re-Fitting XGBoost	0.964	1.023	1.010	0.998	0.965	0.992	0.026
Prediction							
Predicting results	29.738	40.749	42.131	34.473	66.477	42.714	14.188
Display	0.776	0.705	1.043	0.917	0.612	0.811	0.171
Total	152.607	169.508	177.309	166.857	189.175	171.091	13.489

Taking into account the dimensions of the processed hypercube (1400 × 1200) and the initial number of bands (274), it was observed how a great demand of resources was required to open the file itself and calculate spectral indexes, accumulating 61.7 s on average. Similarly, the features filtering process in the training section also demanded considerable time, exceeding 50 s. On the other hand, the elapsed time executing the remaining tasks of the training phase was remarkably short. Specifically, the report highlights the benefits of filtering irrelevant features by comparing the duration of fitting the classifier for the first time with the duration of re-fitting it again with less yet relevant data from 8.74 to 0.99 s. Overall, the application spent 2 min and 51 s to evaluate and map an area of 338 m^2 approximately.

The GSD value of 4.7 cm/pixel from the acquired hyperspectral imagery represented a minor challenge in labelling individual trees, but is still problematic when specific stems or leaflets need to be highlighted. Higher resolution can assist in higher classification rates. As an illustration, the final segmented image of the optimised classifier is shown in Figure 9, where Figure 9a shows the digital labelling of every class region and Figure 9b depicts the generated segmentation output by Algorithm 1. A hypercube covering the entire area flown was also processed using the trained model, with results shown in Figure 10.

Results show a segmentation output using XGBoost as the supervised machine learning classifier that works well for this task. This classifier as well as Algorithm 1 are not only important for their capabilities to offer a pixel-wise classification task, but they also allow a rapid convergence, do not involve many complex mathematical calculations, and filter irrelevant data, compared to other methods. Nevertheless, it is suggested that their prediction performance be revised with new data. Like any model based on decision trees, over-fitting may occur, and misleading results might be generated. In those situations, labelling misclassified data, aggregating them into the features database and rerunning the algorithm is suggested.

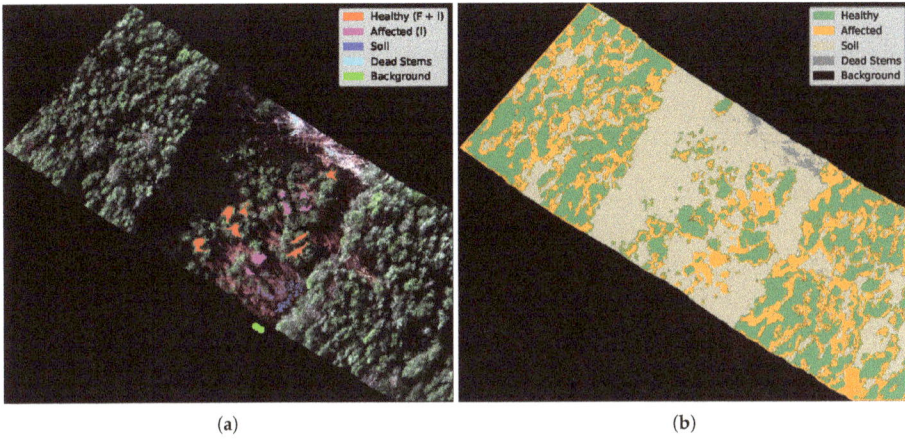

(a)   (b)

Figure 9. Segmentation results of the proposed approach. (**a**) Recovered hyperspectral image in red–green–blue (RGB) colour model; (**b**) Segmentation results.

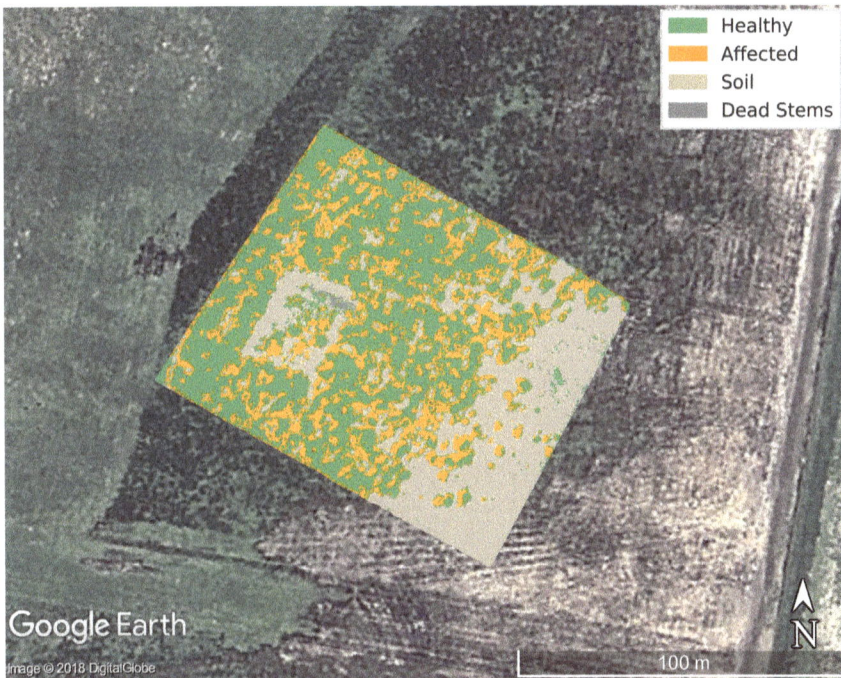

Figure 10. Layer of mapping results of the study area in Google Earth.

The availability to process and classify data with small GSD values demonstrates the potential of UASs for remote sensing equipment compared with satellite and manned aircraft for forest health assessments on forest and tree plantations and with traditional estimation methods, such as statistical regression models. In comparison with similar approaches of non-invasive assessment techniques using UAVs and spectral sensors, this framework does not provide general spectral indexes that can be applied with different classifiers and similar evaluations. In contrast, this presented method boosts the efficiency of the classifier by receiving feedback from the accuracy scores of every feature and transforming the input data in consequence. The more explicit the data for the classifier is, the better the classification rates are. Furthermore, it is also demonstrated that a classifier which processes and combines data from multiple spectral indexes provides better performance than analysing individual components from different source sensors.

5. Conclusions

This paper describes a pipeline methodology for effective detection and mapping of indicators of poor health in forest and plantation trees integrating UAS technology and artificial intelligence approaches. The techniques were illustrated with an accurate classification and segmentation task of paperbark tea trees deteriorated by myrtle rust from an exclusion trial in NSW, Australia. Here, the system achieved detection rates of 97.24% for healthy trees and 94.72% for affected trees. The algorithm obtained a multiclass detection rate of 97.35%. Data labelling is a task that demands many resources from both site-based and remote sensing methods, and, due to human error, affects the accuracy and reliability of the classifier results.

The approach can be used to train various datasets from different sensors to improve detection rates that single solutions offer as well as the capability of processing large datasets using freeware software. The case study demonstrates an effective approach that allows for rapid and accurate indicators, and for alterations of exposed areas at early stages. However, understanding disease epidemiology and interactions between pathogens and hosts is still required for the effective use of these technologies.

Future research should discuss the potential of monitoring the evolution of affected species through time, the prediction of expansion directions and rates of the disease, and how data will contribute to improving control actions to deter their presence in unaffected areas. Technologically, future works should analyse and compare the efficacy of unsupervised algorithms to label vegetation items accurately, integrate the best approaches in the proposed pipeline, and evaluate regression models that predict data based on other biophysical information offered by site-based methods.

Acknowledgments: This work was funded by the Plant Biosecurity Cooperative Research Centre (PBCRC) 2135 project. The authors would like to acknowledge Jonathan Kok for his contributions in co-planning the experimentation phase. We also gratefully acknowledge the support of the Queensland University of Technology (QUT) Research Engineering Facility (REF) Operations team (Dirk Lessner, Dean Gilligan, Gavin Broadbent and Dmitry Bratanov), who operated the DJI S800 EVO UAV and image sensors, and performed ground referencing. We thank Gavin Broadbent for the design, manufacturing, and tuning of a customised 2-axis gimbal for the spectral cameras. We acknowledge the High-Performance Computing and Research Support Group at QUT, for the computational resources and services used in this work.

Author Contributions: Felipe Gonzalez and Geoff Pegg contributed to experimentation and data collection planning. Felipe Gonzalez supervised the airborne surveys, the quality of the acquired data, and logistics. Juan Sandino designed the proposed pipeline and conducted the data processing phase. Felipe Gonzalez, Geoff Pegg, and Grant Smith provided definitions, assistance, and essential advice. Juan Sandino analysed the generated outputs, and validated and optimised the algorithm. All the authors contributed significantly to the composition and revision of the paper.

Abbreviations

The following abbreviations are used in this manuscript:

2D	Two-dimensional
3D	Three-dimensional
F	Fungicides
F + I	Fungicides and Insecticides

GDAL	Geospatial data abstraction library
GPS	Global positioning system
GNDVI	Green normalised difference vegetation index
GSD	Ground sampling distance
I	Insecticides
k-NN	k-nearest neighbours
LDA	Linear discriminant analysis
LiDAR	Light detection and ranging
MDPI	Multidisciplinary digital publishing institute
MSAVI2	Second modified soil-adjusted vegetation index
NDVI	Normalised difference vegetation index
NSW	New South Wales
RGB	Red–green–blue colour model
SAVI	Soil-adjusted Vegetation Index
SVM	Support Vector Machines
TIF	Tagged Image File
UAS	Unmanned Aerial System
UAV	Unmanned Aerial Vehicle
VNIR	Visible Near Infrared
WRIS	White Reference Illumination Spectrum
XGBoost	eXtreme Gradient Boosting

References

1. Smock, L.A.; MacGregor, C.M. Impact of the American Chestnut Blight on Aquatic Shredding Macroinvertebrates. *J. N. Am. Benthol. Soc.* **1988**, *7*, 212–221.

2. Anagnostakis, S.L. Chestnut Blight: The classical problem of an introduced pathogen. *Mycologia* **1987**, *79*, 23–37.

3. Burke, K.L. The effects of logging and disease on American chestnut. *For. Ecol. Manag.* **2011**, *261*, 1027–1033.

4. Rizzo, D.M.; Garbelotto, M.; Hansen, E.M. Phytophthora ramorum: Integrative research and management of an emerging pathogen in California and oregon forests. *Annu. Rev. Phytopathol.* **2005**, *43*, 309–335.

5. Frankel, S.J. Sudden oak death and Phytophthora ramorum in the USA: A management challenge. *Australas. Plant Pathol.* **2008**, *37*, 19–25.

6. Grünwald, N.J.; Garbelotto, M.; Goss, E.M.; Heungens, K.; Prospero, S. Emergence of the sudden oak death pathogen Phytophthora ramorum. *Trends Microbiol.* **2012**, *20*, 131–138.

7. Hardham, A.R. Phytophthora cinnamomi. *Mol. Plant Pathol.* **2005**, *6*, 589–604.

8. Shearer, B.L.; Crane, C.E.; Barrett, S.; Cochrane, A. Phytophthora cinnamomi invasion, a major threatening process to conservation of flora diversity in the South-West Botanical Province of Western Australia. *Aust. J. Bot.* **2007**, *55*, 225–238.

9. Burgess, T.I.; Scott, J.K.; Mcdougall, K.L.; Stukely, M.J.; Crane, C.; Dunstan, W.A.; Brigg, F.; Andjic, V.; White, D.; Rudman, T.; et al. Current and projected global distribution of Phytophthora cinnamomi, one of the world's worst plant pathogens. *Glob. Chang. Biol.* **2017**, *23*, 1661–1674.

10. Pegg, G.S.; Giblin, F.R.; McTaggart, A.R.; Guymer, G.P.; Taylor, H.; Ireland, K.B.; Shivas, R.G.; Perry, S. Puccinia psidii in Queensland, Australia: Disease symptoms, distribution and impact. *Plant Pathol.* **2014**, *63*, 1005–1021.

11. Carnegie, A.J.; Kathuria, A.; Pegg, G.S.; Entwistle, P.; Nagel, M.; Giblin, F.R. Impact of the invasive rust Puccinia psidii (myrtle rust) on native Myrtaceae in natural ecosystems in Australia. *Biol. Invasions* **2016**, *18*, 127–144.

12. Howard, C.; Findlay, V.; Grant, C. Australia's transition to management of myrtle rust. *J. For. Sci.* **2016**, *61*, 138–139.

13. Fernandez Winzer, L.; Carnegie, A.J.; Pegg, G.S.; Leishman, M.R. Impacts of the invasive fungus Austropuccinia psidii (myrtle rust) on three Australian Myrtaceae species of coastal swamp woodland. *Austral Ecol.* **2017**, *43*, doi:10.1111/aec.12534.

14. Dayton, L.; Higgins, E. Myrtle rust 'biggest threat to ecosystem'. Available online: http://www.webcitation.org/6y61T6sI6 (accessed on 19 February 2018).

15. Carnegie, A.J.; Cooper, K. Emergency response to the incursion of an exotic myrtaceous rust in Australia. *Australas. Plant Pathol.* **2011**, *40*, 346–359.

16. Carnegie, A.J. First Report of Puccinia psidii (Myrtle Rust) in Eucalyptus Plantations in Australia. *Plant Dis.* **2015**, *99*, 161, doi:10.1094/PDIS-09-14-0901-PDN.

17. Pegg, G.; Taylor, T.; Entwistle, P.; Guymer, G.; Giblin, F.; Carnegie, A. Impact of Austropuccinia psidii (myrtle rust) on Myrtaceae-rich wet sclerophyll forests in south east Queensland. *PLoS ONE* **2017**, *12*, e0188058, doi:10.1371/journal.pone.0188058.

18. Government of Western Australia. Phytophthora Dieback—Parks and Wildlife Service. Available online: http://www.webcitation.org/6xLA86qjW (accessed on 19 February 2018).

19. U.S. Forest Service. Sudden Oak Death (SOD) | Partnerships | PSW Research Station | Forest Service. Available online: http://www.webcitation.org/6xLDwPURd (accessed on 19 February 2018).

20. State of Hawaii. Department of Agriculture | How to Report Suspected Ohia Wilt/Rapid Ohia Death. Available online: http://www.webcitation.org/6xLCVG70h (accessed on 19 February 2018).

21. Lawley, V.; Lewis, M.; Clarke, K.; Ostendorf, B. Site-based and remote sensing methods for monitoring indicators of vegetation condition: An Australian review. *Ecol. Indic.* **2016**, *60*, 1273–1283.

22. Oliver, I.; Smith, P.L.; Lunt, I.; Parkes, D. Pre-1750 vegetation, naturalness and vegetation condition: What are the implications for biodiversity conservation? *Ecol. Manag. Restor.* **2002**, *3*, 176–178.

23. Lawley, V.; Parrott, L.; Lewis, M.; Sinclair, R.; Ostendorf, B. Self-organization and complex dynamics of regenerating vegetation in an arid ecosystem: 82 years of recovery after grazing. *J. Arid Environ.* **2013**, *88*, 156–164.

24. Ostendorf, B. Overview: Spatial information and indicators for sustainable management of natural resources. *Ecol. Indic.* **2011**, *11*, 97–102.

25. Roux, J.; Germishuizen, I.; Nadel, R.; Lee, D.J.; Wingfield, M.J.; Pegg, G.S. Risk assessment for Puccinia psidii becoming established in South Africa. *Plant Pathol.* **2015**, *64*, 1326–1335.

26. Berthon, K.; Esperon-Rodriguez, M.; Beaumont, L.; Carnegie, A.; Leishman, M. Assessment and prioritisation of plant species at risk from myrtle rust (Austropuccinia psidii) under current and future climates in Australia. *Biol. Conserv.* **2018**, *218*, 154–162.

27. Lausch, A.; Erasmi, S.; King, D.J.; Magdon, P.; Heurich, M. Understanding Forest Health with Remote Sensing -Part I –A Review of Spectral Traits, Processes and Remote-Sensing Characteristics. *Remote Sens.* **2016**, *8*, 1029, doi:10.3390/rs8121029

28. Tuominen, J.; Lipping, T.; Kuosmanen, V.; Haapanen, R. Remote sensing of forest health. In *Geoscience and Remote Sensing*; Ho, P.G.P., Ed.; InTech: Rijeka, Croatia, 2009; Chapter 02.

29. Lausch, A.; Erasmi, S.; King, D.J.; Magdon, P.; Heurich, M. Understanding forest health with remote sensing-Part II–A review of approaches and data models. *Remote Sens.* **2017**, *9*, 129, doi:10.3390/rs9020129.

30. Cui, D.; Zhang, Q.; Li, M.; Zhao, Y.; Hartman, G.L. Detection of soybean rust using a multispectral image sensor. *Sens. Instrum. Food Qual. Saf.* **2009**, *3*, 49–56.

31. Candiago, S.; Remondino, F.; De Giglio, M.; Dubbini, M.; Gattelli, M. Evaluating multispectral images and vegetation indices for precision farming applications from UAV images. *Remote Sens.* **2015**, *7*, 4026–4047.

32. Lowe, A.; Harrison, N.; French, A.P. Hyperspectral image analysis techniques for the detection and classification of the early onset of plant disease and stress. *Plant Methods* **2017**, *13*, 80, doi:10.1186/s13007-017-0233-z.

33. Khanal, S.; Fulton, J.; Shearer, S. An overview of current and potential applications of thermal remote sensing in precision agriculture. *Comput. Electron. Agric.* **2017**, *139*, 22–32.

34. Devadas, R.; Lamb, D.W.; Simpfendorfer, S.; Backhouse, D. Evaluating ten spectral vegetation indices for identifying rust infection in individual wheat leaves. *Precis. Agric.* **2009**, *10*, 459–470.

35. Ashourloo, D.; Mobasheri, M.; Huete, A. Developing two spectral disease indices for detection of wheat leaf rust (Pucciniatriticina). *Remote Sens.* **2014**, *6*, 4723–4740.

36. Wang, H.; Qin, F.; Liu, Q.; Ruan, L.; Wang, R.; Ma, Z.; Li, X.; Cheng, P.; Wang, H. Identification and disease index inversion of wheat stripe rust and wheat leaf rust based on hyperspectral data at canopy level. *J. Spectrosc.* **2015**, *2015*, 1–10.

37. Heim, R.H.J.; Wright, I.J.; Chang, H.C.; Carnegie, A.J.; Pegg, G.S.; Lancaster, E.K.; Falster, D.S.; Oldeland, J. Detecting myrtle rust (Austropuccinia psidii) on lemon myrtle trees using spectral signatures and machine learning. *Plant Pathol.* **2018**, doi:10.1111/ppa.12830.
38. Booth, T.H.; Jovanovic, T. Assessing vulnerable areas for Puccinia psidii (eucalyptus rust) in Australia. *Australas. Plant Pathol.* **2012**, *41*, 425–429.
39. Elith, J.; Simpson, J.; Hirsch, M.; Burgman, M.A. Taxonomic uncertainty and decision making for biosecurity: spatial models for myrtle/guava rust. *Australas. Plant Pathol.* **2013**, *42*, 43–51.
40. Salami, E.; Barrado, C.; Pastor, E. UAV flight experiments applied to the remote sensing of vegetated areas. *Remote Sens.* **2014**, *6*, 11051–11081.
41. Glassock, R.; Hung, J.Y.; Gonzalez, L.F.; Walker, R.A. Design, modelling and measurement of a hybrid powerplant for unmanned aerial systems. *Aust. J. Mech. Eng.* **2008**, *6*, 69–78.
42. Whitney, E.; Gonzalez, L.; Periaux, J.; Sefrioui, M.; Srinivas, K. A robust evolutionary technique for inverse aerodynamic design. In Proceedings of the European Congress on Computational Methods in Applied Sciences and Engineering, Jyvaskyla, Finland, 24–28 July 2004; Volume 2, pp. 1–2.
43. Gonzalez, L.; Whitney, E.; Srinivas, K.; Periaux, J. Multidisciplinary aircraft design and optimisation using a robust evolutionary technique with variable fidelity models. In Proceedings of the 10th AIAA/ISSMO Multidisciplinary Analysis and Optimization Conference, Albany, NY, USA, 30 August–1 September 2004; Volume 6, pp. 3610–3624.
44. Ken, W.; Chris, H.C. Remote sensing of the environment with small unmanned aircraft systems (UASs), part 1: A review of progress and challenges. *J. Unmanned Veh. Syst.* **2014**, *2*, 69–85.
45. Gonzalez, L.; Montes, G.; Puig, E.; Johnson, S.; Mengersen, K.; Gaston, K. Unmanned Aerial Vehicles (UAVs) and artificial intelligence revolutionizing wildlife monitoring and conservation. *Sensors* **2016**, *16*, 97, doi:10.3390/s16010097.
46. Sandino, J.; Wooler, A.; Gonzalez, F. Towards the automatic detection of pre-existing termite mounds through UAS and hyperspectral imagery. *Sensors* **2017**, *17*, 2196, doi:10.3390/s17102196.
47. Vanegas, F.; Bratanov, D.; Powell, K.; Weiss, J.; Gonzalez, F. A novel methodology for improving plant pest surveillance in vineyards and crops using UAV-based hyperspectral and spatial data. *Sensors* **2018**, *18*, e260, doi:10.3390/s18010260.
48. Vanegas, F.; Gonzalez, F. Enabling UAV navigation with sensor and environmental uncertainty in cluttered and GPS-denied environments. *Sensors* **2016**, *16*, 666, doi:10.3390/s16050666.
49. Aasen, H.; Burkart, A.; Bolten, A.; Bareth, G. Generating 3D hyperspectral information with lightweight UAV snapshot cameras for vegetation monitoring: From camera calibration to quality assurance. *ISPRS J. Photogramm. Remote Sens.* **2015**, *108*, 245–259.
50. Nasi, R.; Honkavaara, E.; Lyytikainen-Saarenmaa, P.; Blomqvist, M.; Litkey, P.; Hakala, T.; Viljanen, N.; Kantola, T.; Tanhuanpaa, T.; Holopainen, M. Using UAV-Based photogrammetry and hyperspectral imaging for mapping bark beetle damage at tree-level. *Remote Sens.* **2015**, *7*, 15467–15493.
51. Calderon, R.; Navas-Cortes, J.; Lucena, C.; Zarco-Tejada, P. High-resolution airborne hyperspectral and thermal imagery for early detection of Verticillium wilt of olive using fluorescence, temperature and narrow-band spectral indices. *Remote Sens. Environ.* **2013**, *139*, 231–245.
52. Calderon, R.; Navas-Cortes, J.A.; Zarco-Tejada, P.J. Early detection and quantification of verticillium wilt in olive using hyperspectral and thermal imagery over large areas. *Remote Sens.* **2015**, *7*, 5584–5610.
53. Albetis, J.; Duthoit, S.; Guttler, F.; Jacquin, A.; Goulard, M.; Poilvé, H.; Féret, J.B.; Dedieu, G. Detection of Flavescence dorée Grapevine Disease using Unmanned Aerial Vehicle (UAV) multispectral imagery. *Remote Sens.* **2017**, *9*, 308, doi:10.3390/rs9040308.
54. Pause, M.; Schweitzer, C.; Rosenthal, M.; Keuck, V.; Bumberger, J.; Dietrich, P.; Heurich, M.; Jung, A.; Lausch, A. In situ/remote sensing integration to assess forest health–A review. *Remote Sens.* **2016**, *8*, 471, doi:10.3390/rs8060471.
55. Stone, C.; Mohammed, C. Application of remote sensing technologies for assessing planted forests damaged by insect pests and fungal pathogens: A review. *Curr. For. Rep.* **2017**, *3*, 75–92.
56. Habili, N.; Oorloff, J. Scyllarus™: From Research to Commercial Software. In Proceedings of the ASWEC 24th Australasian Software Engineering Conference, Adelaide, SA, Australia, 28 September–1 October 2015; ACM Press: New York, NY, USA, 2015; Volume II, pp. 119–122.

57. Gu, L.; Robles-Kelly, A.A.; Zhou, J. Efficient estimation of reflectance parameters from imaging spectroscopy. *IEEE Trans. Image Process.* **2013**, *22*, 3648–3663.

58. GDAL Development Team. *GDAL—Geospatial Data Abstraction Library, Version 2.1.0*; Open Source Geospatial Foundation: Beaverton, OR, USA, 2017.

59. Chen, T.; Guestrin, C. XGBoost: A Scalable Tree Boosting System. In Proceedings of the 22nd ACM SIGKDD International Conference on Knowledge Discovery and Data Mining (KDD'16), San Francisco, CA, USA, 13–17 August 2016; ACM Press: New York, NY, USA, 2016; pp. 785–794.

60. Pedregosa, F.; Varoquaux, G.; Gramfort, A.; Michel, V.; Thirion, B.; Grisel, O.; Blondel, M.; Prettenhofer, P.; Weiss, R.; Dubourg, V.; et al. Scikit-learn: Machine Learning in Python. *J. Mach. Learn. Res.* **2011**, *12*, 2825–2830.

61. Bradski, G. The OpenCV library. *Dr. Dobb's J. Softw. Tools* **2000**, *25*, 120, 122–125.

62. Hunter, J.D. Matplotlib: A 2D graphics environment. *Comput. Sci. Eng.* **2007**, *9*, 90–95.

63. Rouse, J.W., Jr.; Haas, R.H.; Schell, J.A.; Deering, D.W. Monitoring vegetation systems in the great plains with Erts. *NASA Spec. Publ.* **1974**, *351*, 309–317.

64. Gitelson, A.A.; Kaufman, Y.J.; Merzlyak, M.N. Use of a green channel in remote sensing of global vegetation from EOS-MODIS. *Remote Sens. Environ.* **1996**, *58*, 289–298.

65. Huete, A. A soil-adjusted vegetation index (SAVI). *Remote Sens. Environ.* **1988**, *25*, 295–309.

66. Laosuwan, T.; Uttaruk, P. Estimating tree biomass via remote sensing, MSAVI 2, and fractional cover model. *IETE Tech. Rev.* **2014**, *31*, 362–368.

67. Australian Government. *Evans Head, NSW–August 2016–Daily Weather Observations*; Bureau of Meteorology: Evans Head, NSW, Australia, 2016.

sensors

MDPI

Article

Secure Utilization of Beacons and UAVs in Emergency Response Systems for Building Fire Hazard

Seung-Hyun Seo *, Jung-In Choi and Jinseok Song

The Division of Electrical Engineering, Hanyang University, ERICA Campus, Gyeonggi-do, Ansan 15588, Korea; peach0206@hanyang.ac.kr (J.-I.C.); bbtam99@naver.com (J.S.)
* Correspondence: seosh77@hanyang.ac.kr

Received: 15 August 2017 ; Accepted: 19 September 2017; Published: 25 September 2017

Abstract: An intelligent emergency system for hazard monitoring and building evacuation is a very important application area in Internet of Things (IoT) technology. Through the use of smart sensors, such a system can provide more vital and reliable information to first-responders and also reduce the incidents of false alarms. Several smart monitoring and warning systems do already exist, though they exhibit key weaknesses such as a limited monitoring coverage and security, which have not yet been sufficiently addressed. In this paper, we propose a monitoring and emergency response method for buildings by utilizing beacons and Unmanned Aerial Vehicles (UAVs) on an IoT security platform. In order to demonstrate the practicability of our method, we also implement a proof of concept prototype, which we call the UAV-EMOR (UAV-assisted Emergency Monitoring and Response) system. Our UAV-EMOR system provides the following novel features: (1) secure communications between UAVs, smart sensors, the control server and a smartphone app for security managers; (2) enhanced coordination between smart sensors and indoor/outdoor UAVs to expand real-time monitoring coverage; and (3) beacon-aided rescue and building evacuation.

Keywords: sensors and beacons; unmanned aerial vehicles; monitoring and emergency response systems; Internet of Things; security

1. Introduction

Fire detection and corresponding safety systems like monitoring and emergency response systems are crucial aspects of any building management system, and billions of dollars are spent annually on the installation and maintenance of such systems [1]. However, with recent advances in IoT technology, intelligent emergency systems using smart sensors offer opportunities to accomplish these tasks more efficiently and economically. Smart sensors can often anticipate emergencies even before they happen, and when emergencies do occur, such sensors can provide more complete and reliable information to building management and emergency personnel. This allows building occupants to be evacuated more quickly and safely than is possible when using conventional systems. For these reasons, it is becoming standard for all new buildings to be equipped with wireless sensors that function as part of the overall building management system. In fact, many proposed alert and response systems already incorporate wireless sensors in their designs [2–6]. However, in these proposals [2–6], sensors are used only in the monitoring of interior locations, which leaves building exteriors insufficiently monitored. Furthermore, since such systems mostly use fixed cameras like closed-circuit television(CCTV) for monitoring, even many areas inside buildings remain vulnerable. Unmanned Aerial Vehicles (UAVs), also known as drones, are another recently-emerging technology that is being used more frequently for monitoring and surveillance in many common public sectors. Specifically, camera-equipped drones have been widely used for visual monitoring in public safety, traffic management and disaster monitoring.

Moreover, UAVs can be used to support search and rescue operations and perform operations that are hard to execute by human operators, at low operating costs. When UAVs are deployed in a disaster area, they can collect evidence of the presence of a victim and report the collected data to a rescue team. Some monitoring and warning and search and rescue systems have already proposed the use of UAVs [7–17], but since most of those systems focus on natural disasters, they are not directly applicable to the monitoring of buildings. Another important weakness of those systems is that they do not account for the secure transmission of data between sensors/drones and control servers. This leaves such systems highly vulnerable to malicious attacks in which people outside of the system could easily alter the data that are being transmitted or even take direct control of various aspects of the system.

In order to address these several constraints, in this paper, we propose a method of property monitoring and emergency response by utilizing beacons and UAVs on an IoT open standard platform such as the AllJoyn platform [18]. We also present a proof of concept prototype, called the UAV-EMOR (UAV-assisted Emergency Monitoring and Response) system. Our UAV-EMOR system offers the following three novel features.

First, our UAV-EMOR system supports end-to-end secure communication between all components of the system, including UAVs, smart sensors, the control server and a smartphone app for security managers. Unfortunately, one aspect of wireless network-based monitoring and emergency response systems that has not been sufficiently addressed thus far is data security. Systems that use wireless networks are vulnerable to external manipulation, impersonation and eavesdropping, etc. In our system, as well, all entities are connected to the control server via wireless networks, and in order to provide secure data transmission, we incorporate the standard authenticated key exchange protocol supported by the Security 2.0 specification of the AllJoyn IoT platform [18]. Since a security manager app is used to remotely monitor the interior and exterior of the building and to access the emergency response systems, it is vital that only a legitimate security manager should be capable of accessing the app and communicating with the emergency response system. To ensure this is the case, we use a bio-metric authentication method using fingerprint identification.

Another key aspect of our system is the integration of UAVs. Our UAV-EMOR system supports enhanced coordination between smart sensors and indoor/outdoor drones in order to expand real-time monitoring coverage. The existing research on UAV-assisted monitoring and warning systems mainly focuses on environmental monitoring and emergency response for natural disasters [7–14]. In our work, we introduce a more dynamic monitoring approach, which uses adaptive indoor/outdoor UAVs to expand monitoring coverage for buildings and territory around them. Indoor UAVs are assigned to each floor of a building, while outdoor UAVs are used to monitor the exterior. This allows our system to monitor the blind spots that the fixed CCTV of conventional systems cannot capture.

Unmanned all-terrain vehicle(ATV) can be utilized for monitoring instead of indoor drones [19]. However, we are only considering devices that can be operated for indoor monitoring without any problem even in an emergency situation. In a disaster situation where buildings are collapsing, ATVs are difficult to get into the room and to move in situations where users are evacuating quickly. ATVs can have visibility issues that the UAV would not have since the UAV would be able to move above and around the flames. ATVs would have benefits in other ways, but their size makes them difficult to stock many in the building itself. More UAVs could be kept on site. While the battery may become an issue, the UAV only has to fly for a short time to collect data and would not necessarily run out of battery. Plus, a large enough supply of smaller UAVs could offset the battery issue. Therefore, an emergency response method using indoor UAVs is more effective than ATVs.

Finally, our UAV-EMOR system provides beacon-aided rescue and building evacuation. There is a distinct need to improve emergency response methods so building occupants can be more efficiently and safely evacuated during emergency situations. There are currently several approaches [20–22] that attempt to track the location of individuals inside of buildings. However, they are not precise enough to be practicable or they require an active response from a user, which may not be possible during certain types of emergencies. For example, some systems use the global positioning system(GPS) of users' smartphones for location, but GPS-based location information cannot be calculated accurately indoors [23]. Other systems use a variety of sensors like ultrasonic, near-field communication and radio frequency signals, but those systems are overly dependent on battery power and require active interaction from users [20–22]. Fixed location beacons are also in use and require very little battery power, though they suffer from the same issue as sensors, as they are dependent on user interaction [24,25]. In our approach, we take advantage of the beacon's low battery consumption and increase its effectiveness by assigning an individual beacon to every person inside a building. This is accomplished by incorporating beacons into smart ID cards that are issued to all building occupants. Using Bluetooth technology, such beacons constantly transmit the unique ID and real-time location of the wearer to all smart sensors and drones. The beacons have no on/off function, so there is never any danger of accidentally disabling the device; nor do they require active input from the user, so they can be used to locate users who are perhaps too badly injured to activate a device or who are unconscious. Furthermore, since beacon signals can transmit through building walls, users can be located even inside of collapsed buildings.

The remainder of this paper is organized as follows: In Section 2, we discuss related works and provide a comparison of our approach with those recent works. In Section 3, we introduce the main components of the UAV-EMOR system and present system protocols. In Section 4, we describe the prototype of our UAV-EMOR system and use scenarios. In Section 5, we evaluate communication and computation overhead for the performance of our system protocols and also the evaluate position accuracy of our UAV-EMOR system. Lastly, in Section 6, we outline our conclusions.

2. Related Works and Backgrounds

In this section, we first explore related works in the field of emergency alert and response systems. Then, we discuss several related works for warning systems that incorporate UAVs. We will also look at works that address how to incorporate sensor-based monitoring systems into buildings that were not designed with this purpose in mind. Table 1 summarizes the comparison of recent related works.

Table 1. Comparison of related works. EMOR, Emergency Monitoring and Response.

	Source		Types of Functionalities				Service Area	User Interface
	Sensor	UAV	Monitoring	Alarm	Rescue	Security		
UAV-EMOR (Our System)	O	O	O	O	O	O	Building, Outside	Smartphone App
[2]	O			O	O		Building	Smartphone App
[3]	O		O	O	O		Building	
[4]	O		O	O	O	O	Building	
[5]	O		O				Building	
[6]	O		O	O	O		Building	Smartphone App
[7]	O	O	O				Nature	
[9]	O	O	O				Nature	
[10]	O		O	O			Nature	Web
[11]	O	O	O				River	

2.1. Emergency Alert and Response System

Conventional emergency and evacuation systems do not provide information about the context of an emergency, nor about the profile of each person in the building. To address these weaknesses, there is a growing interest in alert and response systems using smart sensors that are tailored to the specifics of an emergency situation. Aedo, I. et al. [2] propose a set of criteria related to user profiles, available technologies, kinds of emergency and contextual circumstances for an effective notification and evacuation. The authors present a mechanism that provides personalized alert notifications and evacuation routes to users with smartphones. However, their mechanism cannot automatically identify the building occupant's location. To get the personalized evacuation route, occupants should send their location data to the emergency management server. This system has a disadvantage in that it cannot be used if a part of the building collapses and the occupant cannot grasp the current location.

Liu, J.W.S. et al. [3] propose a data model and architecture of indoor emergency evacuation systems for large public buildings. The authors also present an intelligent indoor emergency evacuation system that utilizes the Building and environment Data-based Indoor Positioning System (BeDIPS) [26] to support indoor positioning service. However, BeDIPS requires installing location beacons in a building, and in the current version of BeDIPS, each location beacon only provides a one-step navigation instruction to the nearest exit. Thus, Liu, J.W.S. et al.'s system could not support a personalized evacuation route for each occupant.

Wu, Z.Y. et al. [4] propose a framework for an intelligent evacuation guidance system for large buildings. They combine fire detection, video surveillance, a mobile terminal and passive RFID tags to create an evacuation guidance system. During regular monitoring, their system notes temperature and atmospheric changes, setting off an alarm when appropriate. During an actual emergency situation, the system monitors population density inside the building, the evacuation status of all entrances and the road capacity outside the building. Furthermore, their system can guide individuals from their specific locations in the building to appropriate exits by sending out evacuation instructions. However, their RFID-based evacuation system requires active input from the occupants to get the evacuation instructions.

Zeng, Y. et al. [5] propose an emergency routing scheme for wireless sensor networks that is designed specifically for building fire emergency. Their scheme is adaptable to several different scenarios and the routing hole problems. Each sensor contains data about its own ID, power status and state and automatically changes its state between safe, low safe, fire and unsafe. Lin, C.Y. et al. [6] propose an active disaster response system for earthquakes, which automatically engages when a natural disaster occurs. Their system consists of two main parts, an emergency broadcast system and an active disaster response system. In the emergency broadcast system, the weather center sends warning messages to embedded boards. Then, the active disaster response system automatically performs specified actions, like cutting off power and gas. During a natural disaster, this system configures embedded boards to construct and maintain a temporary network for sending emergency messages. Using this network, the rescue center can integrate data and enable rescue workers to provide aid where it is needed.

Unlike these previous works, our UAVs-EMOR system supports both indoor and outdoor monitoring by incorporating UAVs and automatically identifies the occupant's location by using beacon technology. Therefore, it can provide evacuation information to occupants.

2.2. UAV-Based Warning System

UAVs deliver timely disaster warnings to public officers and support rescue and recovery operations. Due to their ability to fly autonomously and rapidly acquire sensing data on emergency areas that are inaccessible to humans, UAVs have become a viral candidate platform for emergency monitoring and response systems. Drones are being used by organizations such as customs control, the coast guard, environmental agencies and private businesses for a variety of purposes including safety systems in the gas and chemical industry, fire detection, tactical surveillance, search and rescue, environmental monitoring, and many more [8,12,27–29].

Choi, K. et al. [7] propose a monitoring system for emergency responses. Their system consists of both an aerial and a ground component. The aerial aspect includes a UAV platform, sensors and supporting modules, while the ground aspect includes vehicles and a receiving and processing system. The aerial component receives sensor data from the environment and transmits it to the ground component. Sensor data and control commands are transmitted automatically in real time.

Chen, D. et al. [9] propose a dynamic routing protocol, a method for managing mobile nodes to enable reliable, real-time data transmission. In [9], to support early warning, UAVs monitor the area and collect sensor data, transferring it to the data center for analysis. Frigerio, S. et al. [10] propose a web-based platform for automatic and continuous monitoring of the Rotolon catchment in the Italian

Alps for potential landslides. Ueyama, J. et al. [11] propose the use of UAVs to make their wireless sensor network more resilient to failures for river monitoring.

Most warning systems based on UAVs focus on natural disasters, and such systems are unsuitable for application in security monitoring. In the IoT environment, each individual device and sensor must contain a security module to protect it from malicious interference. Indeed, in an emergency response system for a building, it is most important to prevent malicious attacks on the communications network. Therefore, we customize security modules from the AllJoyn Security Platform to secure every component of our system.

2.3. Incorporating Sensor-Based Monitoring Systems into Existing Buildings

The term "smart building" describes a suite of technologies used to make the design, construction and operation of buildings more efficient and is applicable to both existing and newly-built properties [30]. The smart building in the future will adjust in every situation to provide personalized service to every individual inside the building. To do this, the building must have the ability to sense and to interpret situations automatically. Buildings without smart technology cannot transmit environmental information in real time because they do not incorporate network-connected sensors. However, reconstructing a building to make it "smart" would be an inefficient waste of both time and financial resources.

For energy savings in commercial buildings, Weng, T. et al. [31] have presented a design for converting a conventional building into a smart building. They have developed several mechanisms for activating the HVAC system, IT equipment and lighting depending on occupancy and usage. For monitoring, they attach sensors to the building and install building management systems. They estimate that the cost of converting a building to make it "smart" is offset by the energy savings it produces. Osello, A. et al. [1] also propose using middleware to reduce energy in existing buildings. They use software to monitor and control the lighting and HVAC services. They also design the context framework using an ontology-based knowledge repository and a rule-based context recognition system. Whereas existing smart building systems focus primarily on energy-reduction, our system provides a cost-effective, smart emergency alert and response network that provides situationally-specific and personalized service for users.

3. UAV-Assisted Emergency Monitoring and Response System

In this section, we briefly describe an overview of our UAV-assisted Emergency Monitoring and Response (UAV-EMOR) system for buildings. We then present each of the processes of which our system is composed: (1) day-to-day monitoring; (2) abnormal situation response; (3) emergency response.

3.1. Overview of UAV-EMOR System

There are four novel functionalities of our UAV-EMOR system: (1) provide secure communications between UAVs, smart sensors, a control server and the security manager's smartphone app, which is based on the AllJoyn security platform; (2) support real-time monitoring using smart sensors; (3) support active response using indoor/outdoor UAVs; and (4) support beacon-aided rescue. Figure 1 provides an overview of our UAV-EMOR system. The system consists of six entities: smart sensors, indoor/outdoor monitoring UAVs (Unmanned Aerial Vehicles), a security manager, a control server, users and CCTV. Indoor and outdoor UAVs periodically follow designated routes, monitor conditions inside and outside of the building while interfacing with smart sensors that have been placed in each room. Each smart sensor checks the surrounding temperature and notes potentially abnormal situations inside the room. Each UAV transmits its monitoring data and images directly to the control server, where a building security manager can monitor the images at any time. CCTV is still used to monitor the space for potential dangers, like fire, and stores those recordings on the server. However, the indoor UAVs on each floor can focus on blind spots that CCTV cannot see.

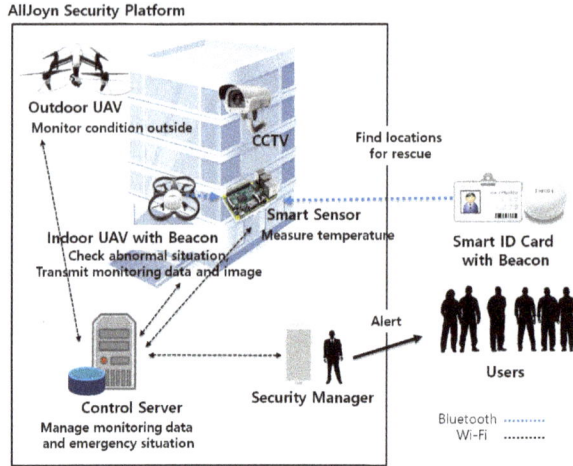

Figure 1. The overview of our UAV-assisted emergency monitoring and response system.

The indoor UAV only monitors when it is given a task, but UAVs currently in commercial use have an average of about 30 min–1 h of battery life, which is sufficient for most tasks. So after use, they must be recharged. To solve this battery issue, we assume that an indoor UAV usually stays in a specific landing spot on each floor. The indoor UAV can remain charged at the landing spot until it receives the command from the control server.

The Bluetooth beacons are attached to or embedded in the indoor UAVs and the users' smart ID cards for indoor location recognition. The indoor UAV's beacon transmits the ID of the indoor UAV to the smart sensor. Only authorized UAVs can get entry into the room. The beacons in the smart ID cards transmit a signal to the smart sensor in an emergency situation so that the smart sensor determines the number of users and the users' information in each room. The smart sensors, UAVs and security manager's smartphone are all connected to the control server via the wireless network. We use the AllJoyn security platform in order to provide secure communications among the various components of our system. Figure 2 shows the main components of UAV-EMOR system.

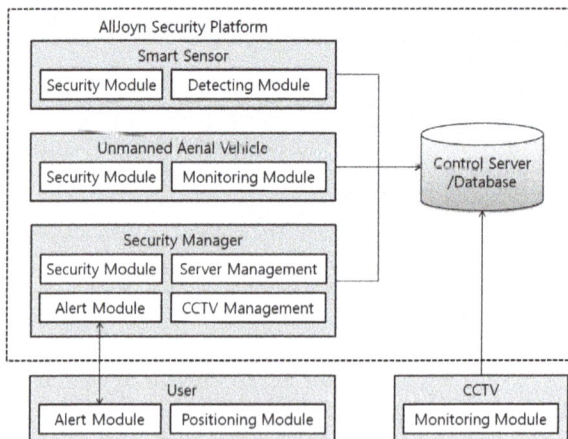

Figure 2. The Main components of the UAV-EMOR system.

3.2. AllJoyn Security Platform for UAV-EMOR System

The AllJoyn system supports a security framework for IoT applications to authenticate each other and send encrypted data between them. Since AllJoyn applications can be installed on the IoT devices, it can be utilized to provide end-to-end application level security between IoT devices, and thus authentication and data encryption are done at the application layer. All of the logic for authentication and encryption except the Auth is implemented in the AllJoyn core library. The Auth Listener is a callback function implemented by the application to provide auth credentials (e.g., PIN or password) or verify auth credentials (e.g., verify the certificate chain in the case of ALLJOYN_ECDHE_ECDSA). The security module manages a key store to save authentication and encryption keys. The AllJoyn security platform makes use of the ECDHE (Elliptic Curve based Diffie–Hellman Key Exchange) [32] algorithm for generating session keys. Furthermore, it utilizes the Simple Authentication and Security Layer (SASL) [33] security framework for authentication and key exchange. It supports the following authentication mechanisms for app-to-app level authentication:

- ALLJOYN_SRP_KEYX: Secure Remote Password (SRP) key exchange.
- ALLJOYN_SRP_LOGON: Secure Remote Password (SRP) log on with username and password.
- ALLJOYN_ECDHE_NULL: Elliptic Curve Diffie–Hellman (ephemeral) key exchange with no authentication.
- ALLJOYN_ECDHE_PSK: Elliptic Curve Diffie–Hellman (ephemeral) key exchange authenticated with a pre-shared key (PSK).
- ALLJOYN_ECDHE_ECDSA: Elliptic Curve Diffie–Hellman (ephemeral) key exchange authenticated with an X.509 ECDSA certificate.

We applied the ALLJOYN_ECDHE_PSK mechanism to support authentication and secure communication between all components. Figure 3 shows an overview of the security architecture of our UAV-EMOR system. This ALLJOYN_ECDHE_PSK mechanism provides the pre-shared key-based authenticated ECDHE [32] to share the communication session encryption key between all components. In the setup phase, the control server generates a Pre-shared Secret Key (PSK) with each of the UAVs, the security manager's smartphone and the smart sensors. Each component shares a unique PSK with the control server. In order to generate a session encryption key, the pre-shared key-based authenticated ECDHE protocol should be performed between all components.

Since all components such as UAVs, smart sensors, etc., have already been authenticated, a PSK is found in the key store. Therefore, the UAV can directly execute the next step of generating a session key and/or a group key. Each component may generate a session key with the control server by using the ECDHE algorithm based on the pre-shared key authentication. Once the session key is generated, it is used along with the AES (Advanced Encryption Standard) [34] algorithm to encrypt communication messages. This is accomplished by installing the AllJoyn security platform on the control server and the security manager's smartphone, as well as on all of the UAVs and smart sensors. Figure 4 shows the high-level message flow of the end to end security communication between the control server and the UAV as an example. This flow consists of following four steps:

- Exchange Auth GUIDs (Authentication Group User Identifications): In this step, two peers such as the control server and the UAV exchange Auth GUIDs. The Auth GUID is used to see if the pre-shared key is present for that Auth GUID in the key store. If no pre-shared key is found, the two peers have not authenticated each other. Therefore, the two peers should execute the next app-to-app authentication step.
- App-to-app authentication: In this step, two peers perform the authentication mechanism supported by the AllJoyn platform. At the end of this step, the two peers have authenticated each other and share a common pre-shared key.

- Generate a session key: In this step, two peers generate a session key for encrypting communication messages between them. The session key is generated independently by both peers based on the pre-shared key. A group key is also generated when the first session key is generated.
- Exchange group keys: In this step, two peers exchange their own group keys with each other via an encrypted AllJoyn message. The group key is used to encrypt the session multicast and broadcast signals. At the end of this step, the two peers share group keys to decrypt secure broadcast signals received from each other.

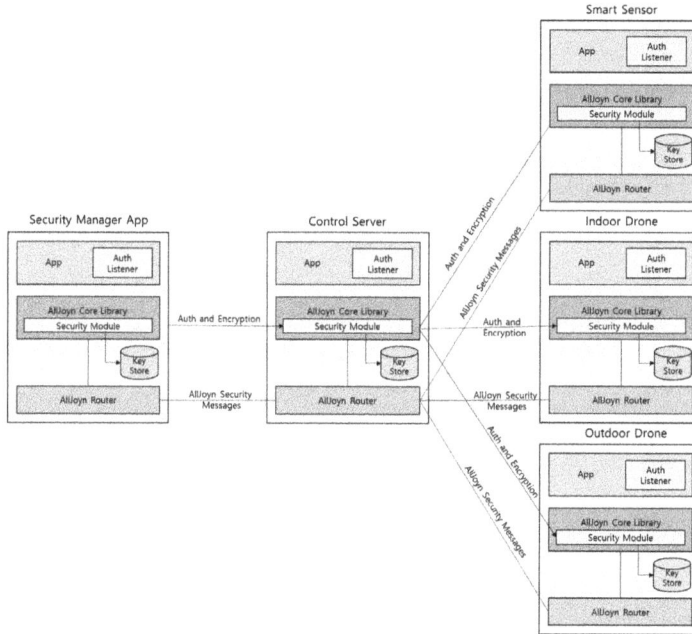

Figure 3. The security architecture of our system.

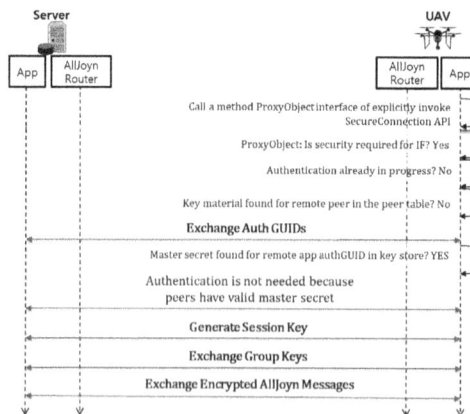

Figure 4. The end to end security flow of our UAV-EMOR system.

3.3. System Protocols

In our UAV-EMOR system, a smart sensor indicates one of four states: (1) "Active": indicates that the sensor is functioning correctly; (2) "Unsafe": indicates when the local temperature is beyond an established threshold; (3) "Fire": indicates when a fire has been detected; (4) "Inactive": indicates that the sensor is no longer functioning. A smart sensor automatically indicates these states, according to the settings chosen by the system administrator, as shown in Figure 5.

- Active: When a smart sensor is functioning correctly or if it detects that a fire has ended, it indicates this by sending an "Active" message to the server.
- Unsafe: If a smart sensor detects an abnormal temperature or the smoke detection system detects smoke , it is indicated with the message "Unsafe". The manager receives the abnormal temperature data or the smoke data and confirms whether there is an abnormality or not.
- Fire: If a manager confirms that there is a fire, the smart sensor enters the "Fire" state. When the control server receives the "Fire" message, it guides UAVs to observe the location of the smart sensor where the fire has been detected and the sprinkler is activated Furthermore, the control server sends an emergency message to everyone who is connected to the system. Via beacons, the smart sensors collect user positions and send them to the control server and manager.
- Inactive: This status indicates that a smart sensor is no longer functioning correctly due to error, loss of power, etc. In such a situation, the smart sensor notifies the control server of its "Inactive" state.

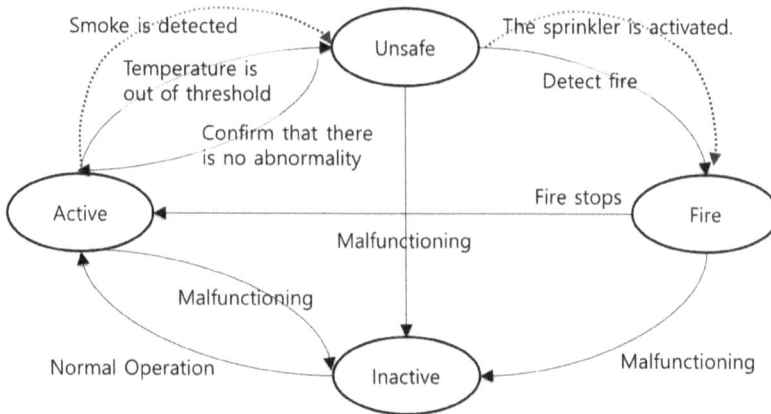

Figure 5. State diagram of a smart sensor.

Algorithm 1 shows the overall emergency management algorithm of the control server.

Algorithm 1: Management Algorithm of Control Server

```
State  :   Active, InActive, Unsafe, Fire
Tt :   Temperature value in time T
Tm :   Normal threshold of temperature
IPhoto :   Photo taken by an indoor drone
OPhoto :   Photo taken by an outdoor drone
CCTVimage :   image taken by CCTV
Monitoring Group :   Indoor Drone, Outdoor Drone, CCTV, Smart Sensor
```

1 Connect to Smart Sensor
2 **if** *Smart Sensor == InActive* **then**
3   Transmit "InActive" message to Security Manager
4 **end**
5 **while** *Smart Sensor == Active* **do**
6   Receive Tt from Smart Sensor
7   **if** *The temperature change == True* **then**
8   Save Tt to DB
9   Transmit Warning Message to Security Manager
10   **if** *Tt < Tm* **then**
11   Receive "Safe" message from Security Manager
12   Run Normal Operation of Day-to-Day Monitoring
13   **end**
14   **if** *Tt >= Tm* **then**
15   State = UnSafe
16   Broadcast "UnSafe" message to Security Manager and Monitoring Group
17   **while** *State == Unsafe of Fire* **do**
18   Run "Abnormal Response" operation
19   Run the fire detection system (or smoke detection system)
20   Update IPhoto in DB
21   Update OPhoto in DB
22   Update CCTVimage in DB
23   Receive Tt from Smart sensors
24   Receive State and location from Security Manager
25   **if** *State == Fire* **then**
26   Broadcast "Fire" message to Users and Monitoring Group
27   Run "Emergency Response" operation
28   Run the sprinkler system
29   **end**
30   **end**
31   **end**
32   **end**
33 **end**

3.3.1. Day to Day Monitoring

The temperature in each room is constantly monitored by a smart sensor attached near the entrance of the room. To maximize efficiency, a smart sensor only transmits a temperature value to the control server if the temperature value changes. The server stores the received temperature value and checks whether it is over a certain range of temperature. If the temperature value goes beyond a certain range, the server broadcasts the message "Unsafe" to an indoor drone located in the same floor, a security manager and all smart sensors in the building. The drone is then directed to the location of the smart sensor that sent the high temperature value to take a photo and video of the zone. Then, the drone sends the photo and video to the control server. The security manager can access the server and monitor these images to determine whether or not the situation is abnormal. If it is an abnormal situation, the "Abnormal Situation Response Process" is engaged. If the situation is safe, then the server broadcasts the message "Safe" to the indoor drone, smart sensors and a security manager's smartphone. In the absence of any "Unsafe" messages, all parts of the system continue with their regular daily monitoring duties.

3.3.2. Abnormal Situation Response

When a smart sensor detects that the temperature has climbed above a set threshold, the server immediately sends a message to the drones and security managers, informing them of the location where the abnormal situation has been detected, as well as the present temperature at that location. Upon receiving that message, the outdoor drone moves to the location under question, where it takes photos of the area and immediately sends them back to the server. Concurrently, inside the building on the floor where the abnormal situation was detected, an indoor drone moves to the location. Since the temperature sensor is located at the entrance way to the room, it cannot determine whether the abnormal temperature has occurred inside the room or outside in the hallway. Therefore, when the drone arrives at the location of the abnormal temperature change, it first photographs the hallway area just outside the room and sends that data to the server. Then, it moves into the room to take photographs there. To enter the room, the drone requires authentication, and it is outfitted with a beacon sensor for this purpose. Each drone has its own unique ID that is stored on the server. Using Bluetooth, the drone sends this ID to the smart sensor, which checks against the data in the server to determine whether or not the drone has permission to enter the room. To accomplish this authentication process, we make use of the AllJoyn security platform, which quickly certifies the permission status of the drone. Once this permission has been certified, the door automatically opens, and the drone moves into the room to take photos there. While all of this is happening, the security manager can monitor the whole process on a smartphone, viewing the photographs taken by the drones, as well as the video provided by the CCTV cameras. In addition, when a suspicious situation such as smoke detection occurs, the existing smoke-alert system is activated. The smart sensor transmits smoke data and the video and photo data of CCTV or UAV to the manager. Using this information, the manager can come to a final decision regarding the status of the situation, whether it is an emergency or not.

3.3.3. Emergency Response

In the event that a situation is determined to be an emergency, the building manager begins the emergency response process by sending a message to the control server. In order to prevent malicious interference from external sources or false reports, only managers authenticated by the Alljoyn security platform have the right to begin this process. Once an authenticated message has been received from an authorized source, the control server broadcasts emergency alert messages to everyone in the building and to all smart sensors, drones and other building personnel. In addition, the sprinkler installed in the building is operated by the server. This emergency response process is shown in Figure 6. CCTV and indoor and outdoor drones continue to monitor the situation both inside and outside the building, sending all visual data to the server. Furthermore, in an emergency situation, the smart sensors utilize the beacon signals in the users' smart ID cards in each room to determine how many users are indoors and transmit the number to the server.

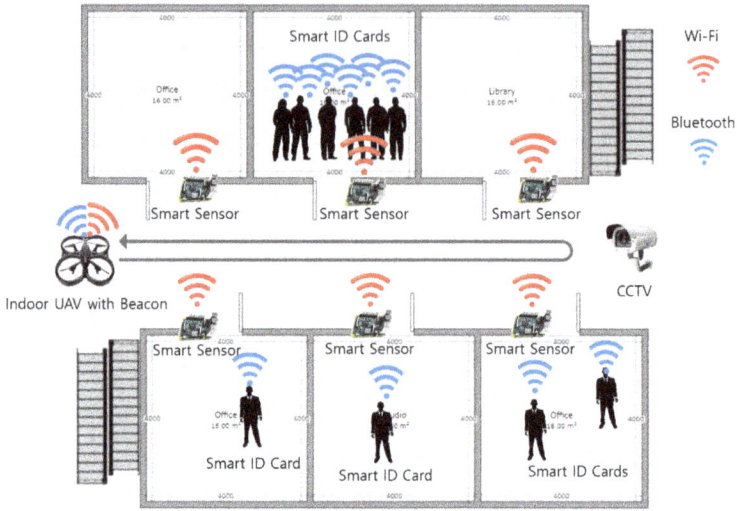

Figure 6. Emergency response process.

To make the above response possible, the system's smart sensors are constantly connected to Bluetooth and therefore always capable of receiving a Bluetooth signal from beacons that are attached to or embedded in users' smart ID cards. In this way, the smart sensors can precisely locate all building occupants at all times. The Bluetooth beacons in the ID cards transmit a constant signal to all the components of the system. This signal includes a time stamp, a TX power setting, an RSSI (dB), a major number and a minor number. The minor number is particularly important, as it is used to provide a unique identification number to each user's ID card, which is what allows the system to specify exactly where each user is located in the building. The location of an ID card, also called its proximity, can be determined by the strength of the signal as it is received by a sensor. Usually, the Friis model formula [35] is used to calculate this distance. Equation (1) is the Friis transmission equation [35].

$$P_r(\omega) = P_t(\omega)\frac{G_t(\omega)G_r(\omega)\lambda^2}{(4\pi r)^2} \tag{1}$$

where P_t and P_r are the transmitted and received powers, respectively, G_t and G_r are the antenna gains of the receiving and transmitting antenna, respectively, λ is the wavelength at the operating frequency and r is the distance between the transmitting and receiving antennas.

By this Friis model formula, the proximity value of beacons r is obtained as Equation (2):

$$r = \frac{\lambda}{4\pi}\sqrt{\frac{P_t}{P_r}G_tG_r} \tag{2}$$

However, the calculated proximity value r has an accuracy of 68% [25]. Therefore, in order to improve the accuracy, we apply the median of the proximity value r for a certain period of time to calculate the user's position.

Therefore, during an emergency situation, the control server informs the occupants in the building of the disaster and, using the proximity data received from their ID cards, determines the optimal emergency route for each person based on his/her current location in the building. In the event that a smart sensor becomes disabled in an emergency situation, by fire, for example, an occupant's proximity data can also be transmitted by the indoor/outdoor UAVs as they are also constantly receiving that information via Bluetooth.

4. Prototype of UAV-EMOR and Use Scenarios

In order to demonstrate the possibility of our system, we implement prototype for our UAV-EMOR system. In this section, we present the main prototype of our UAV-EMOR system and use scenario. Figure 7 shows the prototypes our UAV-EMOR system. Our prototype consists of commercial AR drones, temperature sensors, Raspberry Pi and a standard control server for implementing functionalities such as fire detection, smart notification and fire response services. Moreover, in order to secure communication within the building, the internal Wi-Fi network was protected by applying the AllJoyn security framework based on P2P, which is an open standard platform for the Internet of Things.

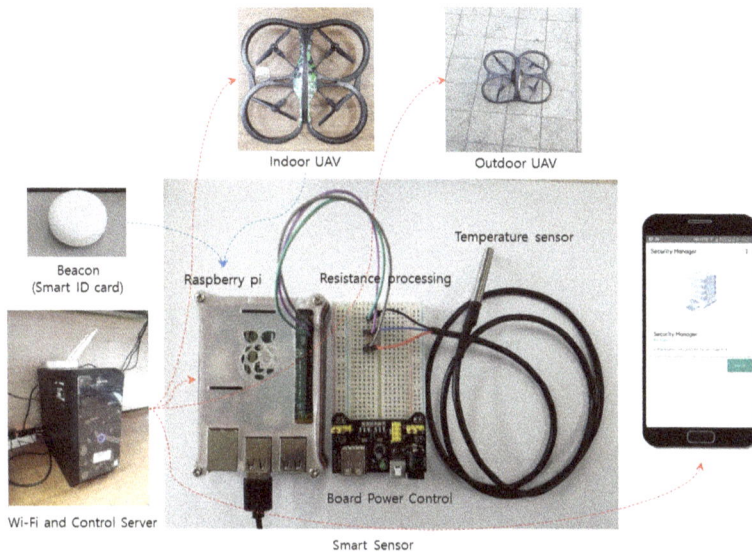

Figure 7. Prototype of UAV-EMOR system.

To construct a smart sensor, we combined a conventional DS18B20 sensor for temperature measurement with a single-board computer, the Raspberry Pi. This computer has a built-in input/output pin (General-purpose input/output, GPIO) for hardware control, which is useful for configuring the system to utilize the sensor. The DS18B20 temperature-sensitive drive circuit responds to between the −55 °C and +125 °C range and detects temperatures within −0.5 °C and +0.5 °C. We used the conventional temperature sensor to implement the prototype, but the temperature sensor could be changed to a high-end sensor for real-world use. In addition, on each floor, there are multiple smart sensors with temperature sensors, so the control server can use data from all over the floor to determine the actual temperature. Therefore, the control server can avoid false alarms. If a room sensor registers a high temperature, the control server sends an indoor UAV to the room to check the situation and that will confirm the temperature.

Temperature-sensitive sensors are used for early fire detection. The sensor module for sensing the initial temperature of a fire is connected via the Raspberry Pi GPIO, and everything is linked via the control server. In addition, source coding is performed for the smart sensor using Razbian (a Linux-based OS), which automates data exchange between the control server, sensors and drones and can also automate certain actions, as well. For example, the control server commands the drones to monitor the inside and outside of the building and sends the monitoring results to the security manager so that any abnormal situations can be dealt with promptly.

Figure 8 shows the execution screen of the security manager app, which is used by a security manager to monitor a building. We use a biometric-based authentication method to implement user authentication of the Security Manager app. Therefore, only users who are specifically granted permission can access and control the app. The security manager's fingerprint is registered on the smartphone, and the only way the app can be accessed and the building monitored is via fingerprint identification. Using this app, the security manager can always check and monitor the inside or outside of the building. Figure 8b shows the user confirmation message, which appears on the screen upon the verification of the security manager's fingerprint. At the same time, the security manger app generates a PSK-based on the AllJoyn security platform to establish a secure connection with the control server. Our smart building monitoring prototype considers three different statuses: 'Active', 'Unsafe' and 'Fire'. In a normal situation, the status of a smart sensor is 'Active', which means that conditions are stable. However, if a smart sensor senses a temperature that is outside of the normal range, the status of the sensor is changed to 'Unsafe'. If the sensor identifies a fire in the building, it changes its status to 'Fire'.

The Security Manager app receives these status updates on the building through a secure connection between the smart sensors and the control server by using the AllJoyn security platform. When the security manager receives notice of an emergency situation on the Security Manager app, he/she can check the extent of the danger and the size of any fires through real-time video and photos taken by indoor and outdoor drones. As shown in Figure 8d, if an abnormal temperature is detected, the control server registers the 'Unsafe' status, then transmits this status information to all the smart sensors in the building, to the indoor and outdoor drones and to the security manger smartphone. Once the security manager checks the detailed information of a fire situation, he/she can declare an emergency command by pressing the 'Emergency' button, which will set in motion the fire emergency response procedure.

Figure 8. The interface of the building Security Manager app.(**a**) first page of security manager app , (**b**) fingerprint authentication process, (**c**) security manager app in a normal situation, (**d**) an abnormal temperature is detected, (**e**) security manager app in abnormal situation, (**f**) rescue and emergency evacuation mode of security manager app

5. Performance Evaluation

In order to evaluate the performance of our UAV-EMOR system protocol, we implemented our day-to-day monitoring protocol on a commercially available augmented reality drone (AR.Drone) 2.0 for both indoor and outdoor monitoring. The AR.Drone 2.0 is a quad-copter equipped with front and ground cameras and Wi-Fi (2.4 GHz). It has a 1-GHz 32-bit ARM cortex A8 CPU and a 1 GB RAM. The main board of the AR.Drone 2.0 runs the BusyBox-based GNU/Linux distribution with the 2.6.32 kernel. We also used a personal computer with Linux OS as the control server and Raspberry

Pi as a smart sensor. Figure 9 shows the experimental flow of the day to day monitoring protocol. In this experiment, we mainly evaluated the performance of the indoor/outdoor UAV and smart sensors during execution of the protocol, as they are resource-constrained devices. We measured the completion time of the protocol for day-to-day monitoring to evaluate the communication overhead according to flight altitude. For evaluating the communication overhead, we placed the server on the ground and periodically increased the drone's flying altitude, measuring the communication time between the UAV and the server at each altitude. The parameters used in the simulation are shown in Table 2. We measured the communication time between the UAV and the server under two different circumstances: (1) communication secured by an encryption algorithm, AES (Advanced Encryption Standard) on the AllJoyn Security platform, and (2) unsecured communication without using the encryption algorithm on the AllJoyn security platform. The communication time was measured from the time that a ping was sent from the server to the time that the drone's response ping was received back at the server. Figure 10 shows the communication overhead according to the distance between the UAV and the server. In Figure 10, we can see that, generally, the communication overhead increases as the distance between the UAV and the server increases. However, it can also be seen that the communication overhead is not significantly increased by the use of an encryption algorithm for a secure communication, as represented on the graph by the orange line.

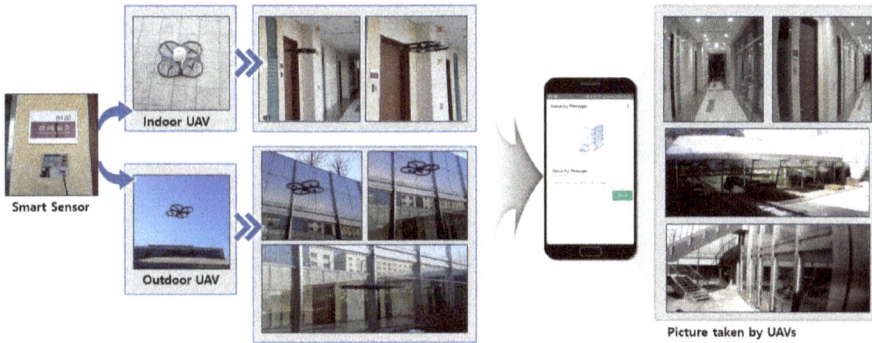

Figure 9. The experimental flow of the day-to-day monitoring process.

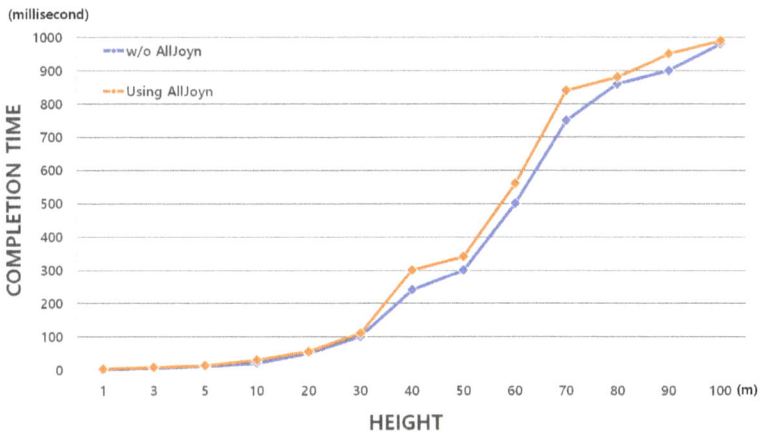

Figure 10. The communication and computation overhead according to flight altitude.

Moreover, in order to evaluate the computational overhead of the AR.Drone 2.0 (which we used as our indoor/outdoor UAV), we measured the encryption/decryption time of messages or images to complete the monitoring mission of UAV. Table 2 shows the time required to execute encryption and decryption algorithms for completion of a mission according to the size of the messages or images. When the size of the text message for transmission was 128 bytes, the UAV with our protocol took 1.1 ms and 0.9 ms to complete the protocol with encryption and decryption, respectively. Even for a large text size such as 10,240 byte, the UAV with our protocol encrypted and decrypted the message relatively quickly. It took only 180 ms to perform the encryption or decryption operation. Generally, images take longer to encrypt and decrypt than text messages do. However, our system is still relatively fast in this regard. It took the UAV 15 ms to encrypt/decrypt 473 bytes (one or two images). For the secure transmission of 10,703 bytes (about six or seven images), the UAV took 306 ms.

Furthermore, in order to evaluate the computational overhead of the smart sensor, we measured the cryptographic operation time such as the encryption or decryption time of messages to securely communicate with the UAV. Table 3 shows the computation time required to execute encryption and decryption algorithms for secure communication to the size of the messages. When the size of the transmitted text message was 256 bytes, the smart sensor took 1.79 ms and 1.77 ms to complete the protocol with encryption and decryption, respectively. Even for a large text size such as 10,240 byte, the smart sensor encrypted and decrypted the message relatively quickly. It took only 164 ms to perform the encryption or decryption operation.

Table 2. Computation time on UAVs.

	Text					Image				
Message Size (Byte)	32	128	512	2048	10,240	473	824	2453	8530	10,703
Encryption (ms)	0.6	1.1	5	40	180	15	36.5	106	220.5	306
Decryption (ms)	0.6	0.9	5	40	180	15	36.5	106	220.5	306

Table 3. Computation time on a smart sensor.

	Text									
Message Size (Byte)	16	32	64	128	256	512	1024	2048	4096	10,240
Encryption (ms)	0.31	0.5	0.7	0.86	1.79	4.1	13.4	36	71	164
Decryption (ms)	0.31	0.49	0.69	0.86	1.77	4	13.3	36	70	164

6. Discussion

6.1. Position Accuracy

The distance measurement using the beacon is based on the received Bluetooth signal strength. However, if signal interference due to surrounding obstacles occurs, the measured distance data may be inaccurate. Therefore, the proximity value of a beacon calculated using the Friis model formula has generally low accuracy such as 68%, due to frequent signal interference [24,36]. To improve the accuracy, Choi et al. [37] used the average value after storing the distance calculation results for 30 s. However, the method of taking the average value shows higher accuracy only when the experiment is performed by constructing the optimal environment without signal interference. If a large distance error is transmitted due to Bluetooth signal interference, the large error value, i.e., the outlier, affects the average value.

In order to reduce the effect of outliers and improve the position accuracy measured by the beacon, we propose to apply the median of the proximity value r for a certain period of time to measure the occupant's precise location. In order to evaluate the accuracy of our system's proximity data, we performed the experiment using raw values, the average of all values and median values in a variety of emergency scenarios. For our experiments, it is assumed that the building is fully outfitted with smart sensors and that each building occupant carries a beacon-embedded smart ID card that is used to locate the occupant during an emergency.

Figure 11 shows four experimental conditions. We attached smart sensors to the front of the door in each room, and the installation radius between sensors was 3–5 m. Each experiment participant carried a beacon and was instructed to walk in a radius of 3 m, 4 m then 5 m away from a smart sensor. Figure 11a–c indicates that a user with a beacon is moving or standing at a distance of 3 m, 4 m and 5 m from the smart sensor, respectively. Figure 11d shows the experimental environment in which the participants were allowed to either remain in one location or to freely change their location while remaining within the 3–5 m radius from the smart sensor.

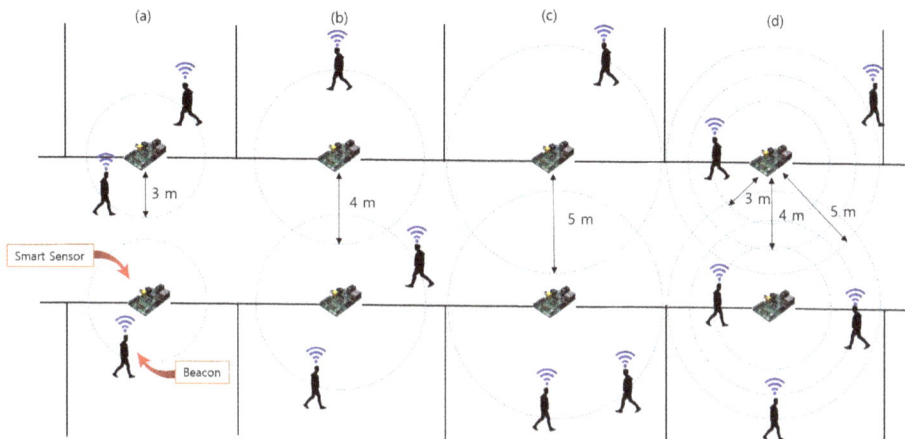

Figure 11. The experimental setting used for localization accuracy evaluation. Participant walks in a (**a**) 3 m radius, (**b**) 4 m radius, (**c**) 5 m radius, (**d**) 3–5 m radius.

Figure 12 shows the results of our position accuracy calculations. To evaluate position accuracy, we used the following three methods: the raw method [24], the average method [37] and the median method. The raw method accepts the proximity value calculated as the distance of the beacon from a single smart sensor. The average method and the median method both refer to methods in which the average or median of multiple proximities is obtained by calculating the location of the beacon over a set period of time. In order to evaluate the accuracy of these methods, we tested different periods of time for collecting the proximity data. These set periods of proximity data collection were set to 1 s, 5 s, 10 s, 15 s, 20 s and 30 s.

One challenge to calculating precise locations with beacons is that smart sensors are affected by where a user holds the beacon, i.e., on the upper or lower body. This is partly responsible for the sometimes wide variance in proximity data that smart sensors receive from beacons. In our study, the goal of position recognition is to locate building occupants and inform them of appropriate evacuation routes in an emergency situation. For this purpose, we considered a small location range of up to 1 m acceptable. Accordingly, we set a tolerance threshold value of 50 cm on the data received from the smart sensor. That is, if the smart sensor calculates the position of the user as being 3 m away, we regard the user's correct position as a range from 2.5–3.5 m.

Experimental results show that the accuracy of proximity data provided by systems that use beacons is significantly low (33.7%, 48.7% and 30%). This is because beacons broadcast inconsistent proximity data. Most of the data broadcasts may be correct, but on occasion, they may also broadcast a highly incorrect distance. For example, if a user is 3 m away from a smart sensor, the raw method may read an incorrect broadcast from a beacon and calculate that the user is at a distance of 12 m. Likewise, the inaccuracy of beacon broadcasts also adversely affects the data calculated using the average method. At distances of 3–4 m, the average method shows decreasing accuracy as time for collecting the proximity data increases; see Figure 11a,b. For these reasons, neither the raw method, nor the average method are suitable for the calculation of precise location using beacons. In contrast to those methods, the median method provided a significantly high level of accuracy. This accuracy was also sustained over long periods of collection time for the proximity data. At all distances, 3 m, 4 m and 5 m, the proximity data gathered in a 30-s period and calculated using the median method showed an accuracy of more than 90%. Thus, in our UAV-EMOR system, we program smart sensors to collect proximity data from a beacon for 30-s periods, then to calculate the median proximity value of that data and send it to the server as the user's location.

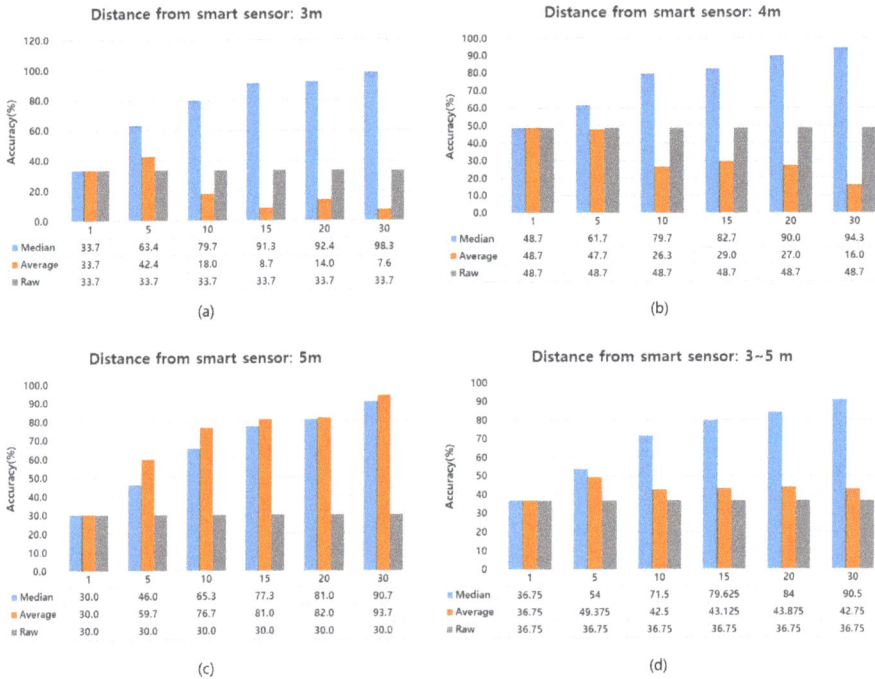

Distance from smart sensor: 3m

	1	5	10	15	20	30
Median	33.7	63.4	79.7	91.3	92.4	98.3
Average	33.7	42.4	18.0	8.7	14.0	7.6
Raw	33.7	33.7	33.7	33.7	33.7	33.7

(a)

Distance from smart sensor: 4m

	1	5	10	15	20	30
Median	48.7	61.7	79.7	82.7	90.0	94.3
Average	48.7	47.7	26.3	29.0	27.0	16.0
Raw	48.7	48.7	48.7	48.7	48.7	48.7

(b)

Distance from smart sensor: 5m

	1	5	10	15	20	30
Median	30.0	46.0	65.3	77.3	81.0	90.7
Average	30.0	59.7	76.7	81.0	82.0	93.7
Raw	30.0	30.0	30.0	30.0	30.0	30.0

(c)

Distance from smart sensor: 3~5 m

	1	5	10	15	20	30
Median	36.75	54	71.5	79.625	84	90.5
Average	36.75	49.375	42.5	43.125	43.875	42.75
Raw	36.75	36.75	36.75	36.75	36.75	36.75

(d)

Figure 12. Accuracy of proximity calculations. The interval between smart sensor and beacon is (**a**) 3 m, (**b**) 4 m, (**c**) 5 m, (**d**) 3–5 m.

6.2. Privacy Considerations

Recently, many studies related to location privacy have gained much attention, because users may want to use location-based services without revealing their location. Location privacy in wireless sensor networks including IoT environments may be classified into content privacy and contextual (e.g., the traffic) privacy. Several approaches for protecting content-oriented privacy have been designed in location tracking systems, which can decide the users' positions for location-based services [38–40]. Spreitzer et al. [38] utilize a location broker residing at the middleware layer to hide users' positions. Gruteser et al. [39] agitate the sensed location information by using the k-anonymity criterion. Al-Muhtadi et al. [40] presented the location privacy protection method through a mist router and a handle-based virtual circuit routing protocol. In addition, to protect the receiver's (i.e., the sink) location, privacy is important, because the receiver collects data from all sensors. Deng et al. [41] first introduced the issue of how to hide the location of the sink (e.g., the base station) and presented techniques of multi-path routing and fake message injection. Jian et al. [42] proposed a location-privacy routing protocol incorporating fake packet injection.

In our UAV-EMOR system, the beacon sensors are used to implement rescue and evacuation functionality. Using Bluetooth technology, such beacons constantly transmit the unique ID and real-time location of the occupants to all smart sensors and indoor-UAVs. Since the building management system can guide the occupants to the evacuation route based on these real-time location data and unique ID, it is useful for evacuating occupants in an emergency situation. The control server of the UAV-EMOR system allows smart sensors and indoor UAVs to enable Bluetooth and receive beacon signals from users only when the emergency situation happens. The system may be connected at all times, but the control server does not collect or store the users' location information at normal times because it has no need to ask for the information. Still, the use of location information can introduce potentially unresolved privacy concerns [43]. Therefore, we need to consider the location privacy issue; simply put, we can make the security policy for our UAV-EMOR to update the beacon ID periodically. To more actively protect the location privacy, we can apply the above mentioned studies about location privacy protection in wireless sensor networks to the UAV-EMOR system.

7. Conclusions

In this paper, we have presented a proof of concept prototype of a monitoring and emergency response method that we call the UAV-EMOR system that is intended to address the weaknesses common in such smart systems, specifically limited monitoring range, inaccuracy of proximity data and data security. In order to enhance the real-time monitoring range, we present an approach that incorporates indoor/outdoor UAVs in coordination with smart sensors; to improve the accuracy of data received from beacons, we introduce the median method of proximity data calculation; and to address security issues, we use the AllJoyn security platform to provide strong encryption for all communications between all components used in the system.

Acknowledgments: This work was supported by the research fund of Hanyang University(HY-2017-N).

Author Contributions: Seung-Hyun Seo, Jinseok Song and Jung-In Choi designed a monitoring and emergency response algorithm for buildings by utilizing beacons and Unmanned Aerial Vehicles (UAVs), called UAV-EMOR system. All authors analyzed the related works for warning systems that incorporate UAVs and discussed the weaknesses and vulnerabilities of the previous works. We then proposed the advanced algorithm for monitoring and emergency response system. Jinseok Song mainly contributed to implementing a proof of concept prototype of UAV-EMOR system and a security manager's smartphone app. He then conducted experiments to evaluate the performance of UAV-EMOR system and analyzed the communication and computation overhead according to flight altitude. Jung-In Choi mainly contributed to designing the beacon-aided rescue and building evacuation mechanism for our UAV-EMOR system and conducted experiments of position accuracy to measure the distance using the beacon. She also contributed to surveying related works in the field of emergency alert and response systems and making an initial version of this manuscript. Seung-Hyun Seo led all the plans for this research work, including the implementation and the overall experiments as an advisor. She wrote this manuscript in its entirety with responsibility as the lead author.

Conflicts of Interest: The authors declare no conflict of interest.

References

1. Osello, A.; Acquaviva, A.; Aghemo, C.; Blaso, L.; Dalmasso, D.; Erba, D.; Fracastoro, G.; Dondre, D.; Jahn, M.; Macii, E.; et al. Energy saving in existing buildings by an intelligent use of interoperable ICTs. *Energy Effic.* **2013**, *6*, 707–723.

2. Aedo, I.; Yu, S.; Diaz, P.; Acuna, P.; Onorati, T. Personalized alert notifications and evacuation routes in indoor environments. *Sensors* **2012**, *12*, 7804–7827.

3. Liu, J.W.S.; Lin, F.T.; Chu, E.T.H.; Zhong, J.L. Intelligent Indoor Emergency Evacuation Systems. In Proceedings of the Future Technologies Conference (FTC), San Francisco, CA, USA, 6–7 December 2016; pp. 600–609.

4. Wu, Z.Y.; Lv, W.; Yu, K. A Framework of Intelligent Evacuation Guidance System for Large Building. In Proceedings of the 5th International Conference on Civil, Architectural and Hydraulic Engineering (ICCAHE 2016), Zhuhai, China, 30–31 July 2016; pp. 695–701.

5. Zeng, Y.; Sreenan, C.J.; Sitanayah, L.; Xiong, N.; Park, J.H.; Zheng, G. An emergency-adaptive routing scheme for wireless sensor networks for building fire hazard monitoring. *Sensors* **2011**, *11*, 2899–2919.

6. Lin, C.Y.; Chu, E.T.H.; Ku, L.W.; Liu, J.W. Active disaster response system for a smart building. *Sensors* **2014**, *14*, 17451–17470.

7. Choi, K.; Lee, I. A UAV based close-range rapid aerial monitoring system for emergency responses. *Int. Arch. Photogramm. Remote Sens. Spat. Inf. Sci.* **2011**, *38*, 247–252.

8. Erdelj, M.; Natalizio, E. UAV-assisted disaster management: Applications and open issues. In Proceedings of the 2016 International Conference on Computing, Networking and Communications (ICNC), Kauai, Hawaii, USA, 15–18 February 2016; pp. 1–5.

9. Chen, D.; Liu, Z.; Wang, L.; Dou, M.; Chen, J.; Li, H. Natural disaster monitoring with wireless sensor networks: A case study of data intensive applications upon low-cost scalable systems. *Mob. Netw. Appl.* **2013**, *18*, 651–663.

10. Frigerio, S.; Schenato, L.; Bossi, G.; Cavalli, M.; Mantovani, M.; Marcato, G.; Pasuto, A. A web-based platform for automatic and continuous landslide monitoring: The rotolon (Eastern Italian Alps) case study. *Comput. Geosci.* **2014**, *63*, 96–105.

11. Ueyama, J.; Freitas, H.; Faiçal, B.S.; Geraldo Filho, P.R.; Fini, P.; Pessin, G.; Gomes, P.H.; Villas, L.A. Exploiting the use of unmanned aerial vehicles to provide resilience in wireless sensor networks. *IEEE Commun. Mag.* **2014**, *52*, 81–87.

12. Marinho, M.A.; De Freitas, E.P.; da Costa, J.P.C.L.; De Almeida, A.L.F.; De Sousa, R.T. Using cooperative MIMO techniques and UAV relay networks to support connectivity in sparse Wireless Sensor Networks. In Proceedings of the 2013 International Conference on Computing, Management and Telecommunications (ComManTel), Ho Chi Minh City, Vietnam, 21–24 January 2013; pp. 49–54.

13. Morgenthaler, S.; Braun, T.; Zhao, Z.; Staub, T.; Anwander, M. UAVNet: A mobile wireless mesh network using unmanned aerial vehicles. In Proceedings of the 2012 IEEE Globecom Workshops (GC Wkshps), Anaheim, CA, USA, 3–7 December 2012; pp. 1603–1608.

14. Tuna, G.; Mumcu, T.V.; Gulez, K. Design strategies of unmanned aerial vehicle-aided communication for disaster recovery. In Proceeding of the 9th International Conference on High Capacity Optical Networks and Enabling Technologies (HONET), Istanbul, Turkey, 12–14 December 2012; pp. 115–119.

15. Waharte, S.; Trigoni, N. Supporting search and rescue operations with UAVs. In Proceeding of the 2010 International Conference on Emerging Security Technologies (EST), Canterbury, UK, 6–7 September 2010; pp. 142–147.

16. Tomic, T.; Schmid, K.; Lutz, P.; Domel, A.; Kassecker, M.; Mair, E.; Grixa, I.L.; Ruess, F.; Suppa, M.; Burschka, D. Toward a fully autonomous UAV: Research platform for indoor and outdoor urban search and rescue. *IEEE Robot. Autom. Mag.* **2012**, *19*, 46–56.

17. Erdos, D.; Erdos, A.; Watkins, S.E. An experimental UAV system for search and rescue challenge. *IEEE Aerosp. Electron. Syst. Mag.* **2013**, *28*, 32–37.

18. Allseen Alliance, AllJoyn Security 2.0 Feature High-Level Design. 2016. Available online: https://allseenalliance.org/ (accessed on 6 December 2016).

19. Papadakis, P. Terrain traversability analysis methods for unmanned ground vehicles: A survey. *Eng. Appl. Artif. Intell.* **2013**, *26*, 1373–1385.

20. Yang, Z.; Zhang, P.; Chen, L. RFID-enabled indoor positioning method for a real-time manufacturing execution system using OS-ELM. *Neurocomputing* **2016**, *174*, 121–133.

21. Gualda, D.; Ureña, J.; García, J.C.; Lindo, A. Locally-referenced ultrasonic–LPS for localization and navigation. *Sensors* **2014**, *14*, 21750–21769.

22. Ozdenizci, B.; Coskun, V.; Ok, K. NFC internal: An indoor navigation system. *Sensors* **2015**, *15*, 7571–7595.

23. Cheng, R.S.; Hong, W.J.; Wang, J.S.; Lin, K.W. Seamless guidance system combining GPS, BLE beacon, and NFC technologies. *Mob. Inf. Syst.* **2016**, *2016*, 5032365.

24. Reco Beacon Homepage. Available online: http://reco2.me/ (accessed on 27 January 2017).

25. iBeacon for Developers. Available online: http://developer.apple.com/ibeacon/(accessed on 27 January 2017).

26. Liu, J.W.S.; Chen, L.J.; Su, J.; Li, C.C.; Chu, E.T.H. BeDIPS: A building environment data based indoor positioning system. In Proceeding of the International Conference on Internet of Things (iThings 2015), Sydney, Australia, 11–13 December 2015.

27. Murphy, R.R.; Stover, S. Rescue robots for mudslides: A descriptive study of the 2005 La Conchita mudslide response. *J. Field Robot.* **2008**, *25*, 3–16.

28. Sardouk, A.; Mansouri, M.; Merghem-Boulahia, L.; Gaiti, D.; Rahim-Amoud, R. Multi-agent system based wireless sensor network for crisis management. In Proceedings of the Global Telecommunications Conference (GLOBECOM 2010), Miami, FL, USA, 6–10 December 2010; Volume 2010, pp. 1–6.

29. Dalmasso, I.; Galletti, I.; Giuliano, R.; Mazzenga, F. WiMAX networks for emergency management based on UAVs. In Proceedings of the 2012 IEEE First AESS European Conference on Satellite Telecommunications (ESTEL), Rome, Italy, 2–5 October 2012; pp. 1–6.

30. Ye, J.; Hassan, T.M.; Carter, C.D.; Zarli, A. *ICT for Energy Efficiency: The Case for Smart Buildings*; Department of Civil and Building Engineering, Loughborough University: Loughborough, UK, 2008.

31. Weng, T.; Agarwal, Y. From buildings to smart buildings—Sensing and actuation to improve energy efficiency. *IEEE Des. Test Comput.* **2012**, *29*, 36–44.

32. Certicom Research, Standards for Efficient Cryptography, SEC 1: Elliptic Curve Cryptography, Version 2.0, 21 May 2009. Available online: http://www.secg.org/sec1-v2.pdf (accessed on 5 April 2017).

33. RFC 4422, Simple Authentication and Security Layer (SASL), June 2006. Available online: https://tools.ietf.org/html/rfc4422 (accessed on 28 March 2017).

34. National Institute of Standards and Technology (NIST). Federal Information Processing Standards Publication 197; Announcing the Advanced Encription Standard (AES). Available online: http://nvlpubs.nist.gov/nistpubs/FIPS/NIST.FIPS.197.pdf (accessed on 1 August 2017).

35. Constantine, A.B. *Antenna Theory: Analysis and Design*; John Wiley and Sons: New York, NY, USA, 1997.

36. Apple Core Location. Available online: https://developer.apple.com/documentation/corelocation (accessed on 28 December 2016).

37. Choi, J.I.; Yong, H.S. Real-time intragroup familiarity analysis model using beacon based on proximity. In Proceedings of the IEEE Annual Ubiquitous Computing, Electronics and Mobile Communication Conference (UEMCON), New York, NY, USA, 20–22 October 2016; pp. 1–5.

38. Spreitzer, M.; Theimer, M. Providing Location Information in a Ubiquitous Computing Environment. In Proceedings of the 14th ACM Symposium on Operating System Principles, Asheville, NC, USA, 5–8 December 1994; Volume 27, pp. 270–283.

39. Gruteser, M.; Schelle, G.; Jain, A.; Han, R.; Grunwald, D. Privacy-Aware Location Sensor Networks. In Proceedings of the 9th USENIX Workshop on Hot Topics in Operating Systems(HotOS IX), Lihue, HI, USA, 18–21 May 2003; Volume 3, pp. 163–168.

40. Al-Muhtadi, J.; Campbell, R.; Kapadia, A.; Mickunas, M.D.; Yi, S. Routing through the mist: Privacy preserving communication in ubiquitous computing environments. In Proceedings of the 22nd International Conference on Distributed Computing Systems, Vienna, Austria, 2–5 July 2002; pp. 74–83.

41. Deng, J.; Han, R.; Mishra, S. Countermeasures against traffic analysis attacks in wireless sensor networks. In Proceedings of the First International Conference on Security and Privacy for Emerging Areas in Communications Networks, SecureComm 2005, Athens, Greece, 5–9 September 2005; pp. 113–126.

42. Jian, Y.; Chen, S.; Zhang, Z.; Zhang, L. Protecting receiver-location privacy in wireless sensor networks. In Proceedings of the INFOCOM 2007, 26th IEEE International Conference on Computer Communications, Barcelona, Spain, 6–12 May 2007; pp. 1955–1963.

43. Beresford, A.R.; Stajano, F. Location privacy in pervasive computing. *IEEE Pervasive Comput.* **2003**, *2*, 46–55.

sensors

MDPI

Article

Calculation and Identification of the Aerodynamic Parameters for Small-Scaled Fixed-Wing UAVs

Jieliang Shen [1], Yan Su [1,*], Qing Liang [2] and Xinhua Zhu [1,*]

[1] School of Mechanical Engineering, Nanjing University of Science and Technology, Nanjing 210094, China; perfect.sjlchg2008@163.com
[2] School of Computer Technologies and Control, ITMO University, St. Petersburg 197101, Russia; liangqing1688@gmail.com
* Correspondence: suyan@njust.edu.cn (Y.S.); zhuxinhua@njust.edu.cn (X.Z.)

Received: 9 October 2017; Accepted: 11 January 2018; Published: 13 January 2018

Abstract: The establishment of the Aircraft Dynamic Model (ADM) constitutes the prerequisite for the design of the navigation and control system, but the aerodynamic parameters in the model could not be readily obtained especially for small-scaled fixed-wing UAVs. In this paper, the procedure of computing the aerodynamic parameters is developed. All the longitudinal and lateral aerodynamic derivatives are firstly calculated through semi-empirical method based on the aerodynamics, rather than the wind tunnel tests or fluid dynamics software analysis. Secondly, the residuals of each derivative are proposed to be identified or estimated further via Extended Kalman Filter (EKF), with the observations of the attitude and velocity from the airborne integrated navigation system. Meanwhile, the observability of the targeted parameters is analyzed and strengthened through multiple maneuvers. Based on a small-scaled fixed-wing aircraft driven by propeller, the airborne sensors are chosen and the model of the actuators are constructed. Then, real flight tests are implemented to verify the calculation and identification process. Test results tell the rationality of the semi-empirical method and show the improvement of accuracy of ADM after the compensation of the parameters.

Keywords: aerodynamic parameters; semi-empirical aerodynamic coefficient modeling; parameters identification; EKF; real flight tests

---

1. Introduction

In contrast to the large-scaled high-altitude long-term Unmanned Aerial Vehicles (UAVs) for remote surveillance and combat, the small-scaled fixed-wing UAVs, like "Raven" RQ-11 and Sand Hawk of the US, or "Rainbow-802" of China, have unique superiority in civilian and military fields. With the light weight, small volume and simple take-off, small UAVs could execute tasks like the close-range monitoring in a covert and flexible way [1]. The establishment of Aircraft Dynamic Model (ADM) should be accomplished first before any task designing. It contains the computation of aircrafts' structural parameters, position of the center of gravity, moment of inertia, aerodynamic parameters and the modeling of actuators.

For small-scaled fixed-wing UAVs, the structural parameters, like the wing span l, mean aerodynamic chord $\bar{c}$ and the reference area S, could be measured precisely. The thrust, driven by the propeller, can be determined by the diameter of the propeller D, rotation rate n_P and airspeed V_a [2]. However, the aerodynamic parameters, covering stability derivatives and the control derivatives, are relatively hard to get, which is crucial in calculating the force and moment. Stability derivatives involve partial derivatives with respect to states and control derivatives involve partial derivatives with respect to control inputs. The stability derivatives could be further divided into static stability derivatives for derivatives associated with air-relative velocity quantities and the dynamic stability derivatives for derivatives associated with angular rates and unsteady aerodynamics [3]. Generally, the aerodynamic parameters are analyzed through wind tunnel tests or CFD software, which are not appropriate for small ones. Wind tunnel test is high-cost and time-consuming, most of all, the results are not accurate in the flight condition of low Reynolds number. And CFD software analysis usually has heavy work on the 3D model meshing, and the gap effect as well as the frictional drag is often neglected [4,5]. DATCOM, introduced by the US Air Force, is another choice for the parameter calculation. On the basis of the huge flight database, all the aerodynamic coefficients could be obtained when the required parameters are imported. But DATCOM does have drawbacks. The input parameters are quite complex and the analysis will somehow be affected by the computational interval of the angle-of-attack α, resulting in limited accuracy for small UAVs [6,7].

Considering the special aerodynamic configuration and flight condition of the small UAVs, the semi-empirical method based on the aerodynamic theory is proposed here, by means of which the longitudinal and lateral derivatives could be calculated step by step according to the semi-empirical formulas or diagrams. Semi-empirical method was also mentioned in Arifianto's research [8]. The reason why it is called "Semi-empirical" is that the formulas and diagrams used are actually the combination of the fundamental aerodynamics and empirical summary from abundant flight tests. For instance, the diagram for calculate the lift curve slope of wing C_W^α is fitted with the flight data according to the theoretical relationship with the geometrical shape of the wing and the Mach number. Monographs [9–14] introduce the basic aerodynamic theory of different aircrafts flying under different circumstances, and describe the detailed procedure of calculating all the aerodynamic derivatives. Specifically, the aerodynamic analysis of small-scaled aircrafts is often simplified for quite small Mach number, plain aerodynamic structure and small angle-of-attack assumption. So, this kind semi-empirical method could rationally calculate the initial parameters in a small amount of computation with low cost and it will be adopted in the paper. Although the semi-empirical analysis is an effectively feasible method, the accuracy of the parameters is sometimes suspicious due to the structure simplification and the changing of the flight condition. After the theoretical calculation, the aerodynamic parameters are nearly all constants, which is understandable in the small Mach number condition and small angle-of-attack assumption. It means that the method owns poor real-time performance. Therefore, the identification process is carried out to further compensate the error of aerodynamic parameters with the real-time flight data.

Monographs [3,15] synthesize the basic issues of system identification for fixed-wing aircrafts or rotorcrafts, including aircraft's mathematical model, estimation theories, classification of identification methods as well as engineering practices, like experiment design, data compatibility check and data analysis. Generally, there exists many approaches to identify or estimate the error of the aerodynamic parameters, and it could be classified into off-line and on-line way. Off-line way consists of the formula error method and the output error method, the former is based on the Least Square (LS) while the latter based on the Maximum Likelihood Method(MLE). And off-line way could be operated either in the the the time domain or the frequency domain [16,17]. Neural network is also chosen as the structure to describe or identify the dynamic characteristics of aircraft [18,19]. All the methods above own a large amount of computation and call for the whole accurate measurements corresponding to the system states. Morelli [20] focuses on the aerodynamic model identification of the combat aircraft like F-16 using MLE in the frequency domain, and the measurements used are the 6 outputs of the IMU and the aerodynamic states like V_a, α and β, which requires high-precision IMUs and expensive sensors for α, β detection. But it is quite an expectation for small-loaded aircrafts. Dorobantu's research [21,22] exhibit the system identification for small, low-cost, fixed-wing UAVs, which usually carry lightweight devices, like the MEMS IMUs, but the accuracy of the measurement is usually hard to guarantee.

On-line identification, or the real-time estimation, could handle the problems above. Based on the linear or non-linear Kalman Filter, errors of the aerodynamic parameters are expanded into the motion states, and then estimated to some extent in a real time. Garcia [23] introduces that EKF method breaks the limitation of the linearity of the system and the estimation results could be variant.

But, the inaccuracy of the statistical characteristics of noise and the observability issue affect the estimating results. According to [24], maneuvering motion, like the changing of the acceleration and angular velocity, will be able to increase the observability of system states. Hence, the specific maneuvers excited by the changing of the inputs of the actuators definitely improve the observability of each expanded states [25]. After the categorization of current identification methods for different kinds of aircrafts equipped with different sensors, Hoffer [26] proposed that the Square-root UKF would be a good choice for the on-line identification. Girish [27] employs both EKF and UKF to estimate the aerodynamic parameters with the flight data, and it is pointed out that although UKF owns a theoretically better precision, EKF shows the same performance with less computation. Therefore, EKF is utilized in the paper to estimate and compensate the error of all the longitudinal and lateral aerodynamic derivatives. And the observations are the relatively accurate airborne integrated INS/GPS attitude and velocity. Meanwhile, maneuvering flight path is designed for the promotion of observability. The identified parameters could be used to calculate the navigation results with the dynamic and kinematic model of aircraft, or the motion model, then aiding the low-grade INS [28]. So the precision of identification results could be evaluated by comparing the navigation results with that of INS/GPS. It is quite different for the evaluation of the identification of second-order transfer function.

The outline of this paper is as follows. In Section 2, the aerodynamic characteristics of the small-scaled fixed-wing UAVs are summarized and all the parameters are calculated semi-empirically. In Section 3, we present the process of the identification of the aerodynamic parameters in details, and also the observability analysis is discussed. Section 4 selects the airborne sensors and the data acquisition board, then describes the modeling of the propeller and the control surfaces as a test preparation. In Section 5, the results of the flight tests are demonstrated as a verification. At last, the paper is concluded in Section 6.

2. Semi-Empirical Calculation

Aerodynamic parameters reside in the aerodynamic forces and moments acting on the aircraft, containing the lift Z, drag X and pitch moment M_y in the longitudinal channel and the side force Y, roll moment M_x and yaw moment M_z in the lateral channel, shown in Figure 1. It is a typical fixed-wing aircraft with a conventional configuration. Table 1 lists the formulas for the calculation of the thrust driven by the propeller and all the aerodynamic forces and moments, which are the product of the dynamic pressure Q ($Q = \rho V_a^2/2$), reference area S, reference size l or $\bar{c}$ and the non-dimensional coefficients C_i or m_i, $i = X, Y, Z$. As seen in the formulas, C_i or m_i consists of several non-dimensional and dimensional parameters, namely the aerodynamic derivatives, which are exactly the targeted parameters in this paper.

Aerodynamic derivatives are determined by the flight condition, aerodynamic layout and structural characteristic. Reynolds number of small UAVs, around 200,000, belongs to low Reynolds number. Laminar separation occurs and the aerodynamic efficiency decreases. For instance, the relationship between the lift coefficient and α, namely C_Z^α, would be nonlinear as the α increases. Hysteresis effect would happen around the stall angle resulting in the abnormal change of C_Z^α [29]. Given the condition of small Mach number Ma, the real-time changing of the aerodynamic derivatives caused by the velocity could be neglected, and the assumption of small α or β could make the derivatives, for example C_Z^α, C_Y^β, constants [12,13]. Therefore, the nonlinear aerodynamic derivatives, like $C_Z^{\alpha^2}$, $C_Z^{\alpha q}$, C_{Z0} (α, β), are neglected for the ignoring of large-amplitude maneuvers and high angle-of-attack, which could be a huge simplification. And dynamic derivatives referenced to the $\dot{\alpha}$, $\dot{\beta}$ could be small and also neglected, just as shown in Table 1. Moreover, small-scaled UAVs usually own a conventional configuration, namely wing-body-horizontal tail-vertical tail structure, and the sweepback angle χ or dihedral angle ψ_d doesn't exist in the wing designing. So the complexity of the aerodynamic analysis is reduced especially for the lateral channel. And the wing of small UAV functions predominately in the aerodynamic performance, so substituting the wing for the whole aircraft could be acceptable for the lift or pitch moment parameter calculation. The detailed process of calculating all the aerodynamic derivatives will be analyzed then.

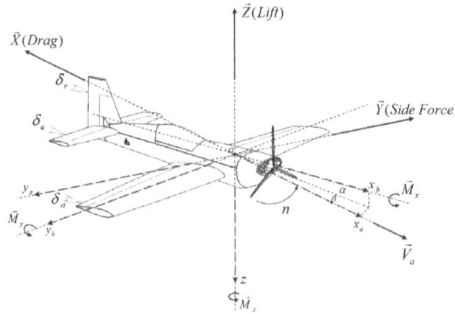

Figure 1. Aerodynamic Forces & Moments.

Table 1. Formulas of the Forces & Moments.

Forces/Moments	Formulas
Thrust F_T	$F_T = \rho n_p^2 D^4 \left[C_{FT1} + C_{FT2} \frac{V_a Ma}{D\pi n_p} + C_{FT3} \left(\frac{V_a Ma}{D\pi n_p} \right)^2 \right]$
Lift Z	$Z = QSC_Z \quad C_Z = C_{Z0} + C_Z^\alpha \alpha + C_Z^{\delta_e} \delta_e$
Drag X	$X = QSC_X \quad C_X = C_{X0} + C_{Xi} = C_{X0} + AC_Z^2$
Pitch Moment M_y	$M_y = QS\bar{c}m_y \quad m_y = m_{y0} + m_y^\alpha \alpha + m_y^{\delta_e} \delta_e + m_y^{\bar{\omega}_y} \frac{\omega_y \bar{c}}{V_a}$
Side Force Y	$Y = QSC_Y \quad C_Y = C_Y^\beta \beta + C_Y^{\delta_r} \delta_r$
Yaw Moment M_z	$M_z = QSlm_z \quad m_z = m_z^\beta \beta + m_z^{\delta_r} \delta_r + m_z^{\bar{\omega}_z} \frac{\omega_z l}{2V_a} + m_z^{\bar{\omega}_x} \frac{\omega_x l}{2V_a}$
Roll Moment M_x	$M_x = QSlm_x \quad m_x = m_x^\beta \beta + m_x^{\delta_r} \delta_r + m_x^{\delta_a} \delta_a + m_x^{\bar{\omega}_z} \frac{\omega_z l}{2V_a} + m_x^{\bar{\omega}_x} \frac{\omega_x l}{2V_a}$

2.1. Longitudinal Channel Analysis

Total lift of an aircraft is the collective effect of the wing, body and horizontal tail. With regard to the small-scaled fixed-wing aircraft, wing constitutes the main source of the lift. And there often exists interaction between the wing and body. As depicted in Table 1, lift coefficient consists of C_{Z0}, C_Z^α and $C_Z^{\delta_e}$. Zero lift coefficient, C_{Z0}, is determined by the airfoil camber, and the incidence angle of the wing φ_W and horizontal tail φ_S, which means that lift still exists when the angle-of-attack is zero. C_Z^α, known as the slope of lift curve, fuses the separate cantilever-wing slope $C_{ZCW_S}^\alpha$, the separate body slope $C_{ZB_S}^\alpha$, the separate horizontal tail slope $C_{ZHT_S}^\alpha$ and all the mutual effect between each of them. In details, $C_{ZCW_S}^\alpha$, $C_{ZHT_S}^\alpha$ could be computed with the same diagram in [13,14], both of which are the function of the structural parameters of the wing and the Mach number Ma. While the lift slope of body could be divided into the conical or ogive head, the cylindrical body and the shrinking tail part. $C_Z^{\delta_e}$ concerns with the aerodynamic efficiency η_e and the size of the elevator.

Lift of the wing-body part Z_{WB} is the sum of the lift of separate body Z_{Bs} and the lift caused by the existence of the cantilever-wing Z_{CW}, while the latter contains the lift of separate cantilever-wing Z_{CWs}, the interference lift from the body to the cantilever-wing Z_{B-CW} and the opposite lift Z_{CW-B}. Interference factors [13,14], $K_{\alpha\alpha}$, $K_{\varphi 0}$, are introduced to describe the mutual effect numerically, both of which concerns with the shape of the wing and the diameter-to-span ratio $\bar{D}$. Generally, when we have the condition that the aspect ratio λ is large enough ($\lambda \geqslant 5$) and $\bar{D}$ is small enough ($\bar{D} \leqslant 0.1$), neglecting the mutual effect hardly matters. In the linear range, the lift slope of the wing-body part are shown in (1) and (2).

$$C_{ZWB}^{\alpha} = C_{ZBs}^{\alpha}\frac{S_B}{S} + C_{ZCWs}^{\alpha}K_{\alpha\alpha}\frac{S_{CW}}{S} \tag{1}$$

$$C_{ZWB}^{\varphi} = C_{ZCWs}^{\alpha}K_{\varphi0}\frac{S_{CW}}{S} \tag{2}$$

The airflow block factor k_{HT} and the down wash angle ε, caused by the wing part, should be taken into account when analyzing the lift of the horizontal tail. The block factor k_{HT} ($k_{HT} = 0.85$) should be multiplied to the dynamic pressure while the down wash angle ε should be eliminated to obtain the real angle-of-attack of the horizontal tail α_{HT}. The down wash angle after the wing-body part ε_{WB} are nearly linear to the α and the incidence angle of the wing, so the related derivatives are calculated in (3) and (4), in which ε_W^{α}, the down wash caused by the separate wing could be calculated in [13,14]. Thus, the lift coefficients could be deduced in Formulas (5)–(7), among which $(K_{\alpha\alpha})_{HT}$ and $(K_{\varphi0})_{HT}$ are the interference factor between the body and the horizontal tail.

$$\varepsilon_{WB}^{\alpha} = K_{\alpha\alpha}\frac{S_{CW}}{S}\frac{C_{ZCWs}^{\alpha}}{C_{ZWs}^{\alpha}}\varepsilon_W^{\alpha} \tag{3}$$

$$\varepsilon_{WB}^{\varphi} = K_{\varphi0}\frac{S_{CW}}{S}\frac{C_{ZCWs}^{\alpha}}{C_{ZWs}^{\alpha}}\frac{\varepsilon_W^{\alpha}}{1-\bar{D}} \tag{4}$$

$$C_{Z0} = C_{ZWB}^{\varphi}\varphi_W + C_{ZHTs}^{\alpha}(K_{\varphi0})_{HT}(\varphi_S - \varepsilon_{WB}^{\varphi}\varphi_W)k_{HT}\frac{S_{HT}}{S} \tag{5}$$

$$C_Z^{\alpha} = C_{ZWB}^{\alpha} + C_{ZHTs}^{\alpha}(K_{\alpha\alpha})_{HT}(1 - \varepsilon_{WB}^{\alpha})k_{HT}\frac{S_{HT}}{S} \tag{6}$$

$$C_Z^{\delta_e} = C_{ZHT}^{\delta_e}k_{HT}\frac{S_{HT}}{S} \tag{7}$$

Generally, aircraft drag contains the parasitic drag and the induced drag, neglecting the wave drag in supersonic flight. Parasitic drag usually consists of friction drag, form drag and interference drag, among which the friction drag are the main source of the parasitic drag in the low speed flight. Induced drag is caused by the lift of the wing. Under the small Mach number and small angle-of-attack assumption, the drag coefficient C_X is separated into zero lift coefficient C_{X0} and the lift-induced drag coefficient C_{Xi}, depicted as the parabolic formula in Table 1. C_{X0} mainly contains the friction part and the due-to-lift part in the parasitic drag is neglected. So C_{X0} is calculated with the surface friction coefficient c_f and correction factor η [13,14]. C_{Xi}, consistent with the formation AC_Z^2, is calculated considering the wing-body part according to [13,14], in which C_{Xi} of wing are closely related to the wing's shape and C_{Xi} of body is proportional to α^2.

Assuming that the vector of thrust points through the center of gravity, the pitch moments are primarily caused by the lift of different parts. Prior to calculating the pitch moment coefficients, the action point of the lift, namely the aerodynamic center, should be analyzed, which covers the analysis of the wing, body, wing-body and the horizontal tail. Similar to the lift, the location of the aerodynamic center of the separate wing and horizontal tail x_{pW}, x_{pHT} could be obtained by the same diagram in [13,14] according to the size and shape. Referenced to the vertex in the front, the location of aerodynamic center is usually positive. But for the body part, x_{pB} could be negative as the tail generates large minus pitch moment. The location of the wing-body part $(x_{pWB})_{\alpha\alpha}$, $(x_{pWB})_{\varphi0}$ could also be computed with the factor $K_{\alpha\alpha}$, $K_{\varphi0}$. For the elevator, $x_{pHT\delta_e}$ locates at the leading edge of the control surface.

Along with the aerodynamic center, the aerodynamic focus is also introduced to reflect the longitudinal stability. The location of aerodynamic focus x_F, only determined by the angle-of-attack, are related to the longitudinal static stability derivative m_y^{α} and the location of the center of gravity x_G, depicting in (8). It is explained that when the m_y^{α} is negative ($x_G < x_F$), the aircraft is statically stable. While the aircraft deviates from its equilibrium position, the pitching moment makes itself back to the equilibrium automatically. Three static aerodynamic derivatives m_{Z0}, m_Z^{α}, $m_Z^{\delta_e}$ are deduced in (9)–(11).

For the dynamic derivative, the damping moment coefficient $m_y^{\bar{\omega}_y}$ caused by the non-dimensional $\bar{\omega}_y$ should be taken into account to strengthen the stability which could be obtained in [14].

$$m_y^\alpha = C_Z^\alpha \frac{x_G - x_F}{\bar{c}} \tag{8}$$

$$m_{y0} = C_{ZWB}^\varphi \varphi_W \frac{x_G - (x_{pWB})_{\varphi 0}}{\bar{c}} + C_{Z0HT} k_{HT} \frac{S_{HT}}{S} \frac{x_G - x_{pHT}}{\bar{c}} \tag{9}$$

$$m_y^\alpha = C_{ZBs}^\alpha \frac{S_B}{S} \frac{x_G - x_{pB}}{\bar{c}} + C_{ZWB}^\alpha \frac{x_G - (x_{pWB})_{\alpha\alpha}}{\bar{c}} + C_{ZHT}^\alpha k_{HT} \frac{S_{HT}}{S} \frac{x_G - x_{pHT}}{\bar{c}} \tag{10}$$

$$m_y^{\delta_e} = C_{ZHT}^{\delta_e} k_{HT} \frac{S_{HT}}{S} \frac{x_G - x_{pHT\delta_e}}{\bar{c}} \tag{11}$$

2.2. Lateral Channel Analysis

The aerodynamic analysis in the lateral channel is comparatively more complicated concerning the coupling between the roll and yaw moments. But there are some easy ways to simplify the process. Firstly, side force Y mainly stems from the body and the vertical tail. When the aircraft rotates around the x-axis for $90°$, Y is just similar to the lift Z. As a result, the angle of sideslip β resembles the angle-of-attack α and the deflection angle of the rudder δ_r resembles that of elevator δ_e. Then, the aerodynamic derivatives C_Y^β, $C_Y^{\delta_r}$ could be deduced. Similarly, the calculation of the yaw moment derivatives could be operated after the rotation of $90°$ and at this time, the side force Y is the main cause of the yaw moment. The static derivatives m_z^β, $m_z^{\delta_r}$ and the dynamic derivative $m_z^{\bar{\omega}_z}$ are computed, among which m_z^β ($m_z^\beta < 0$) determines the stability of yawing channel [13,14].

Among the roll moment derivatives, m_x^β is an important derivative that concerns with the α, β, the aerodynamic configuration and structural size, such as the dihedral angle ψ_d, sweepback angle $\chi_{0.5}$, the wing-body interaction, the wing tip characteristic and so on, depicting in (12). The aileron plays a vital role in rolling the aircraft and the relevant control derivative $m_x^{\delta_a}$ could be obtained in (13), in which η_a is the aerodynamic efficiency of the aileron. The damping roll derivative $m_x^{\bar{\omega}_x}$ could also be obtained just like the damping one in the yaw moment. There does exist coupling effects between the yaw and roll moment, described by the relevant derivatives $m_z^{\delta_a}$, $m_x^{\delta_r}$, $m_z^{\bar{\omega}_x}$, $m_x^{\bar{\omega}_z}$. They are usually small but ought to be considered according to the specific situation [13,14].

$$(m_x^\beta)_* = (m_x^\beta)_{\psi_d} + (m_x^\beta)_{WB} + (m_x^\beta)_{VT} + [(\frac{\partial^2 m_x}{\partial \alpha \partial \beta})_{X0.5} + (\frac{\partial^2 m_x}{\partial \alpha \partial \beta})_{Wtip} + (\frac{\partial^2 m_x}{\partial \alpha \partial \beta})_\varepsilon]\alpha + (\frac{\partial^2 m_x}{\partial \delta_e \partial \beta})_\varepsilon \delta_e \tag{12}$$

$$m_x^{\delta_a} = C_{ZWs}^\alpha \eta_a (\frac{1}{C_Z^\alpha} \cdot \frac{m_x^\beta}{\psi_d}) \tag{13}$$

3. Parameter Identification

After the semi-empirical analysis, the initial aerodynamic parameters are obtained, but they are not accurate enough for further research. An identification process is necessary to correct the parameters. The nonlinear dynamic and kinematic differential equations of V^b, ω^b, and Euler angles are to be used as the system equations in the identification process, shown in (14)–(16) , in which a non-rotating and flat earth is assumed. Considering the possible fast dynamic change of the motion states, the error states are chosen as system states, $X_M = [\delta u, \delta v, \delta w, \delta p, \delta q, \delta r, \delta \phi, \delta \theta, \delta \psi]^T$. The error of aerodynamic parameters X_C are expanded into the system state, depicted in (17). Here, X_C are assumed as constants given the condition of small Mach number and angle-of-attack, and they are exactly the parameters to be estimated and then compensated to the initial value . The linearization of 1st-order Taylor Series approximation of the nonlinear model is operated and the related Jacobian matrix F is formulated in (18), in which F could be partitioned into F_M, F_C and the zero matrix. Adding up the related noise $W_{9\times1}$, the linearized system equation is shown in (19).

$$\dot{V}^b = C_n^b g^n + (F_T + C_a^b F_A)/m - \omega^b \times V^b \tag{14}$$

$$\dot{\omega}^b = I^{-1}(M - \omega^b \times I\omega^b) \tag{15}$$

$$\begin{pmatrix} \dot{\phi} \\ \dot{\theta} \\ \dot{\psi} \end{pmatrix} = \begin{bmatrix} 1 & \tan\theta\sin\phi & \tan\theta\cos\phi \\ 0 & \cos\phi & -\sin\phi \\ 0 & \frac{\sin\phi}{\cos\theta} & \frac{\cos\phi}{\cos\theta} \end{bmatrix} \begin{pmatrix} p \\ q \\ r \end{pmatrix} \tag{16}$$

$$\begin{aligned} X = [&\delta u, \delta v, \delta w, \delta p, \delta q, \delta r, \delta\phi, \delta\theta, \delta\psi, \\ &\delta C_{Z0}, \delta C_Z^\alpha, \delta C_Z^{\delta_e}, \delta C_{X0}, \delta C_{XA}, \delta m_{y0}, \delta m_y^\alpha, \delta m_y^{\delta_e}, \delta m_y^{\bar{\omega}_y}, \\ &\delta C_Y^\beta, \delta C_Y^{\delta_r}, \delta m_z^\beta, \delta m_z^{\delta_r}, \delta m_z^{\bar{\omega}_z}, \delta m_z^{\bar{\omega}_x}, \delta m_x^\beta, \delta m_x^{\delta_a}, \delta m_x^{\delta_r}, \delta m_x^{\bar{\omega}_x}, \delta m_x^{\bar{\omega}_z}]^T_{29\times1} \end{aligned} \tag{17}$$

$$F = \begin{bmatrix} F_{M9\times9} & F_{C9\times20} \\ 0_{20\times9} & 0_{20\times20} \end{bmatrix} \tag{18}$$

$$\dot{X} = FX + W \tag{19}$$

There are 20 aerodynamic parameters in the aircraft dynamic model according to Table 1. All the derivatives are directly fused with the airspeed V_a residing in the dynamic pressure Q. It could be excited through acceleration or deceleration by changing the rotation rate of the propeller. Some derivatives have close relations with the aerodynamic angle α, β, like C_Z^α, m_y^α, C_Y^β, m_x^β, and the damping derivatives like $m_y^{\bar{\omega}_y}$, are concerned with the rotation rate of the aircraft. All of them are supposed to be stimulated by the uniform or non-uniform changing of the attitude. And the rest derivatives are directly related to the deflection angle of the control surfaces δ_e, δ_r, δ_a.

As strong nonlinearity exists among the motion states, depicted in (14)–(16), and the number of invariant targeted derivatives are too many, it is not easy to compensate all of them correctly at the same time. Different parameters may have different impact on the model, and the observability is also different. Lift coefficient C_Z is taken as an example to analyze the observability of each aerodynamic derivatives, depicted in (20). With the observations of non-gravitational acceleration and the thrust, the left hand Z is potentially calculated according to (14). Right hand of the equation is a sum formation, and it is easier to distinguish each element with different characteristics. So the aerodynamic derivative C_{Z0}, C_Z^α and $C_Z^{\delta_e}$ could be observed just when the angle-of-attack and deflection angle δ_e vary differently, namely they have distinguishing frequency spectrum. The analysis could be generalized into the moment coefficients, like the roll moment in (21). Also it is not easy to tell apart ω_x from ω_z, so $m_x^{\bar{\omega}_x}$, $m_x^{\bar{\omega}_z}$ may not be separated totally, same theory for $m_x^{\delta_a}$, $m_x^{\delta_r}$. Therefore, for the similar derivatives, they could be observed when the corresponding actuators are excited at different time.

$$Z = QSC_Z = \frac{1}{2}\rho S(V_a^2 C_{Z0} + V_a^2 C_Z^\alpha \alpha + V_a^2 C_Z^{\delta_e}\delta_e) \tag{20}$$

$$M_x = QSlm_x = \frac{1}{2}\rho Sl\left(V_a^2 m_x^\beta \beta + V_a^2 m_x^{\delta_a}\delta_a + V_a^2 m_x^{\delta_r}\delta_r + \frac{l}{2}V_a m_x^{\bar{\omega}_x}\omega_x + \frac{l}{2}V_a m_x^{\bar{\omega}_z}\omega_z\right) \tag{21}$$

To sum up theoretically, if all the aerodynamic parameters are to be observed, the aerodynamic forces and moments are calculated firstly and the observations of acceleration and angular velocity are necessary, referring to the dynamic Equations (14) and (15). But, for small-scaled fixed-wing aircraft, it is hard for the MEMS-IMU to provide precise observations of acceleration and angular velocity on account of the unpredictable drift or bias. As MEMS-IMU integrated with GPS is the most common airborne navigation devices, the integrated navigation results could be adopted as measurements for further compensation of the aerodynamic coefficients. The acceleration and angular velocity could be deduced by the first order derivative, instead of the outputs of MEMS-IMU. So, it is reasonable to accomplish the identification process just with the observations of velocity and attitude. Now, the differences between the relatively accurate velocity and attitude from the INS/GPS fusion(subscript "SF") and that from the integral of the aircraft motion model of Equations (14)–(16) (subscript "AMM") are used as measurements, shown in (22). u_{SF}, v_{SF}, w_{SF} are deduced from the INS/GPS ground velocity V_N, V_E, V_D through a coordinate transformation by multiplying the rotation matrix C_n^b computed with Euler angles. Then, the identification process will be accomplished by EKF.

$$Z = \begin{pmatrix} \delta u \\ \delta v \\ \delta w \\ \delta \phi \\ \delta \theta \\ \delta \psi \end{pmatrix} = \begin{pmatrix} u_{AMM} - u_{SF} \\ v_{AMM} - v_{SF} \\ w_{AMM} - w_{SF} \\ \phi_{AMM} - \phi_{SF} \\ \theta_{AMM} - \theta_{SF} \\ \psi_{AMM} - \psi_{SF} \end{pmatrix} \tag{22}$$

The maneuvering flight is also indispensable to excite all the motion states and the inputs of the actuators to increase the observability of all the aerodynamic derivatives. Relevant researches [3,8,15] introduce the specific changes of the elevator, rudder and aileron to excite the expecting maneuvering, like dutch roll, bank-to-bank and the short period motion of the aircraft, which is an effective approach to identify the parameters in the time domain. Here in the paper, a maneuvering flight is designed, including acceleration and deceleration, rolling, the rise and fall when circling around.

4. Experimental Preparation

A model UAV called Extra300, shown in Figure 2, is employed to carry out the real flight test. Two goals are about to achieve in the practical tests: (1) Rationality of the empirically calculated aerodynamic coefficients should be validated. (2) Parameters identification is operated and the improvement should be proved compared with the semi-empirical results. For the empirical calculation of the aerodynamic parameters, the 3D model of Extra300 is established and the structural size is measured. The characteristic parameters of the aircraft like mass m, position of center of gravity (x_G, y_G, z_G) and moment of inertia I are tested successively on a Mass & Gravity Center Test Board and a Moment of Inertia Turntable. Firstly, by setting each aircraft's axis along the central axis of the Test Board, shown in the left part of Figure 3, m and (x_G, y_G, z_G) are obtained based on the moment balance principle. It is tested that the thrust from the propeller F_T points nearly through the center of gravity. Then, the aircraft will be installed on the turntable. Make sure that the rotation axis of the turntable points right through the obtained gravity center, shown in the right part of Figure 3. Moment of inertia is calculated from the rotation period tested by an optoelectronic switch. The rotation is generated by the turntable and slowed down by a spring device. Before the whole test, a standard cylinder with known mass (about 3 kg) and moment of inertia (about 0.00058 kg $\cdot$ m^2) is used to calibrate the parameters of the two test tables, like the elastic coefficient of the spring device. Together with the above, the empirical aerodynamic derivatives calculated according to Section 2 are listed in Table 2.

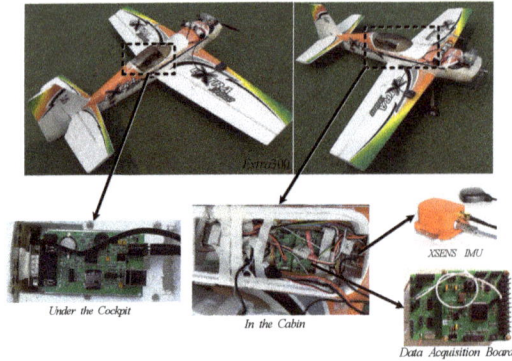

Figure 2. Flight Test Devices.

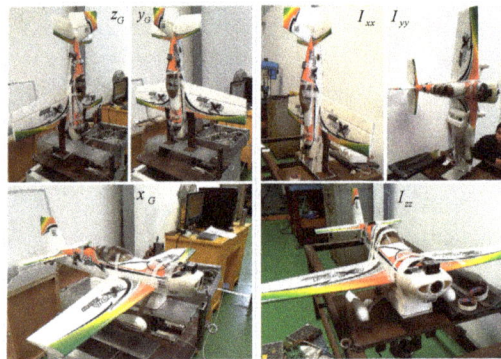

Figure 3. Characteristic Parameter Test.

The airborne sensors and devices are shown in Figure 2. A XSENS MTi-G module is chosen, which owns a mature integrated navigation system of INS/GPS. It also contains an Attitude and Heading Reference System (AHRS) processor. XSENS MTi-G deploys MEMS inertial sensors in 9 axes, in which the outputs of gyroscopes own the $1°/s$ bias stability and $0.05°/s$ noise density, the accelerometers have the 0.02 m/s^2 bias stability and 0.002 m/s^2 noise density, the magnetic sensors has 0.1 m Gauss bias stability and 1.5 m Gauss noise density. MTi-G provides not only the outputs of different inertial sensors but also the integrated navigation results of attitude, velocity and position. The accuracy of the integrated position and velocity is less than 2.5 m and 0.1 m/s respectively. The dynamic accuracy of the pitch/roll angle is less than $1°$ and that of the heading angle is less than $2°$. Thus, the complex initial alignment process for INS or ADM is left out. The data output rate of XSENS is usually set to be 100 Hz [30].

The data of MTi-G is collected through a data acquisition board. As shown in the picture, the collecting board, placed in the cabin, has 2 layers, both of which has the core of DSP-BF506F. The lower board is in charge of collecting the data of all the sensors and transmitting to the upper board. The upper board will collect the data from the lower board and that of the signal receiver of 4 channels transmitted from the controller, then output all of them to a data recorder. The data recorder is recommended, rather than the radio transmission, to avoid the loss of data or electromagnetic interference. Furthermore, the upper board is responsible for the switching of the flight mode. In the tests, the aircraft is controlled remotely by a controller, namely the RC mode [31,32].

Table 2. Extra300 Parameters.

Structural Parameters	$S = 0.285 \text{ m}^2, l = 1.2 \text{ m}, \bar{c} = 0.24 \text{ m}$		
Propeller Parameters	$\rho = 1.22 \text{ kg/m}^3, D = 0.33 \text{ m}$ $C_{FT1} = 0.0842, C_{FT2} = -0.136, C_{FT3} = -0.928$		
Inertia Parameters	$m = 1.72 \text{ kg}, x_G = 0.323 \text{ m}, y_G \approx 0 \text{ m}, z_G \approx 0 \text{ m}$ $I = \begin{bmatrix} 0.001393 & 0 & 0 \\ 0 & 0.003543 & 0 \\ 0 & 0 & 0.003631 \end{bmatrix} \text{ kg·m}^2$		

Aerodynamic Parameters	C_Z	C_X	m_y
	$C_{Z0} = 0.1601$	$C_{X0} = 0.01602$	$m_{y0} = 0.05267$
	$C_Z^{\alpha} = 4.4851 \text{ rad}^{-1}$	$C_{XA} = 0.1347$	$m_y^{\alpha} = -0.1195 \text{ rad}^{-1}$
	$C_Z^{\delta_e} = 0.5901 \text{ rad}^{-1}$		$m_y^{\delta_e} = -2.8648 \text{ rad}^{-1}$
			$m_y^{\bar{\omega}_y} = -6.4099 \text{ rad}^{-1}$
	C_Y	m_z	m_x
	$C_Y^{\beta} = -0.7076 \text{ rad}^{-1}$	$m_z^{\beta} = -0.007517 \text{ rad}^{-1}$	$m_x^{\beta} = -0.05094 \text{ rad}^{-1}$
	$C_Y^{\delta_r} = -0.3893 \text{ rad}^{-1}$	$m_z^{\delta_r} = -0.2152 \text{ rad}^{-1}$	$m_x^{\delta_a} = -0.3022 \text{ rad}^{-1}$
		$m_z^{\bar{\omega}_z} = -0.5053 \text{ rad}^{-1}$	$m_x^{\delta_r} = -0.06492 \text{ rad}^{-1}$
		$m_z^{\bar{\omega}_x} = -0.05272 \text{ rad}^{-1}$	$m_x^{\bar{\omega}_x} = -0.3968 \text{ rad}^{-1}$
			$m_x^{\bar{\omega}_z} = -0.007 \text{ rad}^{-1}$

Aiming at the verification of the accuracy of the aerodynamic parameters, the aircraft dynamic and kinematic model should be calculated with the input data of the actuators in the real flight. So the correspondence between the width of PWM signal , d_{PWM}, transmitted from the controller and the actuators' inputs, namely the rotation rate n and the deflection angles δ, has to be derived. Arifianto [8] establishes the model by testing the force of the propeller corresponding to the rotation rate, but the airspeed is ignored in the static experiment. Herein, the "n-d_{PWM}" relationship is constructed directly, and the thrust will be calculated with the equation in Table 1. An experiment of testing the rotation rate of the propeller is firstly carried out, and the test and the relevant experimental apparatus is shown in Figure 4. The PWM signal width of the channel of the throttle d_{PWM} ranges from 2202 to 3879. The rotation rate of the propeller is measured by a handheld digital tachometer corresponding to the different position of the joystick, which is limited to 70% of the full range for safety. After several tests and second order polynomial fit, the equation could be obtained in (23), and the maximal rotation rate could reach 8966.3 RPM, namely 149.4 r/s. Then, the relationship between the deflection angle of the control surfaces (measured by a digital goniometer) and the PWM width is fitted linearly in (24)–(26).

$$n_P = (-0.001571d_n^2 + 13.18d_n - 21300)/60 \tag{23}$$

$$\delta_e(^\circ) = \begin{cases} 49.2480 - 0.0162d_e & 2202 \leq d_e \leq 3029 \\ 60.1920 - 0.0198d_e & 3029 < d_e \leq 3879 \end{cases} \tag{24}$$

$$\delta_r(^\circ) = \begin{cases} 60.7800 - 0.0200d_r & 2200 \leq d_r \leq 3040 \\ 60.7800 - 0.0200d_r & 3040 < d_r \leq 3879 \end{cases} \tag{25}$$

$$\delta_a(^\circ) = \begin{cases} 55.5840 - 0.0180d_a & 2199 \leq d_a \leq 3085 \\ 58.6720 - 0.0190d_a & 3085 < d_a \leq 3878 \end{cases} \tag{26}$$

Appratus	Type
Propeller	13 ×6.5 inch (0.33m)
Digital Tachometer	SW6234C ±0.05%
Brushless Motor	NO4018 KV950
Brushless ESC	Hobbywing 40A BEC 3A@5V
Battary	3SLi 2200mAh 40C
Receiver	Futaba R6041HS
Controller	Futaba T12ZH
Recorder	LCA3211

Figure 4. Rotation Rate Test.

5. Flight Test and Analysis

Real flight tests were carried out in an open playground in Figure 5. The aircraft Extra300 was basically set to circle around above the field, controlled by a pilot on the ground. Several flight tests were completed and one of the experiment data was chosen to identify the aerodynamic derivatives, and then to verify the improvement compared with that calculated by the semi-empirical method. The 3D flight path is shown in Figure 5. It is truncated from the whole flight, lasting 3 min. To make it more clear to recognize the changing of the path, it is decomposed into three parts, from point A to D. In the flight route, there exists circling, slight climbing and diving, accelerating and decelerating and other maneuvering, which contribute to the observation. Besides, the measured speed is estimated as ground speed not the airspeed as the wind is neglected.

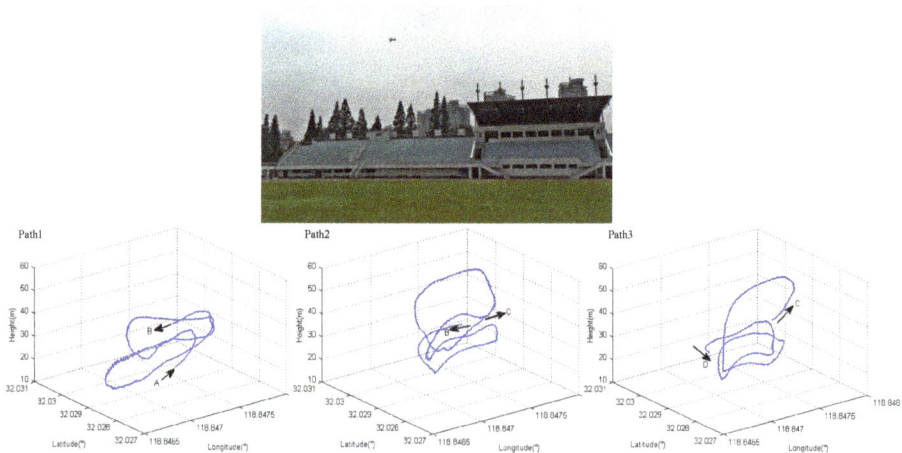

Figure 5. Flight Test.

The identification was operated according to the analysis in Section 3. The data fusion frequency is 10 Hz and the standard deviation of white noise is set according to the precision of the INS/GPS navigation results. The estimation results of the error of aerodynamic parameters are shown in Figure 6a–h, and all of them could converge to a certain range. Some coefficients, like $\delta m_y^{\delta_e}$, $\delta m_y^{\bar{\omega}_y}$, $\delta m_z^{\delta_r}$, $\delta m_x^{\bar{\omega}_x}$, converge piece-wisely. It may be caused by the changing of the real-time flight condition, and it matters slightly in this estimation process. The average of the estimation results in the converging section are calculated and approximated, shown in Table 3. Then, the estimated errors of the aerodynamic derivatives are subtracted or compensated from the initial derivatives calculated by semi-empirical method, listed in Table 2, thus further correcting the aircraft model.

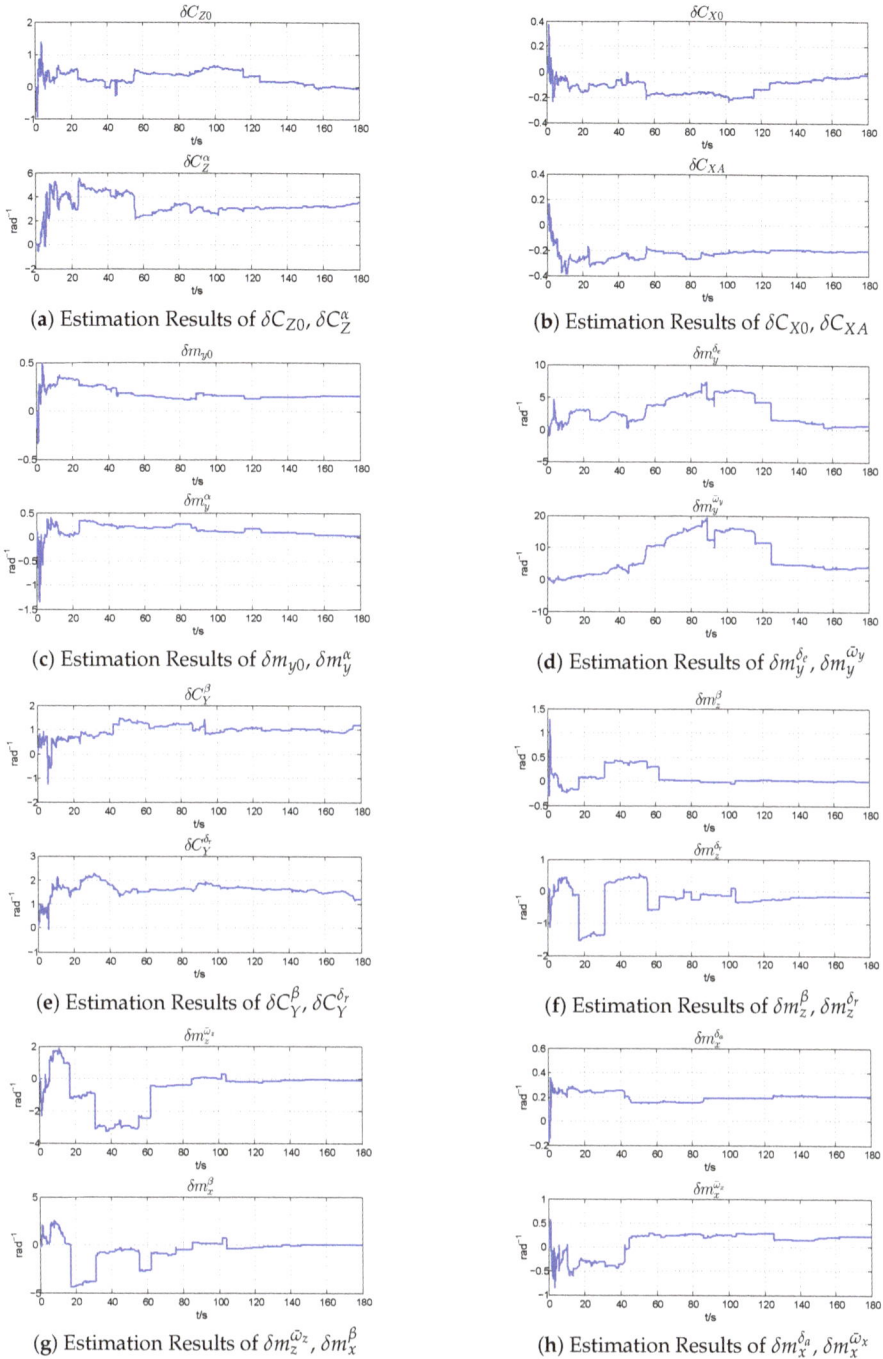

(**a**) Estimation Results of δC_{Z0}, δC_Z^{α}

(**b**) Estimation Results of δC_{X0}, δC_{XA}

(**c**) Estimation Results of δm_{y0}, δm_y^{α}

(**d**) Estimation Results of $\delta m_y^{\delta_r}$, $\delta m_y^{\bar{\omega}_y}$

(**e**) Estimation Results of δC_Y^{β}, $\delta C_Y^{\delta_r}$

(**f**) Estimation Results of δm_z^{β}, $\delta m_z^{\delta_r}$

(**g**) Estimation Results of $\delta m_z^{\bar{\omega}_z}$, δm_x^{β}

(**h**) Estimation Results of $\delta m_x^{\delta_a}$, $\delta m_x^{\bar{\omega}_x}$

Figure 6. Aerodynamic Parameters Error Estimation Results.

Table 3. Parameter Error Estimation Results.

Parameter	Result	Parameter	Result
δC_{Z0}	0.000	δC_Y^{β}	1.000
δC_Z^{α}	3.000	$\delta C_Y^{\delta_r}$	1.600
$\delta C_Z^{\delta_e}$	0.000	δm_z^{β}	0.000
δC_{X0}	−0.100	$\delta m_z^{\delta_r}$	−0.166
δC_{XA}	−0.200	$\delta m_z^{\bar{\omega}_z}$	−0.110
δm_{y0}	0.150	$\delta m_z^{\bar{\omega}_x}$	0.147
δm_y^{α}	0.100	δm_x^{β}	0.000
$\delta m_y^{\delta_e}$	1.000	$\delta m_x^{\delta_a}$	0.200
$\delta m_y^{\bar{\omega}_y}$	5.000	$\delta m_x^{\delta_r}$	0.000
		$\delta m_x^{\bar{\omega}_x}$	0.200
		$\delta m_x^{\bar{\omega}_z}$	0.593

With the initial motion states provided by the INS/GPS integration in XSENS MTi-G, the navigation results from the integration of the aircraft motion equations of (14)–(16) is calculated. By comparing with the INS/GPS navigation results, on one hand, the rationality of the aerodynamic parameters computed by the semi-empirical method will be proved, on the other hand, the effect of latter compensation process is also verified, which is a unique method to validate the correctness of the calculation and identification results. Figures 7–9 show the comparison of the errors of navigation results between the empirically calculated model and the compensated one by subtracting the INS/GPS navigation results.

Firstly, it is demonstrated that the semi-empirically calculated parameters based on aerodynamics are rational and useful, which is the premise for further analysis. After the compensation of the error of the aerodynamic derivatives, there shows an improvement in the navigation results. Specifically, the error of the latitude and longitude of the two processes are similar, and it is calculated that the average error of the latitude and longitude are about −39.1824 m and 5.5703 m respectively. The average of height error decreases to about 177.6206 m from 272.3769 m after the identification. The average errors of the horizontal velocity are alike, namely δV_n, δV_e are about −0.9853 m/s, 0.8024 m/s respectively while δV_d changes from 3.1785 m/s to 1.8770 m/s. Although there exists just a slight improvement in the horizontal velocity, it is obviously figured that the compensated model are more stable and owns a better tendency compared with the INS/GPS results. The average error of roll angle decreases from −44.4609° to−11.1175°. The average error of pitch angle, around 12.6113°, are similar but more stable. Yaw angle, with the average error of −26.4751°, are more rational and precise after the compensation as the wide swing are taken out.

Despite the improvement after the correction process, differences still exist between the compensated aircraft model and the precise integration results. The model of the rotation rate of propeller or the deflection angle of the actuators maybe inaccurate, but it has little effect on the navigation results according to the experiment. It is largely caused by the inaccuracy of the identification of aerodynamic parameters. The flight condition, like the large variation of the aerodynamic angle, will influence the estimation precision. It could be seen from the changing of α, β in Figure 10. As seen from the pictures, it doesn't match the small angle-of-attack assumption sometimes, so some aerodynamic derivatives are no longer constants. Compared with the navigation error in Figure 8, the error just becomes larger when the angle-of-attack increases largely. The phenomenon also take place in other flight tests, and the estimation results of the parameters' error are a little different in different flight sorties concerning the different flight condition. To address the problem, the nonlinear aerodynamic derivatives should be taken into account as the simplified modeling of the coefficients and derivatives leads to the uncertainty. For example, the slope of lift curve C_Z^{α} is variant, and some nonlinear derivatives associated with α, β should be augmented to the states to be estimated. The variant coefficients should be estimated precisely and compensated piece-wisely in a real time.

Furthermore, more sensors measuring the aerodynamic angles α, β and the airspeed should be added to take the changing circumstances into account.

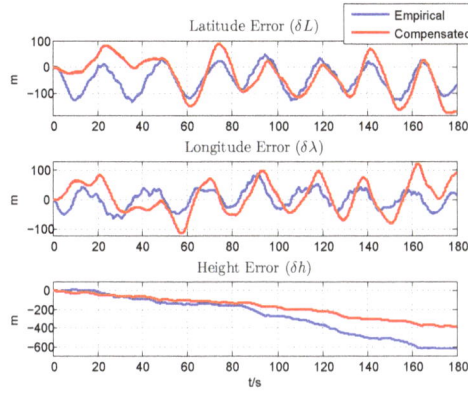

Figure 7. Position Error.

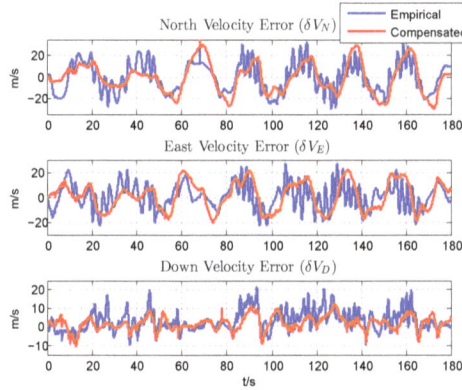

Figure 8. Velocity Error.

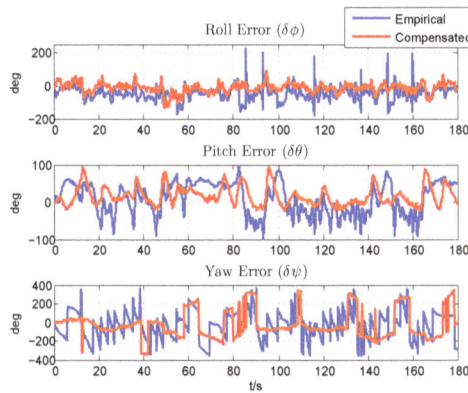

Figure 9. Attitude Error.

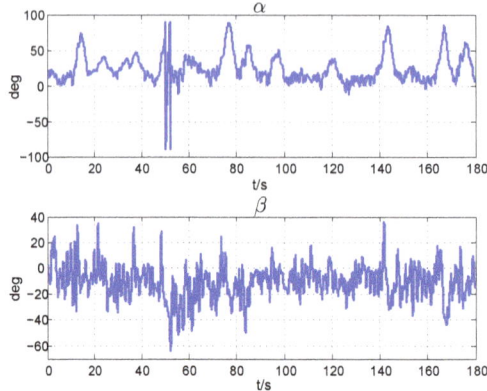

Figure 10. Aerodynamic Angle in the Flight.

6. Conclusions

The paper mainly studied the method of calculating the aerodynamic parameters of the small-scaled fixed-wing aircraft, and it is an important part in the establishment of the aircraft dynamic and kinematic model. It contains 2 sequential processes:

(1) Prepared with the related structural parameters and the position of gravity center, the aerodynamic derivatives are semi-empirically calculated based on fundamental aerodynamics. The simplified structure and the specific flight condition reduce the complexity. So, the aerodynamic analysis could be operated in a fast, effective and low-cost way to originally provide a serial rational parameters. The paper introduces and analyzes the entire course of the calculating of the longitudinal and lateral derivatives, and it is helpful for the modeling of all the other similar small aircraft;

(2) The initially obtained parameters is further identified or compensated with the real flight data. Given the airborne MEMS-based inertial sensors with finite precision, the relatively accurate velocity and attitude from the integration of INS/GPS was employed as observations to estimate the error of the aerodynamic parameters. It may have the problem of lack of observations, but it could be solved with the abundant maneuvering flight. Then, the aerodynamic parameters could be corrected with the estimated error of the derivatives and the aircraft model could be more accurate and reliable.

The real flight test demonstrates a promising result of the above 2 methods. Abide by the traditional aerodynamics and the empiricism from flight database, a rational aircraft model is constructed. Although about 20% or more error exists in the aerodynamic coefficients, it still provides a basic prototype or a baseline model. During the circling flight test, the controller adjusted the model aircraft all the time, thus providing the required maneuvers to increase the observability of all the derivatives in the identification process. And the accuracy of the compensated parameters was verified by integrating the aircraft dynamic and kinematic model, then comparing with the INS/GPS integration results, which is different from the traditional verification. The accuracy of the method synthesizes the precision of the model itself including the structural and aerodynamic part as well as the precision of the actuators' inputs. But the correctness of the aerodynamic parameters affect predominantly.

As discovered from the tests, the ideal aerodynamic parameters, namely the true values, are hard to reach. Firstly, the inaccuracy of the structural parameters, the possible shifting of the position of the center of gravity, and the related assumptions and simplifications could make an effect on the semi-empirical analysis. Then, the variant flight condition makes it more difficult for the filter to identify the variant parameters correctly and instantaneously, especially encountering the disturbance of the wind or gust. In the subsequent study or experiment, more aerodynamic sensors, like the low-cost or homemade vanes measuring the angle-of-attack and the side-slip angle, will be equipped to take the wind into consideration. Then, the relatively accurate aircraft model are expected to be used as an assistance for the fast divergent pure INS in the future research.

Acknowledgments: The study and the related experiment was supported by the MEMS Research Center of Nanjing University of Science and Technology. The work was also supported by a Natural Science Foundation Project (No.51375244), China.

Author Contributions: Jieliang Shen conducted all the research and wrote the paper. Yan Su guided and revised the thoughts of the research. Qing Liang helped conceiving and designing the experiment. Xinhua Zhu contributed to the improvement of the thoughts and the writing of the paper.

Conflicts of Interest: The authors declare no conflict of interest.

Nomenclature

Abbreviation

ADM	Aircraft Dynamic Model
AMM	Aircraft Motion Model
CFD	Computational Fluid Dynamics
EKF	Extended Kalman Filter
GPS	Global Positioning System
IMU	Inertial Measurement Unit
INS	Inertial Navigation System
MEMS	Micro-Electro Mechanical Systems
MLE	Maximum Likelihood Estimation
PMW	Pulse Width Modulation
UAV	Unmanned Aerial Vehicle
UKF	Unscented Kalman Filter

Coordinate

g	Earth-surface inertial reference frame pointing North, East, Down
b	Aircraft-body coordinate frame
a	Wind coordinate frame
C_g^b	Coordinate rotation matrix from g to b
C_a^b	Coordinate rotation matrix from a to b

Aircraft

m	Total mass, kg
I	Moment of inertia, kg·m^2
x_G, y_G, z_G	Position of center of gravity, m
x_F	Aerodynamic focus, m
x_{pW}, x_{pB}, x_{pWB}	Aerodynamic center of wing, body and wing-body, m
$x_{pHT}, x_{pHT\delta_e}$	Aerodynamic center of horizontal tail and elevator, m
l	Wing span, m
$\bar{c}$	Mean aerodynamic chord, m
S	Reference area, m^2
S_{CW}	Cantilever wing area, m^2
S_B	Fuselage maximum cross-sectional area, m^2
S_{HT}	Horizontal tail area, m^2
λ	Aspect ratio

$\bar{D}$	Fuselage diameter-to-wing span ratio
D	Diameter of propeller, m
φ_W, φ_S	Incidence angle of wing, horizontal tail, degree
$K_{\alpha\alpha}, K_{\phi 0}$	Wing-body interference factor
$(K_{\alpha\alpha})_{HT}, (K_{\phi 0})_{HT}$	Horizontal tail-body interference factor
F_T	Thrust of propeller, N
$F_A = [X, Y, Z]^T$	Aerodynamic forces in wind coordinates, Drag, Side Force and Lift, N
$M_b = [M_x, M_y, M_z]^T$	Aerodynamic moments in body coordinates, Roll, Pitch and Yaw moment, N· m
$C_{FT1}, C_{FT2}, C_{FT3}$	Dimensionless thrust coefficients
C_X, C_Y, C_Z	Dimensionless aerodynamic-force coefficients
m_x, m_y, m_z	Dimensionless aerodynamic-moment coefficients
V_a	Airspeed, m/s
α	Angle-of-attack, rad
β	Angle of sideslip, rad
ε	Angle of downwash, rad
ρ	Air density, kg/m^3
Q	Dynamic pressure, Pa
k_{HT}	Airflow block factor
$\delta_e, \delta_r, \delta_a$	Deflection angle of elevator, rudder and aileron, rad
n_P	Revolutions per second of propeller, r/s
d_n, d_e, d_r, d_a	PWM width corresponding to the channel of thrust, elevator, rudder and aileron
ϕ, θ, ψ	Roll, pitch, yaw angle rad
$V^b = [u, v, w]^T$	Body axis velocities, m/s
$\omega^b = [p, q, r]^T$	Body axis angular rates, rad/s

References

1. Austin, R. *Unmanned Aircraft Systems: UAVS Design, Development and Deployment*, 1st ed.; John Wiley & Sons: Chichester, UK, 2011.
2. Ducard, G.J. *Fault-Tolerant Flight Control and Guidance Systems: Practical Methods for Small Unmanned Aerial Vehicles*; Springer Science & Business Media: London, UK, 2009.
3. Klein, V.; Morelli, E.A. *Aircraft System Identification: Theory and Practice*; American Institute of Aeronautics and Astronautics: Reston, VA, USA, 2006.
4. Marqués, P.; Da Ronch, A. *Advanced Uav Aerodynamics, Flight Stability and Control: Novel Concepts, Theory and Applications*; John Wiley & Sons: Southport, UK, 2017.
5. Wisnoe, W.; Nasir, R.E.M.; Kuntjoro, W.; Mamat, A.M. Wind tunnel experiments and CFD analysis of Blended Wing Body (BWB) Unmanned Aerial Vehicle (UAV) at mach 0.1 and mach 0.3. In Proceedings of the 13th International Conference on Aerospace Sciences & Aviation Technology, ASAT-13, Cairo, Egypt, 26–28 May 2009.
6. Williams, J.E.; Vukelich, S.R. *The USAF Stability and Control DATCOM, Users Manual*; McDonnell Douglas Astronautics Company: Berkeley, MO, USA, 1999; Volume 1.
7. STABILITY, USAF. Control Datcom. In Air Force Flight Dynamics Laboratory, Wright-Patterson Air Force Base, Ohio; 1972.
8. Arifianto, O.; Farhood, M. Development and modeling of a low-cost unmanned aerial vehicle research platform. *J. Intell. Robot. Syst.* **2015**, *80*, 139.
9. Etkin, B. *Dynamics of Atmospheric Flight*; Dover Publications: New York, NY, USA, 2005.
10. Nelson, R.C. *Flight Stability and Automatic Control*, 2nd ed.; WCB/McGraw Hill: New York, NY, USA, 1998.
11. Phillips, W.F. *Mechanics of Flight*; John Wiley & Sons: Chichester, UK, 2004.

12. Schmidt, L.V. *Introduction to Aircraft Flight Dynamics*; American Institute of Aeronautics and Astronautics (AIAA): Reston, VA, USA, 1998.

13. Lebedev, A. *Flight Dynamics of the Unmanned Aerial Vehicles*; Zhang, B., Translator; National Defence Industry Press: Beijing China, 1964.

14. Xu, M.; An, X. *Analysis and Calculation of Aerodynamic Characteristics of Aircraft*; Northwestern Polytechnic University Press: Xi'an, China, 2012.

15. Tischler, M.B.; Remple, R.K. *Aircraft and Rotorcraft System Identification*; AIAA Education Series; American Institute of Aeronautics and Astronautics: Reston, VA, USA, 2006.

16. Ljung, L. System identification. In *Signal Analysis and Prediction*; Springer: New York, NY, USA, 1998; pp. 163–173.

17. Guo, M.; Su, Y.; Gu, D. System identification of the quadrotor with inner loop stabilisation system. *Int. J. Model. Identif. Control* **2017**, *28*, 245–255.

18. Ghosh, A.; Raisinghani, S. Frequency-domain estimation of parameters from flight data using neural networks. *J. Guid. Control Dyn.* **2001**, *24*, 525–530.

19. Chen, S.; Billings, S. Neural networks for nonlinear dynamic system modelling and identification. *Int. J. Control* **1992**, *56*, 319–346.

20. Morelli, E.A. Real-time aerodynamic parameter estimation without air flow angle measurements. *J. Aircr.* **2012**, *49*, 1064–1074.

21. Dorobantu, A.; Ozdemir, A.A.; Turkoglu, K.; Freeman, P.; Murch, A.; Mettler, B.; Balas, G. Frequency domain system identification for a small, low-cost, fixed-wing uav. In Proceedings of the AIAA Guidance, Navigation, and Control Conference, Portland, OR, USA, 8–11 August 2011, p. 6719.

22. Dorobantu, A.; Murch, A.; Mettler, B.; Balas, G. System identification for small, low-cost, fixed-wing unmanned aircraft. *J. Aircr.* **2013**.

23. Garcia-Velo, J.; Walker, B.K. Aerodynamic parameter estimation for high-performance aircraft using extended Kalman filtering. *J. Guid. Control Dyn.* **1997**, *20*, 1257–1260.

24. Tang, Y.; Wu, Y.; Wu, M.; Wu, W.; Hu, X.; Shen, L. INS/GPS integration: Global observability analysis. *IEEE Trans. Veh. Technol.* **2009**, *58*, 1129–1142.

25. Jategaonkar, R. *Flight Vehicle System Identification: A Time Domain Methodology*; American Institute of Aeronautics and Astronautics: Reston, VA, USA, 2006; Volume 216.

26. Von Hoffer, N. *System Identification of a Small Low-Cost Unmanned Aerial Vehicle Using Flight Data from Low-Cost Sensors*; Utah State University: Logan, UT, USA, 2015.

27. Chowdhary, G.; Jategaonkar, R. Aerodynamic parameter estimation from flight data applying extended and unscented Kalman filter. *Aerosp. Sci. Technol.* **2010**, *14*, 106–117.

28. Koifman, M.; Bar-Itzhack, I. Inertial navigation system aided by aircraft dynamics. *IEEE Trans. Control Syst. Technol.* **1999**, *7*, 487–493.

29. Yang, Z.; Igarashi, H.; Martin, M.; Hu, H. An experimental investigation on aerodynamic hysteresis of a low-Reynolds number airfoil. In Proceedings of the 46th AIAA Aerospace Sciences Meeting and Exhibit, Reno, NV, USA, 7–10 January 2008.

30. Xsens, T. *MTi-G User Manual and Technical Documentation*; Xsens Technologies B.V.: Enschede, The Netherlands, 2010.

31. Jung, D.; Levy, E.; Zhou, D.; Fink, R.; Moshe, J.; Earl, A.; Tsiotras, P. Design and development of a low-cost test-bed for undergraduate education in UAVs. In Proceedings of the 44th IEEE Conference on Decision and Control, and 2005 European Control Conference, Seville, Spain, 12–15 December 2005, pp. 2739–2744.

32. Owens, D.B.; Cox, D.E.; Morelli, E.A. Development of a low-cost sub-scale aircraft for flight research: The FASER project. In Proceedings of the 25th AIAA Aerodynamic Measurement Technology and Ground Testing Conference, Fluid Dynamics and Co-located Conferences, San Francisco, CA, USA, 5–8 June 2006.

sensors

MDPI

Article

Olive Actual "on Year" Yield Forecast Tool Based on the Tree Canopy Geometry Using UAS Imagery

Rafael R. Sola-Guirado *, Francisco J. Castillo-Ruiz, Francisco Jiménez-Jiménez, Gregorio L. Blanco-Roldan, Sergio Castro-Garciaand Jesus A. Gil-Ribes

Department of Rural Engineering, University of Cordoba, E.T.S.I. Agronomos y Montes, Campus de Rabanales, Ctra. Nacional IV Km 396, 14014 Cordoba, Spain; g62caruf@uco.es (F.J.C.-R.); francisjimenez2@gmail.com (F.J.-J.); ir3blrog@uco.es (G.L.B.-R.); scastro@uco.es (S.C.-G.); gilribes@uco.es (J.A.G.-R.)
* Correspondence: ir2sogur@uco.es

Received: 2 June 2017; Accepted: 26 July 2017; Published: 30 July 2017

Abstract: Olive has a notable importance in countries of Mediterranean basin and its profitability depends on several factors such as actual yield, production cost or product price. Actual "on year" Yield (AY) is production (kg tree^{-1}) in "on years", and this research attempts to relate it with geometrical parameters of the tree canopy. Regression equation to forecast AY based on manual canopy volume was determined based on data acquired from different orchard categories and cultivars during different harvesting seasons in southern Spain. Orthoimages were acquired with unmanned aerial systems (UAS) imagery calculating individual crown for relating to canopy volume and AY. Yield levels did not vary between orchard categories; however, it did between irrigated orchards (7000–17,000 kg ha^{-1}) and rainfed ones (4000–7000 kg ha^{-1}). After that, manual canopy volume was related with the individual crown area of trees that were calculated by orthoimages acquired with UAS imagery. Finally, AY was forecasted using both manual canopy volume and individual tree crown area as main factors for olive productivity. AY forecast only by using individual crown area made it possible to get a simple and cheap forecast tool for a wide range of olive orchards. Finally, the acquired information was introduced in a thematic map describing spatial AY variability obtained from orthoimage analysis that may be a powerful tool for farmers, insurance systems, market forecasts or to detect agronomical problems.

Keywords: olive; production forecast; manual canopy volume; individual crown area; tree mapping

---

1. Introduction

The olive crop covers more than 10 Mha in the world and has a notable social and economic importance in countries of Mediterranean basin, such as Spain that constitutes 44% and 22% of the global olive oil and table olive production, respectively [1]. Olive orchard profitability is highly influenced by yield which depends on orchard category [2] as well as harvesting season, due to alternate bearing [3]. This fact leads to discern between "on" and "off" years. Actual "on year" Yield (AY) is the potential yield in a forecasting model, so the yield actually achieved considering all limiting factors, and it is smaller than potential yield [4]. Thus, AY is the tree productivity, in kg tree^{-1}, obtained in an "on year" and orchard actual "on year" Yield (OAY) is the orchard yield, in kg ha^{-1}.

AY depends on several factors, such as environmental limitations, plant material or intercepted radiation, which is determined by canopy volume, leaf area index and leaf area density. The relationship between production and the canopy volume of a tree has been studied [5]. Tools to forecast AY would be highly useful to facilitate olive orchard management, policy making or to predict supply chain behavior, improving price stability.

It is possible to measure the canopy volume of trees using manual measurements or electronic devices [6]. Nonetheless, most applications in olive growing operations do not require accurate information of canopy volume because of the fact that manual canopy volume measurement with surveying rod is widely used as one goal of looking at overall tree shape and using a minimum number of measurements. Furthermore, aerial imagery from unmanned aerial systems (UASs) is a very promising method to characterize olive tree canopies [7]. It can be hypothesized that individual crown area obtained from an orthoimage may be highly correlated with canopy volume [8] assuming that, within an orchard, tree heights are rather uniform. Then, individual contour area of trees would be a useful parameter for other application that traditionally had been used, the tree canopy volume eliminating the dimension of tree height. The use of a 2D approach by ortho-mosaics would allow for the creation of site-specific management maps that could be related with geographic information systems (GIS) given an extra value. Furthermore, for farmers it is easier to use individual crown area than "apparent canopy volume" calculated by digital surface and terrain models; requiring less technical knowledge of aerial imagery with UAS combined with a simple image processing would be a valid tool to get individual crown area from the orthoimages generated, faster and easier than calculate canopy volume manually or digitally with point cloud.

The objective of this research was to assess tools for olive AY forecast based on tree canopy measurements for a wide range of olive orchards in southern Spain. AY was correlated with manual canopy volume. To provide a more useful method, we propose relating the manual canopy volume of trees to their individual crown area calculated from orthoimages acquired by UASs. This allowed for the creation of maps which could predict the AY for a particular use in a specific farm and enforcing the results understandability.

2. Material and Methods

2.1. Measurement and Relationship between AY and Manual Canopy Volume

Ten orchards were randomly selected from a research database among the most common orchard categories and layout [2]. Orchards are classified attending two variables: Irrigation (irrigated/rainfed) and orchard category (traditional/intensive/large hedgerow). Trees from the main olive cultivars in south Spain were included, such as Picual, Hojiblanca, Arbequina and Manzanilla. Super high-density orchards (more than 800 trees ha^{-1}) were not considered in this study, considering that canopy volume profiles may be rectangular, truncated rectangular, or triangular [9] instead of ellipsoid volume assumed for the rest of orchard categories.

The manual canopy volume of 518 trees located in different orchards was measured. The trees were harvested in four different harvesting seasons of "on years from 2011 to 2014" and fruits were weighed to obtain AY (kg tree^{-1}). Manual canopy volume (MCV) in m^3 was calculated following Equation (1) [10].

$$MCV = \frac{1}{6}\pi D_1 \times D_2 \times \frac{1}{2}\left((Ht_1 - Hs_1) + (Ht_2 - Hs_2)\right) \tag{1}$$

where D_1 and D_2 are crown diameters, Ht is tree height, Hs is skirt height, as the lowest canopy height from the ground. Heights were measured in two perpendicular positions. To obtain the diameters and heights of canopy geometry, two operators using a surveying rod were required using the next procedure (Figure 1): Firstly, operator A placed a surveying rod vertically at the olive canopy center. Then, operator B was located 10 m away from the surveying rod and took measurements of the outer canopy parts (Ht$_1$ and Hs$_1$). Afterward, operator A placed the surveying rod horizontally 1.5 m above the ground. The rod was extended from the two furthest points of the crown in that direction while operator B took the measurement of the canopy width (D$_1$). Finally, these steps were performed again in a position located at a 90-degree angle from the first position.

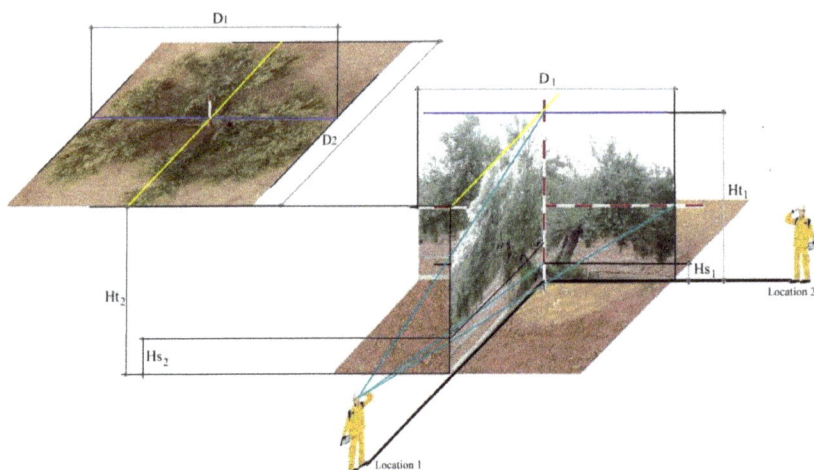

Figure 1. Manual Canopy Volume (MCV) measurement procedure for an olive tree.

A linear regression was described between manual canopy volume and AY. Two linear regressions were built separating data from irrigated orchards and rainfed ones. OAY, in kg ha^{-1} was calculated multiplying AY by planting density, in trees ha^{-1}. Multiple regression models were avoided considering that AY forecasting tool was targeted to be used by farmers.

2.2. Measurement and Relationship between Manual Canopy Volume and Individual Crown Area

A total of 36 trees from the most representative orchards studied (irrigated intensive, irrigated traditional and rainfed traditional) [1], were randomly selected (supplementary material). The trees were measured manually to determine manual canopy volume per tree (Figure 1), and digitally using the information acquired from a UAS to determine individual crown area (Figure 2). A linear regression was obtained to set a relationship between both variables.

Individual crown area was obtained from orthoimages acquired using a UAS. The UAS operated fully autonomously using waypoint navigation guidance for automatic image acquisition. The flight plan was set at 90 m height, 8 m s^{-1} cruising speed, and images were overlapped 85% and 70% in longitudinal and transversal directions respectively. Imagery was synchronized using a global positioning system (GPS) and triggering time was recorded for each image. There was no user interaction required for processing the acquired images during the flight. The pictures obtained from the UAS enabled orchard orthoimages creation using Pix4D (Ecublens). The matching image was calculated using the options "capture time", "triangulation of image geolocation", and "image similarity". An older map of the area was used to set the ground control points. Aerial pictures were acquired using a digital camera (NEX7, Sony) with a shutter speed of 1:4000 s, an aperture of 3.617, a focal length of 18 mm and an ISO velocity index of 100, without a flash on windless days close to 12:00 p.m. The camera was placed in a gimbal (Zenmuse Z15, Dji) that maintained the camera position using 3 axes that were programmed using the aerial position. This device was fixed to a UAS frame (S800, Dji) controlled using a multirotor autopilot system (Wookong-M, Dji).

Orthoimages analysis (Figure 2) was performed using the open-source Java-based ImageJ package (ImageJ, National Institutes of Health). In the first stage, the image size was adjusted to the set measurement scale to convert pixel^2 into m^2 using a known measurement reference in the original image. Secondly, the original red, green, blue (RGB) values of the image were transformed into monochromatic grey-scale according to algorithm based on the lightness [11]. The green band was used to perform the segmentation procedure of the tree crown from the ground, based on the discrimination of undesirable elements using the RGB values of the individual pixels [12]. Manual segmentation of the image was made by applying a threshold as a function of the intensity values of the RGB image between 0 and 255 and labeling each pixel as black or white, depending on whether the pixel value was greater or less than the threshold selected. Two thresholds were selected, one to remove the ground from the image and another to remove the tree shadow. Once the tree canopy was separated by this binary process, the tree crown area was calculated using the automated routine tool known as "analyze particles" based on edge detection algorithms, which numbered and outlined each tree [13].

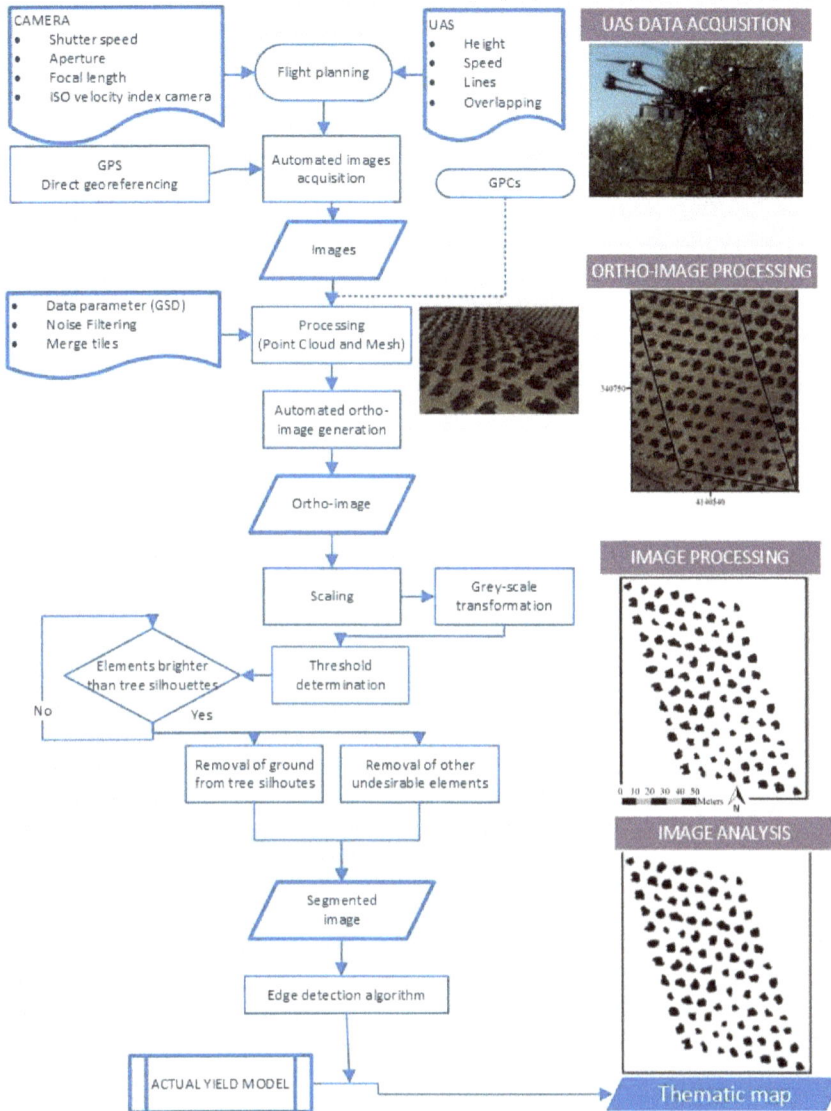

Figure 2. Orthoimage analysis procedure in a traditional olive orchard. UAS (Unmanned Aerial System), GPS (Global Positioning System), GSD (Ground Sample Distance), GPC (Ground Point Control).

Finally, AY was estimated with individual crown area by the relation between individual crown area and manual canopy volume. A plot of one orchard was mapped for predicting AY with the relationship obtained by using the individual crown area measured.

3. Results and Discussion

3.1. Orchard Actual "on Year" Yield and Other Features

Olive yield was influenced by many factors such as canopy dimensions, orchard layout, water availability or other stresses. Farmers are aware of these complex relations. However, all these variables are often summed up in orchard yield as the main factor for orchard profitability. Within the same orchard category, OAY increased when the planting density increased, while AY exhibited the opposite trend (Table 1). Fruit set and fruit fresh weight may influence AY [14] along with planting density. The ratio between production and canopy volume (kg m^{-3}) was generally higher for denser plantations, ranging from 1.6 to 3.1 kg m^{-3} for orchards with 70 trees ha^{-1} and 408 trees ha^{-1}, respectively, though there were exceptions due to specific orchard constraints. Other authors describe lower production efficiency ratios, i.e., 1 to 0.26 kg m^{-3}, depending on the tree cultivar [15].

Table 1. Orchards and trees parameters measured in "on year" season. Parameters are the mean ± standard deviation.

Orchard Category	Planting Density (tree ha^{-1}) – (Planting Distance)	Tree Production or Actual Yield (kg tree^{-1})	Orchard Actual Yield (kg ha^{-1})	Manual Canopy Volume (m^3 tree^{-1})	Production Per Canopy Volume (kg m^{-3})	Orchard Canopy Volume (m^3 ha^{-1})
Irrigated large hedgerow	555 (6 × 3 m)	31.4 ± 9.3	17,463	11.7 ± 3.4	2.8 ± 0.8	6723
	408 (7 × 3.5 m)	24.2 ± 11.4	9883	7.6 ± 2.5	3.1 ± 0.9	3099
	312 (8 × 4 m)	29.3 ± 6.1	9171	21.6 ± 5.3	1.4 ± 0.3	6760
Irrigated intensive	285 (7 × 5 m)	53.3 ± 17.3	15,251	19.9 ± 9.5	2.9 ± 0.6	5690
	208 (6 × 8 m)	39.0 ± 16.8	7479	21.4 ± 9.3	1.9 ± 0.6	4655
	204 (7 × 7 m)	45.1 ± 15.4	9190	23.9 ± 8.1	1.9 ± 0.5	4879
Rainfed intensive	158 (7 × 9 m)	45.2 ± 11.5	7181	21.4 ± 5.1	2.2 ± 0.7	3395
	138 (8 × 9 m)	31.4 ± 9.6	4278	22.3 ± 9.6	1.5 ± 0.4	3023
Irrigated traditional	70 (12 × 12 m)	162.9 ± 27.9	11,241	96.4 ± 15.6	1.7 ± 0.3	6652
Rainfed traditional	70 (12 quincunx)	81.2 ± 23.6	6496	61.2 ± 30.6	1.6 ± 0.8	4893

The mean OAY in the irrigated orchards was higher than that in the rainfed ones. The mean OAY in the irrigated large hedgerows (12,172 kg ha^{-1}) was slightly higher than that for irrigated traditional orchards (11,241 kg ha^{-1}) and irrigated intensive orchards (10,640 kg ha^{-1}). By contrast, the rainfed orchards had much lower mean "on year" yields, from 6496 kg ha^{-1} produced by the traditional orchards to 5729 kg ha^{-1} for intensive orchards. Yield values were in accordance with those described in Spain for new plantations (aged 3 to 7 years old) with 408 trees ha^{-1}, which produced 9540 kg ha^{-1}, and with 816 trees ha^{-1}, which produced 13,898 kg ha^{-1} [10]. Previous research also stated that the mean yield was influenced by tree training, planting density and location, with yields ranging between 6380 kg ha^{-1} and 10,580 kg ha^{-1} [3].

It was notable that high OAY values could be obtained in any irrigated orchard category with any tree training system. The lack of differences indicated that the profitability of orchard categories is not due only to their OAY, but other limiting factors, such as operational cost, inputs requirements, orchard size, or lower harvesting machinery performance. In this way, the number of high-density and hedgerow orchards is increasing, although traditional orchards are still the most widely used category of the total cultivated olive area. The linear regression (Equation (2)) predicted actual "on year" yield depending on manual canopy volume for a wide range of olive orchard categories (Table 1).

3.2. Actual "on Year" Yield Forecasting Tool Based on Manual Canopy Volume

A general regression (Equation (2)) to predict actual "on year" yield (AY) in kg per tree depending on manual canopy volume (MCV) was built for both irrigated and non-irrigated orchards.

The regression showed a highly linear trend ($r^2 = 0.76$, $p \leq 0.01$) and a standard error of 18.6 kg tree^{-1} for the mean estimation. The provided forecast equation was representative for south Spain considering the wide range of trees, orchard categories, locations and harvesting seasons tested. Further research is needed to adjust the model coefficients for other locations with different soil, climate constraints, or cultivars [16]. All data represented the yields from "on years", considering that production in "off year" did not provide a significant regression to describe tree production according to manual canopy volume (data not shown). The general regression showed the relationship between tree size and yield which have been described relating olive tree productivity to trunk girth [17]. This trend has also been proven in other crops, such as apple, in which orchard yield and photosynthetically active radiation were correlated with tree training system or tree architecture, canopy volume and the trunk cross-sectional area [18].

$$AY = 15.928 + 1.215 \times MCV \tag{2}$$

The general regression can be separated into two different specific forecast equations, one for irrigated and one for rainfed orchards.

$$Irrigated : AY = 10.642 + 1.541 \times MCV \tag{3}$$

$$Rainfed : AY = 25.932 + 0.781 \times MCV \tag{4}$$

Linear regression was obtained for irrigated orchards ($r^2 = 0.89$, $p \leq 0.01$) and 13.1 kg tree^{-1} as the standard error of the mean estimated (Figure 2, Equation (3)), whereas the regression for the rainfed orchards showed a slightly worse adjustment ($r^2 = 0.62$, $p \leq 0.01$), and 17.1 kg tree^{-1} as the standard error of the mean estimate (Equation (4)) (Figure 3).

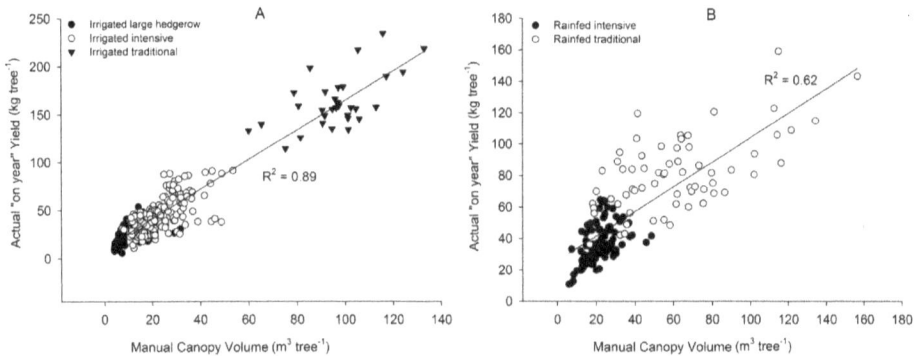

Figure 3. Linear model between the Manual Canopy Volume and the Actual "on year" Yield (AY) for the irrigated orchards (**A**) and for the rainfed orchards (**B**).

Irrigation improved AY by reducing water-limiting factor [4], although AY in rainfed orchards was close to that obtained in the irrigated orchards in rainy years. Accordingly, the soil variability resulted in higher data scattering for the rainfed orchards, while the irrigated orchards showed a scatter pattern that was similar to the general regression. These results agreed with previous research showing that olive production is strongly influenced by irrigation, although the response gradually decreases when the water applied approaches the maximum demand [19].

Not only canopy volume per tree has a significant influence on AY, but the canopy volume per hectare determined OAY, which should be maximized in order to increase crop profitability. It is advisable to adapt pruning intensity to reach optimal orchard canopy volume [10], considering that tree density, canopy size and soil management are strategic decisions for olive water relations [20] and then, for olive AY. For high-density orchards under climate conditions in southern Spain, the optimal canopy volume should be around 8000 m^3 ha^{-1} for rainfed orchards, while it should be between 11,000 and 13,000 m^3 ha^{-1} for irrigated ones [19]. For all studied orchards, the orchard canopy volume was below optimal values so it might be advisable to increase orchard canopy volume to increase OAY, although in some cases, vigorous trees may decrease harvesting performance reducing olive orchard profitability [21]. Similarly, it is important to choose when and how to perform the pruning according to the harvesting method to enhance the harvest effectiveness based on the branches' vibrations [22].

3.3. Actual "on Year" Yield Forecasting Tool Based on Tree Crown Area

Manual canopy volume was an adequate predictor for AY (Figure 3), although the method introduces several errors. Firstly, it does not consider the high variability between crown shapes [23] and estimates the canopy volume assuming that all trees had an ellipsoid volume. Secondly, manual measurements might show a lower resolution than other available systems, because of the accuracy of the operator related to how well he can identify the crown contour. Furthermore, this method requires remarkable time consumption for taking measurements. In this work, the time spent for measuring 1 ha (96 trees) of a traditional orchard was approximately 4 h, with a mean value of 2.5 min tree^{-1}. However, although the method may be tedious and have high labour requirements for large areas, it could be valid for smaller ones.

Significant differences ($p < 0.05$) for AY, manual canopy volume and individual tree crown area matched between orchard categories (Table 2). Intensive orchard had smaller canopy volume, and the irrigated orchard volumes were less scattered than the rainfed ones. Therefore, irrigation not only increases AY but also reduces orchard variability.

Table 2. Tree parameters of common orchards measured in "on year" season used for AY estimation. Parameters are the mean ± standard deviation values. Different letters within a column show significant differences between orchard categories (Duncan's post hoc test, $p < 0.05$).

Orchard Category	Production in or Actual Yield (kg tree^{-1})	Manual Canopy Volume (m^3 tree^{-1})	Individual Crown Area (m^2 tree^{-1})
Irrigated intensive	38.6 ± 4.3 a	12.1 ± 1.7 a	10.3 ± 1.3 a
Rainfed traditional	65.8 ± 29.8 b	73.6 ± 27.6 b	24.2 ± 13.4 b
Irrigated traditional	155.2 ± 15.0 c	98.0 ± 9.9 c	34.8 ± 3.1 c

A linear regression was built to determine the relationship between manual canopy volume (MCV) in m^3 and the individual tree crown area (ICA) in m^2 for different orchard categories (Figure 4) ($r^2 = 0.83$, $p \leq 0.01$) (Equation (5)) because tree height was rather similar in commercial orchards, which was 3.8 ± 0.2 m for the irrigated traditional orchards and 3.9 ± 0.4 m for the rainfed orchards (mean ± standard deviation). This relationship may simplify the process of measurements and tree characterization for this purpose and does not need to be carried out using complex methods. Results confirmed that it is possible to estimate canopy volume from orthoimagery obtained from a commercial digital camera mounted on an inexpensive UAS by applying Equation (5) which resulted in a standard error of the mean estimate of 16.77 m^3 tree^{-1}.

$$MCV = -7.906 + 3.080 \times ICA \qquad (5)$$

Canopy volume estimation from aerial imagery may include tree height measurements obtained from Digital Surface Models (DSMs), though this does require data processing that increases the computation time. However, for determining individual tree crown and density arrangements, this information is not necessary. UAS orthoimagery has been appropriated for determining the individual crown area of the trees, although this can be estimated by obtaining data from other aerial remote method. In this way, satellite or UAS imagery could be useful sources for estimating these parameters using very high-resolution satellites such as Pleiades, WorldView-2 or OrbView, which provide images with a resolution of up to 0.4 m that allows for the recognition, identification and delineation of individual tree crowns using object-based image analysis [24]. Digital elevation models using traditional airborne platforms can also be used for tree delineation [25], although the images obtained might make the process more expensive and unaffordable for small- and medium-sized orchards, considering that they are adapted to cover large areas. Therefore, UASs is a flexible tool for the acquisition of digital images to get thematic maps in any moment without depending on cloud condition.

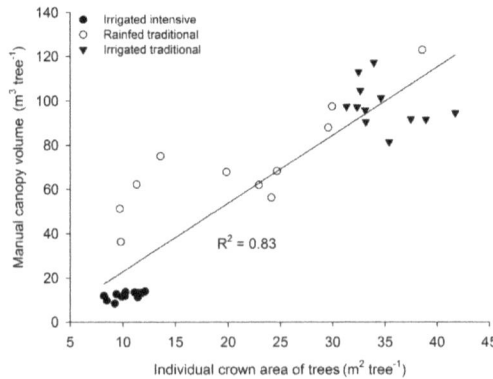

Figure 4. Linear model between the individual crown area (ICA) of trees and the manual canopy volume (MCV) for three different orchard categories.

The resolution of orthoimages obtained has been acceptable, having a resolution up 0.02 m, and, therefore, the procedure has supposed a promising alternative to characterize the canopy of olive trees [7]. These models describe the surfaces of both terrain and trees, so the detection and delineation of trees represents an effective technique for imagery analysis. The processing result demonstrates that the proposed method was not affected by the structure of the vegetation or by ground unevenness. Moreover, it was not necessary to use any radiometric normalization because the mosaic created did not comprise a large number of images collected in the same period of time under the same conditions. In this work, the threshold values for segmentation were set manually and the tree delineation was achieved automatically. Nonetheless, the assessment of individual crown area may be performed automatically using specific algorithms and image-processing software [26]. Further advances should accomplish a fully automated process.

Manual canopy volume was replaced by individual crown area in the specific regressions for irrigated and rainfed orchards (Equations (3) and (4)). Estimated value of AY using individual crown area was compared to measured yield, providing good adjustment slope close to 1:1 (Figure 5). These equations could be used in a wide range of olive orchard categories from large hedgerow orchards to traditional ones.

Further studies should include a real-time system to detect single trees and tree canopies measures might be integrated [27] into new agricultural machinery for different labour requirements, such as pest and disease control or to determine harvesting labour using specific mapping. It could be used to develop innovations for olive harvesting machinery; for instance, they could adapt vibration parameters (e.g., vibration power, frequency, amplitude) to tree canopy volume in order to enhance harvesting efficiency [28].

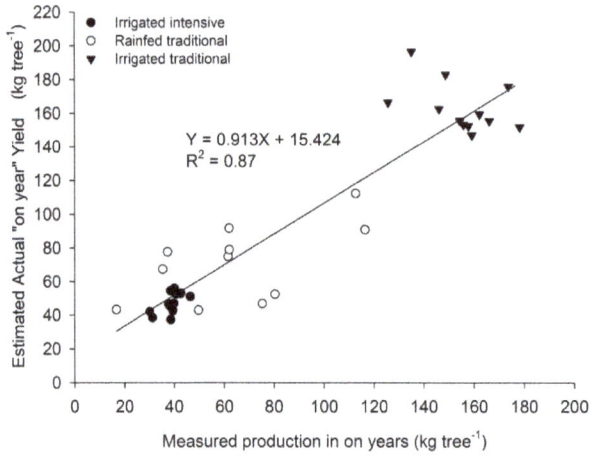

Figure 5. Predicted AY and measured production in an "on-year".

Orthoimages analysis may provide valuable information for farmers representing acquired information on thematic maps. A plot of 1 hectare of rainfed traditional olive trees was mapped using this technique (Figure 6). As a result, AY was calculated based on each tree individual crown area. Therefore, canopy volume or individual crown area mapping could be useful tools to describe spatial yield variability. The information provided by these maps should be integrated into precision agriculture decision-making processes to enhance orchard management efficiency.

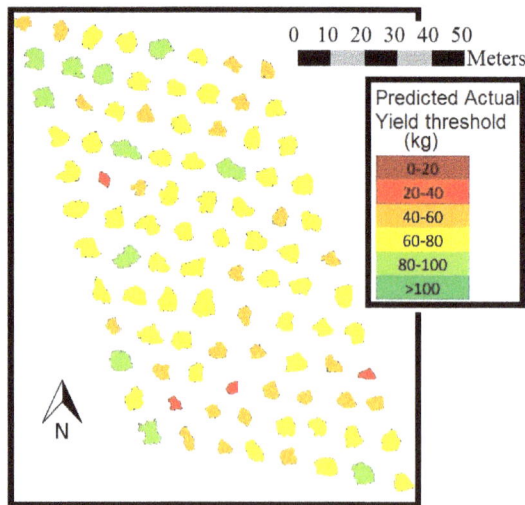

Figure 6. Mapping of the expected Actual Yield in a rainfed traditional olive orchard related to the variability in the individual crown area of the trees.

4. Conclusions

This research provides new advances for the description of broadly applicable methodology to describe tree and orchard actual "on year" yield for both oil ant table olives and for all olive orchard

Sensors **2017**, *17*, 1743

categories except super high-density ones. Significant relationships were described between tree crown features and actual yield (AY), so forecasts for AY can be performed using manual canopy volume, but also by using individual crown area, simplifying the method to calculate the canopy volume over large areas, and avoiding other tree crown measurements. Once AY could be forecasted, OAY might be estimated for a farm, area, or even for a region using quick and inexpensive methodology. Nonetheless, these relationships are valid for southern Spain conditions, thus, further research is required to obtain the adjusted coefficients for other geographical areas. Irrigation was a key factor that should be considered to assess AY. However, there were no big differences in OAY between the studied orchard categories. An optimal canopy volume per hectare should be achieved to reduce yield gap by tree training, taking into account the enhancements in harvesting operations. Crown features measurements and yield forecasting provide very valuable maps for farmers, agricultural insurance systems and researchers to characterize an orchard, to enhance orchard management and to predict economic, agronomical or social aspects.

Acknowledgments: The authors gratefully acknowledge financial support from 'Interprofesional del Aceite de Oliva Español 'and 'Organización Interprofesional de la Aceituna de Mesa'. The authors also acknowledge all involved farms that help to gather in-field data.

Author Contributions: All the authors made significant contributions to the manuscripts. R.R.S.G. and F.J.C.R. designed the experiment and wrote the manuscript. R.R.S.G. performed the UAS data acquisition. F.J.C.R. and R.R.S.G. performed the field component. F.J.R.C. performed the statistical analysis. F.J.J., G.L.B.R., J.A.G.R. and S.C.G. scientifically supported and reviewed the paper. J.A.G.R. managed the project.

Conflicts of Interest: The authors declare no conflict of interest.

References

1. International Olive Council [IOC]. World Table Olive Figures 2014. Available online: http://www.internationaloliveoil.org (accessed on 12 December 2014).
2. Rallo, L.; Barranco, D.; Castro-García, S.; Connor, D.J.; Gómez del Campo, M.; Rallo, P. High-density olive plantations. *Hortic. Rev.* **2013**, *41*, 303–384.
3. Lavee, S.; Haskal, A.; Avidan, B. The effect of planting distances and tree shape on yield and harvest efficiency of cv. Manzanillo table olives. *Sci. Hortic.* **2012**, *142*, 166–173. [CrossRef]
4. Van Ittersum, M.K.; Cassman, K.G.; Grassini, P.; Wolf, J.; Tittonell, P.; Hochman, Z. Yield gap analysis with local to global relevance—A review. *Field Crops Res.* **2013**, *143*, 4–17. [CrossRef]
5. Castro-García, S.; Blanco-Roldán, G.L.; Gil-Ribes, J. Suitability of Spanish 'Manzanilla' table olive orchards for trunk shaker harvesting. *Biosyst. Eng.* **2015**, *129*, 388–395. [CrossRef]
6. Llorens, J.; Gil, E.; Llop, J.; Escolà, A. Ultrasonic and LiDAR sensors for electronic canopy characterization in vineyards: Advances to improve pesticide application methods. *Sensors* **2011**, *11*, 2177–2194. [CrossRef] [PubMed]
7. Dandois, J.P.; Ellis, E.C. High spatial resolution three-dimensional mapping of vegetation spectral dynamics using computer vision. *Remote Sens. Environ.* **2013**, *36*, 259–276. [CrossRef]
8. Redei, K.; Veperdi, I. Study of the relationships between crown and volume production of black locust trees (*Robinia Pseudoacacia* L.). *For. J.* **2001**, *47*, 135–142.
9. Connor, D.J. Towards optimal designs for hedgerow orchards. *Aust. J. Agric. Res.* **2006**, *57*, 1067–1072. [CrossRef]
10. Muñoz-Cobo, M.P.; Humanes-Guillén, J. (Eds.) Poda de producción. In *La Poda Del Olivo: Moderna Olivicultura*; Editorial Agrícola Española: Madrid, Spain, 2006; pp. 77–139.
11. Kanan, C.; Cottrell, G.W. Color-to-grayscale: Does the method matter in image recognition? *PLoS ONE* **2012**, *7*, e29740. [CrossRef] [PubMed]
12. Meyer, G.E.; Neto, J.C. Verification of colour vegetation indices for automated crop imaging applications. *Comput. Electron. Agric.* **2008**, *63*, 282–293. [CrossRef]
13. Igathinathane, C.; Pordesimo, L.O.; Columbus, E.P.; Batchelor, W.D.; Methuku, S.R. Shape identification and particles size distribution from basic shape parameters using ImageJ. *Comput. Electron. Agric.* **2008**, *63*, 168–182. [CrossRef]

14. Barone, E.; La Mantia, M.; Marchese, A.; Marra, F.P. Improvement in yield and fruit size and quality of the main Italian table olive cultivar 'Nocellara del Belice'. *Sci. Agric.* **2014**, *71*, 52–57. [CrossRef]
15. Farinelli, D.; Ruffolo, M.; Boco, M.; Tombesi, A. Yield efficiency and mechanical harvesting with trunk shaker of some international olive cultivars. *Acta Hortic.* **2012**, *949*, 379–384. [CrossRef]
16. Simkeshzadeh, N.; Etemadi, N.; Mobli, M.; Baninasab, B. Use of olive cultivars in landscape planning regarding form and texture. *J. Agric. Sci. Technol.* **2015**, *17*, 717–724.
17. Fernández-Escobar, R. A model to estimate tree size when trunk girth cannot be used. *J. Am. Pomol. Soc.* **2014**, *68*, 80–88.
18. Aggelopoulou, A.D.; Bochtis, D.; Fountas, S.; Swain, K.C.; Gemtos, T.A.; Nanos, G.D. Yield prediction in apple orchards based on image processing. *Precis. Agric.* **2011**, *12*, 448–456. [CrossRef]
19. Moriana, A.; Orgaz, F.; Pastor, M.; Fereres, E. Yield responses of a mature olive orchard to water deficits. *J. Am. Soc. Hortic. Sci.* **2003**, *128*, 425–431.
20. Connor, D.J. Adaptation of olive (*Olea europaea* L.) to water-limited environments. *Aust. J. Agric. Res.* **2005**, *56*, 1181–1189. [CrossRef]
21. Villalobos, F.J.; Testi, L.; Hidalgo, J.; Pastor, M.; Orgaz, F. Modelling potential growth and yield of olive (*Olea europaea* L.) canopies. *Eur. J. Agron.* **2006**, *24*, 296–303. [CrossRef]
22. Tombesi, S.; Poni, S.; Palliotti, A.; Farinelli, D. Mechanical vibration transmission and harvesting effectiveness is affected by the presence of branch suckers in olive trees. *Biosyst. Eng.* **2017**, *158*, 1–9. [CrossRef]
23. Castillo-Ruiz, F.J.; Castro-Garcia, S.; Blanco-Roldan, G.L.; Sola-Guirado, R.R.; Gil-Ribes, J.A. Olive crown porosity measurement based on radiation transmittance: An assessment of pruning effect. *Sensors* **2016**, *16*, 723. [CrossRef] [PubMed]
24. Gougeon, F.A.; Leckie, D.G. The individual tree crown approach applied to Ikonos images of a coniferous plantation area. *Photogramm. Eng. Remote Sens.* **2006**, *2*, 1287–1297. [CrossRef]
25. Koch, B.; Heyder, U.; Weinacker, H. Detection of individual tree crowns in airborne lidar data. *Photogramm. Eng. Remote Sens.* **2006**, *72*, 357–363. [CrossRef]
26. Ke, Y.H.; Quackenbush, L.J. A review of methods for automatic individual tree-crown detection and delineation from passive remote sensing. *Int. J. Remote Sens.* **2011**, *32*, 4725–4747. [CrossRef]
27. Turner, D.; Lucieer, A.; Watson, C. An automated technique for generating georectified mosaics from ultra-high resolution unmanned aerial vehicle (UAV) imagery, based on structure from motion (SfM) point clouds. *Remote Sens.* **2012**, *4*, 1392–1410. [CrossRef]
28. Sola-Guirado, R.R.; Jimenez-Jimenez, F.; Blanco-Roldan, G.L.; Castro-Garcia, S.; Castillo-Ruiz, F.J.; Gil-Ribes, J.A. Vibration parameters assessment to develop a continuous lateral canopy shaker for mechanical harvesting of traditional olive trees. *Span. J. Agric. Res.* **2016**, *14*, 0204. [CrossRef]

sensors

MDPI

Article

Vicarious Calibration of sUAS Microbolometer Temperature Imagery for Estimation of Radiometric Land Surface Temperature

Alfonso Torres-Rua

Utah Water Research Laboratory, Utah State University, Logan, UT 84322, USA; alfonso.torres@usu.edu;
Tel.: +1-435-797-0397

Received: 20 May 2017; Accepted: 20 June 2017; Published: 26 June 2017

Abstract: In recent years, the availability of lightweight microbolometer thermal cameras compatible with small unmanned aerial systems (sUAS) has allowed their use in diverse scientific and management activities that require sub-meter pixel resolution. Nevertheless, as with sensors already used in temperature remote sensing (e.g., Landsat satellites), a radiance atmospheric correction is necessary to estimate land surface temperature. This is because atmospheric conditions at any sUAS flight elevation will have an adverse impact on the image accuracy, derived calculations, and study replicability using the microbolometer technology. This study presents a vicarious calibration methodology (sUAS-specific, time-specific, flight-specific, and sensor-specific) for sUAS temperature imagery traceable back to NIST-standards and current atmospheric correction methods. For this methodology, a three-year data collection campaign with a sUAS called "AggieAir", developed at Utah State University, was performed for vineyards near Lodi, California, for flights conducted at different times (early morning, Landsat overpass, and mid-afternoon") and seasonal conditions. From the results of this study, it was found that, despite the spectral response of microbolometer cameras (7.0 to 14.0 μm), it was possible to account for the effects of atmospheric and sUAS operational conditions, regardless of time and weather, to acquire accurate surface temperature data. In addition, it was found that the main atmospheric correction parameters (transmissivity and atmospheric radiance) significantly varied over the course of a day. These parameters fluctuated the most in early morning and partially stabilized in Landsat overpass and in mid-afternoon times. In terms of accuracy, estimated atmospheric correction parameters presented adequate statistics (confidence bounds under ± 0.1 for transmissivity and ± 1.2 W/m^2/sr/um for atmospheric radiance, with a range of RMSE below 1.0 W/m^2/sr/um) for all sUAS flights. Differences in estimated temperatures between original thermal image and the vicarious calibration procedure reported here were estimated from $-5\,°$C to $10\,°$C for early morning, and from 0 to $20\,°$C for Landsat overpass and mid-afternoon times.

Keywords: sUAS; vicarious calibration; thermal calibration; surface temperature; atmospheric correction; microbolometer cameras; thermal remote sensing

---

1. Introduction

Spatially distributed estimates of surface temperature can be useful in water resources research for applications in agriculture, geology, riparian habitat, and river corridor analysis [1]. Current efforts to monitor surface temperature using remote-sensing instruments vary in scale from continental (km/pixel) to plant (cm/pixel) and in instrumentation type (satellites, airborne/unmanned sensors) [2] and on-ground infrared radiometer sensors [3,4]. Surface temperature is of value in water resources studies due to its direct impact on processes such as evapotranspiration, soil moisture, open water evaporation, soil/water temperature profiles, climate change, drought monitoring, fish habitat, and others [1,5–9]. Satellite sensors can commonly provide easily accessible temperature information with worldwide coverage. Common satellites with thermal sensors are GOES, MODIS, and Landsat, while others are country-specific solutions, such as CBERS (China-Brazil Earth Resources Satellite) [10]. The imagery provided by these satellites ranges from 30 m/pixel/16 days at its finest resolution (Landsat ETM+/TIRS), to 375 m/pixel/1 day (VIIRS) to 500 m/pixel/1 day (MODIS Terra/Aqua), to 5 km/pixel/1 day (GOES). While the satellite information is used for large-scale surface processes (entire farm to sub-basin and basin scales), the information is of limited value for fine-scale processes that require sub-meter scale measurements and/or multiple measurements on the same day (e.g., sunrise, solar noon, mid-afternoon, night). For these requirements, manned aircraft and sUAS equipped with temperature sensors have been used. Examples of airborne and sUAS thermal applications can be found in [1,11–23]. Despite its importance in satellite temperature related research, little attention has been paid to atmospheric calibration of thermal imagery from manned aircraft and sUAS. One reason may be the often expensive meteorological sondes (to measures air temperature and relative humidity that must be used along with atmospheric profile models such MODTRAN and 6S [1,24–30]. By contrast, the technology implemented in manned aircraft/sUAS (lightweight, relative low-cost) thermal cameras is affected by local weather and flight elevation conditions. Therefore, the absence of standards or recommended procedures for referential calibration and atmospheric correction of thermal cameras for deployment on manned aircraft or sUAS can introduce uncertainty and systematic error in the data they collect and limit its synergistic use in combination with available satellite thermal imagery. The objective of this study was to develop standard procedures for vicarious (sUAS-specific, time-specific, flight-specific, and sensor-specific) atmospheric calibration of thermal cameras used in sUAS platforms.

1.1. Microbolometers UAS Temperature Cameras

In terms of weight limitations, sUAS (under 25 kg) have only one available radiometric temperature sensor solution: microbolometer infrared sensors (below 200 gr), which have a typical spectral response from ~7 um to ~14 um, using vanadium oxide (VOx) or amorphous silicon (A-Si) as the sensor core [31–33]. Figure 1 shows a microbolometer camera that is suitable for thermal remote sensing applications, as well as the spectral response of the sensor. Microbolometer technology uses the responsiveness of the sensor core material to changes in surface temperature that are larger than those of the sensor itself [31]. These sensors are an alternative to the cryogenically cooled thermal technology used in NASA and ESA satellites [34–36]. Miniaturized cryogenic temperature sensors exist, but are still too heavy for sUAS (over 4.0 Kg) [37]; thus, they are used mainly for manned aircraft. The manufacturer's absolute radiometric calibration plays a major role in data quality, with reported laboratory accuracies of ±5 °C (FLIR) [37] and ±°C (ICI). Figure 1 shows the specifications of an ICI microbolometer camera with a typical spectral response [38].

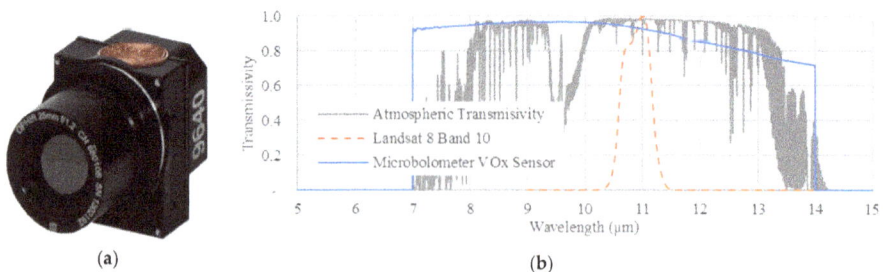

(a)   (b)

Figure 1. Example of the dimensions of a microbolometer ICI Camera 9640 Series, used in this study (**a**). VOx Spectral Response in the 7 to 14 μm Filter, Landsat 8 Band 10 spectral response and average atmospheric transmissivity in the long infrared region (**b**).

1.2. Atmospheric Correction of Surface Temperature

All imaging sensors are affected by atmospheric conditions, as indicated by the atmospheric correction models available for the optical and thermal sensors in satellites such as Landsat, Sentinel-3, MODIS, and others [25,27,39–41]. For temperature sensors, the largest sources of distortion are water content in the atmospheric path between the sensor and the surface, in addition to sensor technology and payload integration (i.e., the unit's accuracy and camera attachment options such as gimbal or frame fitting with/without casing, etc.). For temperature-capable satellites, solutions include an onboard thermal blackbody [27] with a gas-coolant or cryogenic sensor design [42] and minimization of temperature waveband(s) [43] (Figure 1, Landsat 8 spectral response). Nevertheless, these solutions are not available for microbolometer temperature technology (under 200 gr). It is important to compare the spectral response from microbolometer technology for satellites against the atmospheric transmission on the spectral wavebands used to measure surface temperature (7 to 14 um). In this spectral region, water vapor and atmospheric gasses will differentially affect the transmissivity per wavelength and use narrow wavebands: between 8 and 9, and between 10.5 and 12 µm is recommended. However, current microbolometer technology cannot selectively access these recommended narrow wavebands- because the technology itself would decrease the sensing capability of the microbolometer with a narrow band (signal to noise ratio) [38].

For satellites and sUAS, the temperature sensor observes the radiation originally emitted from the surface (L_G), but reduced or attenuated by atmospheric factors such as the amount of water vapor and other gasses in the atmosphere column between the ground and the sensor, along with weather conditions, sensor view geometry, etc. The measured radiation at the temperature sensor is called "radiance at sensor" (L_S). The radiation at ground and sensor levels, along with the atmospheric conditions between the sensor and the ground, can be related by using a radiative transfer model [25,44] as presented in Equation (1):

$$L_S = \tau \, \varepsilon \, L_G + L_U + (1 - \varepsilon) \, L_D \qquad (1)$$

where τ is the atmospheric transmissivity, ε is the emissivity of the surface, L_G is the radiance of a blackbody target of kinetic temperature T at ground level, L_U is the upwelling or atmospheric path radiance, LD is the downwelling or sky radiance, and LS is the radiance measured by the temperature sensor on board the satellite or manned/sUAS. Radiance is in units of $W/m^2/sr/\mu m$, and τ and ε do not have dimensions. If only brightness or radiometric temperature is required, ε can be considered as 1.0, simplifying Equation (1) to Equation (2):

$$L_S = \tau \, L_G + L_U \qquad (2)$$

For satellite sensors, L_U and τ can be determined for the specific image date using a radiative transfer model such as MODTRAN [25,45,46] and 6SV [28–30,47] to calculate the scattering and transmission of radiance through the entire earth atmosphere. These models are time-consuming and require input data that is not often available, such as the vertical profile of atmospheric water vapor and other gasses [1]. The quantification of atmospheric water vapor is needed because, while the atmosphere is a mix of gases (nitrogen, oxygen, carbon dioxide, etc.), these can be considered to be present in constant quantities (with resulting constant effect), but water vapor changes continuously in time and space [48]. To determine the radiance of an object from temperature measurements, Planck's Law allows the nonlinear relationship of the total emittance as a blackbody, at a specific wavelength, to be determined from its temperature and vice versa [49]. When expressed per unit wavelength, the simplified form of Planck's Law is Equation (3):

$$W(\lambda, T) = c_1 (\lambda^5 (\exp(c_2 \times T)^{-1}) - 1)^{-1} \qquad (3)$$

where $W(\lambda, T)$ is the total spectral radiant emittance at a temperature per unit area of emitting surface at wavelength (λ) in meters ($W \cdot m^{-2} \cdot sr^{-1} \cdot um^{-1}$), T is the temperature in Kelvins, c_1

is 1.1910×10^{-22} W·m^{-2}·µm^{-1}·sr^{-1}, and c_2 is 1.4388×10^{-2} m·K. To obtain surface brightness temperature (without emissivity correction) the equation is inverted as follows to Equation (4):

$$T(\lambda, W) = c_2 (\lambda \times \ln(c_1 (\lambda^5 W)^{-1}) - 1)^{-1} \tag{4}$$

Equations (3) and (4) use the weighted band center from the specific spectral response of the sensor [1]. It is important to note the linear relationship among the radiance at ground and sensor levels in Equations (1) and (2), while Equations (3) and (4) indicate a nonlinear relationship between temperature and radiance.

2. Materials and Methods

2.1. Study Site

The study area is a commercial vineyard near Galt, CA (field center located at 38°17'7.40'' N, 121°6'58.11'' W), operated by E&J Gallo [50]. The farm is equipped with a drip irrigation system and covers an area of approximately 77 hectares (188 acres). Figure 2 shows the location of the farm. The location is also an intensive experimental field by the USDA-ARS Hydrological and Remote Sensing Laboratory—GRAPEX (Grape Remote Sensing Atmospheric Profiling and Evapotranspiration Project).

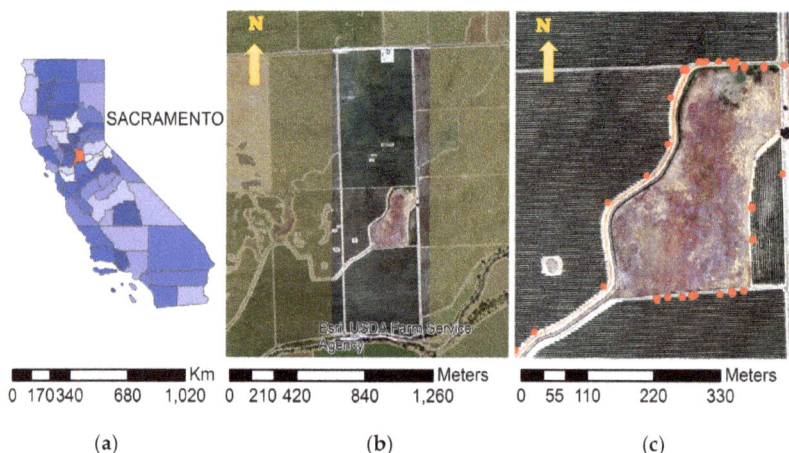

Figure 2. Location of the study area: County location in California (**a**); AggieAir sUAS coverage area in RGB mosaic for all flights (**b**); and close view of sUAS RGB mosaic along with ground temperature sampling locations (dots) (**c**).

2.2. AggieAir sUAS

The AggieAir sUAS platforms and payloads developed by Utah State University have been widely used for remote sensing assignments in support of research in natural resources, water resources, and agricultural applications. The system incorporates a collection of sUAS remote sensing equipment, including multiple platforms and interchangeable sensor packages. The customizable payload includes short, medium, and long waveband sensors. The extended flight times of AggieAir platforms have incorporated continuous improvements (3.0 h on a single battery charge, up to 12,000 ft MSL, weather sensors, etc.). To achieve scientific accuracy, intensive ground data collection efforts have been conducted to produce reflectance estimation protocols, address camera vignetting, assure accurate image orthorectification, etc. In addition, the optical and thermal cameras are located within a payload frame to minimize atmospheric effects (chilling) on the sensor due to flight elevations (up to 1000 m above ground) and speeds (~50 mph). Figure 3 shows details of the AggieAir "Minion" sUAS and payloads used in this study.

(a)   (b)

Figure 3. An example of the AggieAir "Minion" sUAS Fixed Wing Aircraft (**a**); and AggieAir custom payload detail (**b**).

2.3. Methods

To develop a vicarious calibration procedure for a microbolometer sensor, the AggieAir sUAS was employed to fly over the area of study and collect thermal imagery during a 3-year campaign (2014–2016) in agricultural lands (vineyards) in California (Figure 2). Temperature information was collected at ground level during each sUAS flight. The flight altitude was 450 m above ground level (AGL) and was constant for all flights. Measurements (and flights) were made at early morning (approximately a half-hour after sunrise), Landsat 8 overpass time (close to solar noon), and mid-afternoon. The AggieAir sUAS navigated over the area of interest based on a pre-programmed flight plan with total flight times of less than 30 min.

The thermal cameras included in this study are described in Table 1. Both microbolometer cameras were acquired from ICI [38]. These instruments were selected partly on the basis of their reported laboratory calibration accuracy and ease of integration with the AggieAir payload [38]. In addition, cameras from this manufacturer have been used by other research groups mentioned in the scientific literature [51–54]. A National Institute of Standards and Technology NIST traceable temperature camera calibration ambient blackbody was acquired from Palmer Wahl [55]. The "ambient" notation indicates that the blackbody can be used in exterior locations and it does not require cryogenic or external cooling for absolute temperature measurement. Table 1 specifies the technical characteristics of the temperature instruments used in this study.

Table 1. Instruments used to collect temperature information in this study.

Instrument	Blackbody	2014	2015–2016
Brand/Model	Wahl Palmer/WD1042	ICI/7640-P	ICI/9640-P
Weight (gr)	1000	148	141
Image Size (pixel)	–	640 by 480	640 by 480
Spectral Range (µm)	–	7 to 14	7 to 14
Spectral Band Centre (µm)	–	10.35	10.35
Operating Range	-40 to $70\,^{\circ}$C	-20 to $100\,^{\circ}$C	-40 to $140\,^{\circ}$C
Reported Accuracy	$\pm0.2\,^{\circ}$C	$\pm1.0\,^{\circ}$C or $\pm1.0\%$	$\pm1.0\,^{\circ}$C
Reported Emissivity	0.95 ± 0.02	1.0	1.0
NIST Traceable?	YES	NOT REPORTED	NOT REPORTED

For this study, the AggieAir sUAS was equipped with visual, near-infrared, and thermal cameras. It was flown over the study area on four different dates and times (early morning, Landsat overpass and mid-afternoon) (Table 2). These flights acquired thermal imagery at 60-cm/pixel resolution at an elevation of 450 m (1476 ft.) AGL for less than 30 min flight time. The three daily flight times were selected to compare sUAS information with specialized algorithms for evapotranspiration alongside Landsat satellite imagery, which are not part of this study. Agisoft Photoscan software [56] was used to create temperature imagery mosaics, while custom MATLAB code and ground control points collected with an RTK GPS system [57] were used to orthorectify the AggieAir imagery [11].

Table 2. AggieAir sUAS flights included in this study (Times in Pacific Daylight Time zone).

Date	Early Morning Flights		Landsat Overpass Flights		Mid-Afternoon Flights	
	Launch	Landing	Launch	Landing	Launch	Landing
09 August 2014	7:10 AM	7:30 AM	11:30 AM	11:50 AM	No UAS flight	
02 June 2015	6:51 AM	7:32 AM	11:21 AM	12:06 PM	2:54 PM	3:20 PM
11 July 2015	6:37AM	7:11 AM	11:26 AM	12:00 PM	2:58 PM	3:31 PM
02 May 2016	8:13 AM	8:35 AM	12:53 PM	1:17 PM	3:52 PM	4:16 PM
03 May 2016	8:40 AM	9:06 AM	No UAS flight		1:35 PM	2:00 PM

The vicarious calibration methodology used in this study is initially based on the earlier work by [11], which compared georeferenced ground and sUAS temperature pixels for water pools. The present study considered three major vicarious calibration steps as presented in Table 3.

Table 3. Followed vicarious calibration methodology used in this study.

Steps	Activity Description
Before Flight	• Camera—blackbody temperature measurement • GPS survey of ground temperature sampling locations
During Flight	• Temperature ground sampling
After Flight	• Ground/sUAS temperature pixel extraction • Calibration of Radiative Transfer Model • sUAS temperature image correction

The three main steps of the vicarious calibration methodology (Table 3) are as follows:

- **Before Flight**

   Two activities had to be accomplished before the flight: (1) a measurement of the ambient temperature blackbody using the sUAS and ground temperature cameras, and (2) a selection and RTK-GPS survey of the locations to be used for ground data collection during the sUAS flight. The first activity allowed the bias to be determined between the temperature cameras and a NIST-traceable instrument. Given that both instruments include reported accuracies, this activity also allowed the bias source (e.g., instrument or environmental) to be determined [58]. The second activity identified areas of interest in the area of study. In agricultural lands, for example, a range of locations was considered that included bare soil (wet and dry), short vegetation (green, dry), tall canopy, and open water surface. The sub-meter pixel resolution of the sUAS thermal images made it necessary to establish the selected locations with temporary or permanent ground markers and perform GPS surveys with sub-centimeter accuracy, thus the need for RTK-GPS equipment.

- **During Flight**

   The previously selected locations were measured with a ground level temperature camera simultaneously with the sUAS flight over the study area. It was important to complete the sUAS flight and the ground data collection in a short amount of time, generally much less than 30 min. This was to avoid the introduction of measurement errors due to diurnal surface temperature changes. A tall, portable frame was erected on a truck to enable a large number of ground temperature images (and pixels) to be collected quickly.

- **After Flight**

After the sUAS and ground temperature data were collected, the sUAS temperature map was developed using mosaicking software (Agisoft Photoscan) and custom MATLAB code to georeference the temperature images from the ground data collection. Temperature pixels were then extracted from both the ground and sUAS images at the resolution of the sUAS image. Temperature pixel data was then transformed into radiance using Equation (3), and the radiometric model (proposed by [25,44]) shown in Equation (2) was applied. Finally, atmospheric transmissivity and atmospheric path radiance were applied to the entire sUAS radiance image (converted from a temperature map) and transformed back into an atmospherically corrected temperature image.

3. Results and Discussion

3.1. Before Flight

Camera–Blackbody Temperature Imagery Measurement: A total of 213 individual temperature images of the blackbody were compared to determine bias in the microbolometer cameras used in this 3-year study. Figure 4 shows an example of the visual and temperature images of the NIST-traceable ambient temperature blackbody in the field.

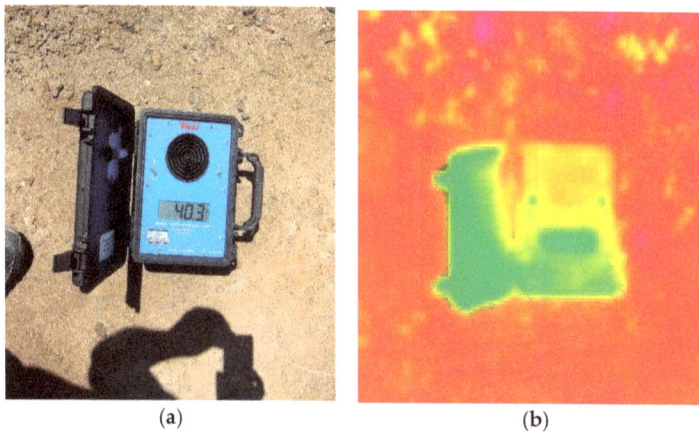

(a)   (b)

Figure 4. Visual (**a**) and temperature (**b**) images of the NIST traceable ambient temperature blackbody used in this study. Black disk (**a**) is the blackbody temperature sensor.

As specified in Table 1, the temperature blackbody works at an emissivity value of 0.95 ± 0.02, while the microbolometer camera worked at a value of 1.0. Therefore, the blackbody temperature was adjusted to the camera emissivity using the Stefan-Boltzmann Law (5):

$$T_{blackbody \cdot corrected}^{4} = (\varepsilon_{blacbody} / \varepsilon_{camera}) \times T_{blackbody}^{4} \tag{5}$$

where $T_{blackbody}$ corrected is the emissivity adjusted temperature, ε_{camera} (1.0) and $\varepsilon_{blackbody}$ (0.95) are the emissivities of the microbolometer cameras and the blackbody, respectively, and $T_{blackbody}$ is the temperature reported by the ambient blackbody. Once the blackbody temperature was corrected for emissivity, the temperatures (blackbody and cameras) were modeled as shown in Figure 5. In addition, a sensitivity analysis of the camera and blackbody instrument accuracies (Table 1) were performed.

As shown in Figure 5 and Table 4, a strong linear relationship exists between the data from the microbolometer cameras provided by ICI and the ambient blackbody. The linear response to a 1:1-line slope indicates that the ICI microbolometer cameras needed only a constant bias correction expressed by an independent term ($-2.67\,^{\circ}$C) in the equation shown in Figure 5a. The relationship thus identified was not affected by weather conditions or seasonality (air temperature, wind, humidity, etc.). In addition, Figure 5b shows a residual analysis of the camera-blackbody linear model. The bias residuals have a distribution similar to a Gaussian curve (mean = $0.0\,^{\circ}$C, and standard deviation $\pm 1.22\,^{\circ}$C). In addition, up to 48% of the residual variability around the mean can be explained by the accuracy of the blackbody ($\pm 0.2\,^{\circ}$C, $\pm 0.02\,\varepsilon$ or $\pm 0.35\,^{\circ}$C when both accuracies are combined), and up to 68% of the residual variability around the mean is explained by the accuracy of the thermal camera ($\pm 1.00\,^{\circ}$C). Therefore, 32% of the variability in the residuals of the modeled bias seemed to be caused by measurement factors (e.g., optical characteristics of the ICI camera such as the point spread function and selection of blackbody pixels from the temperature image).

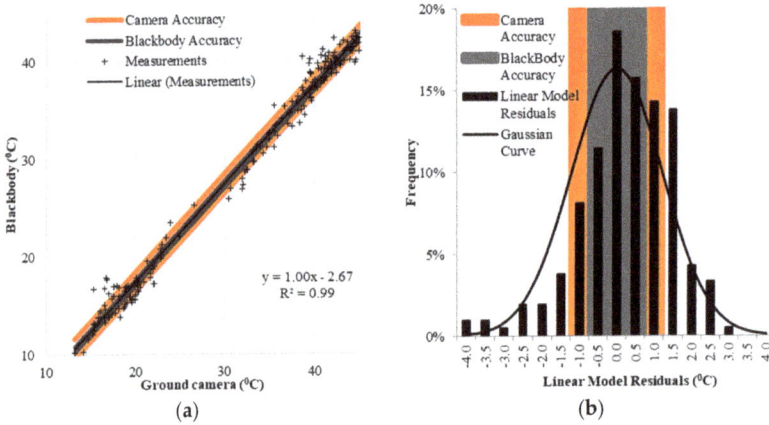

Figure 5. Temperature camera–blackbody comparison for 213 individual measurements. A linear model with a unit slope (1:1) and a constant bias ($-2.67\,^{\circ}$C) fitted the camera bias value over 3 years of study. Temperature values below 30 °C for the ground camera axis belong to early morning measurements, higher values are for Landsat overpass and mid-afternoon times (**a**); Reported accuracies from blackbody and camera manufacturers can explain up to 48% (gray region) and 68% (orange region) of linear model residuals variability, respectively (**b**).

Table 4. Statistics for Temperature camera—blackbody linear model analysis.

Model	Slope (95% Confidence Bounds)	Bias (95% Confidence Bounds)	R^2	RMSE (C°)	Reported Camera Accuracy (C°)	Reported Blackbody Accuracy (C°)
Linear	1.00 (0.99 1.02)	−2.67 (−3.19–2.22)	0.99	1.23	±1.00	±0.35

Ground temperature sampling locations: Different land surfaces were considered during the three-year data collection effort for this project. Examples of different surface types are presented in Figures 2 and 6. These locations were visually homogeneous and covered the range of possible land surfaces in the area of study. An RTK GPS system was used to survey perimeters made with PVC and aluminum tape (1.6 by 1.6 m and 0.8 by 0.8 m) so that sUAS and ground temperature pixels could be accurately located.

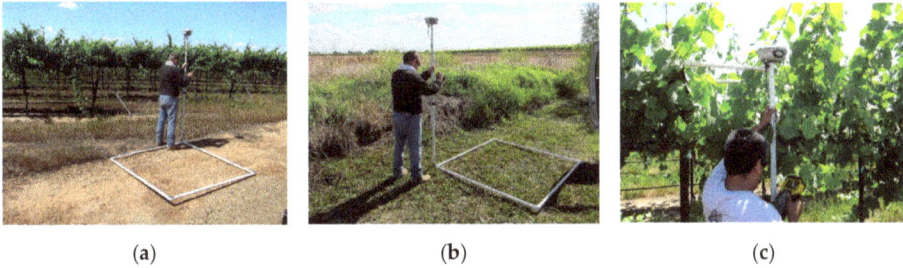

Figure 6. Examples of ground locations for temperature sampling the day before the sUAS flight shown in Figure 2. Locations were delimited using PVC and surveyed using RTK GPS. Note the diversity of locations: bare dry soil (**a**), short green canopy (**b**), and tall canopy (**c**).

3.2. During Flight

Temperature Ground Sampling: Simultaneous examples of AggieAir sUAS flights and ground temperature captures are presented in Figure 7:

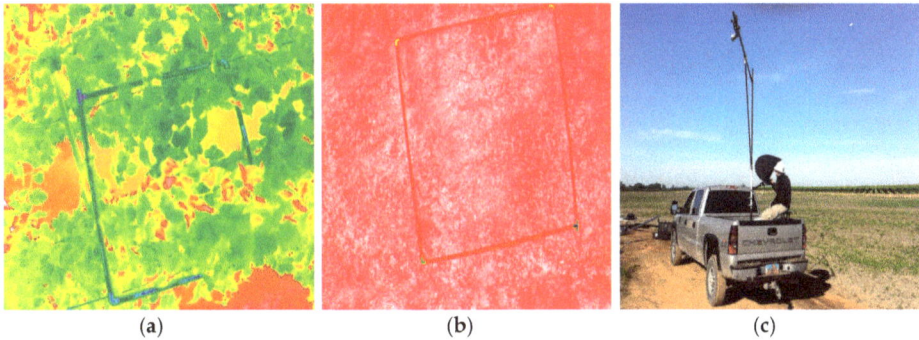

Figure 7. An example of ground temperature samples taken using the temperature camera mounted on a tripod (3 m height) and connected to a laptop loaded with ICI software IR FLASH on 3 May 2016, at Landsat overpass time. Temperature images of the top of vine canopy (**a**) and bare soil (**b**). Temperature camera and tripod mounted over a truck for fast ground temperature sampling (**c**).

3.3. After Flight

Ground and sUAS Pixel Extraction: After each flight, sUAS temperature maps were developed using data from the sUAS onboard inertial measurement unit (IMU) and GPS receiver. This provided the sUAS location (x, y, z coordinates) and orientation (pitch, roll, yaw) to the Agisoft Photoscan version 1.3 software [56]. RTK-GPS surveyed ground control points, specifically for thermal cameras (aluminum-based blankets), were also included. For the ground temperature camera images, these points were registered using their respective RTK-GPS coordinates and ground PVC frame dimensions using ESRI ArcGIS software. An example of georeferenced ground temperature images is shown in Figure 8.

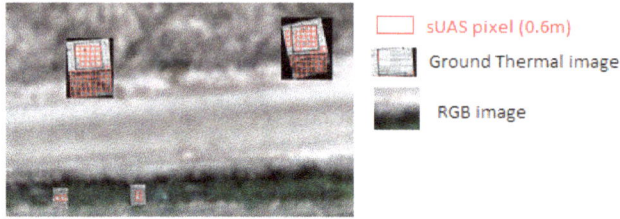

Figure 8. Example of ground temperature images georeferenced from locations shown in Figure 6 and others over a visual image using ArcGIS. The figure showcases thermal temperature images of two bare soil locations (top) and two tall canopies (vines) (bottom). Square grids indicate sUAS and ground pixels that can be extracted for atmospheric radiance calibration.

Calibration of Atmospheric Radiance Model: Once extracted, data from the sUAS thermal images and ground temperature values were transformed into radiance Equation (3). The atmospheric radiance model presented in Equation (2) was calibrated for every sUAS flight, and the results of the calibration are shown in Table 5. Figure 9 shows the radiance comparison from sUAS and ground pixels.

Table 5. Statistical results from atmospheric radiance model (τ and Lu) using ground and sUAS pixels.

Date	Flight Time	τ (95% Confidence Bounds)	L_u (95% Confidence Bounds) W/m²/μm/sr	r^2	RMSE W/m²/μm/sr	Used Pixels
8/9/2014	Early Morning	0.40 (0.51 0.33)	−5.54 (−9.38–3.03)	0.65	0.37	48
	Landsat Overpass	0.69 (0.88 0.57)	−2.94 (−6.75–0.45)	0.58	0.84	63
6/2/2015	Early Morning	0.35 (0.38 0.33)	−4.28 (−5.10–3.56)	0.71	0.26	336
	Landsat Overpass	0.57 (0.62 0.53)	−3.61 (−4.88–2.53)	0.62	0.88	330
	Mid Afternoon	0.53 (0.58 0.50)	−5.17 (−6.49–4.04)	0.68	0.97	330
7/11/2015	Early Morning	0.81 (0.88 0.76)	−0.94 (−1.58–0.39)	0.73	0.12	283
	Landsat Overpass	0.49 (0.52 0.47)	−5.08 (−5.86–4.38)	0.83	0.77	352
	Mid Afternoon	0.47 (0.48 0.45)	−5.21 (−5.71–4.74)	0.94	0.47	278
5/2/2016	Early Morning	0.29 (0.33 0.27)	−5.23 (−6.66–4.05)	0.53	0.21	343
	Landsat Overpass	0.62 (0.63 0.60)	−2.21 (−2.49–1.94)	0.95	0.46	299
	Mid Afternoon	0.65 (0.69 0.62)	−2.25 (−2.94–1.62)	0.79	0.55	299
5/3/2016	Early Morning	0.43 (0.45 0.41)	−4.74 (−5.49–4.07)	0.8	0.13	326
	Mid Afternoon	0.52 (0.55 0.50)	−3.38 (−4.01–2.82)	0.83	0.43	349

In Figure 9 and Table 5 the Equation (2) linear model assumption is confirmed by the linearity of the ground and sUAS pixel comparison. Not all Equation (4) regression calibrations (Table 5) have a high R2 value, due to the scattering of the compared pixels, but small average errors (RMSE) were observed for early morning (<0.5 W/m²/μm/sr) and Landsat overpass and mid-afternoon (RMSE < 1 W/m²/μm/sr) flights. In all early morning flights, the temperature radiance ranged from 0 to 6 W/m²/μm/sr, and for Landsat overpass and mid-afternoon flight times, these ranged from 0 to 16 W/m²/μm/sr.

In terms of atmospheric correction parameters, the transmittance (τ) ranged from 0.29 to 0.81 for all flights. The early morning values do not seem to concentrate on a given range (0.29 to 0.81). For Landsat overpass and mid-afternoon, τ values were within the 0.47 to 0.69 range. LU values varied from −0.94 to −5.54 W/m²/μm/sr for early morning flights and −2.21 to −5.21 W/m²/μm/sr for Landsat overpass and mid-afternoon flights, with no evidence of a preferred value range. The 95% confidence bounds included in Table 5 indicate that even in the best atmospheric conditions (7/11/2015 early morning, $\tau = 0.81$) an atmospheric correction (confidence upper bound = 0.88) is still needed. Furthermore, the confidence bound estimates for LU in Table 5 indicate that there is no record where this parameter can be omitted (zero or positive values). In terms of confidence bound ranges, all τ estimates are within the ±0.1 range and ±1.2 W/m²/μm/sr for LU for all measurement times (early morning, Landsat overpass, and mid-afternoon).

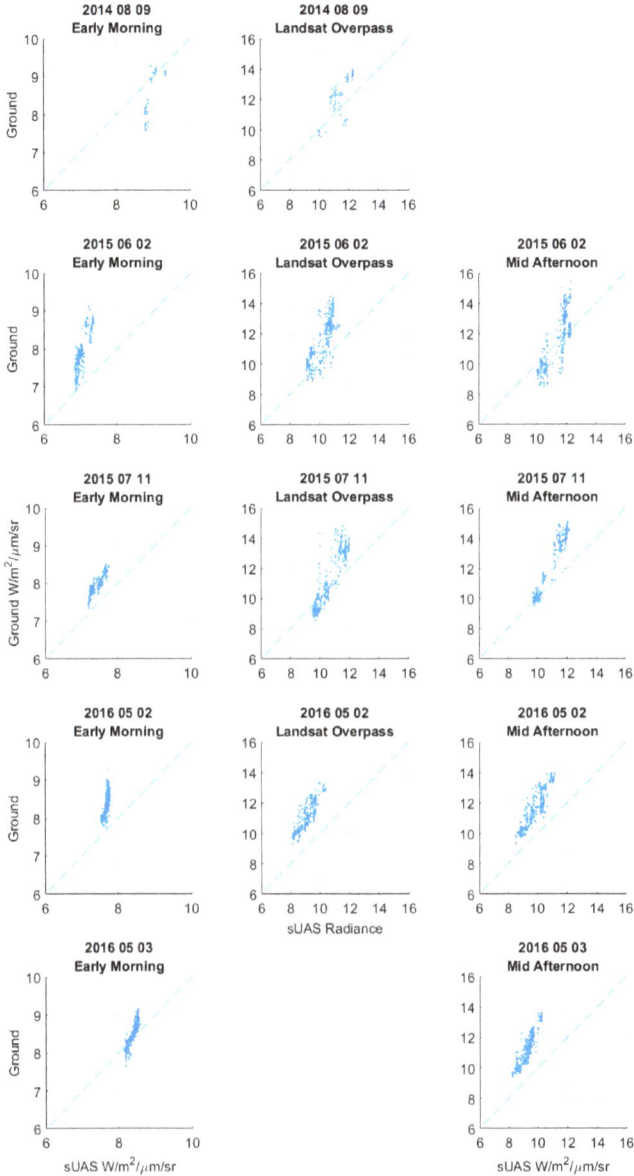

Figure 9. Radiance (W/m²/sr/μm) from sUAS (horizontal axis) and ground pixels (vertical axis) for all the sUAS flights in the site of study. The left column is for early morning, a center column for Landsat overpass and the right column for mid-afternoon times respectively.

sUAS Temperature Image Correction: When the estimation of the atmospheric correction model with the calculation of atmospheric transmissivity τ and radiance LU was completed, the sUAS images were processed by converting them to radiance (W/m^2/µm/sr) using Equation (3) and then back to temperature using Equation (4). Figure 10 demonstrates the differences in surface temperature due to the atmospheric correction model for a given date. This example shows the sUAS flights for early morning, Landsat overpass and mid-afternoon for 2 May 2016.

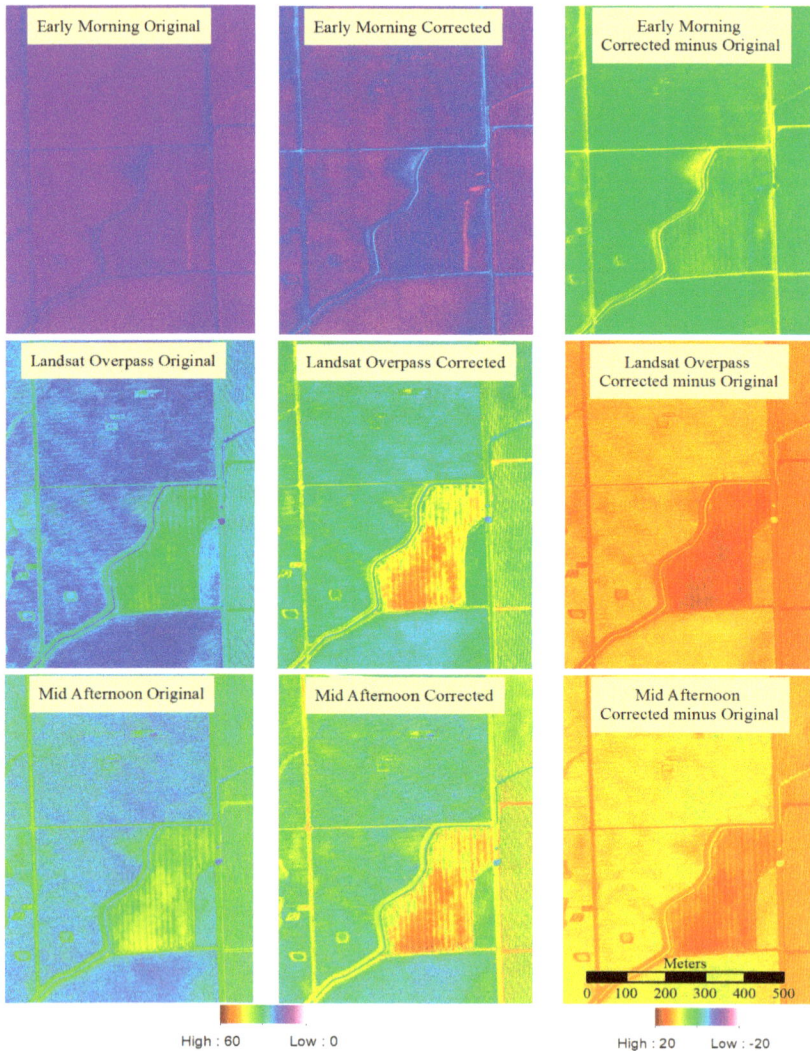

Figure 10. An example of temperature image correction for 2 May 2016 using the vicarious calibration for early morning (**top row**), Landsat overpass (**center row**) and mid-afternoon (**bottom row**). Units in Celsius. The left column is the sUAS original temperature image, the center column is the original sUAS image after the atmospheric radiance correction, and the right column is the difference between the corrected minus original sUAS imagery.

The results of the application of τ and L$_U$ to a sUAS image, as shown in Figure 10, indicate a significant change in the estimation of the surface temperatures. These changes range from −5 to 10 °C for the early morning images, and 0 to 20 °C for both the Landsat overpass and mid-afternoon images. The significant changes in temperature estimates can be explained by the Stefan-Boltzmann Law, which indicates changes in radiances relate to changes in temperatures at the 4th power, as indicated in Equation (5). Therefore, variations in radiance by the atmospheric radiance correction will translate into significant variation in radiance temperature.

4. Conclusions

This study proposes a vicarious calibration methodology for atmospheric correction of microbolometer temperature sensors used on sUAS platforms, such as those of the AggieAir sUAS Research Group at USU. This methodology uses NIST-traceable ambient temperature blackbody and ground level temperature images from different land surfaces from a second temperature camera of the same model and manufacturer. This methodology avoids the use of atmospheric models such as MODTRAN and 6SV, while referring to local measurements during the sUAS flight. The procedure is applicable to any sUAS flight elevation, time, camera spectral response, and camera set up, although the procedure demands additional human effort and surveying equipment for ground data collection. The results of this study indicate that, overall, microbolometer temperature cameras, despite the impact of atmospheric conditions and sUAS setup, can be related to a NIST-traceable temperature device.

The advantage of the proposed vicarious calibration methodology is the accountability of atmospheric and other factors that can affect the acquisition of land surface temperatures, such as daytime and weather conditions, spectral response, camera operational temperature, and others. This methodology requires the use of a microbolometer camera with a laboratory calibration accuracy adequate to the expected posterior analysis. The ICI cameras used in this study have a laboratory accuracy of ±1 °C.

The atmospheric radiance correction of the sUAS thermal imagery requires adequate conversion of the temperature maps from sUAS and georeferenced ground imagery into thermal radiance using the Planck equation and an estimate of the thermal central waveband. The atmospheric radiance calibration provides two main parameters: (1) atmospheric transmissivity (dimensionless), and (2) atmospheric radiance (W/m^2/μm/sr). These two parameters can be directly applied to the sUAS radiance image. It is important to note that the selection of ground sampling locations plays an important role in the calibration of the atmospheric radiance correction model. It is recommended that different flat locations be considered based on their temperature response (dry/wet, bare, and vegetation-covered soils, top of the canopy for tall crops, open water, snow, etc.), thus creating a temperature gradient necessary for the regression analysis. Shadowed areas are not recommended because sun elevation changes continuously and will affect the temperature conditions of shadow-covered locations during the sUAS flight.

The results from the dates where the vicarious calibration methodology was applied in this study demonstrated that the atmospheric correction model parameters are different for each date and flight time (early morning, Landsat overpass, and mid-afternoon). These results indicate that instantaneous atmospheric conditions (air temperature, water vapor, etc.) between the sUAS and the ground may play a major role in the atmospheric correction parameters values. A simplification of the proposed vicarious calibration methodology might be possible by mounting additional sensors (air temperature, relative humidity, atmospheric pressure, incident radiation, wind speed, etc.) on the sUAS to collect data about air column conditions during periods when the aircraft is climbing or descending. These onboard sensors could build an "atmospheric profile" from the ground to the targeted elevation at the beginning and end of the flight with a continuous monitoring of the weather conditions during the flight. This is an important source of information to support radiometric calibration of thermal imagery for flights that extend for longer time periods (larger than 30 min), or are conducted over time intervals that experience changing atmospheric and/or sunlight conditions.

A variant on the vicarious calibration presented in this study would be ground temperature collection using a second sUAS, which would carry the ground thermal camera. The expected sUAS elevation should be less than 10 m AGL. Keeping the recommendations about the ground temperature sampling locations as described in this study, the number of pixels available for the radiance atmospheric model can increase by an order of magnitude while keeping the necessary sUAS flight time as described in the procedure presented here.

Future work involves the comparison and cross-calibration of other temperature sensors and image sources, such as atmospherically corrected Landsat, temperature canopy sensors, and hemispherical radiometers towards the integration of multiple thermal measurements and emissivity estimation.

Acknowledgments: The author wishes to acknowledge the AggieAir sUAS Research Group at Utah State University for imagery collection and initial processing of the temperature maps. Also, acknowledgments are given to the E&J Gallo Scientific Team, and USDA-ARS-Hydrology and Remote Sensing Laboratory—GRAPEX (Grape Remote Sensing Atmospheric Profiling & Evapotranspiration Experiment) for access to the commercial farm and recommendations for the development of this study. The present study was supported by Mineral Lease Funds WR-2188 provided by the Utah Water Research Laboratory at Utah State University.

Conflicts of Interest: The author declares no conflict of interest.

References

1. Handcock, R.N.; Torgersen, C.E.; Cherkauer, K.A.; Gillespie, A.R.; Klement, T.; Faux, R.N.; Jing, T. Thermal Infrared Remote Sensing of Water Temperature in Riverine Landscapes. In *Carbonneau/Fluvial Remote Sensing for Science and Management*; Carbonneau, P.E., Piegay, H., Eds.; Wiley-Blackwell: Hoboken, UK; Hoboken, NJ, USA, 2012; pp. 85–113.
2. Petropoulos, G.P. *Remote Sensing of Energy Fluxes and Soil Moisture Content*; CRC Press: Boca Raton, FL, USA, 2013; p. 546.
3. Blonquist, J.M., Jr.; Norman, J.M.; Bugbee, B. Automated measurement of canopy stomatal conductance based on infrared temperature. *Agric. For. Meteorol.* **2009**, *149*, 2183–2197. [CrossRef]
4. Apogee Instruments. Available online: http://www.apogeeinstruments.com/infraredradiometer/ (accessed on 24 February 2017).
5. Anderson, M.C.; Allen, R.G.; Morse, A.; Kustas, W.P. Use of Landsat thermal imagery in monitoring evapotranspiration and managing water resources. *Remote Sens. Environ.* **2012**, *122*, 50–65. [CrossRef]
6. Jackson, R.D.; Pinter, P.J. Detection of Water Stress in Wheat by Measurement of Reflected Solar and Emitted Thermal IR Radiation. In Proceedings of the Spectral Signatures of Objects in Remote Sensing, Avigion, France, 8–11 September 1981; pp. 399–406.
7. Kogan, F.N. Application of vegetation index and brightness temperature for drought detection. *Adv. Space Res.* **1995**, *15*, 91–100. [CrossRef]
8. Kogan, F.N. Operational space technology for global vegetation assessment. *Bull. Am. Meteorol. Soc.* **2001**, *82*, 1949–1964. [CrossRef]
9. UMACCI Committee. *Abrupt Impacts of Climate Change. Committee on Understanding and Monitoring Abrupt Climate Change and Its Impacts*; National Academies Press: Washington, DC, USA, 2013; Volume 16, p. 201.
10. Câmara, G. The Future of the CBERS Program: A View from Brazil. In Proceedings of the CBERS Chinese Users Conference, Beijing, China, 21 October 2003.
11. Jensen, A.M.; McKee, M.; Chen, Y. Calibrating thermal imagery from an unmanned aerial system—AggieAir. In Proceedings of the 2013 IEEE International Geoscience and Remote Sensing Symposium (IGARSS), Melbourne, Australia, 21–26 July 2013; pp. 542–545.
12. Cherkauer, K.A.; Burges, S.J.; Handcock, R.N.; Kay, J.E.; Kampf, S.K.; Gillespie, A.R. Assessing Satellite-Based and Aircraft-Based Thermal Infrared Remote Sensing for Monitoring Pacific Northwest River Temperature. *J. Am. Water Resour. Assoc.* **2005**, *41*, 1149–1159. [CrossRef]
13. Hoffmann, H.; Nieto, H.; Jensen, R.; Guzinski, R.; Zarco-Tejada, P.; Friborg, T. Estimating evaporation with thermal UAV data and two-source energy balance models. *Hydrol. Earth Syst. Sci.* **2016**, *20*, 697–713. [CrossRef]

14. Elarab, M.; Ticlavilca, A.M.; Torres-Rua, A.F.; Maslova, I.; McKee, M. Estimating chlorophyll with thermal and broadband multispectral high resolution imagery from an unmanned aerial system using relevance vector machines for precision agriculture. *Int. J. Appl. Earth Obs. Geoinf.* **2015**, *43*, 32–42. [CrossRef]
15. Berni, J.A.J.; Zarco-Tejada, P.J.; Sepulcre-Cantó, G.; Fereres, E.; Villalobos, F. Mapping canopy conductance and CWSI in olive orchards using high resolution thermal remote sensing imagery. *Remote Sens. Environ.* **2009**, *113*, 2380–2388. [CrossRef]
16. Hassan-Esfahani, L.; Torres-Rua, A.; Jensen, A.; McKee, M. Assessment of Surface Soil Moisture Using High-Resolution Multi-Spectral Imagery and Artificial Neural Networks. *Remote Sens.* **2015**, *7*, 2627–2646. [CrossRef]
17. Torres-Rua, A.; Al Arab, M.; Hassan-Esfahani, L.; Jensen, A.; McKee, M. Development of unmanned aerial systems for use in precision agriculture: The AggieAir experience. In Proceedings of the 2015 IEEE Conference on Technologies for Sustainability (SusTech), Ogden, UT, USA, 30 July–1 August 2015; pp. 77–82.
18. Hassan-Esfahani, L. High Resolution Multi-Spectral Imagery and Learning Machines in Precision Irrigation Water Management. Ph.D. Dissertation, Utah State University, Logan, UT, USA, 2015.
19. Paul, G.; Gowda, P.H.; Prasad, P.V.V.; Howell, T.A.; Staggenborg, S.A.; Neale, C.M.U. Lysimetric evaluation of SEBAL using high resolution airborne imagery from BEAREX08. *Adv. Water Resour.* **2013**, *59*, 157–168. [CrossRef]
20. Xia, T.; Kustas, W.P.; Anderson, M.C.; Alfieri, J.G.; Gao, F.; McKee, L.; Prueger, J.H.; Geli, H.M.E.; Neale, C.M.U.; Sanchez, L.; et al. Mapping evapotranspiration with high-resolution aircraft imagery over vineyards using one- and two-source modeling schemes. *Hydrol. Earth Syst. Sci.* **2016**, *20*, 1523–1545. [CrossRef]
21. Neale, C.M.U.; Jayanthi, H.; Wright, J.L. Irrigation water management using high resolution airborne remote sensing. *Irrig. Drain. Syst.* **2005**, *19*, 321–336. [CrossRef]
22. Glenn, E.P.; Neale, C.M.U.; Hunsaker, D.J.; Nagler, P.L. Vegetation index-based crop coefficients to estimate evapotranspiration by remote sensing in agricultural and natural ecosystems. *Hydrol. Process.* **2011**, *25*, 4050–4062. [CrossRef]
23. Prueger, J.H.; Alfieri, J.G.; Hipps, L.E.; Kustas, W.P.; Chavez, J.L.; Evett, S.R.; Anderson, M.C.; French, A.N.; Neale, C.M.U.; McKee, L.G.; et al. Patch scale turbulence over dryland and irrigated surfaces in a semi-arid landscape under advective conditions during BEAREX08. *Adv. Water Resour.* **2012**, *50*, 106–119. [CrossRef]
24. Allen, R.G.; Tasumi, M.; Morse, A.; Trezza, R.; Wright, J.L.; Bastiaanssen, W.; Kramber, W.; Lorite, I.; Robison, C.W. Satellite-Based Energy Balance for Mapping Evapotranspiration with Internalized Calibration (METRIC)—Model. *J. Irrig. Drain. Eng.* **2007**, *133*, 395–406. [CrossRef]
25. Barsi, J.A.; Barker, J.L.; Schott, J.R. An Atmospheric Correction Parameter Calculator for a single thermal band earth-sensing instrument. In Proceedings of the IEEE International Geoscience and Remote Sensing Symposium (IGARSS) (IEEE Cat. No.03CH37477), Toulouse, France, 21–25 July 2003; Volume 5, pp. 3014–3016.
26. McCarville, D.; Buenemann, M.; Bleiweiss, M.; Barsi, J. Atmospheric correction of Landsat thermal infrared data: A calculator based on North American Regional Reanalysis (NARR) data. In Proceedings of the American Society for Photogrammetry and Remote Sensing Conference, Milwaukee, WI, USA, 1–5 May 2011; p. 12.
27. Barsi, J.; Schott, J.; Hook, S.; Raqueno, N.; Markham, B.; Radocinski, R. Landsat-8 Thermal Infrared Sensor (TIRS) Vicarious Radiometric Calibration. *Remote Sens.* **2014**, *6*, 11607–11626. [CrossRef]
28. Kotchenova, S.Y.; Vermote, E.F. Validation of a vector version of the 6S radiative transfer code for atmospheric correction of satellite data. Part II. Homogeneous Lambertian and anisotropic surfaces. *Appl. Opt.* **2007**, *46*, 4455–4464. [CrossRef] [PubMed]
29. Kotchenova, S.Y.; Vermote, E.F.; Matarrese, R.; Klemm, F.J., Jr. Validation of a vector version of the 6S radiative transfer code for atmospheric correction of satellite data. Part I: Path radiance. *Appl. Opt.* **2006**, *45*, 6762–6774. [CrossRef] [PubMed]
30. Kotchenova, S.Y.; Vermote, E.F.; Levy, R.; Lyapustin, A. Radiative transfer codes for atmospheric correction and aerosol retrieval: Intercomparison study. *Appl. Opt.* **2008**, *47*, 2215–2226. [CrossRef] [PubMed]
31. Liddiard, K.C. The active microbolometer: A new concept in infrared detection. In *Microelectronics, MEMS, and Nanotechnology*; International Society for Optics and Photonics: Bellingham, DC, USA, 2004; pp. 227–238.

211

32. Green, W.J.; Maurer, D.E. Merlin microbolometer camera calibration. In Proceedings of the Infrared Imaging Systems: Design, Analysis, Modeling, and Testing XII, Orlando, FL, USA, 16 April 2001.

33. Behnken, B.N.; Karunasiri, G.; Chamberlin, D.R.; Robrish, P.R.; Faist, J. Real-time imaging using a 2.8 THz quantum cascade laser and uncooled infrared microbolometer camera. *Opt. Lett.* **2008**, *33*, 440–442. [CrossRef] [PubMed]

34. Irish, R.R. *Landsat 7 Science Data Users Handbook*; NASA Contract. Rep. NASA CR 2000; National Aeronautics and Space Administration: Houston, TX, USA, 2000.

35. United States Geological Survey (USGS). *Landsat 8 Data Users Handbook*; USGS: Reston, VA, USA, 2016.

36. Drinkwater, M.; Rebhan, H. *Sentinel-3: Mission Requirements Document*; EOP-SMO/1151/MD-md; European Space Agency (ESA): Paris, France, 2007; pp. 19–22.

37. FLIR Systems, I. FLIR Systems | Thermal Imaging, Night Vision and Infrared Camera Systems. Available online: http://www.flir.com/home/ (accessed on 22 December 2016).

38. Infrared & Thermal Camera Specialists. Available online: http://www.infraredcamerasinc.com/ (accessed on 22 December 2016).

39. Srivastava, P.K.; Han, D.; Rico-Ramirez, M.A.; Bray, M.; Islam, T.; Gupta, M.; Dai, Q. Estimation of land surface temperature from atmospherically corrected LANDSAT TM image using 6S and NCEP global reanalysis product. *Environ. Earth Sci.* **2014**, *72*, 5183–5196. [CrossRef]

40. Liang, S.; Fang, H.; Chen, M. Atmospheric correction of Landsat ETM+ land surface imagery—Part I: Methods. *IEEE Trans. Geosci. Remote Sens.* **2001**, *39*, 2490–2498. [CrossRef]

41. Liang, S.; Fang, H.; Chen, M.; Shuey, C.J.; Walthall, C.; Daughtry, C.; Morisette, J.; Schaaf, C.; Strahler, A. Validating MODIS land surface reflectance and albedo products: Methods and preliminary results. *Remote Sens. Environ.* **2002**, *83*, 149–162. [CrossRef]

42. Gilmore, D.G.; Donabedian, M. *Spacecraft Thermal Control Handbook: Cryogenics*; American Institute of Aeronautics and Astronautics (AIAA): Reston, VA, USA, 2003.

43. Vollmer, M.; Möllmann, K.-P. *Infrared Thermal Imaging: Fundamentals, Research and Applications*; John Wiley & Sons: Hoboken, NJ, USA, 2011.

44. Schott, J.R.; Brown, S.D.; Barsi, J.A. Calibration of thermal infrared sensors. In *Thermal Remote Sensing in Land Surface Processing*; Luvall, J.C., Ed.; CRC PRESS: Boca Raton, FL, USA, 2004.

45. Berk, A.; Conforti, P.; Hawes, F. An accelerated line-by-line option for MODTRAN combining on-the-fly generation of line center absorption within 0.1 cm-1 bins and pre-computed line tails. In *SPIE Defense + Security*; International Society for Optics and Photonics: Baltimore, MD, USA, 20 April 2015; pp. 947217-1–947217-11.

46. Berk, A.; Conforti, P.; Kennett, R.; Perkins, T.; Hawes, F.; van den Bosch, J. MODTRAN6: A major upgrade of the MODTRAN radiative transfer code. In *SPIE Defense + Security*; International Society for Optics and Photonics: Baltimore, MD, USA, 5 May 2014; pp. 90880H-1–90880H-7.

47. Vermote, E.F.; Tanre, D.; Deuze, J.L.; Herman, M.; Morcette, J.J. Second Simulation of the Satellite Signal in the Solar Spectrum, 6S: An overview. *IEEE Trans. Geosci. Remote Sens.* **1997**, *35*, 675–686. [CrossRef]

48. Sabatini, R.; Richardson, M.A.; Jia, H.; Zammit-Mangion, D. Airborne laser systems for atmospheric sounding in the near infrared. In *SPIE Photonics Europe*; International Society for Optics and Photonics: Brussels, Belgium, 16 April 2012; pp. 843314-1–843314-40.

49. Evans, H.; Lange, J.; Schmitz, J. *The Phenomenology of Intelligence-Focused Remote Sensing: Volume 1: Electro-Optical Remote Sensing*; Riverside Research: New York, NY, USA, 2013.

50. E. & J. Gallo Winery. Available online: http://www.gallo.com/ (accessed on 21 June 2017).

51. Harvey, M.C.; Luketina, K. Thermal Infrared Cameras and Drones: A Match Made in Heaven for Cost-Effective Geothermal Exploration, Monitoring and Development. In Proceedings of the 37th New Zealand Geothermal Workshop, Taupo, New Zealand, 18–20 November 2015; Volume 18, p. 20.

52. Cassis, L.A.; Urbas, A.; Lodder, R.A. Hyperspectral integrated computational imaging. *Anal. Bioanal. Chem.* **2005**, *382*, 868–872. [CrossRef] [PubMed]

53. Harvey, M.; Harvey, C.; Rowland, J.; Luketina, K. Drones in Geothermal Exploration: Thermal Infrared Imagery, Aerial Photos and Digital Elevation Models. In Proceedings of the 6th African Rift Geothermal Conference, Addis Ababa, Ethiopia, 2–4 November 2016; p. 12.

54. Nugent, P.W. Wide-angle Infrared Cloud Imaging for Cloud Cover Statistics. Master's Thesis, Montana State University, Bozeman, USA, 2008.

55. PalmerWahl. Available online: http://www.palmerwahl.com/ (accessed on 21 June 2017).

56. AgiSoft, L. *Agisoft PhotoScan Professional Edition*, version 1.0.3; Agisoft PhotoScan: St. Petersburg, Russia, 2014.

57. Trimble. Available online: http://www.trimble.com/ (accessed on 21 June 2017).

58. Torres-Rua, A.F.; Ticlavilca, A.M.; Walker, W.R.; McKee, M. Machine Learning Approaches for Error Correction of Hydraulic Simulation Models for Canal Flow Schemes. *J. Irrig. Drain. Eng.* **2012**, *138*, 999–1010. [CrossRef]

sensors

MDPI

Article

UAV-Assisted Dynamic Clustering of Wireless Sensor Networks for Crop Health Monitoring

Mohammad Ammad Uddin [1,2,*], Ali Mansour [1], Denis Le Jeune [1], Mohammad Ayaz [2] and el-Hadi M. Aggoune [2]

[1] Lab STICC, ENSTA Bretagne, Brest 29200, France; ali.mansour@ensta-bretagne.fr (A.M.); denis.le_jeune@ensta-bretagne.fr (D.L.J.)
[2] Sensor Networks and Cellular Systems Research Center, University of Tabuk, Tabuk 71491, Saudi Arabia; ayazsharif@ut.edu.sa (M.A.); hadi.aggoune@gmail.com (e.-H.M.A.)
* Correspondence: mohammad.ammad@gmail.com

Received: 11 December 2017; Accepted: 5 February 2018; Published: 11 February 2018

Abstract: In this study, a crop health monitoring system is developed by using state of the art technologies including wireless sensors and Unmanned Aerial Vehicles (UAVs). Conventionally data is collected from sensor nodes either by fixed base stations or mobile sinks. Mobile sinks are considered a better choice nowadays due to their improved network coverage and energy utilization. Usually, the mobile sink is used in two ways: either it goes for random walk to find the scattered nodes and collect data, or follows a pre-defined path established by the ground network/clusters. Neither of these options is suitable in our scenario due to the factors like dynamic data collection, the strict targeted area required to be scanned, unavailability of a large number of nodes, dynamic path of the UAV, and most importantly, none of these are known in advance. The contribution of this paper is the formation of dynamic runtime clusters of field sensors by considering the above mentioned factors. Furthermore a mechanism (Bayesian classifier) is defined to select best node as cluster head. The proposed system is validated through simulation results, lab and infield experiments using concept devices. The obtained results are encouraging, especially in terms of deployment time, energy, efficiency, throughput and ease of use.

Keywords: dynamic clustering; cluster head selection; IoT for agriculture; UAVs for agriculture

1. Introduction

The use of Internet of Things (IoT) technology to collect up-to-date information from crop fields to protect it from any kind of damage is the main objective of this research. Dynamic clustering is used to collect versatile data under harsh conditions. Clustering is an important way of increasing network lifetime and reliability. Many clustering techniques are proposed, which can be classified into four broad categories: static-sink static-nodes (Low-energy Adaptive Clustering Hierarchy (LEACH) and Hybrid Energy-Efficient Distributed clustering (HEED) [1,2]), mobile-sink static-nodes (Rendez-vous base routing [3–6]), static-sink mobile-nodes (cellular network [7]), and mobile-sink mobile-nodes (ad-hoc routing [8]). This research focuses on the mobile sink static nodes clustering method as all agriculture sensors are assumed to be static and we consider a mobile UAV to collect data from crop fields. For static sensor nodes, researchers propose predefined clusters and cluster head schemes to collect data. This type of clustering is not feasible in our case because of the frequent fluctuations of the sensor nodes. The situation becomes more critical, if the cluster head (CH) is down and the whole network becomes unfunctional. In addition, the path of the UAV is dynamic and sensor nodes are unaware of it; in this case, it rarely happens that a predefined CH lays on the path of the UAV and has a good link to it. Network-defined and Rendez-vous base clustering are also proposed in the literature, where all the nodes send periodic updates to maintain up-to-date CH locations or

Rendez-vous Points (RVPs) from where an UAV can collect data. In this situation, the main drawback is the overhead for all nodes to update the CH locations continuously which results in battery drainage and reduces the network lifetime. Besides, UAVs have to search for and track the network-assigned CHs (Rendez-vous), which will affect the throughput of the system and deflect the UAV from its path. To the best of our knowledge, none of previously published clustering schemes considers the UAV path as a clustering criterion.

We proposs a dynamic clustering scheme. All the field sensor nodes are initially considered indistinguishable (no potential CH), so the the UAV sends a beacon message to activate all nodes residing in its vicinity, forming a cluster by considering its path and type of required data. The next step is to choose one node as the CH, merge all the cluster data on this point, and locate and connect it with the UAV at some reasonable height and distance.

IoT for agriculture is proposed in many research articles like [9–11], where they elaborate how sensors are useful in this field, what are the potential areas of application and how to collect and process information. To the best of our knowledge none of them provide in depth details like how to form clusters, select a CH, proper utilization of UAVs, independent movement of UAVs and energy preservation of the nodes. The most important aspect of our article is the evaluation of the proposed system through intensive simulations and real life experiments.

The rest of the paper is organized as follows: Section 2 provides a short background of the developed system. Sections 3 and 4 present our dynamic cluster formation and cluster head selection techniques. The proposed three layer architecture is described in Section 5, while the system algorithm is given in Section 6. Different system parameters and their characteristics are defined in Section 7. The developed system in evaluated by using simulation models and IoT devices in Sections 8 and 9, respectively. Finally Section 10 concludes this article.

2. Background

Data collection from Wireless Sensor Networks (WSNs) installed in farm fields faces many challenges like extreme climatic conditions characterized by high temperatures, dry air, dust/sand storms, lack of infrastructure, remote and huge geographical locations, etc. We propose a dynamic data collection mechanism to overcome the above mentioned challenges. In this study, smartness in agriculture is achieved in four steps:

1. Deployment of heterogeneous sensor nodes: a large number and wide range of heterogeneous sensors are used to monitor different parameters related to crop, soil and environment.
2. Use of Unmanned Aerial Vehicles (UAVs): UAVs are utilized to build the communication infrastructure between sensing devices and end-users.
3. Dynamic clustering: Clustering is the most important aspect of our study. It is the process used to arrange different sensing devices in groups according to geographical area, required data, path of the UAV, communication limitations, similarities or any other criterion.
4. Dynamic cluster head selection: Once available alive sensors have arranged themselves in a cluster formation then the next challenge is how to select a node as CH which will collect all the data from neighboring nodes and transmit it to the UAV. Selection of the CH is a tricky task, as the node having the best specifications and more suitable for the UAV (i.e., near to the UAV path) has to be selected as CH.

The developed system has many advantages over conventional ones. First of all it is a quickly deployed handy system which can be employed anywhere without any existing infrastructure. A farmer can collect needed data from an area of interest by using this system as all the data from the whole area is rarely needed. The mission is established for the UAV in advance in the form of waypoints to collect data from selective sensors and area. Network life can be improved massively as we propose the use of dynamic clustering, hence there is no need to discover and maintain data routes continuously. In WSNs, the node energy is mostly exhausted in broadcasting periodic updates to keep the network alive all the time. In our system, the network is formed only when data really needs to be collected and only selected nodes will participate. The use of UAVs also helps us to prolong the network life, as it enables us to collect data by visiting the IoT nodes at a suitable pre-arranged height. We are introducing a dual frequency communication system to further reduce the energy initialization. A low frequency (e.g., 433 MHz) is used to locate, navigate and shake-hand with sensor nodes while the high bandwidth and more energy demanding transceiver (e.g., 2.4 GHz) is switched on only when data need to be transmitted. Sensor nodes don't have GPS modules and their reduced working loads are two other factors helping to optimize the node energy. Another important benefit of our proposed system is that it can survive and work well under harsh conditions, as sensor nodes are often fluctuating due to bad weather, and becoming covered by sand, water, mud or plant follicles. Pre-defined CHs and routes are not feasible, and in dynamic clustering only available nodes participate and the best among them will act as CH and formulate a temporary network to deliver the required data to the UAV.

The life cycle of the proposed system is composed of seven steps, as shown in Figure 1. The UAV, which acts like data mule and the means of communication among the sensors, is the main part. The system life cycle starts when the UAV initiates the process of data gathering by sending a beacon message in step-1. Type of nodes, suggested data collection height and threshold to limit number of UAV-node connections are mentioned in this beacon. The nodes addressed in the UAV beacon are activated in step-2. In step-3 clusters are formed in the path of the UAV to preserve its predefined path. There is a possibility that none of the desired sensors have the capability to communicate with the UAV because of limited resources, while on the contrary, the UAV may get many responses from activated ground sensors and be unable to handle them at the same time. To tackle both of these conditions, step-4 is introduced to select some reasonable amount of field sensors for further processing. We name "shunting" this process of pushing or pulling some nodes from the process to make sure they are in a reasonable range. Localization (step-5) is to find out sensor nodes installed in crop field with the UAV, and a special lightweight energy-efficient antenna is designed for this purpose. We have given full details about this virtual phase array antenna in a previous article [12]. The best node among all will be selected as CH in step-6. Many parameters like energy, antenna size, energy consumption rate and distance to the UAV are investigated before selecting a node as CH. The final step-7 is data collection, in which the CH collects data from all neighboring nodes and the aggregated data is transmitted to the UAV by using a point to point dedicated link. This lifecycle keeps on until the whole or the selected area of the crop field is scanned and data is harvested successfully. The whole process spans three dimensions: dynamic clustering, dynamic cluster head selection and localization of field sensors by the UAV.

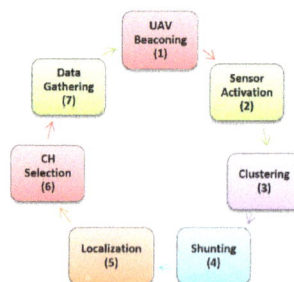

Figure 1. Proposed System lifecycle.

3. Proposed UAV Assisted Dynamic Clustering Architecture

As described above, the developed dynamic clustering consists of two components: sensor nodes and UAVs. No special CHs are installed/defined and all sensor nodes are considered indistinguishable.

3.1. UAV

In the developed system, the UAV plays a vital role in cluster formation, localization, communication and data gathering and the whole system is trigged by the UAV beacon. The UAV under consideration is a quad copter equipped with a localization system with the following specifications:

- Minimum operating flight height = 20 m,
- Maximum operating flight height = 500 m,
- Maximum speed = 7 m/s.

The role of UAV is very critical in the developed system hence it is designed in such way that it bears most of the processing burden to simplify the tasks of the sensor nodes. It performs many key functions throughout its working cycle: activating the sensor nodes, shunting them for further processing, getting nodes' initial information, locating all activated nodes, evaluating them, nominating the CH and finally collecting data from the selected CH. The working cycle of the UAV is given in Figure 2.

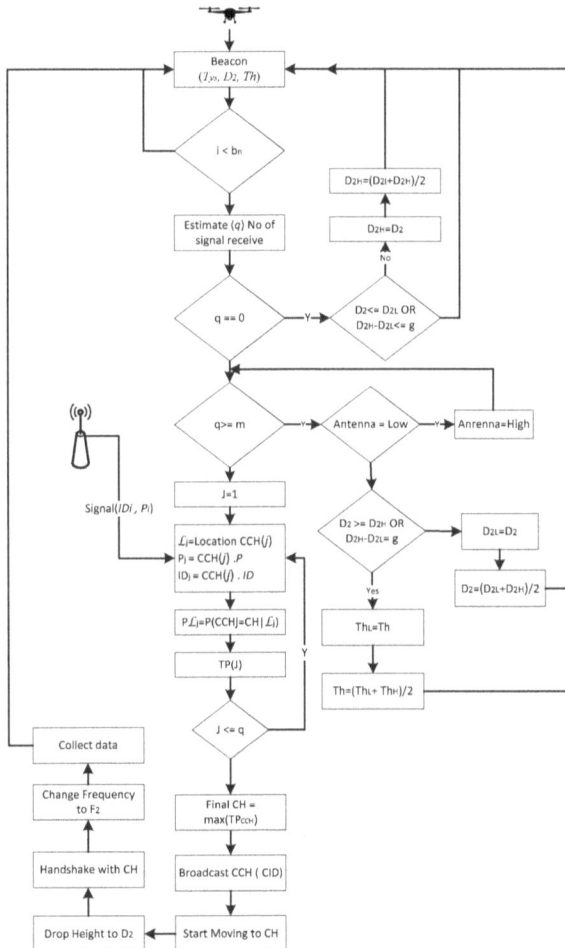

Figure 2. Flow chart of the UAV working.

The UAV sends a beacon message (see Table 1) to activate sensor nodes, and the parameters set in the beacon are as follows:

- *Sensor type*: Type of sensor nodes or combination need to be activated in response of this beacon.
- *Height*: it is current data collection height of UAV not the flying height. The sensor nodes that are capable to communicate at this distance will be considered as candidate for a CH.
- *Threshold*: This threshold will be used to limit the number of sensor nodes contesting for CH.
- *Trailer*: A trailer contains Error Detection Code (EDC) or any other information.

Table 1. UAV beacon message.

Type (6 bytes)	Height (2 bytes)	Threshold (1 byte)	Other (2 bytes)
Header		**Payload**	**Trailer**

3.2. Sensor Node

Field sensors installed in the field to monitor a specific parameters about crop, soil or the environment have the following properties:

- Location-unaware, cheap in cost and left unattended.
- Support for multiple frequencies (at least two frequencies—433 MHz and 2.4 GHz—for localization and data transmission.
- By default, 433 MHz is switched on to hear the beacon from the UAV, afterward it will be used for localization and synchronization, while the 2.4 GHz transceiver will be activated on demand for communication only.
- Can hear the UAV if in range, but all might not be able to connect with it because of their internal parameters.
- Maximum communication range = 500 m.
- Processing and memory enabled.

All sensor nodes are considered to have a unique ID of the format given below:

- *Type*: We consider three basic types: crop, soil and environment, but we allocate 1 byte for further and future extensions.
- *Subtype*: Leaf, stem, root and any combination.
- *Purpose*: Temperature, humidity, thickness, flow and all possible combinations.
- *Unique ID*: Unique number of each sensor node.

The field sensor nodes will have 6 bytes of unique ID as shown in Table 2. The UAV will select any particular type or combination with the help of this ID, also called prefix. Sensor nodes will also use it to send beacon acknowledgement/reply messages.

Table 2. Sensor noder unique ID.

Circle No.	Type	Sub Type	Purpose	Unique No.	Total Size
1 byte	1 byte	1 byte	1 byte	2 bytes	6 bytes
Prefix					

In our proposed system, initially all sensor nodes are considered as undistinguishable (no special CH). A sensor node has to maintain five parameter values about its health:

1. *Energy*: How much energy is remaining,
2. *Consumption rate*: The consumption rate of energy to perform its job,
3. *Renewable energy*: Whether renewable energy is available or not,
4. *Antenna size*: Antenna size to estimate its communication range,
5. *Data size*: How much data should be transmitted.

All activated sensor nodes will calculate a probability value to become a CH based on these health indicators. The sensor nodes' reply to the UAV will consist of 9 Bytes containing a prefix (node ID) in the packet header, a probability value (the probability to be selected as a CH, calculated by the Bayesian formula by putting its entire health indicator's values) in payload and trailer. The 1 trailer byte is allocated for Cyclic Redundancy Check (CRC) and future use. A sensor node activated in response to an UAV beacon will send a reply shown in Table 3:

Table 3. Sensor node reply.

ID (6 bytes)	Probability (2 bytes)	Trailer (1 byte)
Header	**Payload**	**Trailer**

The most important component of our developed system is the sensor node. The whole system is developed in such a way to facilitate the tasks of this component and let it optimize its resources to monitor the crop parameters for a longer time. In our developed system, the daily routine working of sensor node is limited to sense and sleep. It becomes only active for communication after receiving a beacon message from the UAV. The sensor node working process is expressed in an algorithmic form as shown in Figure 3. The process starts when the node receives a beacon then it calculates its Bayesian probability to be a CH and informs the UAV about it. The UAV then selects one of the activated nodes as CH, which should collect data from the whole cluster, aggregate and forward it to the UAV. The list of variables used in the rest of this article is given in Table 4.

Table 4. The list of variables used in state diagram.

Variable	Meaning
CCH	Set of cluster heads
CID	Id of CH broadcast by UAV
D_1	Distance for localization at least 120 m we are taking it 200 m
D_2	Distance for data collection (95 m)
D_{2L}	Minimum possible distance for data collection (170 m)
D_{2H}	Maximum possible distance for data collection (170 m)
E_i	Energy of node i
F_1	1st Frequency 433 MHz
F_2	2nd Frequency 2.4 GHz
g	GPS accuracy
ID_i	Id of crop sensor i
δ_i	Data of node i
$\mathcal{L}_i$	Location of node i
m	Number of antenna elements
n	Number of Sensor nodes ($S1, S2, \ldots, Sn$)
$P\mathcal{L}_i$	Probability of node i by known its location
q	Number of CCH nodes
$T\delta$	Total data of a cluster
Th	Threshold set by UAV, default = 0
TE_{Tx}	Total energy requires to send $T\delta$ bits to D_2 distance
T_{ys}	Type of sensor node announce by UAV
P_i	Probability of node i to become a cluster head

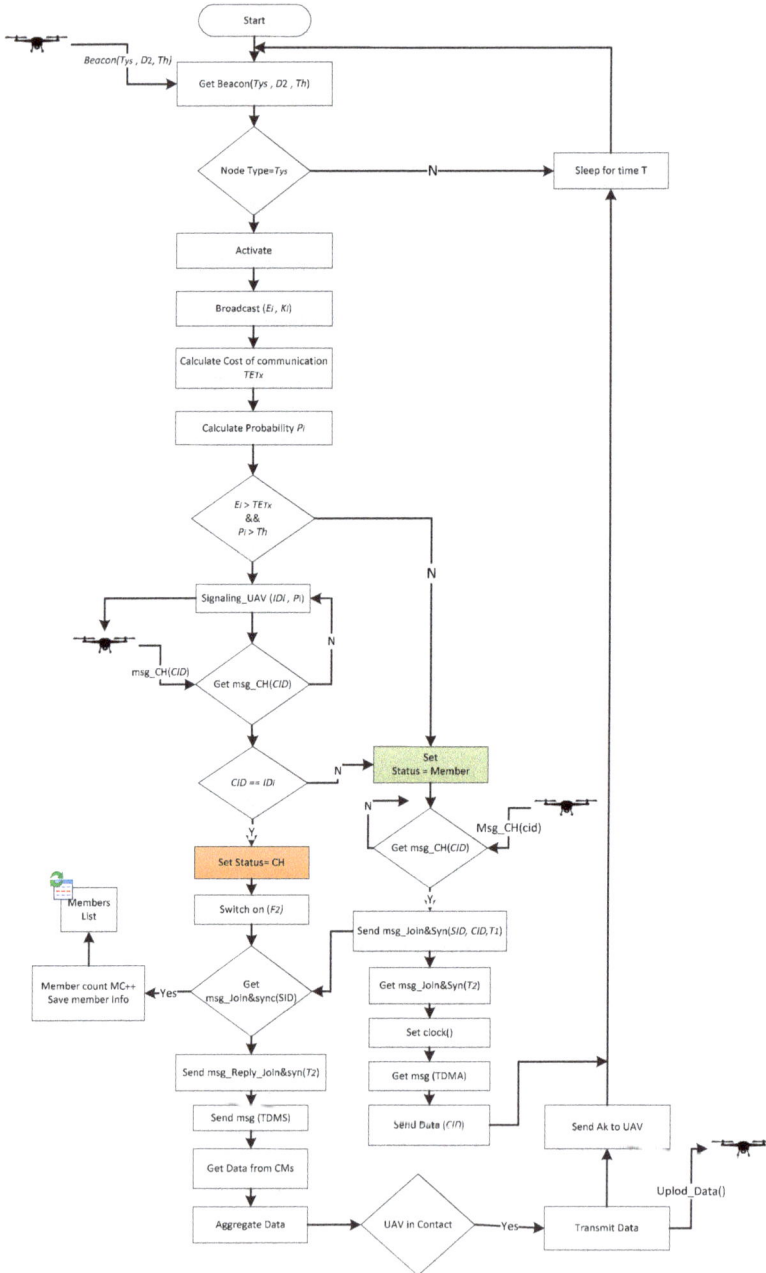

Figure 3. Working flow of sensor node.

4. Dynamic Cluster Head Selection

Once the UAV assisted cluster is formed, the developed system will grade cluster nodes into two types: the first type contains the nodes that don't have the capability to approach the UAV, called cluster members (CMs); candidate cluster heads (CCHs) are the other type. The CCHs are further shunted by the developed system to keep them in a range from 1 to N (where N is the maximum capacity of the UAV to locate sensor nodes). All CCHs and the UAV will collectively take part in the selection process to nominate a node as CH.

Many parameters (like remaining energy, available renewable energy, energy consumption rate, antenna size and distance to the UAV) are considered in this selection process. In the proposed system, the tasks of cluster formation and the CH selection are conducted dynamically at runtime according to the context and then a reliable point to point backbone connection is established between the CH and the UAV to collect all required data for further processing and decision making. The proposed dynamic clustering scheme is illustrated in Figure 4.

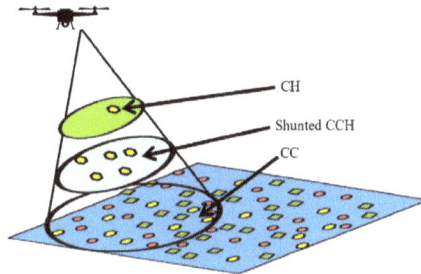

Figure 4. Dynamic clustering scheme.

Each node will participate in the CH selection process depending upon its probability calculated by using a Bayesian classifier. Bayesian probability has been studied for two centuries and many researchers are using it for different purposes [13–16]. We are inspired by Bayesian spam filtering [17] as it is more close to our problem domain. We derived our probabilities in as follows:

Let us suppose there are n sensor nodes $S = (s_1, s_2, \ldots, s_n)$ and each sensor node s_i has z attributes (independent variables) represented by a vector $A = (a_1, a_2, \ldots, a_z)$. A sensor node s_i can be in one of two states:

(a)   Cluster Head (CH)
(b)   Cluster Member (CM)

represented by $State = (CH, CM)$

$P(s_i = CH|a_{ij})$ is the posterior probability of a given sensor node s_i to be a CH knowing its attribute a_{ij}.

$P(s_i = CH)$ is the prior probability of given node to be a CH.

$P(a_{ij}|s_i = CH)$ is the likelihood probability that the highest value of the attribute a_{ij} is in CH node.

$P(s_i = CM)$ is the prior probability that a given node s_i is a member node.

$P(a_{ij}|s_i = CM)$ is the probability that the highest value of attribute a_{ij} is in CM and not in CH.

$P(a_{ij})$ is the prior probability of the attribute value a_{ij} is the highest one.

Equation (1) is the probability of the node s_i to be a cluster head by knowing only one parameter a_{ij}. If all parameters $A_i = (a_1, a_2, \ldots, a_z)$ of node s_i are independent in nature then the conditional probability of this node, by considering whole set A_i can be calculated as:

$$P(a_{i1}, a_{i2}, \ldots, a_{iz}|s_i = CH) = \prod_{j=1}^{z} P(a_{ij}|s_i = CH) = \prod_{j=1}^{z} \frac{P(s_i = CH|a_{ij})\,P(a_{ij})}{P(s_i = CH)} = \frac{\prod_{j=1}^{z} P_{ij}\,P(a_{ij})}{[P(s_i = CH)]^z} \quad (1)$$

$$P(a_{i1}, a_{i2}, \ldots, a_{iz}|s_i = CM) = \prod_{j=1}^{z} P(a_{ij}|s_i = CM) = \prod_{j=1}^{z} \frac{P(s_i = CM|a_{ij})\,P(a_{ij})}{P(s_i = CM)} = \frac{\prod_{j=1}^{z}(1 - P_{ij})\,P(a_{ij})}{[1 - P(s_i = CH)]^z} \quad (2)$$

If:

P_i is the probability $P(s_i = CH|\mathcal{A}_i)$ of the *i*th node, s_i to be a CH by knowing a set of all its parameters $\mathcal{A}_i$, P_{ij} is the probability $P(s_i = CH|a_{ij})$ of the *i*th node, s_i to be a CH by knowing its *j*th parameter a_{ij} from the parameters set $\mathcal{A}_i$ and a_{ij} is the *j*th attribute of the *i*th sensor node.

Then:

$$P(s_i = CH|\mathcal{A}_i) = \frac{P(s_i = CH)\,P(\mathcal{A}i|s_i = CH)}{P(\mathcal{A}_i)} = \frac{P(s_i = CH)\,P(\mathcal{A}_i|s_i = CH)}{P(s_i = CH)\,P(\mathcal{A}_i|s_i = CH) + P(s_i = CM)\,P(\mathcal{A}_i|s_i = CM)} \tag{3}$$

$$P(s_i = CH|a_{ij}) = \frac{P(s_i = CH)\,P(a_{ij}|s_i = CH)}{P(a_{ij})} \tag{4}$$

$$P(s_i = CH|a_{ij}) = \frac{P(s_i = CH)\,P(a_{ij}|s_i = CH)}{P(s_i = CH)\,P(a_{ij}|s_i = CH) + P(s_i = CM)\,P(a_{ij}|s_i = CM)} \tag{5}$$

Putting the values of (1) and (2) in Equation (5):

$$P(s_i = CH|\mathcal{A}_i) = \frac{P(s_i = CH)\,\frac{\prod_{j=1}^{z} P_{ij}\,P(a_{ij})}{[P(s_i=CH)]^z}}{P(s_i = CH)\,\frac{\prod_{j=1}^{z} P_{ij}\,P(a_{ij})}{[P(s_i=CH)]^z} + [1 - P(s_i = CH)]\,\frac{\prod_{j=1}^{z}(1-P_{ij})\,P(a_{ij})}{[1-P(s_i=CH)]^z}}$$

$$P(s_i = CH|\mathcal{A}_i) = \frac{\frac{1}{P(s_i=CH)^{z-1}}\,\prod_{j=1}^{z} P_{ij}\,P(a_{ij})}{\frac{1}{P(s_i=CH)^{z-1}}\,\prod_{j=1}^{z} P_{ij}\,P(a_{ij}) + \frac{1}{[1-P(s_i=CH)]^{z-1}}\,\prod_{j=1}^{z}(1-P_{ij})\,P(a_{ij})}$$

Let us consider that $P(a_{ij})$ are constant then the previous equation can be simplified as follows:

$$P(s_i = CH|\mathcal{A}_i) = \frac{\prod_{j=1}^{z} P_{ij}}{\prod_{j=1}^{z} P_{ij} + \left(\frac{P(s_i=CH)}{1-P(s_i=CH)}\right)^{z-1}\prod_{j=1}^{z}(1-P_{ij})} \tag{6}$$

If considering "not biased" condition where all nodes in the network have the same probability to become cluster head then Equation (6) can be written as:

$$P(s_i = CH|a_{ij}) = \frac{P(a_{ij}|s_i = CH)}{P(a_{ij}|s_i = CH) + P(a_{ij}|s_i = CM)} \tag{7}$$

and Equation (7) can be written as:

$$P_i = P(s_i = CH|\mathcal{A}_i) = \frac{P_{i1}.P_{i2}\ldots\ldots P_{iz}}{P_{i1}.P_{i2}\ldots\ldots P_{iz} + (1 - P_{i1}).(1 - P_{i2})\ldots\ldots(1 - P_{iz})} \tag{8}$$

As Equation (8) may cause a floating point underflow problem, we convert the equation into the log domain:

$$\frac{1}{P_i} - 1 = \frac{(1-P_{i1}).(1-P_{i2})\ldots\ldots(1-P_{iz})}{P_{i1}.P_{i2}\ldots\ldots P_{iz}} \Rightarrow$$

$$\ln\left(\frac{1}{P_i} - 1\right) = \sum_{j=1}^{z}\left(\ln(1 - P_{ij}) - \ln P_{ij}\right) \tag{9}$$

$\ln(P_{ij})$ can produce a problem when P_{ij} is close to zero; the results will be asymptotically correct. Let $\mu = \sum_{j=1}^{z} \ln(1 - P_{ij}) - \ln P_{ij}$:

$$\mu = \sum_{j=1}^{z} \ln\left(\frac{1 - P_{ij}}{P_{ij}}\right) = \sum_{j=1}^{z} \ln\left(\frac{1}{P_{ij}} - 1\right)$$

Finally, we can write that:

$$P_i = \frac{1}{e^{\mu} + 1} \tag{10}$$

5. Three-Layer Architecture of the Developed System

The proposed UAV assisted routing and data gathering scheme is developed in multi-layers and multi-phases (see Figure 5):

1. Layer-1 UAV: UAV is the main part and the top layer,
2. Layer-2 CH: One node per cluster will be selected as CH and it will form a 2nd middle layer,
3. Layer-3 CM: Composed of ground sensor nodes.

Each layer further is divided into three phases as shown in Figure 5. The main part of this system is the UAV acting like a data mule, and the process is started by its beacon message. The UAV can operate at two different frequencies: very low power and long-range UHF F_1 used for localization and high data rate WiFi F_2 for communication. The UAV is performing localization at distance D_1 which is considered as constant and should be high enough to satisfy the far field conditions of F_1. Data collection is conducted at distance D_2 which can be adjusted from 20 to 170 m by considering the CH transmission ability, safe flight distance and WiFi limitations. Top layer (UAV) layer is composed of three phases discuded in detail in next subsection.

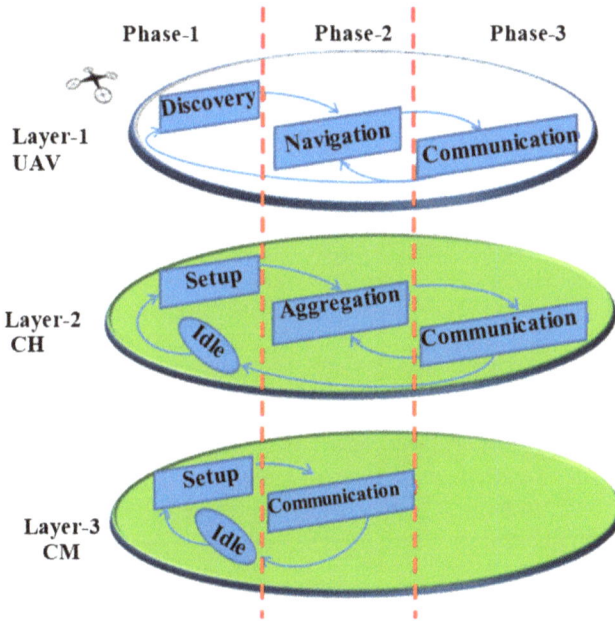

Figure 5. Multi-layer and multi-phase proposed system architecture.

5.1. Discovery

The UAV performs five main tasks in this phase including: beaconing, estimating the number of connected nodes, shunting, localizing and CH nomination, as explained in Figure 6.

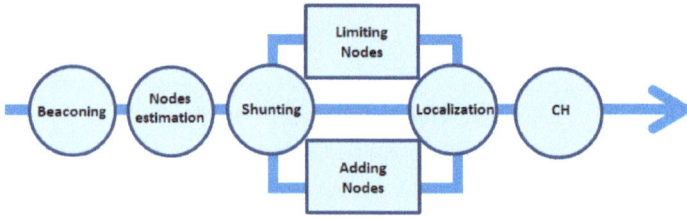

Figure 6. UAV discovery phase.

Let us suppose that the UAV is flying at a height D_1 and initiates its data gathering process by sending a beacon $B(T_{ys}, D_2, Th)$ containing the type T_{ys} of nodes required to activate for data collection, distance D_2 at which it will collect data and Th which is the threshold to limit the activated nodes who will contest for the role of CH. Initially $Th = 0$ means all nodes can participate in the CH selection process. Suppose that there are n sensor nodes $S = \{s_1, s_2, \dots, s_n\}$ installed in a crop field. Assume O nodes become activated in response of UAV beacon such that $1 \le O \le n$ and $SS = \{SS \subseteq S | SS \in T_{ys}\} = \{ss_1, ss_2, \dots, ss_O\}$. SS is the set of nodes that will make cluster in response of this beacon.

If each active node SS_i has $\&_i$ bit data to transmit then $T\&$ represents the total size of cluster data, that is expected to be transmitted by the CH to the UAV:

$$T\& = T\&(SS) = \sum_{i=0}^{O} \&_i \qquad (11)$$

The energy required to transmit $T\&$ to UAV for distance D_2 is:

$$TE_{Tx} = (E_{elec} \times T\&) + \left(E_{amp} \times T\& \times (D_2)^2\right) \qquad (12)$$

where E_{elec} is the energy being dissipated to run the transmitter and E_{amp} is the energy dissipation to amplify the message up to D m.

While UAV is in the discovery phase, activated ground sensors SS will be in setup phase to form a cluster. Every member of SS will calculate a Bayesian probability P_i considering all its parameters (energy, consumption, antenna size, data size, etc.). A detailed discussion on Bayesian probability is given in Section 4. Based on E_i, the energy of the node SS_i, and P_i, the probability to be selected as CH, a further subset called candidate cluster heads of CCH. $CCH = \{cch_1, cch_2, \dots, cch_q\}$, where $1 \le q \le O$, is formed such that:

$$CCH = \{CCH \subseteq SS \, | \vee E_{i_{current}} \, T\& \, \& \, P_i \rangle Th\} \qquad (13)$$

where $E_{i_{current}}$ is the current energy of node i.

As described earlier, $Th = 0$ means any activated node having enough energy can be a member of the CCH.

All CCH nodes start sending beacon replies to the UAV in the form of narrowband signals. At this point, UAV estimates the number of CCH ($|CCH| = q$). If the UAV estimates that $q = 0$ or $q \ge m - 1$, where m is the number of antenna elements on board, then the shunting process is started to keep the candidate cluster heads within a reasonable range; Otherwise, the UAV will locate all CCH nodes by using a special virtual phase array antenna developed for said purpose (a detailed model is given in Section 4). Based on CCH location and P_i values, the UAV nominates a final CH and broadcasts a message to all nodes of set SS to inform them about the CH nomination.

Shunting is the important process of discovery phase to handle the situation when:

$$q = 0 \text{ or } q \geq m$$

If $q = 0$ or $CCH = \varnothing$, it means no CCH member has the capability to send aggregated data to the UAV; In this case, shunting decreases D_2 in steps down to a minimum height depending upon safe flight constraints.

If $q \geq m$ means there are many good CCH nodes and UAV cannot locate all at once; in this case, shunting can take three steps:

- Increase antenna capacity from m to $2m$,
- Increase D_2 in steps up to the limit of F_2,
- Increase Th.

Once a cluster head is selected the next phase of the UAV is navigation.

5.2. Navigation

Once a CH is selected the UAV enters the next phase called navigation. In similar way, the CH and CM switch their phases from setup to aggregation and communication, respectively. In this phase, all active nodes will switch on their frequencies from F_1 to F_2 (433 MHz to 2.4 GHz).

Only the CH will operate on both frequencies. CH will use F_2 to collect data from CMs and F_1 for UAV navigation. While the UAV approaches the CH and attains an agreed height D_2 all CM nodes must transmit their data to the CH which will aggregate it and get ready to make a link with the UAV. As soon as the UAV approaches D_2 and starts handshaking, the CH will switch off its F_2 module and the navigation phase is over. The navigation diagram is elaborated in the diagram of Figure 7.

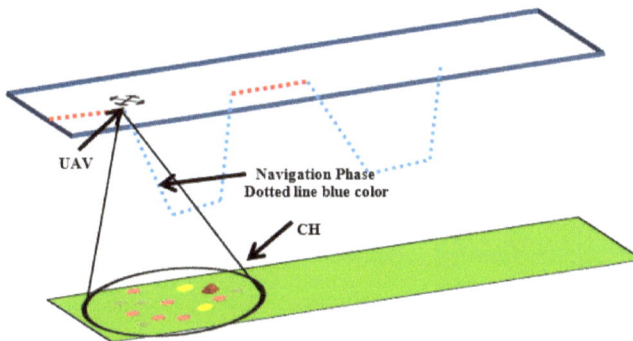

Figure 7. Navigation phase of the proposed system.

5.3. Communication

In the last phase, only the top two layers (UAV and CH) of our developed system will participate. The CH will transmit the whole data to the UAV at the frequency F_2; once transmission is done, the CH will go to sleep for a specific time T, while all CMs had already been shifted to sleep mode. To elaborate the exact sequence of steps the flow chart diagram of the developed system is shown in Figure 8.

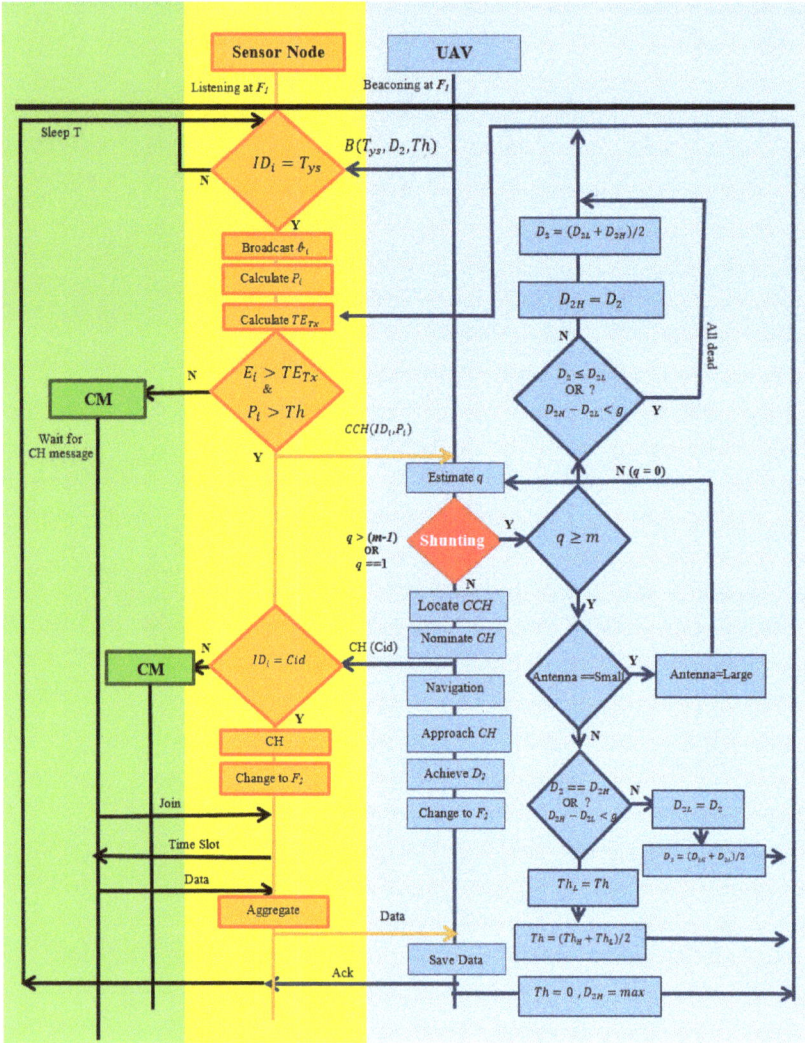

Figure 8. State diagram of the developed system.

6. Developed System Algorithm

The overall working of the developed system is explained in the state diagram (Figure 8) where the homological sequences of steps are shown. The variables used in the algorithm are listed in Table 4. As shown in Figure 8, blue color represents the top layer, Yellow is the middle one and green represents the last layer.. The UAV needs the following inputs to start the mission and data collection procedure: the path of the UAV consists of waypoints $\{(x1, y1), (x2, y2), (x3, y3)....\}$, D_1 height, the UAV normally flies at this height and conducts localization and navigation (e.g., 200), D_{2h} is maximum possible height for data collection (e.g., 170 m), D_{2L} is the lowest possible height for the data collection (e.g., 20 m), D_2 is the data collection height, and the UAV only obtains this height to establish a backbone high data rate link with the CH to collect data, after that UAV goes back to its normal height D_1.

Initially D_2 should mainly be set to an average height (e.g., 170 + 20/2 = 95 m), Th the threshold value $0 \le Th \le 1$, Th_L denotes minimum threshold, Th_H stands for maximum threshold and g is the GPS accuracy (e.g., 5 m).

The UAV is equipped with a virtual phase array antenna which can operate at two different modes Low/High, where Low means with the least specification that can locate fewer targets and High means maximum specifications [12].

The UAV sends beacon messages to activate the ground sensors. The beacon is composed of three basic kinds of information: T_{ys} is the type of nodes must be activated, D_2 distance at which UAV will collect data and Th is threshold. The process steps are as below:

1. The nodes not mentioned in this beacon message continue sleeping, while others will become activated.
2. The activated nodes will broadcast a message to all their neighbors to let them know how much data and energy they have.
3. All activated nodes will calculate the cost of transmission TE_{Tx} as per Equation (12) and their probability to be a cluster head P_i as per Equation (10). All the nodes having energy greater than TE_{Tx} and get a higher probability than the threshold will declare themselves candidate cluster heads (CCHs) and remaining ones will be cluster members (CM).
4. CCHs will proceed further and start sending a narrowband signal to UAV having P_i and their IDs.
5. When the UAV gets a reply from ground sensor nodes, it will use its on board virtual antenna and estimate the number of replying sensor nodes q. If the UAV finds $0 < q < m$ then it will go into localization mode, to estimate the location of these CCHs, otherwise it will start the shunting process.
6. Shunting is a process to pull or push some sensor nodes from the process to keep their number within some reasonable range $(1 \ to \ m-1)$, m is the total number of virtual antenna elements. The shunting process is shown by a red diamond in the state diagram (see Figure 6 For shunting, UAVs have three options that it will be used in steps:

   a. The first option: the UAV is equipped with an adjustable virtual antenna that can operate in two different modes: fast mode with minimum localization ability and high specification mode. If the UAV finds many CCH nodes contesting for cluster head (CH), it will switch its mode to high performance mode.
   b. If the UAV finds that the number of replying CCHs is even greater than the high performance antenna's capacity then it will try to reduce the number of CCHS by increasing the data collection height D_2. The D_2 can also be used in reverse order, if no sensor node is contesting for CH, it means none has the ability to send aggregated data to the UAV at this distance and in this case, D_2 can be decreased up to a minimum level (safe flight).
   c. D_2 is always kept in the middle of highest height D_{2H} and the lowest D_{2L} as $D_2 = (D_{2H} + D_{2L})/2$. We are shrinking and expanding low and high values as per the requirements. If many CCHs were found and we want to limit them by increasing the height the low value is shifted to the middle and the middle value is calculated again. The same treatment in reverse order is used if no CCH is found, and the data collection height is decreased by shifting the high value to the middle and the middle is calculated again.

7. This D_2 height tuning process continues till:

   a. The number of CCHs falls in the range (1 to m − 1)
   b. D_2 reaches the boundary condition extreme high or low
   c. The distance between high and low becomes less than the GPS accuracy of let's say 5 m.

8.   The third option is the variation of the threshold *Th*; by increasing *Th* from 0 to 0.5 roughly half of the nodes will withdraw themselves from CH selection. The UAV can increase *Th* at every iteration.

9.   The UAV will localize the activated candidate cluster heads and select the best one as CH on the basis of P_i information received and the distance to the next waypoint. The node having highest P_i value and closer to the waypoint will be selected as CH.

10. The UAV will send CH messages to all activated nodes.

11. The nodes receiving a CH message will send a join request. The CH receives the join request and allocates a time slot.

12. The nodes receive the time slot and send its data.

13. The CH aggregates the data and transmits it to the UAV.

14. The UAV moves to next cluster. All active nodes finish their transmission, and go to sleep for a fixed time.

7. Characteristics of the Developed System

The characteristics of the proposed system including link budget and communication range are calculates to evaluate the system and to build a simulation model based on these real values.

7.1. Link Budget

The critical parameter in our application is the energy, in particular the energy required for the cluster heads to transmit their beacon and useful information to the UAV. The relationship among energy, distance and frequency is given in Equation (14) [18]:

$$Path\ loss = 32.45\ dB + 20\ Log_{10}\ (frequency\ in\ MHz) + 20\ Log_{10}\ (distance\ in\ Km) \tag{14}$$

We realize that at a large distance from the UAV, it is preferable to use a lower frequency to minimize the power consumption of the cluster head.

In our developed system, the UAV and sensor nodes are performing two major types of communication: one is a long range communication for identification/localization/navigation and other is a short range for the data exchange. For the long range communication, we don't need a high bandwidth. We are using a bit lower frequency that can penetrate to larger distances with very low energy. High frequency (the required one is some standard such as WiFi) is only used for data exchange, as the data size may be large enough so we need high bandwidth. Short-range communication is more energy demanding and increases exponentially with increasing distance. In our developed system, we care about this transmitter receive distance. Link budgets for both the frequencies are defined in the next section.

7.2. Long Range Communication (433 MHz UHF Frequency)

A low power transceiver operating at 433 MHz frequency is suggested for long range communication activities like node localization, synchronization and handshaking. For all these activities, we don't need high bandwidth, as only a few bits of data are needed. Let us assume a small radiated power (100 μW) is considered for transmission:

$$PW = 100\ \mu W = 10\ log_{10}\ (PW\ mW)dBm = 10\ log_{10}(0.1)dBm = -10dBm \tag{15}$$

Masking factor (by objects like vegetation or other) for UHF long distance is considered very small $MF = 5\ dB$.

If receiver sensitivity of a UAV is $\xi = -130\ dBm$ and a signal to noise ratio necessary to demodulate is $SNR = 15\ dB$.

The maximal path loss constraint is calculated as:

$$M_{PL} = PW - \xi - SNR - MF = -10 + 130 - 15 - 5 = 100 \text{ dB}$$

Based on Equation (14), we can calculate the maximum distance covered by UHF communication:

$$20\,log_{10}(D_1) = 100 - 32.45 - 20\log(F_1) = 100 - 32.45 - 52.7 = 14.85$$

$$D_1 = 10^{0.7425} \cong 5.527 \text{ Km}$$

The maximum range between the UAV and the cluster head node is $D_1 = 5.527$ Km.

In our developed system, we don't need to communicate over a long distance of 5 Km. If we are considering a UAV that is flying at a height of 300 m and it wants to activate nodes in 500 m area then only 0.825 µW of power is required by the UAV to transmit the F_1 signal to sensor nodes to communicate.

7.3. Short Range Communication (2.4 GHz WiFi Frequency)

The high power and short range (e.g., WiFi) module is only activated only when data needs to be transmitted. In a similar way, we can find the maximum distance covered by a WiFi signal taking all parameters as considered for F_2 frequency, except for a higher masking factor that is 20.

Receiver sensitivity of a UAV is $\xi = -130$ dBm

A signal to noise ratio necessary to demodulate is $SNR = 15$ dB

The maximum path loss constraint is:

$$M_{PL} = PW - \xi - SNR - MF = 85 \text{ dB}$$

Based on Equation (15), we can calculate the maximum distance covered by WiFi communication:

$$20\,log_{10}(D_2) = 85\,dB - 32\,dB - 20\,log_{10}(F_2) = (85 - 32.45 - 67.6)dB = -15.05\,dB$$

$$Log_{10}(D_2) = -\frac{15.05}{20} = -0.7525 \Rightarrow$$
$$D_2 = 10^{-0.7525} \cong 0.176 \text{Km}$$

The maximum range between the UAV and the cluster head node for data collection is 0.177 Km. We will then consider a maximum distance in the (x, y) plane of the order of $xy_{UAV} = 150$ m.

8. Simulation

Simulations are conducted in OMNeT++ (Objective Modular Network Test bed in C++) and MatLab to evaluate the performance of the developed system.

8.1. Simulation Model

As per our designed protocol, a wireless sensor node is equipped with two different frequencies: 433 MHz for localization and identification of nodes and 2.4 GHz for data transmission. We developed a sensor node simulation model using two NICs CC2420 [19] and CC1021 [20] as shown in Figure 9a. Both NICs are composed of MAC and physical layer and the sensor node consists of three layers (Application, Network and NIC). CC2420 is a 802.15.4 compliant Network Interface Card (NIC) operating at 2.4 GHz frequency and having built-in CSMA/CA in MAC layer, while CC1021 is a low power RF transceiver for narrowband systems operating at 433 MHz. The sensor node is considering as smart enough that it can negotiate a communication height with the UAV by considering the amount of data that must be transmitted and the remaining energy power level. It utilizes only 0.825 µW of energy while operating with the CC1021 for localization. The largest energy depletion factor is data transmission between the CH and UAV using a higher frequency, e.g., WiFi. If the considered node is selected as CH then the energy consumption of the CC2420 is optimized by limiting the agreed CH-UAV height. The UAV receives data at a lower height suitable for the CH. The UAV model is created with same NICs as in the sensor nodes. The only difference is the addition of mobility components in the main module as shown in Figure 9b.

The UAV is simulated to fly with a constant speed of 20 m/h and a height of 200 m. Initially, it switches on its CC1021 module and starts sending the beacon message. Once the UAV identifies the location of the selected CH, it starts moving towards it and reduces its height up to an agreed level. When the UAV approaches the CH, it will change its communication module from CC1021 to CC2420. All these simulation are available on YouTube links [21–25].

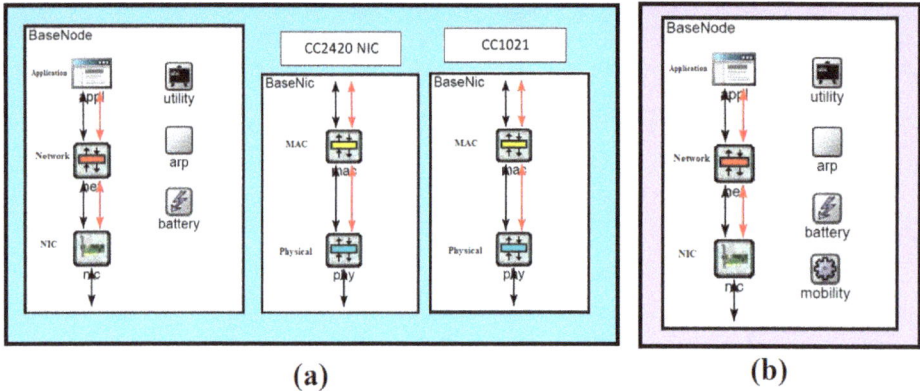

Figure 9. Developed OMNet++ simulation model (**a**) Sensor node simulation model (**b**) Mobile sink extra module other than sensor node model.

8.2. Simulation Cases

As described earlier, our case scenario belongs to the category of a Mobile Sink Static Node (MSSN) network where a sink is traversing the network to collect data while all sensor nodes are kept constant. If we are taking the sink as a mobile node then there are three possibilities:

(1) Direct data collection,
(2) Cluster-based data collection with a controlled path,
(3) Cluster-based data collection with an independent UAV path.

In this scheme, the UAV can move freely and instruct the ground sensor nodes to form clusters and help them to select a cluster head. To evaluate this developed system, we studies three cases, one for each type:

1. Direct Data Collection (DDC) by UAV, as described in [26],
2. Network Assisted Data Collection (NADC), as given in [27]
3. Our developed UAV Routing Protocol (URP).

One hundred sensor nodes are deployed randomly with a uniform distribution in a 2 Km^2 area.

8.3. Simulation Results

Different simulation models and cases are executed a number of times (at least five times) and the average results are considered for more accuracy. Different simulation cases are described one by one as given below:

8.3.1. Number of Dead Nodes vs. Simulation Time

In this simulation run, we wanted to check how energy is utilized and how fast the sensor nodes' energy becomes completely exhausted. We simulate each case (DDC, NADC, URP) separately with same parameters such as number of nodes, simulation time, initial node energy, node/UAV communication ranges, data collection height, etc.

We can conclude that using our developed system, the network lifetime can be improved by 20% or more on the expense of UAV energy, adding more intelligent sensor nodes and dual frequency support in the UAV and nodes (as Figure 10a).

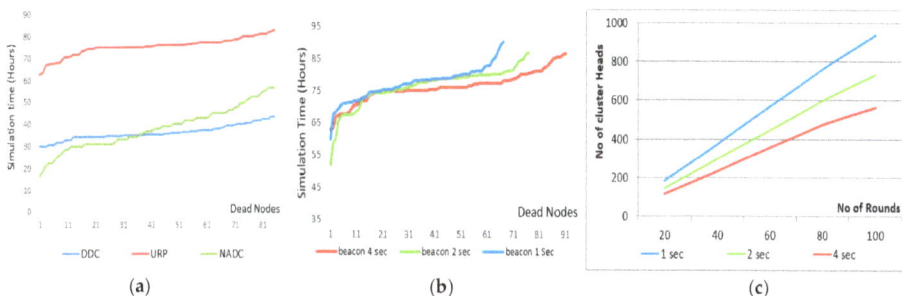

(a)   (b)   (c)

Figure 10. Simulation results of different cases (**a**) Number of dead nodes vs. simulation time (**b**) effect of varying beacon sending period.

8.3.2. Effect of Beacon Sending Period

The UAV starts its data collection process by sending beacon messages to activate the sensor nodes. This beaconing period decides the size of clusters, and the longer the beaconing period the larger the cluster size. This phenomenon can also affect the performance of the system. The developed system is tested against varying beacon sending periods and the results are shown hereinafter.

Figure 10b represents the total number of dead nodes on the X-axis, simulation time on the Y-axis and different curves show the performance of system using different beacon periods. The lower the curve the better is the performance. It is observed in Figure 10c that by increasing the beaconing period the clusters become larger in size and fewer in number. Number and size of clusters may also effect the energy utilization of the system, which is further investigated in Figure 11.

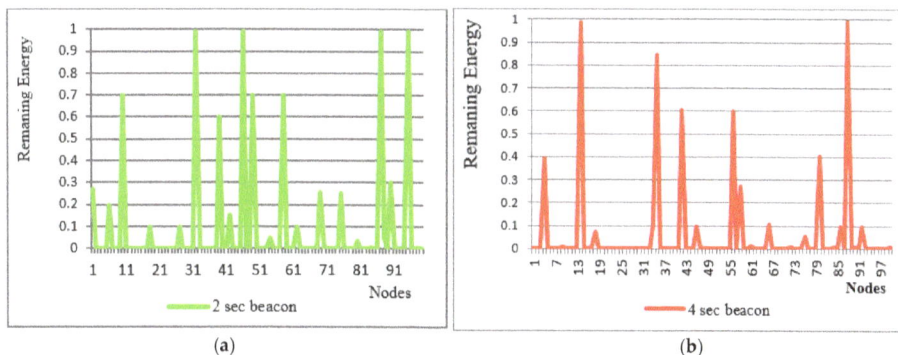

(a)   (b)

Figure 11. Effect of beaconing period on energy utilization (**a**) 2 s beaconing (**b**) 4 s becoming.

Figure 11a shows the remaining energy of each node at the end of simulation while using 2 s beacons. More high peaks mean more nodes are unutilized. Figure 11b shows similar results as Figure 11a but we considered have 4 s beacons.

The system is evaluated by changing the beaconing period from 1 to 4 s. If a 4 s beacon is used about 580 clusters are made in 100 rounds. Further, in the case of a1 s beacon the number of clusters is 980. Figure 11a shows that about 10 nodes are left unattended or unutilized at the end of the simulation, while in Figure 11b the energy utilization is improved in the sense that only two nodes are observed unutilized and few are underutilized.

8.3.3. Dead Nodes Investigation in Matlab

Figure 12a shows the comparison of dead nodes vs. UAV rounds in different routing schemes. The number of rounds is shown on the X-axis and number of dead nodes is shown on the Y-axis. Different colored curves represent different data gathering schemes. The lower the curve, the higher the performance. Red and purple lines show the performance of proposed UAV Routing Protocol (URP). We tested the developed system at constant and fixed height (the UAV is not going down to take data) as in case of other networks (LEACH and HEED) and the results are shown in the red curve. When the UAV is flying at a fixed constant height even then our proposed system is performing quite well. The purple curve URP-adopted height is the best case scenario when the UAV negotiates the best height with the CH in advance and respect it while collecting the data. Thanks to the fact that there are no periodic updates, no flooding of information, better CH selection and duel frequency use, that helps us to optimize the node energy up to a maximum extent. The UAV and CH negotiate a suitable height for data collection. As long-range communication is the main source of energy depletion, adjusting has in a good impact on overall system lifetime.

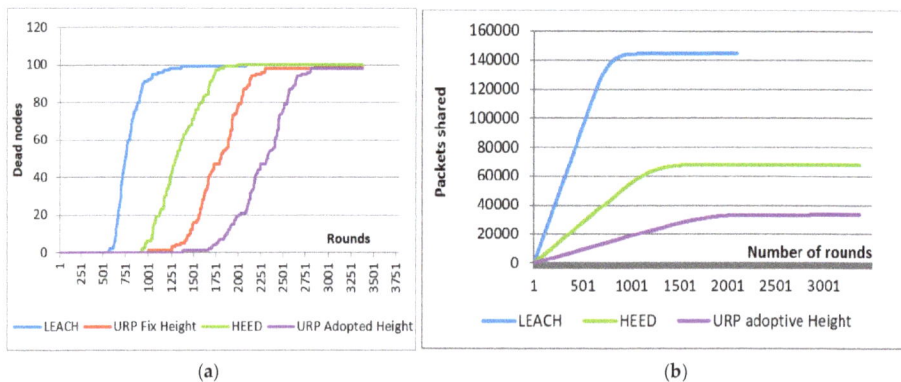

(a)

(b)

Figure 12. Comparison of existing routing protocols with the proposed system (**a**) Number of rounds vs. dead nodes; (**b**) comparison of the amount of inter-cluster communication in different routing algorithms that is required to formulate a cluster.

8.3.4. Inter-Cluster Communication Assessment

Number of packets exchanged among clusters to build a network is shown in Figure 12b. The number of rounds is shown on the X-axis, packets delivered are represented by the Y-axis and the colored curves shows different protocols. The lower the curve the better the system is. It is observed that about 140 k packets are exchanged among cluster members and cluster heads in 1000 rounds just only to keep network live and updated in the form of periodic updates, clustering information and CH notifications. In the proposed system, only 20 k messages are exchanged, which is a great factor that improves the system performance a lot.

9. Proof of Concept

We used an Arduino microcontroller to build two components, an UAV module and ground sensor nodes. A specialized UAV is also made to carry this equipment and can operate as per instructions given onboard by our developed system. IoT and UAV sink nodes are developed using same hardware shown in Figure 13, the only difference being the software uploaded because they have different functionality.

Figure 13. Proof of concept sensor and UAV node.

Both devices are built using a nRF24L01 transceiver, which is a low power consumption transceiver operating at 2.4 GHz frequency and capable of transmitting data at rates up to 2 Mbps. A circuit diagram for the nRF24L01 [28] and wiring information are shown in Figure 14.

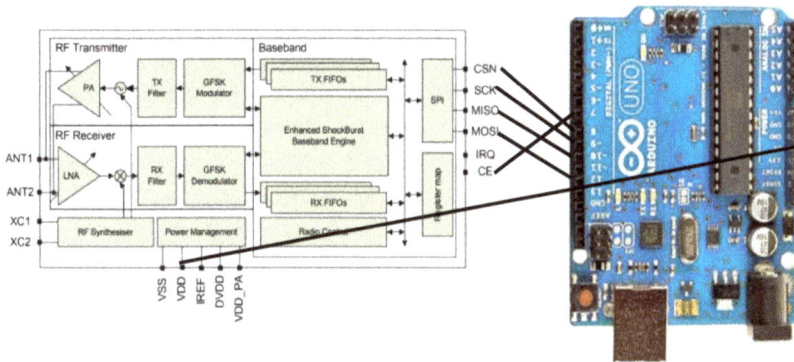

Figure 14. NRF24L01 and Aduino wiring diagram.

Initially, we loaded all sensor nodes with 20 Kb (Kilo Bytes) of data and five nodes are deployed in the field. When the UAV approaches the field, one of them is elected as CH dynamically and it collects all the data from neighboring nodes and finally transmits 100 Kb data to the UAV.

The localization components, including the virtual antenna, UHF 433 MHz transceiver for the UAV and sensor nodes are still in development and not installed so far. The first test run results are given below. All these results do not include the UAV navigation time.

9.1. Working of Cluster Member Nodes

The algorithm described in Section 6 is developed in Arduino language which is mostly based on C++. The developed code then builds on Arduino Uno board shown in Figure 13 that is developed to make a field sensor node. We choose Arduino mini to keep the node size small. Figure 15a represents the activity graph of a ground sensor node that acts as cluster member. It is observed that it takes about 5.4 s to complete the cluster formation and data delivery. In that figure, the X-axis represents the CM sequence of activities, while the Y-axis shows the time in seconds and the curves represent the relationship betweentime and activities. The parameters of this node were set to assume it doesn't have good specification and its CH probability is almost 0, so it decides immediately to set its status as CM. Cluster head selection takes about 4.5 s and finally data transmission takes less than a second to complete its process.

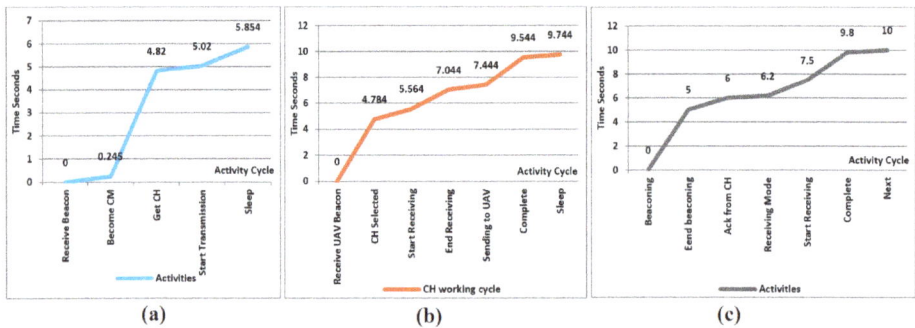

Figure 15. Activities of proof of concept devices related to time (**a**) sensor node (**b**) cluster head (**c**) UAV.

9.2. Cluster Head Node Activities

Figure 15b shows the graph of CH working. It takes about 10 s to complete its working cycle. It is observed that this node becomes a CH in 5 s, during this period, it becomes activated, contacts neighboring nodes and shares information with the UAV. As this node has been selected as a CH, it has to collect data from all other members and transmit the aggregated data to the UAV. This whole procedure is conducted in 5 s.

9.3. Working Cycle of the UAV

Life cycle of the UAV is shown in Figure 15c. It sends a beacon message in the first 5 s, then switches to discovery phase to search for a suitable CH. Once a CH is selected, it will navigate to approach it and collect data at some reasonable height. The whole procedure takes about 10 s.

9.4. Combined Activities Analysis

Comparison of UAV activities with respect to the CM and CH are shown in Figure 16a. The X-axis represents UAV activities, while the Y-axis shows time in seconds and different color curves represent different components (UAV, CH and CM).

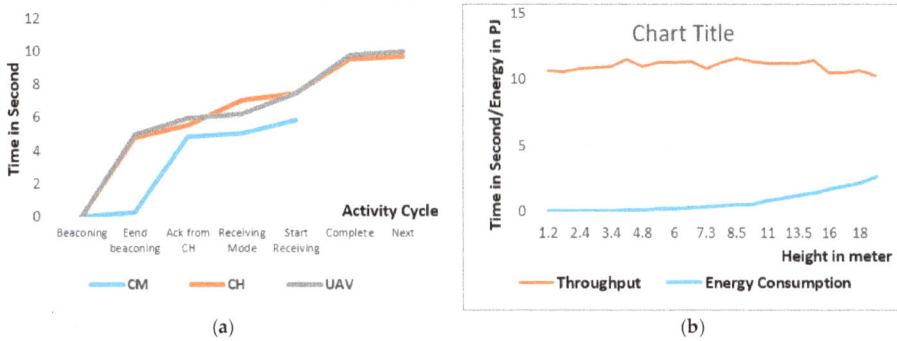

Figure 16. Performance analysis of proof of concept devices (**a**) activities of different components with respect of time (**b**) effect of varying height on throughput and energy consumption.

9.5. Effect of Varying Height on Data Collection

We evaluate the effect of changing height on the system performance and results are shown in Figure 16b. The X-axis shows the height of the UAV from 1.2 m to 20 m. and the Y axis represents both time consumed in seconds and energy utilization in PJ (Peta Joules).

By increasing the height of the UAV, we could not notice any effect on the data collection time but it may have a massive impact on the CH energy utilization.

10. Conclusions

We have developed and tested a dynamic data collection method where data can be collected from selected nodes in a targeted area. In this project, an UAV can move freely without considering the ground network topology. To tackle this, it is proposed that ground nodes will form clusters according to the UAV's path and the nature of data. Further we introduced a Bayesian probability- based cluster head selection process to select the best node as a cluster head. The developed system is evaluated through simulation and proof of concept devices. From the simulation results, we can conclude that the proposed system can drastically increase the network lifetime while maintaining a good throughput. For proof of concept, sensor and sink nodes are developed by using an Arduino microcontroller where the sink node is integrated with the UAV so that it can control the UAV's flight. Real time field data is collected and different parameters are monitored for performance evaluation. The measured results showed that this system is practically feasible while offering many benefits over traditional approaches. Critical data about crops like nutritional stress, bug attacks or the spread of diseases can be detected well in time so they can be cured before the crop is affected or destroyed. By using our developed system, farmers do not need the regular visits to their large farm field and they can rely on this fully automatic system to get accurate, up-to-date and precise information. Furthermore this system can help the farmers to plan the optimum utilization of resources as the system can take appropriate action on their behalf, if some threshold is achieved. Considering all this we can say that developed system will facilitate the use of IoT technology for agriculture to take care of crop health to ensure the quality and quantity of foods. Moreover, the proposed system can be extended in the future in many ways; multiple UAVs can be used in a a coordinated way to further increase the throughput. Single hop clusters can be extended to multi-hop ones at the same time that more professional sensors and sink devices can be made from the proof of concept.

Acknowledgments: The authors gratefully acknowledge the support of SNCS Research Center at the University of Tabuk, Saudi Arabi.

Author Contributions: Mohammad Ammad Uddin originated the idea, concept and algorithms as well as performed simulation study, proof of concept development and real field testing of the system; Ali Mansour participated in mathematical models and algorithm development. Also perform results verification and over all supervision; Denis Le Jeune contributed in developing signaling method, localization, performance measurement, system validation. Also validate results and system working; Manuscript proof reading, editing, change validation and rechecking is done by Muhammad Ayaz; el-Hadi M. Aggoune supervised the project Over all working, evaluation, integration and testing.

Conflicts of Interest: The authors declare no conflict of interest.

References

1. Fu, C.; Jiang, Z.; Wei, W.; Wei, A. An Energy Balanced Algorithm of LEACH Protocol in WSN. *Int. J. Comput. Sci.* **2013**, *10*, 354–359.
2. Kaixin, X.; Gerla, M. A Heterogeneous Routing Protocol Based on a New Stable Clustering Scheme. In Proceedings of the Military Communication Conference MILCOM, San Jose, CA, USA, 31 October–3 November 2010; Volume 2, pp. 838–843.
3. Shangguan, L.; Mai, L.; Du, J.; He, W.; Liu, H. Energy-efficient Heterogeneous Data Collection in Mobile Wireless Sensor Networks. In Proceedings of the 20th International Conference on Computer Communication Networks (ICCCN'2011), Maui, HI, USA, 31 July–4 August 2011.
4. Khan, A.W.; Abdullah, A.H.; Razzaque, M.A.; Bangash, J.I. VGDRA: A Virtual Grid-Based Dynamic Routes Adjustment Scheme for Mobile Sink-Based Wireless Sensor Networks. *IEEE J. Sens.* **2015**, *15*, 526–534. [CrossRef]
5. Okcu, H.; Soyturk, M. Distributed Clustering Approach for UAV Integrated Wireless Sensor Networks. *Int. J. Ad Hoc Ubiquitous Comput.* **2014**, *15*, 106–115. [CrossRef]
6. Khan, A.; Abdullah, A.; Anisi, M.; Bangash, J. A Comprehensive Study of Data Collection Schemes Using Mobile Sinks in Wireless Sensor Networks. *IEEE J. Sens.* **2014**, *14*, 2510–2548. [CrossRef] [PubMed]
7. Chang, J.-Y.; Ju, P.-H. An Efficient Cluster-based Power Saving Scheme for Wireless Sensor Networks. *EURASIP J. Wirel. Commun. Netw.* **2012**, *2012*, 172–179. [CrossRef]
8. Sharef, B.T.; Alsaqour, R.A.; Ismail, M. Vehicular Communication Ad-hoc Routing Protocols: A Survey. *J. Netw. Comput. Appl.* **2014**, *40*, 363–396. [CrossRef]
9. Duan, Y.E. Design of Intelligent Agriculture Management Information System Based on IoT. In Proceedings of the 2011 Fourth International Conference on Intelligent Computation Technology and Automation, Shenzhen, China, 28–29 March 2011; pp. 1045–1049.
10. Fan, T.K. Smart Agriculture Based on Cloud Computing and IOT. *J. Converg. Inf. Technol.* **2013**, *8*, 210–216.
11. Bo, Y.; Wang, H. The Application of Cloud Computing and the Internet of Things in Agriculture and Forestry. In Proceedings of the 2011 International Joint Conference on Service Sciences, Taipei, Taiwan, 25–27 May 2011; pp. 168–172.
12. Uddin, M.A.; le Jeune, D.; Mansour, A.; el Aggoune, H.M. Direction of Arrival of Narrowband Signals Based on Virtual Phased Antennas. In Proceedings of the IEEE 23rd Asia-Pacific Conference on Communication, Perth, Australia, 11–13 December 2017; pp. 627–632.
13. Koutsias, J.; Chandrinos, K.V.; Paliouras, G.; Spyropoulos, C.D. An evaluation of Naive Bayesian anti-spam filtering. In Proceedings of the 11th European Conference on Machine Learning, Barcelona, Spain, 31 May–2 June 2000.
14. Mokhesi, L.; Bagula, A. Context-aware handoff decision for wireless access networks using Bayesian networks. In Proceedings of the Annual Research Conference of the South African Institute of Computer Scientists and Information Technologists, Vanderbijlpark, South Africa, 12–14 October 2009; pp. 104–111.
15. Domingos, P.; Pazzani, M. On the Optimality of the Simple Bayesian Classifier under Zero-One Loss. *Mach. Learn.* **1997**, *29*, 103–130. [CrossRef]
16. Wang, Q.; Garrity, G.M.; Tiedje, J.M.; Cole, J.R. Naive Bayesian Classifier for Rapid Assignment of rRNA Sequences into the New Bacterial Taxonomy. *Appl. Environ. Microbiol.* **2007**, *73*, 5261–5267. [CrossRef] [PubMed]
17. Androutsopoulos, I.; Koutsias, J.; Chandrinos, K.V.; Spyropoulos, C.D. An experimental comparison of naive Bayesian and keyword-based anti-spam filtering with personal e-mail messages. In Proceedings of the 23rd Annual International ACM SIGIR Conference on Research and Development in Information Retrieval, Athens, Greece, 24–28 July 2000; pp. 160–167.

18. Faruque, S. Chapter 2 Free Space Propagation. In *Radio Frequency Propagation Made Easy*; Springer: Berlin, Germany, 2015; pp. 19–26.

19. 2.4 GHz IEEE 802.15.4/ZigBee-Ready RF Transceiver. Texas Instruments Incorporated. Available online: http://www.ti.com/lit/ds/symlink/cc2420.pdf (accessed on 2 May 2017).

20. CC1021 Single Chip Low Power RF Transceiver for Narrowband Systems. Texas Instruments Incorporated. Available online: http://www.ti.com/lit/ds/swrs045e/swrs045e.pdf (accessed on 6 October 2017).

21. Ammad-udDin, M. Field Test of Agriculture IoT Devices. 2017. Available online: https://www.youtube.com/watch?v=Bma2qE7f1gk (accessed on 3 December 2017).

22. Ammad-udDin, M. UAV Connectivity with Sensor Nodes. 2017. Available online: https://www.youtube.com/watch?v=xzf4SkhDteE (accessed on 3 December 2017).

23. Ammad-udDin, M. STK Simulation of Dynamic Data Collection by UAV. 2017. Available online: https://www.youtube.com/watch?v=OXtNn94l1xA (accessed on 3 December 2017).

24. Ammad-udDin, M. MatLab 3D Simulation of Dynamic Clustering. 2017. Available online: https://www.youtube.com/watch?v=zVM42i9BQFY (accessed on 3 December 2017).

25. Ammad-udDin, M. OMNet ++ Simulation of Dynamic Clustering and Crop Health Monitoring. 2017. Available online: https://www.youtube.com/watch?v=pYx-Z0chWoA (accessed on 3 December 2017).

26. Li, X.; Nayak, A.; Stojmenovic, I. Exploiting Actuator Mobility for Energy-Efficient Data Collection in Delay-Tolerant Wireless Sensor Networks. In Proceedings of the 2009 Fifth International Conference on Networking and Services, Valencia, Spain, 20–25 April 2009; pp. 216–221.

27. Rao, J.; Biswas, S. Joint Routing and Navigation Protocols for Data Harvesting in Sensor Networks. In Proceedings of the 5th IEEE International Conference on Mobile Ad Hoc and Sensor Systems, Atlanta, GA, USA, 29 September–2 October 2008; pp. 143–152.

28. RF24L01 2.4GHz Radio/Wireless Transceivers How-To. Available online: https://arduino-info.wikispaces.com/Nrf24L01-2.4GHz-HowTo (accessed on 6 October 2017).

sensors

MDPI

Article

A Novel Methodology for Improving Plant Pest Surveillance in Vineyards and Crops Using UAV-Based Hyperspectral and Spatial Data

Fernando Vanegas [1,*], Dmitry Bratanov [1], Kevin Powell [2,3,†], John Weiss [3,4] and Felipe Gonzalez [1,3]

1   Institute for Future Environments, Robotics and Autonomous Systems, Queensland University of Technology, 2 George St, Brisbane, QLD 4000, Australia; dmitry.bratanov@qut.edu.au (D.B.); felipe.gonzalez@qut.edu.au (F.G.)
2   Agriculture Victoria Research, Victorian Department of Economic Development, Jobs, Transport and Resources, Rutherglen, VIC 3083, Australia; kpowell@sugarresearch.com.au
3   Plant Biosecurity Cooperative Research Centre, Bruce, ACT 2817, Australia; john.weiss@ecodev.vic.gov.au
4   Agriculture Victoria Research, Victorian Department of Economic Development, Jobs, Transport and Resources AgriBio Centre, 5 Ring Road, Bundoora, VIC 3083, Australia
*   Correspondence: f.vanegasalvarez@qut.edu.au; Tel.: +61-7-3138-4593
†   Current address: Sugar Research Australia, Meringa, QLD 4865, Australia.

Received: 30 November 2017; Accepted: 14 January 2018; Published: 17 January 2018

Abstract: Recent advances in remote sensed imagery and geospatial image processing using unmanned aerial vehicles (UAVs) have enabled the rapid and ongoing development of monitoring tools for crop management and the detection/surveillance of insect pests. This paper describes a (UAV) remote sensing-based methodology to increase the efficiency of existing surveillance practices (human inspectors and insect traps) for detecting pest infestations (e.g., grape phylloxera in vineyards). The methodology uses a UAV integrated with advanced digital hyperspectral, multispectral, and RGB sensors. We implemented the methodology for the development of a predictive model for phylloxera detection. In this method, we explore the combination of airborne RGB, multispectral, and hyperspectral imagery with ground-based data at two separate time periods and under different levels of phylloxera infestation. We describe the technology used—the sensors, the UAV, and the flight operations—the processing workflow of the datasets from each imagery type, and the methods for combining multiple airborne with ground-based datasets. Finally, we present relevant results of correlation between the different processed datasets. The objective of this research is to develop a novel methodology for collecting, processing, analysing and integrating multispectral, hyperspectral, ground and spatial data to remote sense different variables in different applications, such as, in this case, plant pest surveillance. The development of such methodology would provide researchers, agronomists, and UAV practitioners reliable data collection protocols and methods to achieve faster processing techniques and integrate multiple sources of data in diverse remote sensing applications.

Keywords: remote sensing; unmanned aerial vehicle; phylloxera; multispectral; hyperspectral; RGB; digital elevation model; digital vigour assessment

---

1. Introduction

Recent advances in remote sensed imagery and geospatial image processing using unmanned aerial vehicles (UAVs) have enabled the creation of rapid and ongoing monitoring tools for crop management [1–4] and the detection/surveillance of insect pests. However, there are still challenges in remote sensing applications, such as detecting early incursions of cryptic pest species such as grape phylloxera (*Daktulosphaira vitifoliae* Fitch) in vineyards. Grape phylloxera is currently present in most grape-growing countries, but relatively localised in wine districts in southeastern Australia. Grape phylloxera is a very small insect that primarily lives underground, feeds on the roots of the grapevines, and consequently damages the root system. This creates impairing water and nutrient uptake to the plant, which causes stress that is expressed above-ground by impairment and changes in photosynthesis, changes in pigment ratios, reduced canopy, slow stunted growth, and reduced yield [5]. The symptoms of infestation appear usually after two to three years, although in some instances this can be longer [6].

The current surveillance practice for growers is to visually inspect their grapevines during December to April for signs of grapevine damage attributable to a phylloxera infestation [7]. A more intensive method of monitoring utilises phylloxera emergence traps at a density of one trap every fifth panel of every third grapevine row (standard trapping method) [8]. Traps are deployed in the spring and summer months, when immature (first instar) insects move from the grapevine roots onto the soil surface [7]. The traps are usually placed at the base of suspected infested grapevines and inspected after two to four weeks. In addition, on suspected infested grapevines, the roots can be inspected for yellow galls, swellings on the older roots, and the yellow insects [9]. Although these practices have been extensively used as standard methods, they are time consuming, season dependent, labour intensive and require taxonomic expertise [10].

Ground-based spectral observations using hyperspectral imagery have been conducted to characterise the spectral response of phylloxera-infested stressed grapevines [11]. However, this study focused only on leaf-level observations. In addition, several aerial studies have been conducted for phylloxera detection and monitoring [12–14]. A canopy-level characterisation of phylloxera has been studied [15], but further research needs to be conducted to develop reliable predictive detection methods. Typical indices for analysis of stress vegetation include the normalised difference vegetation index (NDVI) and colour infrared composite (CIR) [7], but they are not able to distinguish the stress caused by phylloxera infestation from the stress caused by other sources, and may only be useful in monitoring temporal changes in phylloxera distribution over known infested vineyards [16]. Aerial thermal imagery can be used to identify grapevines showing indications of yellowing or decline that require closer inspection [17].

A spatial and temporal analysis methodology has been applied for grapevine vigour analysis to assess the health of a vineyard [18]; however, hyperspectral imagery was not considered. Micro UAV have been used for generating georeferenced orthophotos for precision agriculture purposes, with the aim of producing vegetation indices in the visible (VIS) and near-infrared (NIR) spectrum [19], yet there is a need for higher spectral resolution imagery to produce different vegetation indices. UAV-based cloud point and hyperspectral imagery systems have been developed and used for a biomass estimation of wheat and barley crops [20,21] but this new technology still needs to be tested in broader applications. Low altitude aerial imagery has been trialed to detect diseases in avocado trees using different vegetation indices from RGB images [22]. UAV and aerial RGB imagery have also been applied to the detection of infestation symptoms on olive trees (canopy discolouration) and palm trees mapping [23], but only visible symptoms are considered, and further research needs to be to done to include vegetation indices and spectral signatures analysis.

This paper presents the methodology for developing a predictive detection method for pests using processed airborne RGB, multispectral, and hyperspectral data combined with ground-collected data. Among the methods used for processing the data are the photogrammetry of RGB imagery, the development of an airborne RGB-based digital elevation model in order to assess the canopy vigour, the georeferencing of different datasets, the processing of multispectral and hyperspectral imagery to generate vegetation indices, and the extraction of mean spectral signatures of grapevines for different levels of pest infestation at the canopy level.

Section 3 describes the UAV and sensors used for airborne imagery collection. Section 4 describes the field experiments carried out in December 2016 and February 2017.

Finally, we show results for the correlation of the most relevant vegetation indices to expert vigour assessment and a digital vigour model, and provide conclusions on these preliminary findings and insights on the ongoing and further work required to generate a predictive model for pest infestation in crops (in this case grapevine phylloxera). A second paper will focus on verifying and validating a predictive model of a case study of a serious pest in vineyards. We envision the title to be Improving Plant Health Surveillance in Vineyards and Row Crops: a Predictive Model for the Early Detection, Extent, and Impact of Grape Phylloxera.

2. Methods

2.1. Predictive Detection Model Workflow

Our approach for generating a predictive model for pest and diseases in vineyards or other row crops e.g., avocado trees, macadamia trees, etc., employs multiple stages (Figure 1). The first stage of the process is data collection. This includes the collection of airborne RGB, multispectral and hyperspectral imagery, and ground data in the form of ground control points (GCP), reflectance references, expert visual vigour assessment, EM-38 soil conductivity, and ground traps counts.

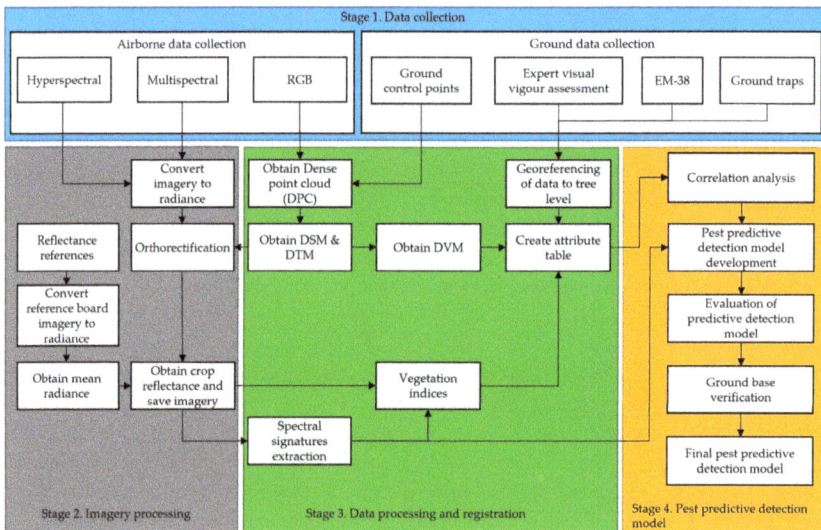

Figure 1. Predictive detection model workflow.

In the second stage, multispectral and hyperspectral images are processed to obtain radiance. All of the imagery was created with data input from ground control points, and using structure from motion algorithms is orthorectified. This process generates orthomosaics of the entire crop. Multi and hyperspectral orthomosaics are processed to obtain reflectance data cubes using mean white references obtained from boards and imagery taken in pre and post-flight operations.

The next stage uses these reflectance images to extract spectral signatures of the crop at the canopy level, with different levels of infestation, for the calculation of numerous vegetation indices that are selected based on the symptoms of infestation (e.g., reduced chlorophyll content, leafs yellowing). Digital surface models (DSM) and digital terrain models are obtained from the RGB orthomosaics in order to produce a digital vigour model (DVM).

The next step in this stage is to combine all of the multiple sources of data into a single information system. In order to do this, there is a segmentation process to obtain data from individual plants, with the aim of creating a table that contains different attributes for a single plant within the crop.

The result of the previous step is an attribute table containing the extracted values of several vegetation indices (see Table 1), expert vigour assessment classes, and values estimated from a DVM. The table also consists of traps data that is only used for the verification of suspected infestation, and EM38 soil conductivity data that was collected on the field, which could highlight differences in soil moisture and thus possible regions where phylloxera could establish.

The following is the list of the attributes used in the georeferenced table:

1. Tree number
2. Latitude
3. Longitude
4. Block
5. Row
6. Panel
7. Tree variety
8. Expert visual vigour assessment
9. Digital vigour model
10. Multispectral derived indices and bands
11. Hyperspectral derives indices and bands
12. EM38 data

Once the table is populated with georeferenced data, the next stage in the process is the development of a pest predictive detection model. The first step in this stage is to carry out a correlation analysis between the expert vigour assessment as ground truth, and the vegetation indices calculated using the multispectral and hyperspectral data, the EM38 data, and the DVM.

The results obtained in the correlation analysis are the foundation for the development of a preliminary phylloxera detection model, which is followed by an evaluation and a ground-based verification to obtain a final phylloxera detection model. This final stage is out of the scope of this paper, and is part of ongoing research work.

2.2. Orthorectification, Hyperspectral Imagery Processing, and GIS Tools

Data orthorectification of colour and multispectral data was conducted using Agisoft Photoscan (Agisoft LLC., St. Petersburg, Russia). The software allowed us to perform the photogrammetric processing of digital images and to generate three-dimensional (3D) spatial data including 3D models, orthomosaics, and digital elevation models (DEM).

The photogrammetry orthorectification process starts with the alignment of photos, where the software refines the camera positions of each photo and builds a sparse and dense point cloud model of objects in the multiple collected airborne images. Then, key points are projected onto the selected reference system. Finally, ground control points (GCP) with highly accurately observed geolocations can then improve the sparse point cloud.

We processed the hyperspectral data with Headwall SpectralViewer, MATLAB and Scyllarus® DATA61 Matlab open source toolbox. Scyven software from DATA61 was used for analysing and cropping the white reference (this process is explained in Section 2.4) for transforming radiance data cubes into reflectance [24]. Matlab was used to produce the different indices, run a first approach for pixel classification based on these indices, generate plots of the phylloxera spectral response, and process the hyperspectral data cubes to obtain radiance.

The ArcMap 10.5 (ESRI, Redlands, CA, USA) software was used to incorporate the different sources of data into several layers, in order to analyse and visualise the data. It was also used to extract the spectral signatures from the grapevines that are next to the phylloxera traps, and to generate the attribute table generating vegetation indices for every plant.

Google Earth enabled us to visualise overlaid imagery obtained from indices and phylloxera traps, as well as to distinguish the boundaries of the different grapevine varieties in the crops.

2.3. Georeferencing

The georeferencing using GCP was conducted and reported using a free online AUSPOS service managed by Geoscience Australia, and took less than 10 min for a GCP. We included GCPs into the georeferencing of multispectral and colour data after the generation of a sparse points cloud, followed by camera locations optimisation and dense point cloud generation routines.

2.4. Ground Control Points (GCP) and White Reference Boards

We refined the georeferencing process of the collected imagery using custom-made GCPs with high contrast ground markers and applied precise point positioning (PPP) localisation methods to geolocate them. We used a triple frequency Novatel DL-V3 GPS receiver with a Novatel 703 antenna and Geoscience Australia online service in order to get the localisation of GCPs to a 3 cm accuracy level; whereas standard single frequency GPS chip in Canon/MicaSense sensors typically provided positioning accuracy within 5–10 m.

Additionally, we deployed a set of ground reflectance references in order to calculate reflectance in the multispectral and hyperspectral imagery. A MicaSense reflectance reference board (grey reference) and a 100 percent white reference Spectralon target were used before and after each flight as primary calibration tools for the cameras. We also deployed a set of nine white reference boards to the sites to be able to adjust reflectance in case of changing lighting conditions during the flight. Figure 2 shows the GCP, white reference boards located aside, and the grey and white reflectance references.

Figure 2. Ground control point (GCP) target and white reference board (**left**), MicaSense reflectance reference board (**top right**), and Spectralon white reference (**bottom right**).

Images of the white reference boards were taken using the hyperspectral camera before and after the flights on the field ground. The hyperspectral images of the white reference board were first verified to avoid using images that have saturation in some of their bands (Figure 3a,b), and the portion of the image with the higher radiance values is cropped (Figure 3c). The aerial images were then divided by the mean radiance value of the bands extracted from the cropped white reference image to obtain the reflectance of the surveyed field.

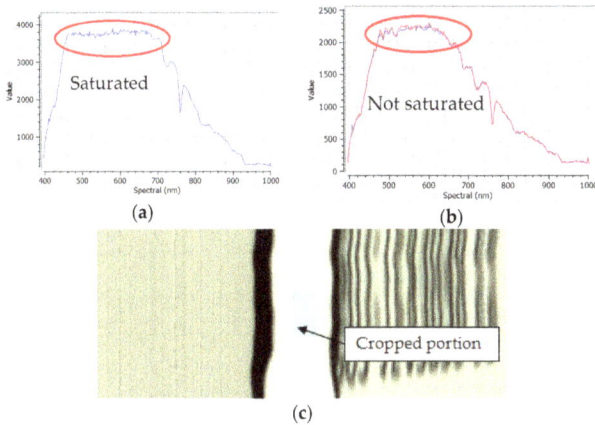

Figure 3. Hyperspectral white reference processing. (**a**) Saturated values around the highlighted area; (**b**) Maximum values are not saturated and this white target portion can be used for obtaining reflectance; (**c**) Cropped portion of the white reference board image with higher radiance values.

2.5. Vigour Assessment

Vigour assessment is conducted both visually by an expert and using orthorectified images to generate a digital vigour model. Estimating and analysing crop and grapevines vigour from orthorectified imagery when the crop or vineyard is on sloping terrain where elevation changes significantly over a small area is a challenging task. Even within one row of grapevines, it is possible to see elevation changes that are much more intense than the height of the grapevines. This means that using the terrain's flat surface as a reference to estimate the vigour and height of individual grapevines is not accurate, and a better method is required.

Our approach was to generate dense point clouds (DPC), and classify these dense points to at least two classes: grapevines and terrain. We ran an optimisation of the DPC (*camera = 10 px, marker accuracy = 0.001 m, marker placement = 0.1 px*). We additionally controlled the reprojection error of generated DPC at level below 2 px. As a segmentation criteria in classification, we used a set of geometrically defined restrains, which we optimised experimentally. We set *max distance = 0.6 m*, i.e., limited a distance between the point in question and terrain model, and set *max angle = 70 deg*, which limited the maximum slope of the ground within the scene. Then, we built two types of meshes and DEM. One set will form a digital surface model (DSM). The second set uses the rest of the DPC augmented with inferred surfaces constructed by closing the gaps under the grapevines. The latter forms a bare-earth or digital terrain model (DTM). We then subtracted the elevations of the DTM from the DSM to get a digital vigour model (DVM) in order to analyse and better represent differences of vigour.

In this way, we generated unbiased models of grapevines' height or vigour. Then, we localised grapevine trunks based on ground data, and performed a zonal analysis of each grapevine within a 0.2 m radius from the trunk.

2.6. Hyperspectral Processing

We implemented the workflow described in Figure 1 (Stage 2) to pre and post-process the airborne collected hyperspectral data. First, the hyperspectral data cubes are transformed from digital numbers into radiance using the Headwall hyperspec software. Second, we orthorectify each of the scans using the GPS time-stamps from the flight. We then generated orthomosaics by stitching together multiple scans. Finally, the reflectance for the orthomosaic data cubes was determined by dividing the radiance data cubes by the mean radiance of the white target boards that are located in the field during the survey flights.

2.7. Mean Spectral Signatures for Different Grapevine Types

We extracted the mean spectral signature for plants with different values of the Modified Cab Absorption in Reflectance Index (MCARI) index. In ArcMap 10.5, we selected polygons containing the pixels of a single crop or grapevine. Figure 4 shows an example of the process where different regions are selected for the signature extraction. Grapevines within the red, orange, and yellow areas show signs of phylloxera infestation, whereas grapevines within the green region are healthy.

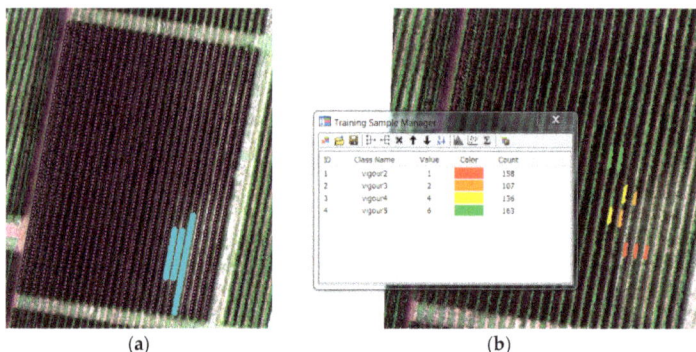

(a)   (b)

Figure 4. Selection of individual plants or grapevines for spectral signature extraction based on vigour assessment into four classes. (**a**) Selected grapevines from an attribute table where vigour is 2/5 (**b**) Four different polygon areas for each class based on grapevine location and vigour assessment.

2.8. Vegetation Indices

We calculated vegetation indices from multispectral and hyperspectral imagery. The indices used were selected in order to evaluate symptoms of infestation such as the premature yellowing of leaves and a reduction in chlorophyll content. A list of the indices used, as well as their equations, is found in Table 1, where index H stands for hyperspectral data and M stands for multispectral data, respectively.

We also created six new indices based on the analysis of the spectral reflectance of infested and uninfested crop grapevines. The two distinct spectral responses were subtracted to highlight the main differences and explore relevant bands with high differences and bands with equal reflectance. In this way, we detected seven relevant spectral bands, from which we created indices PI1 to PI6, as described in Appendix A, Equations (A1)–(A6).

Table 1. Vegetation indices used in the analysis of the phylloxera infestation in vineyards.

Vegetation Index	Equation	Reference
Normalised Difference Vegetation Index (NDVI)	$NDVI_H = (R_{800} - R_{670}/(R_{800} + R_{670})$ $NDVI_M = (R_{840} - R_{668}/(R_{840} + R_{668})$	[25]
Normalised Difference Vegetation Index (NDVI$_{Green}$) (Green band)	$NDVI_{Green\ H} = (R_{800} - R_{551}/(R_{800} + R_{551})$ $NDVI_{Green\ M} = (R_{840} - R_{560}/(R_{840} + R_{560})$	
Normalised Difference Red Edge (NDRE)	$NRDE_M = (R_{840} - R_{717})/(R_{840} + R_{717})$	[26]
Modified C$_{ab}$ Absorption in Reflectance Index (MCARI)	$MCARI_H = [(R_{700} - R_{670}) - 0.2(R_{700} - R_{551})]\,(R_{700}/R_{670})$ $MCARI_M = [(R_{717} - R_{668}) - 0.2(R_{717} - R_{560})]\,(R_{717}/R_{668})$	[27]
Modified Chlorophyll Absorption in Reflectance Index (MCARI$_1$)	$MCARI_{1\ M} = 1.2\,[2.5\,(R840 - R668) - 1.3(R840 - R560)]$	[28]
Modified Chlorophyll Absorption in Reflectance Index (MCARI$_2$)	$MCARI_{2M} = \dfrac{1.5[2.5(R_{840}-R_{668})-1.3(R_{840}-R_{560})]}{\sqrt{(2R_{840}+1)^2-(6R_{840}-5\sqrt{R_{668}})-0.5}}$	[28]
Transformed CARI (TACRI)	$TCARI_H = 3[(R_{700} - R_{670}) - 0.2(R_{700} - R_{551})(R_{700}/R_{670})]$ $TCARI_M = 3[(R_{717} - R_{668}) - 0.2(R_{717} - R_{560})(R_{717}/R_{668})]$	[29]

Table 1. *Cont.*

Vegetation Index	Equation	Reference
Optimised Soil-Adjusted Vegetation Index (OSAVI)	$OSAVI_H = (1 + 0.16)(R_{800} - R_{670})/(R_{800} + R_{670} + 0.16)$ $OSAVI_M = (1 + 0.16)(R_{840} - R_{668})/(R_{840} + R_{668} + 0.16)$	[30]
Blue/Green and Blue/Red Pigment indices	$BGI_{2\,M} = R_{475}/R_{560}$ $BRI_{2\,M} = R_{475}/R_{668}$	[31]
Phylloxera index 1 (PI1) Phylloxera index 2 (PI2) Phylloxera index 3 (PI3) Phylloxera index 4 (PI4) Phylloxera index 5 (PI5) Phylloxera index 6 (PI6)	$PI1 = (R_{522} - R_{504})/(R_{522} + R_{504})$ $PI2 = (R_{551} - R_{562})/(R_{551} + R_{562})$ $PI3 = (R_{700} - R_{680})/(R_{700} + R_{680})$ $PI4 = (R_{782} - R_{700})/(R_{782} + R_{700})$ $PI5 = (R_{782} - R_{671})/(R_{782} + R_{671})$ $PI4 = (R_{680} - R_{563})/(R_{680} + R_{563})$	(This study)

3. UAV and Sensors

The methodology described in Section 2 is platform or sensor agnostic if the UAV platform or sensor has similar characteristics. For the purpose of this work, we demonstrate the methods using the following:

3.1. UAV

The UAV used is an S800 EVO Hexacopter (DJI Ltd., Shenzhen, China) weighing 6.0 kg and capable of taking an additional payload up to 2.0 kg. The frame is fitted with a retractable undercarriage, providing a sensor field of view clear of obstacles. The UAV uses a 16,000 mAh LiPo six-cell battery, which provides a maximum hover time of approximately 20 min with no sensor payload. A WooKong-M flight controller forms the navigation and control system of the UAV, and comes with a stabilisation controller, a GPS unit with an inbuilt compass, and an inertial measurement unit (IMU). The flight controller has multiple autopilot modes to enable both remote control by operator and autonomous go home/landing with an enhanced fail-safe operation following a payload-specific predefined flight path, position and altitude hold.

3.2. High Resolution RGB Camera

A Canon 5DsR camera (Canon Inc., Tokyo, Japan) was integrated into the S800 UAV to capture ultra HD colour (RGB) GPS-stamped images. The camera features the latest full-frame 50.6-megapixel CMOS sensor with Dual DiGIC 6 processors and a 28-mm Canon lens. Specific flight patterns were flown to enable processing the images using photogrammetry software (e.g., Agisoft Photoscan) in order to generate comprehensive geospatial orthomosaics, 3D models, and DEM. The sensor brings to our study an ability to generate remarkable spatially detailed models of the crop in order to look after the spatial component of the infestation symptoms.

3.3. Multispectral Camera

We used a multispectral MicaSense RedEdge camera (MicaSense Inc., Simi Valley, CA, USA) to capture five discrete spectral bands: 475 nm (blue), 560 nm (green), 668 nm (red), 717 nm (red edge), and 840 nm (NIR). The sensor combines 1.2-megapixel CMOS spatial capability and two additional NIR and red edge spectral bands. The images produced by this camera were orthorectified and stitched together to produce rasters that were used to generate vegetation indices e.g., NDVI. Additionally, this camera supported the generation of DEM with resolutions of 3.26 cm/px for 60 m flights and 6.74 cm/px for 100 m flights.

3.4. Hyperspectral Sensor

Hyperspectral imagery was acquired using a Headwall Nano-Hyperspec (Headwall Photonics Inc., Bolton, MA, USA). The hyperspectral sensor recorded data cubes of 274 spectral bands in the visible and near-infrared (VNIR) range (400–1000 nm) with a ~2.2 nm spectral interval and a 5-nm spectral resolution (full width at half maximum (FWHM) with 20 µm slit). This camera is equipped with a calibrated f/1.8 4.8 mm Schneider lens, which results in a 50.7 deg field of view over 640 pixels. The collected hyperspectral data cubes are synchronised with GPS/inertial navigation system (INS) positioning and orientation information in order to perform data cubes orthorectification and multiple data cubes mapping. The hyperspectral sensor was integrated into a S800 UAV using a custom designed gimbal done by the Queensland University of Technology (QUT) Research Engineering Facility (REF). This gimbal has two axes, which ensures the seamless operation of the push-broom hyperspectral scanner in windy conditions. The QUT REF gimbal design features carbon reinforcement over a 3D printed structure, advanced dampening, brushless motors, and a BaseCam SimpleBGS 32 bit gimbal controller, with the total weight below 1 kg. Mounting the push-broom scanner on the gimbal enhances the camera performance in high turbulence environments by ensuring consistent and minimal required overlaps between consecutive data cubes over the large study area. This leads to an increased overall flight efficiency in open environments. Figure 5a shows the S800 UAV with the hyperspectral sensor on the gimbal during one of the missions. The gimbal computer assisted design (CAD) model is presented in Figure 5b.

(a)   (b)

Figure 5. (**a**) Headwall Nano hyperspectral sensor on-board a S800 unmanned aerial vehicle (UAV); (**b**) Custom-made two-axis gimbal hosting the hyperspectral camera (SolidWorks 3D model).

3.5. Expert Visual Vigour Assessment and EM-38 Data

In addition to the UAV airborne data and GCP, the methods use ground truth data for confirming the presence of grape phylloxera. The ground truth data included ground traps for insect presence and abundance, and, when necessary, digging to confirm the presence of the insect on the roots, expert visual vigour assessment, and EM38 soil conductivity. Table 2 shows an example of the vigour assessment classification for phylloxera-infested grapevines used in the visual assessment performed by the expert on the field.

Table 2. Phylloxera vigour classes from an expert visual assessment.

Class	Vigour	Criteria	Phylloxera Presence Conjecture
5	High	Plant or grapevines to or above a given height e.g., top supportive wire	Healthy (probably no infestation e.g., phylloxera)
4	Medium-high	Plant just below a given height e.g., top supportive wire	Mild symptoms (probably no infestation or early stages of impact e.g., phylloxera)
3	Medium	Plant height below middle wire and above bottom wire	Intermediate impact (probably low levels of infestation e.g., phylloxera)
2	Low	Short plants e.g., grapevines. Plants below bottom wire	Severe symptoms of infestation (e.g., phylloxera, surrounding (3–4) plants also likely to be infested)
1	No vigour	Dead plant	Extreme symptoms of infestation (e.g., phylloxera has been affecting the plant for years)

4. Field Experiments

Two major aerial surveys were carried out at two locations during December 2016 and February 2017. Different sensors were deployed to collect comprehensive datasets. Our objective was to evaluate diverse state of the art remote sensing capabilities, and its combinations, in order to evaluate the methodology for the early detection of phylloxera infestations. This section describes the different aerial platforms and sensors used, and how the data was collected.

Two phylloxera-infested vineyards with multiple grapevine varieties in Yarra Valley, Victoria, Australia were surveyed (see Figure 6). These two sites were selected because they have been continuously monitored by Victorian Department of Economic Development, Jobs, Transport and Resources (DEDTR) personnel over several grapevine-growing seasons. Site one included 104 rows of ungrafted *V. vinifera* cultivars Chardonnay, Pinot Noir, Shiraz, and Merlot over an area of 8.5 ha containing 160 to 266 grapevines per row. Site two included 80 rows of ungrafted *V. vinifera* cultivars Cabernet Sauvignon, Pinot Noir, Merlot, and Roussanne, with 59 to 63 grapevines per row over 3.16 ha. In this paper, we present our analysis on a single block within one vineyard containing the Chardonnay variety, and some remarkable results from different blocks for method verification purposes.

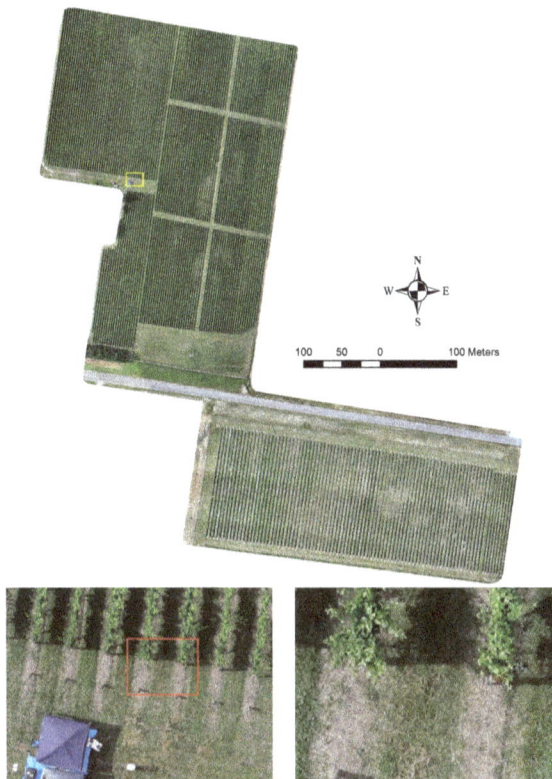

Figure 6. Generated orthomosaic of studied sites; two vineyards in the Yarra valley, Victoria, Australia. Bottom images show enlarged regions of the vineyard with clearly visible plants.

We placed eight GCP over the sites for an hour, which lead to a horizontal positional uncertainty below 0.014 m, and a vertical uncertainty below 0.046 m (95% confidence levels, GDA94) over the array of points. Figure 7 denotes the location of four GCP as marker flags distributed over site 1.

In the studied vineyard, the top, middle, and bottom supportive wires were positioned at approximately 1.75 m, 1.55 m, and 1.2 m (average values for the two studied vineyards) above the ground, respectively. They were used as an approximate guidance to generate the canopy vigour classes (see Table 2).

The imagery was acquired on two consecutive days during the grapevine phenological cycle: post-flowering (14–15 December 2016) and veraison (14–15 February 2017). This approach is common [7,11], and enables temporal studies on the identification of phylloxera symptoms and population development, as well as validation.

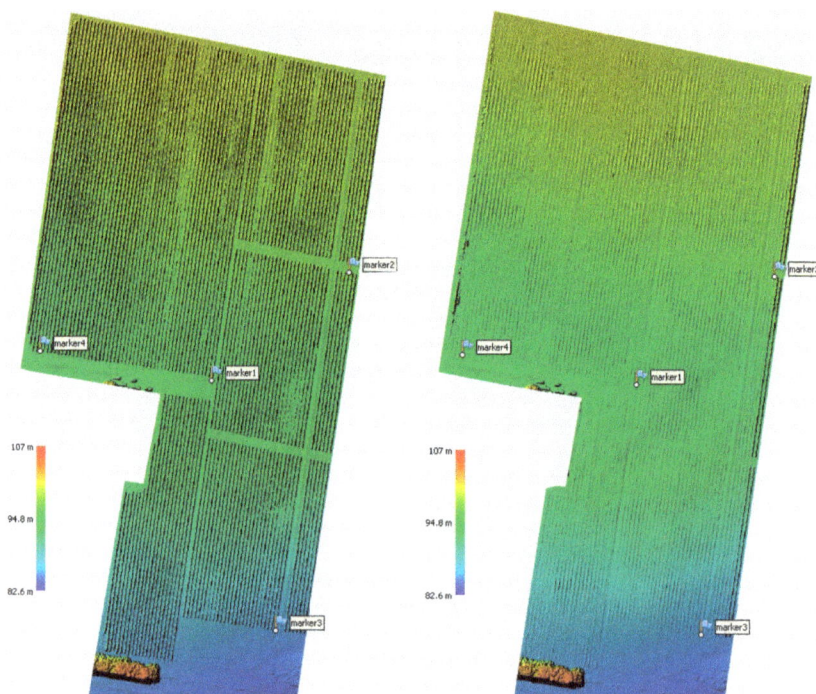

Figure 7. Result of the digital surface model (DSM) and digital terrain model (DTM) of Site 1. Markers are the ground control points (GCP).

5. Results and Discussion

5.1. Visual Vigour Assessment Results

The result of the expert's visual vigour assessment (Figure 8) is a matrix of vigour health, with the cell size equal to the size of the panel and distance between rows. Therefore, the expert-based dataset is spatially sparse, and not aligned geographically. Ground-based vigour assessment is time consuming, and only evaluated every 3–5 m. The expert observations indicated the start of another weak spot (low vigour) at row 48–50, panels 26–28.

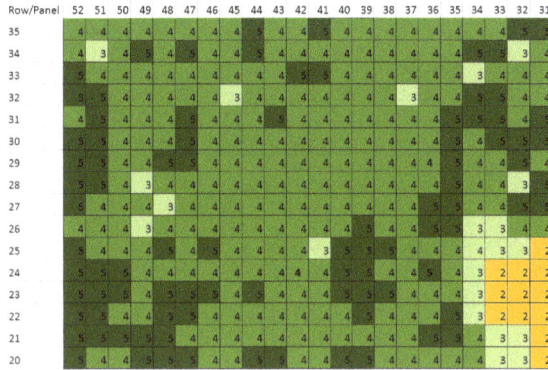

Figure 8. Results of traditional expert-based assessment of grapevines vigour per panel (groups of four to six grapevines, fragment transposed to cardinal directions).

5.2. Digital Vigour Model (DVM)

The results of the generated DVM allowed us to classify the grapevines according to its vigour using classes from 2 to 5 from Table 2. In this study, the two surveyed vineyards did not contain plants with low vigour (class 1).

The result of the DSM and DTM show a range of elevations on the surface and terrain of approximately 24 m (see legend of DSM and DTM for site 1 on Figure 7). The result of the technique described in the methods section for the DVM shows that we reduced that range of elevations within the studied area to 2.5 m (Figures 9–11), which corresponds well to tree heights.

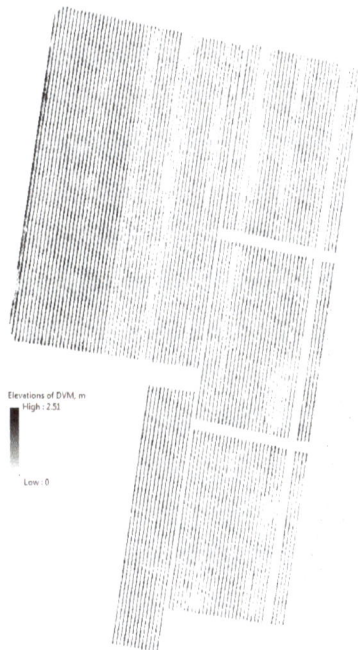

Figure 9. Result of the unbiased DVM for Site 1.

Figures 9–11 show the georeferenced results of an expert's vigour assessment Figures 10a and 11a) and derived DVM (Figures 10b,c and 11b) (inverse distance weighted interpolation) for the two sites. The season average Pearson coefficients of a grapevine's expert-assigned vigour versus DVM remains at 0.396, but mostly correlated at the time of actual vigour assessment (0.414) for the area at site 1 and (0.52) site 2. We note a specific feature in the infestation extension, which tends to happen along the rows from one infested grapevine to another, whereas the transition of the disease between rows is less likely.

The zonal statistics method combines simplicity and numerous statistical parameters for the buffered zones. The Pearson correlation coefficient $r = 0.52$ between an expert-based assessment (Figure 8) and max values of the DVM for individual grapevines as a result of remote sensing for the described area (Figure 11b) for site 2 demonstrates an adequate correlation and validity of the used method.

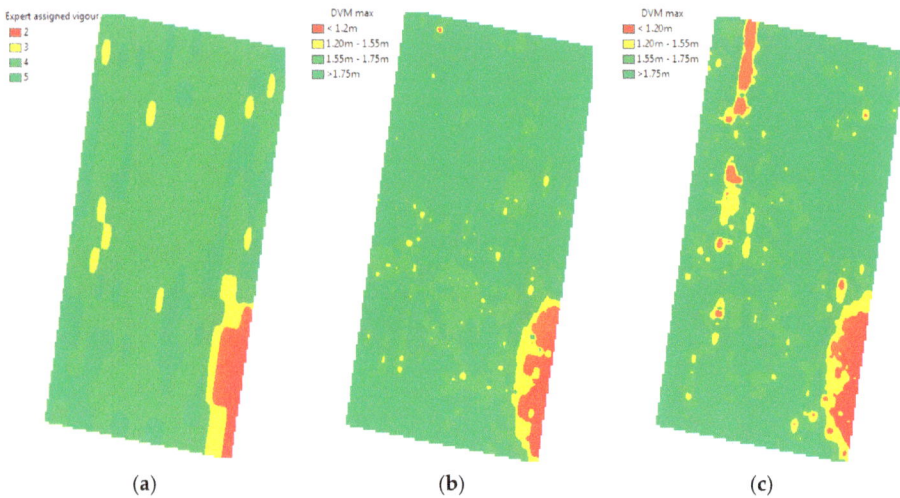

(a)   (b)   (c)

Figure 10. Expert assigned classes of grapevines vigour per panel (**a**) and the results of a remotely-sensed vigour assessment of individual grapevines for December 2016 (**b**) and February 2017 (**c**) for a block with the Chardonnay variety.

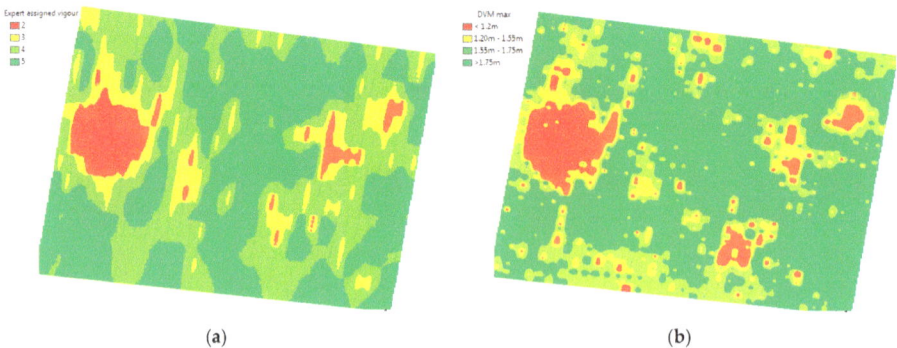

(a)   (b)

Figure 11. Expert-assigned classes of grapevines vigour per panel (**a**) and the results of a remotely-sensed vigour assessment of individual grapevines for February 2017 (**b**), site 2 with the *V. vinifera* Roussanne variety.

Geolocalised results (Figure 10b,c) give a number of advantages over standard expert-based assessments: much more spatially accurate localisation (<0.02 m), individual tree assessment rather than panel approximation, and the ability to detect subtle trends that are unnoticeable due to inaccurate references used for visual assessment, such as the height of the poles. One of the most important advantages is the shorter time required to assess a large area. However, the inability to detect early symptoms of the disease and occasional irregular behaviour of the affected grapevines (early infested grapevines sometimes demonstrate abnormal intense vigour for a short time) are the known limitations of the DVM method. Nevertheless, the DVM method could provide insight to the expert to select candidate zones to revisit for closer and more detail inspections.

5.3. Hyperspectral Analysis

Different vegetation indices were computed on the vineyard imagery in order to highlight the symptoms of the phylloxera infestation, which include a reduction in plant vigour and the reduction of chlorophyll content.

Figure 12 shows examples of four vegetation indices calculated from hyperspectral data for block 3, namely PI2, PI5, NDVI, and $OSAVI_H$.

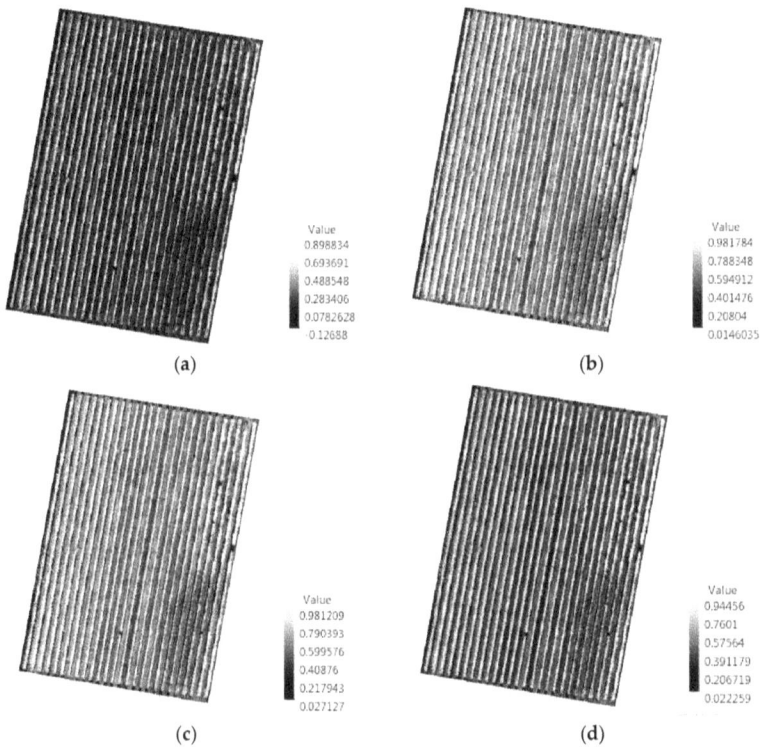

Figure 12. Vegetation indices for block 3 with the highest correlation to vigour assessment. (**a**) PI2; (**b**) PI5; (**c**) NDVI; and (**d**) $OSAVI_H$. All of the indices are based on hyperspectral imagery collected in February 2017.

The result of the mean spectral signature extraction is shown in Figure 13a,b. Notice the different signatures of the grapevines that are affected by phylloxera. The infested grapevines show higher reflectance in the visible region, and lower reflectance in the NIR region. Furthermore, infested vines have higher levels of reflectance at the chlorophyll well around 670 nm, with the healthy grapevines absorbing more light around this wavelength (Figure 14a,b). Spectral signatures also show higher differences between infested and uninfested grapevines for the February 2017 imagery.

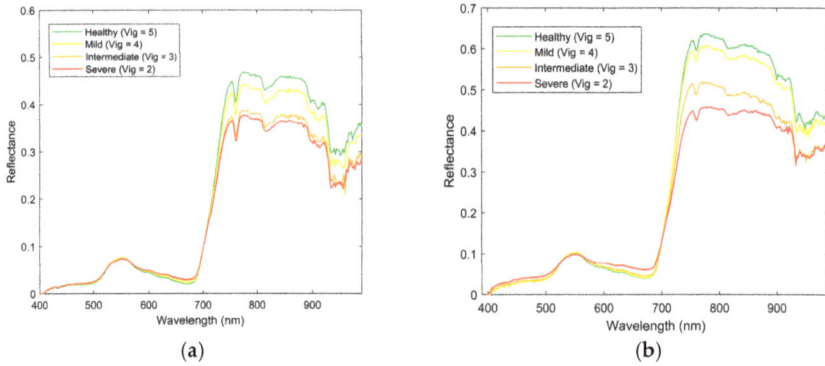

Figure 13. Mean spectral signature for different levels of vigour of the grapevine for the *V. vinifera* Chardonnay variety measured in (**a**) December 2016; and (**b**) February 2017 for wavelengths from 400 to 1000 nm.

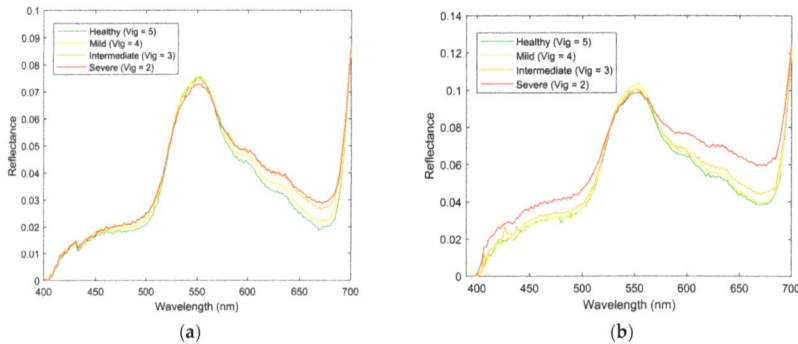

Figure 14. Mean spectral signature for different levels of vigour of the grapevine for the *V. vinifera* Chardonnay variety measured in (**a**) December 2016 and (**b**) February 2017 for wavelengths from 400 nm to 700 nm.

Figure A1 in Appendix A shows the difference in spectral responses for infested and uninfested grapevines for the *V. vinifera* Chardonnay variety. Spectral bands of interest are marked with arrows. These bands correspond to local points where the difference in the reflectances is either high or zero.

5.4. Correlation Analysis of Different Variables Using Attribute Tables

Figures 15 and 16 show the results of calculating Pearson's correlation to indices and bands extracted from multispectral imagery. In particular, we calculated correlation among the following data: blue, green, red, NIR, and red edge bands, and $NDVI_M$, $NDVI_{GreenM}$, Normalised Difference Red Edge ($NDRE_M$), $OSAVI_M$, $MCARI_M$, Transformed Chlorophyll Absorption in Reflectance Index ($TCARI_M$), $MCARI_{1M}$, $MCARI_{2M}$, Blue/Green Index (BGI_{2M}), Blue/Red Index (BRI_{2M}), EM-38, DVM, and expert visual vigour assessment (Vigour).

February	BLUE	GREEN	RED	NIR	RE	NDVI	NDVI_GR	NDRE	OSAVI	MCARI	TCARI	MCARI_1	MCARI_2	BGI_2	BRI_2	EM38	DVM
BLUE	1																
GREEN	0.54	1															
RED	0.81	0.19	1														
NIR	-0.39	0.34	-0.62	1													
RE	-0.04	0.69	-0.33	0.85	1												
NDVI	-0.70	0.05	-0.93	0.96	0.61	1											
NDVI_GR	-0.78	-0.31	-0.79	0.78	0.40	0.87	1										
NDRE	-0.70	-0.33	-0.73	0.67	0.18	0.76	0.90	1									
OSAVI	-0.68	0.07	-0.91	0.88	0.64	1.00	0.87	0.78	1								
MCARI	-0.45	0.34	-0.74	0.86	0.82	0.87	0.64	0.46	0.88	1							
TCARI	-0.45	0.34	-0.74	0.86	0.82	0.87	0.64	0.46	0.88	1.00	1						
MCARI_1	-0.50	0.32	-0.81	0.93	0.79	0.96	0.75	0.66	0.97	0.93	0.93	1					
MCARI_2	-0.56	0.26	-0.87	0.89	0.73	0.96	0.76	0.67	0.98	0.93	0.93	0.99	1				
BGI_2	0.75	-0.14	0.80	-0.73	-0.59	-0.86	-0.68	-0.58	-0.86	-0.79	-0.79	-0.84	-0.86	1			
BRI_2	0.00	0.41	-0.57	0.47	0.49	0.59	0.24	0.24	0.59	0.67	0.67	0.65	0.68	-0.30	1		
EM38	-0.01	-0.02	-0.01	0.05	0.02	0.04	0.07	0.07	0.04	0.00	0.00	0.03	0.03	-0.01	0.00	1	
DVM	-0.16	-0.09	-0.23	0.22	0.07	0.24	0.27	0.30	0.24	0.17	0.17	0.21	0.21	-0.11	0.14	-0.10	1
VIGOUR	-0.18	-0.15	-0.17	0.06	-0.08	0.13	0.16	0.23	0.12	0.04	0.04	0.09	0.09	-0.08	0.02	-0.06	0.30

Figure 15. Pearson's correlation coefficient matrix showing the strength of the relationship between the February 2017 multispectral vegetation indices and expert and digital vigour assessment. Orange colours indicate negative correlation, blue colours indicate positive correlation, and intensity of colour indicates relative strength.

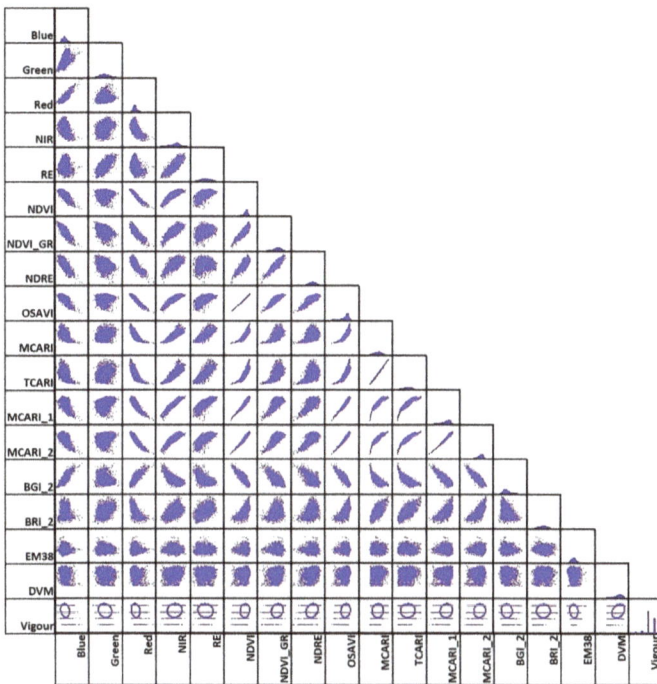

Figure 16. Scatter plot of correlation for data presented in Figure 15, generated from multispectral imagery in February 2017.

We found the following cross-correlations between analysed multispectral-based indices:

1. Vigour in both December 2016 and February 2017 has the highest positive correlation with DVM, and to $NDRE_M$, but only to $NDVI_M$, $NDVI_{GreenM}$, and $OSAVI_M$ in December. However, these are relatively minor relationships, with the linear relationship between 0.23 and 0.3.
2. EM38 has no relationship to any of the vegetation indices, Vigour, or DVM.
3. Certain indices are extremely positively correlated for December and February (indicated by dark blue boxes in Figure 15). These are: BLUE, GREEN, RED, and Red Edge (RE) multispectral bands; NIR with RE bands, $OSAVI_M$, $MCARI_M$, $TCARI_M$, $MCARI_{1M}$, and $MCARI_{2M}$; $NDVI_M$ with

GREEN, NDRE, OSAVI$_M$ and MCARI$_{M1}$; NDVI$_{GreenM}$ with NDRE and OSAVI$_M$; and TCARI$_M$ with MCARI$_M$, MCARI$_{1M}$, and MCARI$_{2M}$.

Most of the newly created hyperspectral vegetation indices showed higher correlations to vigour and to DVM (Figures 17 and 18) compared with the multispectral indices. This might be due to the difference in the spectral resolution of the cameras. Using the hyperspectral camera with a higher spectral resolution, we were able to distinguish specific bands or points of interest such as the ones shown in Figure A1, and thus generate indices that are not possible to create with the multispectral camera.

December	PI1	PI2	PI3	PI4	PI5	PI6	NDVI	NDVI Green	MCARI	TCARI	OSAVI	TCARI /OSAVI	MCARI /OSAVI	Band 670	Band 800	Band 504	Band 551	EM 38	DVM
PI1	1																		
PI2	0.62	1																	
PI3	0.94	0.64	1																
PI4	0.84	0.60	0.89	1															
PI5	0.89	0.60	0.95	0.98	1														
PI6	-0.93	-0.63	-0.99	-0.88	-0.93	1													
NDVI	0.89	0.60	0.94	0.98	1.00	-0.92	1												
NDVI Green	0.75	0.47	0.81	0.96	0.93	-0.76	0.93	1											
MCARI	0.85	0.61	0.89	0.68	0.75	-0.90	0.76	0.55	1										
TCARI	0.76	0.56	0.81	0.56	0.67	-0.84	0.67	0.40	0.93	1									
OSAVI	0.92	0.64	0.97	0.93	0.90	-0.97	0.95	0.85	0.88	0.82	1								
TCARI/OSAVI	0.40	0.33	0.43	0.07	0.22	-0.48	0.21	-0.11	0.66	0.86	0.41	1							
MCARI/OSAVI	0.83	0.60	0.88	0.63	0.72	-0.89	0.73	0.50	1.00	0.94	0.85	0.73	1						
Band 670	-0.83	-0.55	-0.88	-0.96	-0.96	0.85	-0.96	-0.93	-0.66	-0.53	-0.88	-0.07	-0.62	1					
Band 800	0.77	0.58	0.83	0.72	0.75	-0.83	0.75	0.64	0.87	0.82	0.87	0.51	0.84	-0.67	1				
Band 504	-0.75	-0.48	-0.76	-0.90	-0.88	0.72	-0.88	-0.92	-0.49	-0.33	-0.76	0.15	-0.45	0.95	-0.51	1			
Band 551	0.08	0.15	0.10	-0.22	-0.12	-0.16	-0.13	-0.37	0.43	0.58	0.10	0.81	0.46	0.26	0.43	0.48	1		
EM 38	0.09	-0.02	0.07	0.04	0.06	-0.07	0.07	0.01	0.06	0.10	0.07	0.11	0.08	-0.04	0.04	0.00	0.06	1	
DVM	0.33	0.22	0.36	0.38	0.41	-0.35	0.41	0.36	0.29	0.32	0.38	0.16	0.28	-0.40	0.30	-0.34	-0.01	-0.01	1
Vigour	0.40	0.25	0.43	0.49	0.49	-0.43	0.48	0.46	0.35	0.33	0.45	0.11	0.33	-0.49	0.34	-0.45	-0.10	0.02	0.39

Figure 17. Pearson's correlation coefficient matrix showing correlation between the hyperspectral vegetation indices and expert vigour assessment for data collected in December 2016. Orange colour indicates negative correlation; blue colour indicate positive correlation with the intensity of colour indicating relative strength and highlighted red font are the indices that correlate positively with the digital vigour model (DVM) and expert vigour assessment.

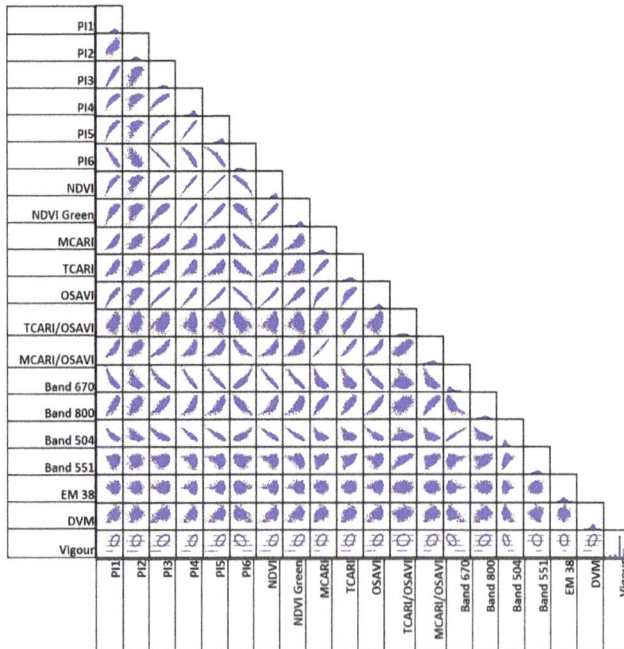

Figure 18. Scatter plot of correlation for data presented in Figure 17, generated from hyperspectral imagery in December 2016.

1. Vigour showed the highest positive relationship ($r > 0.4$) with the vegetation indices PI1, PI3, PI4, PI5, NDVI, NDVI$_{GreenH}$, and OSAVI, as well as with DVM ($r = 0.4$).
2. Similar to the multispectral data, certain vegetation indices were positively correlated for both December and February; PI1, PI3, PI4, PI5, NDVI, NDVI$_{GreenH}$, MCARI$_H$, TCARI$_H$, and OSAVI$_H$ with each other; MCARI/OSAVI with BAND800; and BAND670 with BAND504.
3. The EM38 data showed no relationship with any of the vegetation indices, vigour, or DVM.

Overall, the correlation values of the vegetation indices compared with the vigour assessment are moderate, but some important relationships were found. This was expected, because those are different symptoms that a stressed plant manifests in response to a phylloxera infestation, but in different ways depending on the different stages of infestation and environmental factors. Moreover, some indices might better highlight specific changes in the leave reflectance than others, depending on the specific bands used. Furthermore, the values of the indices are calculated averaging the reflectance for all of the pixels of the tree canopy coverage. This technique allows us to sample every plant within a block in the vineyard.

6. Conclusions and Further Research

In this work, we presented the methodology for processing airborne-collected imagery using RGB, multispectral, and hyperspectral cameras for assessing the condition of vineyards with the aim of generating a predictive model for the detection of pests in crops and vineyards. The methods are demonstrated for detection of phylloxera in vineyards. In particular, we presented the implementation of a digital vigour model, the evaluation of several vegetation indices, and the creation and evaluation of new indices based on hyperspectral signatures to highlight symptoms of grapevine phylloxera infestation.

We presented the results of comparing a digital vigour model of the vineyard to an expert visual assessment. We found that the two assessments correlate positively indicating that the developed method is a correct approach for generating vigour assessments in vineyards.

We generated vegetation indices to highlight possible symptoms of phylloxera infestation, such as changes in the reflectance/absorption of light indicating a reduction in the chlorophyll content in the grapevines. This could be done remotely using both the multispectral and hyperspectral collected and treated imagery.

Furthermore, we identified mean spectral signatures for different levels of infestation for the Chardonnay variety at two different times of the year which helped us find regions of interest in the spectrum in order to generate new vegetation indices to highlight grape phylloxera infestation.

The development and use of a specific phylloxera or existing vegetation index is important to vineyard managers for two reasons. Initially, such an index could be used on imagery collected from an infested vineyard to determine the extent (area) and severity of the plant pest and its impact on grape production. This would also aid in vineyard management decisions, such as where to implement hygiene protocols. The second reason is that this index can improve the potential for early detection of the pest. At present, the existing surveillance methods (visual vine and root inspection and emergence traps) are ineffectual at detecting early infection of the vines by grape phylloxera.

We have shown that hyperspectral imagery has the potential to detect grape phylloxera before it is apparent to visual inspection. The next stage is to determine if these hyperspectral indices have the potential to be adapted to the multispectral data without a loss in correlative power. Future work will focus on deriving predictive models (utilising these indices) and testing their accuracy in predicting the presence/abundance of the plant pest. A paper describing the development and validation of the predictive model (Stage 4 of Figure 1) is in preparation as a complement of this paper.

The methods, workflow, results and analysis presented in this research will contribute to the generation of valuable information for plant pest surveillance. The presented methodology could also be extrapolated to other areas of research in remote sensing, such as minerals exploration, biodiversity, and ecological assessment.

Acknowledgments: This work was funded by the Plant Biosecurity Cooperative Research Centre PB CRC 2135 project, Agriculture Victoria Research and Queensland University of Technology. The authors gratefully acknowledge the support of the REF Operations Team (Dirk Lessner, Dean Gilligan, Gavin Broadbent and Dmitry Bratanov) who operated the DJI S800 EVO, sensors and performed ground referencing. We thank Gavin Broadbent for 2-axis gimbal for hyperspectral camera design, manufacturing and tuning. We also acknowledge the High Performance Computing and Research Support Group at Queensland University of Technology, for the computational resources and services used in this work.

Author Contributions: J.W., K.P., F.G. and D.B. conceived and designed the field and airborne experiments and undertook the relevant field surveys. D.B. and F.G. supervised and quality assessed the collection of the UAV imagery, D.B. and F.V. with input from F.G. developed and implemented the post-processing imagery workflow and analysis, D.B. and J.W. undertook the relevant comparative statistical analysis of vigour and vegetation indices respectively, F.V. was the principal author whilst all authors contributed significantly to the whole paper.

Conflicts of Interest: The authors declare no conflict of interest. The founding sponsors had no role in the design of the study; in the collection, analyses, or interpretation of data; in the writing of the manuscript, and in the decision to publish the results.

Appendix

New Phylloxera Indices Creation

A new phylloxera index was created by first selecting some interest points are chosen for creating indices looking at the difference between the spectral signatures of the infested and the uninfested grapevines (Figure A1). We then selected the wavelengths that have local minimum and maximum and zero values for reflectance in the difference spectral signature.

Seven bands of interest were found which wavelengths are: R_{504}, R_{523}, R_{553}, R_{563}, R_{680}, R_{701} and R_{782}.

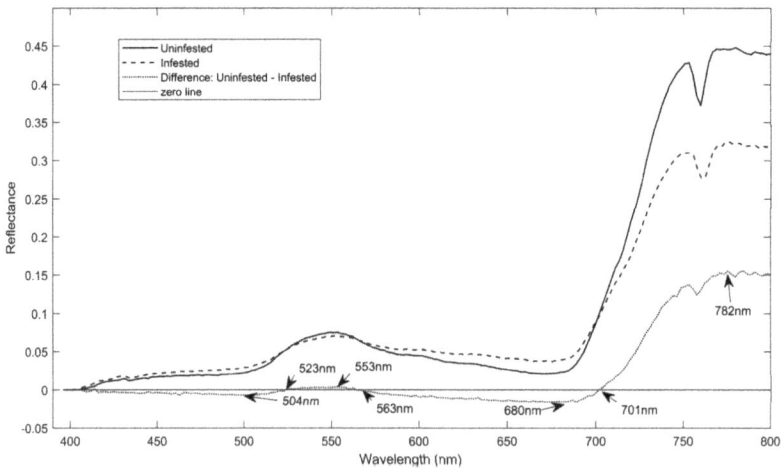

Figure A1. Mean spectral signature for different levels of vigour of the grapevine for the Chardonnay variety measured in December 2016 for wavelengths from 400 nm to 700 nm.

From these seven spectral bands, the following indices were created to highlight the main differences between the uninfested and infested spectral signatures.

The phylloxera infestation indices concentrates on highlighting the differences between the pairs of spectral bands R_{504} and R_{523} (Equation (A1)), R_{551} and R_{562} (Equation (A2)), R_{700} and R_{680} (Equation (A3)), R_{782} and R_{700} (Equation (A4)), R_{782} and R_{671} (Equation (A5)) and bands R_{680} and R_{563} in Equation (A6):

$$PI1 = \frac{R_{522} - R_{504}}{R_{522} + R_{504}}, \tag{A1}$$

$$PI2 = \frac{R_{551} - R_{562}}{R_{551} + R_{562}}, \tag{A2}$$

$$PI3 = \frac{R_{700} - R_{680}}{R_{700} + R_{680}}, \tag{A3}$$

$$PI4 = \frac{R_{782} - R_{700}}{R_{782} + R_{700}}, \tag{A4}$$

$$PI5 = \frac{R_{782} - R_{671}}{R_{782} + R_{671}}, \tag{A5}$$

$$PI6 = \frac{R_{680} - R_{563}}{R_{680} + R_{563}}. \tag{A6}$$

References

1. Mulla, D.J. Twenty five years of remote sensing in precision agriculture: Key advances and remaining knowledge gaps. *Biosyst. Eng.* **2013**, *114*, 358–371. [CrossRef]
2. Marciniak, M. Use of remote sensing to understand the terroir of the Niagara Peninsula. Applications in a Riesling vineyard. *J. Int. Sci. Vigne Vin* **2015**, *49*, 1–26. [CrossRef]
3. Mathews, A.J. A Practical UAV Remote Sensing Methodology to Generate Multispectral Orthophotos for Vineyards: Estimation of Spectral Reflectance. *Int. J. Appl. Geospat. Res.* **2015**, *6*, 65–87. [CrossRef]
4. Zhang, C.; Kovacs, J.M. The application of small unmanned aerial systems for precision agriculture: A review. *Precis. Agric.* **2012**, *13*, 693–712. [CrossRef]
5. Blanchfield, A.L.; Robinson, S.A.; Renzullo, L.J.; Powell, K.S. Phylloxera-infested grapevines have reduced chlorophyll and increased photoprotective pigment content—Can leaf pigment composition aid pest detection? *Funct. Plant Biol.* **2006**, *33*, 507–514. [CrossRef]
6. Skinkis, P.; Walton, V.; Kaiser, C. *Grape Phylloxera, Biology and Management in the Pacific Northwest*; Oregon State University, Extension Service: Corvallis, OR, USA, 2009; pp. 7–9.
7. Benheim, D.; Rochfort, S.; Robertson, E.; Potter, I.D.; Powell, K.S. Grape phylloxera (*Daktulosphaira vitifoliae*)—A review of potential detection and alternative management options. *Ann. Appl. Biol.* **2012**, *161*, 91–115. [CrossRef]
8. Powell, K. *A Holistic Approach to Future Management of Grapevine Phylloxera*; Springer: Dordrecht, The Netherlands, 2012; ISBN 9789400740327.
9. Powell, K.S.; Korosi, G.A.; Mackie, A.M. Monitoring grape phylloxera populations using simple non-destructive trapping systems. *Acta Hortic.* **2009**, *816*, 29–34. [CrossRef]
10. Giblot-Ducray, D.; Correll, R.; Collins, C.; Nankivell, A.; Downs, A.; Pearce, I.; Mckay, A.C.; Ophel-Keller, K.M. Detection of grape phylloxera (*Daktulosphaira vitifoliae* Fitch) by real-time quantitative PCR: Development of a soil sampling protocol. *Aust. J. Grape Wine Res.* **2016**, *22*, 469–477. [CrossRef]
11. Renzullo, L.J.; Blanchfield, A.L.; Powell, K.S. Insights into the early detection of grapevine phylloxera from in situ hyperspectral data. *Acta Hortic.* **2007**, *733*, 59–74. [CrossRef]
12. Wildman, W.E.; Nagaoka, R.T.; Lider, L.A. Monitoring spread of grape phylloxera by color infrared aerial photography and ground investigation. *Am. J. Enol. Vitic.* **1983**, *34*, 83–94.
13. Lobitz, B.; Johnson, L.; Hlavka, C.; Armstrong, R.; Bell, C. *Grapevine Remote Sensing Analysis of Phylloxera Early Stress (GRAPES): Remote Sensing Analysis Summary*; NASA Ames Research Center: Mountain View, CA, USA, 1997; pp. 1–36.
14. Powell, K.S.; Hackworth, P.; Lewis, M.; Lamb, D.W.; Edwards, J. Identification of phylloxera from high resolution infrared aerial imagery: A comparative study between airborne imagery types. *Aust. N. Z. Grapegrow. Winemak.* **2004**, *488*, 51–54.
15. Renzullo, L.J. Characterizing grapevine spectral response to phylloxera infestation. *Boll. Comunità Sci. Australas.* **2004**, *1*, 28–32.
16. Frazier, P.; Whiting, J.; Powell, K.; Lamb, D. Characterising the development of grape phylloxera infestation with multi-temporal near-infrared aerial photography. *Aust. N. Z. Grapegrow. Winemak.* **2004**, *485*, 133–136.
17. Bellvert, J.; Zarco-Tejada, P.J.; Girona, J.; Fereres, E. Mapping crop water stress index in a Pinot-noir vineyard: Comparing ground measurements with thermal remote sensing imagery from an unmanned aerial vehicle. *Precis. Agric.* **2014**, *15*, 361–376. [CrossRef]

18. Mathews, A.J. Object-based spatiotemporal analysis of vine canopy vigor using an inexpensive unmanned aerial vehicle remote sensing system. *J. Appl. Remote Sens.* **2014**, *8*, 085199. [CrossRef]
19. Bachmann, F.; Herbst, R.; Gebbers, R.; Hafner, V.V. Micro Uav Based Georeferenced Orthophoto Generation in Vis + Nir for Precision Agriculture. *ISPRS Int. Arch. Photogramm. Remote Sens. Spat. Inf. Sci.* **2013**, *XL-1/W2*, 11–16. [CrossRef]
20. Honkavaara, E.; Kaivosoja, J.; Pellikka, I.; Pesonen, L.; Saari, H.; Salo, H.; Hakala, T.; Marklelin, L.; Solutions, M. Hyperspectral Reflectance Signatures and Point Clouds for Precision Agriculture by Light Weight Uav Imaging System. *ISPRS Ann. Photogramm. Remote Sens. Spat. Inf. Sci.* **2012**, *I*, 353–358. [CrossRef]
21. Honkavaara, E.; Saari, H.; Kaivosoja, J.; Pölönen, I.; Hakala, T.; Litkey, P.; Mäkynen, J.; Pesonen, L. Processing and assessment of spectrometric, stereoscopic imagery collected using a lightweight UAV spectral camera for precision agriculture. *Remote Sens.* **2013**, *5*, 5006–5039. [CrossRef]
22. De Castro, A.I.; Ehsani, R.; Ploetz, R.C.; Crane, J.H.; Buchanon, S. Detection of laurel wilt disease in avocado using low altitude aerial imaging. *PLoS ONE* **2015**, *10*, 1–13. [CrossRef] [PubMed]
23. Psirofonia, P.; Samaritakis, V.; Eliopoulos, P.; Potamitis, I. Use of Unmanned Aerial Vehicles for Agricultural Applications with Emphasis on Crop Protection: Three Novel Case—Studies. *Int. J. Agric. Sci. Technol.* **2017**, *5*, 30–39. [CrossRef]
24. Habili, N.; Oorloff, J. Scyllarus: From Research to Commercial Software. In Proceedings of the ASWEC 2015 24th Australasian Software Engineering Conference, Adelaide, Australia, 28 September–1 October 2015; pp. 119–122.
25. Rouse, J.W.; Haas, R.H.; Schell, J.A. *Monitoring the Vernal Advancement and Retrogradation (Greenwave Effect) of Natural Vegetation*; NASA's Goddard Space Flight Center: Greenbelt, MD, USA, 1974; pp. 1–8.
26. Rodriguez, D.; Fitzgerald, G.J.; Belford, R.; Christensen, L.K. Detection of nitrogen deficiency in wheat from spectral reflectance indices and basic crop eco-physiological concepts. *Aust. J. Agric. Res.* **2006**, *57*, 781–789. [CrossRef]
27. Daughtry, C. Estimating Corn Leaf Chlorophyll Concentration from Leaf and Canopy Reflectance. *Remote Sens. Environ.* **2000**, *74*, 229–239. [CrossRef]
28. Haboudane, D.; Miller, J.R.; Pattey, E.; Zarco-Tejada, P.J.; Strachan, I.B. Hyperspectral vegetation indices and novel algorithms for predicting green LAI of crop canopies: Modeling and validation in the context of precision agriculture. *Remote Sens. Environ.* **2004**, *90*, 337–352. [CrossRef]
29. Haboudane, D.; Miller, J.R.; Tremblay, N.; Zarco-Tejada, P.J.; Dextraze, L. Integrated narrow-band vegetation indices for prediction of crop chlorophyll content for application to precision agriculture. *Remote Sens. Environ.* **2002**, *81*, 416–426. [CrossRef]
30. Rondeaux, G.; Steven, M.; Baret, F. Optimization of soil-adjusted vegetation indices. *Remote Sens. Environ.* **1996**, *55*, 95–107. [CrossRef]
31. Zarco-Tejada, P.J.; Berjón, A.; López-Lozano, R.; Miller, J.R.; Martín, P.; Cachorro, V.; González, M.R.; De Frutos, A. Assessing vineyard condition with hyperspectral indices: Leaf and canopy reflectance simulation in a row-structured discontinuous canopy. *Remote Sens. Environ.* **2005**, *99*, 271–287. [CrossRef]

![sensors logo] *sensors*

MDPI

Article

Curvature Continuous and Bounded Path Planning for Fixed-Wing UAVs

Xiaoliang Wang [1,†], Peng Jiang [2,†], Deshi Li [1,3,*,†] and Tao Sun [1]

1   Electronic Information School, Wuhan University, Wuhan 430072, China;
    xiaoliangwang@whu.edu.cn (X.W.); suntao@whu.edu.cn (T.S.)
2   GNSS Research Center, Wuhan University, Wuhan 430072, China; jiangp@whu.edu.cn
3   Collaborative Innovation Center of Geospatial Technology, 129 Luoyu Road, Wuhan 430072, China
*   Correspondence: dsli@whu.edu.cn; Tel.: +86-136-0718-2565
†   These authors contributed equally to this work.

Received: 6 August 2017; Accepted: 14 September 2017; Published: 19 September 2017

Abstract: Unmanned Aerial Vehicles (UAVs) play an important role in applications such as data collection and target reconnaissance. An accurate and optimal path can effectively increase the mission success rate in the case of small UAVs. Although path planning for UAVs is similar to that for traditional mobile robots, the special kinematic characteristics of UAVs (such as their minimum turning radius) have not been taken into account in previous studies. In this paper, we propose a locally-adjustable, continuous-curvature, bounded path-planning algorithm for fixed-wing UAVs. To deal with the curvature discontinuity problem, an optimal interpolation algorithm and a key-point shift algorithm are proposed based on the derivation of a curvature continuity condition. To meet the upper bound for curvature and to render the curvature extrema controllable, a local replanning scheme is designed by combining arcs and Bezier curves with monotonic curvature. In particular, a path transition mechanism is built for the replanning phase using minimum curvature circles for a planning philosophy. Numerical results demonstrate that the analytical planning algorithm can effectively generate continuous-curvature paths, while satisfying the curvature upper bound constraint and allowing UAVs to pass through all predefined waypoints in the desired mission region.

Keywords: Catmull-Rom curves; continuous-curvature; curvature upper bound; local regulation; path planning; unmanned aerial vehicles

1. Introduction

Due to recent advances in communication, sensing and battery technologies, Unmanned Aerial Vehicles (UAVs) have drawn much attention for both military and civilian applications. Due to their portability and flexibility, UAVs are employed for numerous civilian applications, including industrial monitoring, scientific data collection and public safety (search and rescue). Soon, autonomous aircraft equipped with various sensors will be routinely surveilling our cities, neighborhoods and rural areas. They promise far reaching benefits for the study and understanding of public collaboration systems.

There are two main categories of UAVs: fixed-wing aircraft and multi-rotor vehicles. Compared with multi-rotor vehicles, fixed-wing aircraft are more advanced in many respects: they tend to be more stable in the air in the face of both piloting and technical errors as they have natural gliding capabilities even without power, and they are able to travel longer distances on less power. More importantly, they have the advantages of being able to fly at high speeds for a long time using a simpler structure. These characteristics make fixed-wing vehicles still widely popular, despite requiring a runway or launcher for takeoff and being unable to hover. For these reasons, we consider the path planning problem for fixed-wing UAVs in this work.

An important technical problem must, however, be addressed: How can we plan feasible and optimal paths that are effective are increasing the mission success rates and reducing the time taken? Path planning aims to determine a path from the initial position to the final position through a complicated space with or without obstacles. This is still an open problem in the field of autonomous systems research as it involves physical constraints due to the UAVs themselves and constraints due to their operating environment, as well as other operational requirements.

Several quintessential methods have been developed for path planning such as Voronoi diagrams [1], triangular-cell-based maps [2], as well as related grid-like maps [3,4], the potential function method [5], the A* algorithm [6,7], probability roadmaps [8], evolutionary algorithms [9–12] and mixed integer-linear programming [13,14]. Although these methods can provide optimal or near-optimal paths that pass through all predefined waypoints, they cannot guarantee smoothness, i.e., curvature continuity. Curvature discontinuities would cause sudden changes in the vehicle's heading and threaten its safety. To generate suitable paths for UAVs, several problems need to be tackled.

(1) Path curvature continuity: This directly influences the vehicles' aerodynamics and kinematics.
(2) Path optimization: We want to find the shortest path that passes through all of the expected waypoints. This is especially critical for missions such as reconnaissance and aerial delivery where vehicles are expected to fly over the specified target precisely, as any deviation would cause target information lost during reconnaissance and drop failures in transportation tasks.
(3) Local path adjustment: Dynamic environments require path replanning, which would consequently consume additional time. Accordingly, the planning method should be able to support local modifications.

The construction of feasible paths has been actively investigated in robotic research fields, because non-smooth motions can cause wheel slippage and degrade the robots' dead-reckoning abilities [15]. Dubins curves are one practical method [16–18], as they can give the shortest path between given initial and final points with a fixed change rate of turn [16]. In these algorithms, two points are connected by concatenating arcs and tangents. Sahingoz et al. [18] and Shanmugavel et al. [19] apply Dubins curves to UAV path planning in 2D space after task allocation. Dubins curves have been extended to include other spline curves such as helix to generate more complex vehicle trajectory models in 3D space [17,20]. However, there are still curvature discontinuities at the junctions between lines and arcs, leading to yaw angle errors for UAVs. As a result, the corresponding applications are limited to straight flights.

To deal with the curvature discontinuity problem, two categories of curves are utilized in [21–26]. The first category involves parameterizing the curvature by arc length using, for example, clothoids [21,22]. Clothoid pairs can provide the shortest curves for a given turn angle [27]. However, there is no closed-form expression for position along the path or its approximation. Moreover, look-up tables are usually required, increasing the operational complexity. The second category can provide closed form solutions, using, for example, B-splines, Bezier curves or spatial Pythagorean hodographs [23–25,27–30]. These curves are used for trajectory generation for both UAVs [23,24] and autonomous robots [25,28]. However, path curvature is a high-order function of curve parameters, and it is a formidable challenge to analyze its monotonicity. Generated paths have continuous curvature when using Bezier curves [31] or B-splines [32], but little attention has been given to the maximum curvature constraint. Although it has been demonstrated that curvature extrema can be obtained numerically for any planar cubic curve, this might be a challenge for parametric polynomial curves [29,33], since it is difficult to solve the formulated curvature extremum model. An objective function has been designed to search for proper Bezier curve parameters that satisfy the maximum curvature constraint, but the optimization problem remains unsolved [34]. Additionally, combined Bezier curves can be used to generate continuous curvature paths with bounded curvature extrema in both 2D and 3D space [24]. Shanmugavel et al. [30] have proposed a Pythagorean hodograph curve for 3D path planning, achieving bounded curvature and equal path lengths. Unfortunately, the generated path only passes through some of the waypoints, accordingly leading to failure of some specified tasks.

Despite all of these efforts, the challenge of generating feasible and optimal trajectories has been partially solved in the literature above. Furthermore, local path adjustment has been ignored, which is important in real applications. Replanning the whole path would be inefficient due to additional time and energy required. By contrast, local adjustment can reduce replanning complexity because the adjustment is used only where needed. Parametric curves with local adjustment, including B-splines and Beta-splines, can potentially be used for path planning. In particular, the construction of geometric continuous Beta-splines has been achieved by DeRose T D et al. In our work, the path planning problem is solved using a three-stage decomposition, which allows curvature continuity and curvature adjustment to be completely independent of each other. This strategy was inspired by the approach in [35], where circular orbits and Bezier curves are concatenated to create more complicated paths.

The objective of this paper is to develop path planning strategies that explicitly account for curvature and waypoint constraints. Previous work has primarily addressed path feasibility, whereas in this paper, we also consider path accuracy. In this paper, a continuous-curvature and bounded path planning algorithm for fixed-wing UAVs is proposed to solve the problem of passing through all of the waypoints under aerodynamic constraints. To this end, a parametric Catmull–Rom curve is employed.

The specific contributions of this paper are given below.

(1) A path planning framework is proposed to generate a feasible path that satisfies both curvature continuity and upper bound constraints.
(2) Path accuracy, which is absolutely necessary for aerial photography and air drop, is explicitly satisfied using the path planning algorithm.
(3) Geometric shaping of the replanned path can be systematically performed with the designed minimum curvature circle transition mechanism.
(4) The proposed path planning algorithm can also be able to solve obstacle avoidance problems to some extent.

The rest of this paper is organized as follows. Section 2 presents the problem formulation. In Section 3, a three-stage trajectory architecture is presented. In Section 4, a cross-value-based interpolation algorithm and a key-point shift algorithm are elaborated and the planar transition algorithm introduced. In Section 5, a path replanning algorithm using minimum curvature circle transitions is proposed. Simulation and experimental results are shown in Section 6, followed by the conclusion in Section 7.

2. Problem Statement

Let $WP = \{W_1, \cdots, W_n\} = \{(x_1, y_1), \cdots, (x_n, y_n)\}$ be a sequence of waypoints in a 2D space returned by a higher planner. The path planning problem is to find a mission flight trajectory that fits all of the waypoints smoothly, satisfying curvature continuity and maximum curvature constraints.

For a feasible path curve, the curvature continuity and maximum curvature requirements must be satisfied simultaneously. Curvature discontinuity would lead to infinite changes in lateral acceleration, which can obviously not be allowed. Likewise, the curvature is proportional to the centripetal force at fixed velocity, and thus, unbounded curvature could result in roll angles that exceed the safety limits.

The smoothness of curve junctions can be measured by geometric and parametric continuity. Different orders of smoothness in terms of geometric and parametric continuity are defined in [36].

Parametric continuity: Let $P(s_0, s_1 : s)$ and $Q(t_0, t_1 : t)$ be parametric curves, such that $P(s_1) = Q(t_0) = m$. They meet at m with k-th order parametric continuity if:

$$\frac{d^k P}{ds^k} = \frac{d^k Q}{dt^k}, k = 1, \cdots, n \tag{1}$$

Geometric continuity: Let $P(s_0, s_1 : s)$ and $Q(t_0, t_1 : t)$ be parametric curves, such that $P(s_1) = Q(t_0) = m$. They meet at m with k-th order geometric continuity if the natural parameterizations of P and Q meet at m with parametric continuity.

These definitions are also used in this paper. Barsky et al. [36] point out that parametric continuity prevents parameterizations from generating geometrically smooth curves due to the strict constraints imposed by the derivatives. By contrast, geometric continuity accommodates differences between the parameterizations of adjoining curves, and thus, it is used to smooth the piecewise path in our work.

The UAV navigation limitations are due to its own kinematic constraints. As it has been widely proven that vertical motions consume additional energy, we also assume that the UAV is flying at constant altitude. The aerodynamics features of a fixed-wing UAV include lift force L and air friction D.

$$L = \frac{1}{2}\rho V^2 S C_L \qquad (2)$$

where ρ is the air density, V is the flying speed, S is the wing area and C_L represents the lift coefficient.

$$D = \frac{1}{2}\rho V^2 S C_D \qquad (3)$$

where ρ is the air density, V is the flying speed, S is the maximum section area of an UAV and C_D represents the resistance coefficient.

From the above two equations, we can see that the flying speed directly determines the lift force and air friction for a UAV under stable atmospheric conditions. In parallel, acceleration is generated from the corresponding component force, and this in turn changes the velocity of a UAV. Due to the designs themselves, fixed-wing UAVs have a minimum speed (the stall speed V_{min}), below which they risk falling out of the sky. In addition, UAVs are limited to a maximum speed of V_{max} by power constraints. In practice, parameters including air density, wing area, lift coefficient, the maximum section area and the resistance coefficient were fixed. Thus, the dynamic constraint is the velocity: $v_{min} \leq v \leq v_{max}$.

To acquire the relation between curvature, velocity and lateral acceleration, a typical circular motion model is used to illustrate the turning motion for fixed-wing UAVs.

$$F = \frac{mv^2}{R} \qquad (4)$$

where F represents the centripetal force, v is the UAV's velocity and R denotes the turning radius. Equation (4) can be further rewritten as: $|\kappa| = \frac{F}{m} = \frac{v^2}{R}$.

Since the curvature is inversely proportional to the turning radius R, the relation among curvature, velocity and lateral can be described as follows:

$$|a| = |v|^2 \kappa \propto \kappa \qquad (5)$$

Due to the total centripetal force, which keeps constant during the turning, the curvature gets its minimum value (κ_{min}) when the flying speed reaches the extreme value (v_{max}). Consequently, the relation between maximum curvature, maximum velocity and lateral acceleration is:

$$|a| = |v_{max}|^2 \kappa_{min} \qquad (6)$$

As far as the dynamics is concerned, (6) shows that the curvature is directly proportional to the vehicle's lateral acceleration. The maximum curvature (κ_{max}) is inversely proportional to the minimum curvature radius (ρ_{min}) of the curves that it is able to follow. A path is valid if the curvature is smaller than the specified maximum value,

$$|\kappa(t)| \leq \kappa_{max} \qquad (7)$$

where κ_{max} denotes the maximum curvature determined by the UAV's dynamic constraints.

The path planning problem in this paper is independent of the UAVs' dynamic models. The single or double integrator model is commonly used to characterize an individual UAV for state analysis. Since the power for a fixed-wing UAV is supplied by motors and direct user interactions are operations on the motor speed that change the state of a UAV, consequently, it is difficult to control the velocity and acceleration directly in engineering. For fixed-wing UAVs, which have a six degree of freedom movement, translational and rotational motions are achieved by adjusting propeller speed and the servomotor controller. Although different dynamic models are employed, the aerodynamic constraints are identical. To be specific, the curvature must be continuous and bounded in the 2D environment.

3. Three-Stage Trajectory Architecture

To deal with problems involving planning, a three-stage path planning algorithm with a transition selection mechanism for piecewise paths is proposed in this paper. The operational workflow of this algorithm is presented in Figure 1.

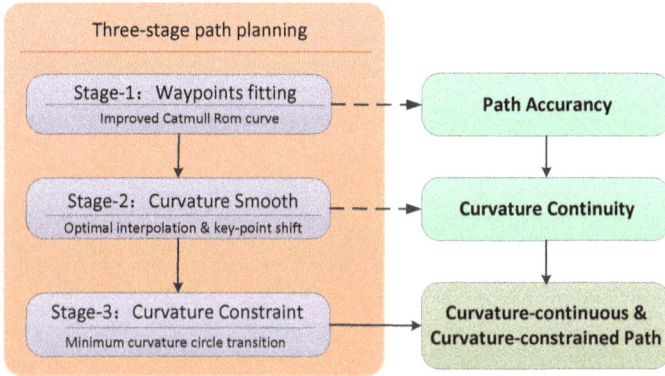

Figure 1. Workflow of the continuous-curvature path planning algorithm.

Our algorithm has two key components: curvature smoothing and curvature constraining. The Catmull-Rom curve is advanced in its local-control ability and owns the property that it passed through the control points. Thus, a Catmull-Rom curve is used to fit the waypoints firstly, to achieve path accuracy. Then, an optimal position or key point is extrapolated to smooth the curvature on the base of Phase 1. After that, to ensure that the UAV flies over each waypoint obeying the aerodynamic constraints, a minimum curvature circle transition scheme is used to replan those piecewise curves whose curvature exceeds the upper boundary.

Obstacle avoidance increases complexity for path planning and leads to path replanning. For off-line path planning, path curves should be designed to avoid obstacles using their prior knowledge such as positions and shapes. This results in constraints on path curves and consequently increases the algorithm complexity of path planning. As for on-line path planning, new paths must be planned rapidly to avoid dynamic obstacles, meaning that the whole path or local path should be replanned.

Our method was advanced in the organic coordination of path planning and obstacle avoidance. The method proposed can potentially solve some obstacle avoidance problems, while generating a continuous curvature and bounded path that passes through all waypoints.

In fact, the obstacle avoidance mechanism in Figure 2 is to navigate around obstacles by designing a path via path re-planning. By adding or moving a waypoint, the predefined path is locally changed. This path re-planning helps to solve the obstacle avoidance problem at the path level.

Specific to multiple waypoint path planning, there are mainly two types of obstacles as shown in Figure 2.

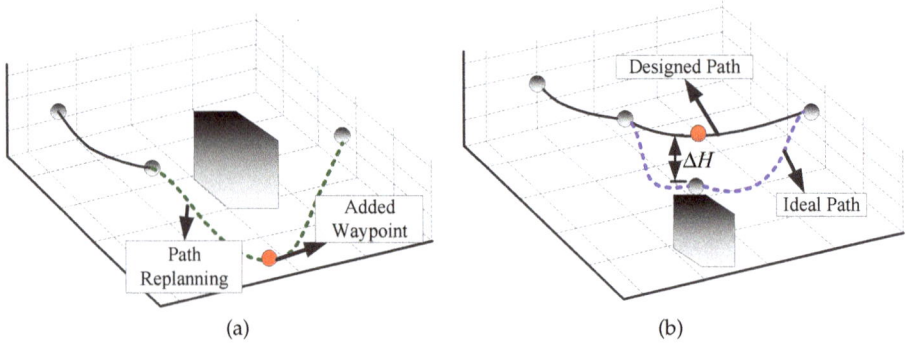

Figure 2. Two typical obstacle scenarios: (**a**) Obstacles are located between two waypoints (**b**) Obstacles are located below a waypoint.

In the first scenario, we believe the collision avoidance problem can be solved by our proposed optimal interpolation method as demonstrated in Figure 2a. By adding a new waypoint, the path is replanned, and UAVs navigate around obstacles successfully. The research background of this paper is mainly ground observation using fixed wings UAVs where aeronautics and astronautic sensors are equipped on the vehicles for surface and human activities explorations. Under such background, the obstacle avoidance problem shown in Figure 2b can be tackled by driving the UAVs to fly over some waypoints. Although this operation results in range deviation between the flight path and waypoint and consequently reduces image quality, we believe that this quality decrease is allowable compared with the UAVs' safety. Additionally, a PTZ (pan/tile/zoom) camera can also help to eliminate or mitigate the above negative influence.

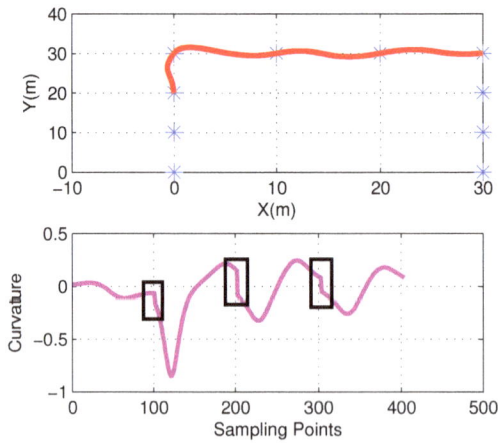

Figure 3. Discontinuity of the curvature value at the junction points: The red line is original path curve and the purple one represents its curvature value.

4. G^2 Continuous Path Smoothing with Multiple Waypoints

It has been proven that Beta-constraints are necessary and sufficient conditions for G^n continuity [37]. As an example, the Beta-constraints for G^2 continuity take the following form:

$$r^{(1)}(0) = \beta_1 q^{(1)}(1)$$
$$r^{(2)}(0) = \beta_1^2 q^{(2)}(1) + \beta_2 q^{(1)}(1) \tag{8}$$

where r and q are abbreviations for the parametric curves $r(s0, s1 : s)$ and $q(t0, t1 : t)$ and β_1 and β_2 are real numbers to represent the linear relationship between the deviation of parameter curves (β_2 can take arbitrary values, but β_1 must be positive.)

Our goal has been to provide a path curve that passes through all waypoints and simultaneously satisfies Beta-constraints. It would be elegant to obtain an algebraic solution demonstrating the curve's adequacy and completeness, but it is arduous to solve sets of second order differential equations with two degrees of freedom since it is formidably difficult to find closed-form solutions. We instead present a practical algorithm based on the idea that a geometric method can be used to design a quantic interpolating spline, which is a subclass of the Catmull–Rom splines with two shape control parameters [38]. In this section, a Catmull–Rom-based continuous-curvature condition is derived, and a relevant path smoothing algorithm is presented.

4.1. Curvature Continuity Condition

In order to get a better and clearer understanding of the work below, the construction method of the Catmull–Rom curve is briefly introduced in this section. The method consists of three parts: set shape parameters, calculate Bezier polygon, determine Bezier control points. The effect of shape parameters was studied in [38], and they were recommended to be set as $\beta_1 = 1$ and $\beta_2 = 0$. After iteration calculation, the polygon of an auxiliary Bezier curve was figured out. Finally, the expected Bezier control points were determined by the linear weight of the auxiliary Bezier polygon.

For predefined waypoints, curvature discontinuities may exist at junction points, as shown in Figure 3. Further investigation indicates that the shape parameters and positions of the corresponding waypoints are the decisive factors in determining the degree of discontinuity. For simplicity's sake, we set the shape parameters as follows: $\beta_1 = 1$ and $\beta_2 = 0$. It should be noted that although the results in Equation (9) would change when the parameters are different, the curvature smoothing algorithm proposed still holds.

According to the Catmull–Rom curve construction method given in [38], the curvature change at the point (x_i, y_i) is given by:

$$\Delta\kappa = (x_{i-2}y_i - x_i y_{i-2} - 9x_{i-1}y_i + 9x_i y_{i-1} - 7x_i y_{i+1} \cdots$$
$$\cdots + 7x_{i+1}y_i + x_i y_{i+2} - x_{i+2}y_i)/900 \tag{9}$$

To ensure curvature continuity, the above condition is used to calculate point position in optimal interpolation and key-point shift algorithms.

4.2. Path Smoothing Algorithm

The curvature continuity condition essentially puts constraints on the relative positions of waypoints. If the local distribution of the original points is properly adjusted, curvature continuity can be achieved. Although (9) is the strict mathematical condition for curvature continuity, this theoretical requirement is difficult to satisfy in practice due to project errors such as measurement or mechanical errors. Thus, we regard the curve as continuous as long as the curvature change is less than a given limit ε. Various tasks have different requests with respect to time cost. For example, a loose time constraint is needed in aerial photography tasks. By contrast, missions such as rescue and reconnaissance require that the time taken must be as little as possible. To meet different demands on time cost while satisfying the continuous-curvature condition, two algorithms (optimal interpolation and key-point shift algorithm) were proposed.

The core idea of the optimal interpolation algorithm is to find a reasonable position to insert a new waypoint. Equation (9) shows that a new waypoint would influence the curvature of five neighboring junctions. Consequently, the goal is to find a point that makes the associated curvature changes tend to zero. By considering how the control points of nearby piecewise curves are affected by the inserted waypoint, the optimization objective is to minimize the length of these curves. As a result, the proposed algorithm solves the following optimization problem.

$$\min f(x; y) = \sum_{1}^{10} \ell_j$$

$$s.t. \begin{cases} |\Delta \kappa_i| \leq \varepsilon \ (i = 1, 2, \cdots, 5) \\ x \in X \\ y \in Y \end{cases} \tag{10}$$

where l_i represents the related path length and ε denotes a user-defined continuity threshold. Numerical methods such as steepest descent and Newton's method are usually employed to solve such optimization problems, but in practice, the integral for calculating the arc length cannot be solved. By contrast, the Gaussian quadrature method only needs the polynomial form of the integral and selects the expected number of points on the integral interval, reducing the computational complexity dramatically. To increase the computational efficiency, both the Gaussian quadrature method and a Genetic Algorithm (GA) are adopted. The input and output of the algorithm are the set of waypoints and the coordinates of the objective point, respectively. The main steps of the optimal interpolation are described in Figure 4.

The optimal interpolation algorithm mainly consists of two phases: construct a cost function and solve the optimization problem. Since the task cycle is a key factor that determines the optimization of flight paths, the path length has also been taken into consideration in the cost function. To obtain the explicit formulation of the cost involved, numerical computation is conducted via Gaussian Quadrature (GQ). Considering that the lengths of adjacent piecewise paths are effected by the newly-added waypoint, those related sequence numbers of waypoints are figured out firstly. Then, the length of each segment is estimated by GQ. As for the curvature change value, the result from Equation (9) is used.

The GA algorithm that was used to solve the optimal interpolation problem was a function encapsulated in a toolbox by MATLAB. The function is: $X = ga(fitnessfcn,nvars,A,b,Aeq,beq,LB,UB)$. In the GA algorithm, the input is curve length and curvature change values with corresponding weight coefficients, and the output is the coordinate of the objective point. Some key parameters in the optimal interpolation algorithm are given as follows: $ctlflag$ is the sequence number of the point needed to be interpolated; Seg_Num is the total number of piecewise curve; seg_num_ahead represents the number of segments that are located ahead of the interpolated point; seg_num_after denotes the number of segments that are located behind the interpolated point; $isnode, ienode$ are the start and end sequence number of the points affected by the interpolation operations, respectively; $data\_len$ is the length of the new waypoint set; fun_len is the total path length related to the interpolation point; CPt are the control point of Bezier curves; and fun_cuv is the total curvature change value.

The key idea of the key-point shift algorithm is to displace as few waypoints as possible and then replan the trajectory using these new points. Due to curve shape changes, the two proposed algorithms both result in path length increment. Thus, curvature smoothing and curve length are two main factors considered in path smoothing. Consequently, we set the curvature change value as the constraint and the curve length as a target, as shown in Equation (10), in both the optimal interpolation and key-point shift algorithms. The optimization problem can be formulated as in Equation (10).

To solve the above optimization problem, weight coefficients (w_1, w_2) were set for the curvature change value and path length, respectively, in the optimal interpolation and key-point shift algorithms.

However, curvature continuity is achieved through curve shape changes by the added piecewise path in the optimal interpolation algorithm, while the key-point shift algorithm is to displace as few waypoints as possible, avoiding the introduction of new waypoints. This results in a much greater increment in path length for the optimal interpolation algorithm. To obtain satisfactory results, different weight coefficient were assigned ($w_1 = 800-3000$, $w_2 = 500-3000$) in the optimal interpolation algorithm and ($w_1 = 800-3000$, $w_2 = 100-500$) in the key point shift algorithm. The results of w_1 and w_2 are empirical values based on a large number of numerical experiments.

if $ctlflag - 5 \geq 0$ **then** $\qquad$ $seg\_num\_ahead = 3;$ **end if** **else if** $ctlflag \geq 3 \& ctlflag < 5$ **then** $\qquad$ $seg\_num\_ahead = mod(ctlflag + 1, 3)$ /* **end if** **if** $data\_len - ctlflag - 4 \geq 3$ **then** $\qquad$ $seg\_num\_after = 3;$ **end if** $\quad$ **else** $\qquad$ $seg\_num\_after = data\_len - ctlflag - 4;$ **end**	*Figure out the influenced waypoints*
$seg\_num = seg\_num\_ahead + seg\_num\_after$ **if** $ctlflag < 5$ **then** $\qquad$ $isnode = 1$ **end if** $\quad$ **else** $\qquad$ $isnode = crlflag - 4$ **end** $\qquad$ $ienode = isnode + seg\_num - 1$	*Find out the index of starting and ending points*
while($j = isnode : ienode$) $\qquad$ $x0 = data(j : j+5, :)$ $\qquad$ $x0(seg\_num\_ahead + 3 - (j - isnode), :) = x$ $\qquad$ $CPt = getCtlPt(x0)$ /* *Determine the control points of Bezier curves* */ $\qquad$ $fun\_len = fun\_len + qgaus(gausfun(CPt))$ /* *qgaus () calculates arc length* */ **end while**	*Using Gaussian quadrature to calculate arc length*
$fun\_cuv = \sum\limits_{i=1}^{Cur\_T} CurChagCacu(data0(i))$ $fun = fun\_len + WT * fun\_cuv$ $\qquad\qquad$ /* *WT is a weight coefficient* */	*Establish optimization objective for GA algorithm*

Figure 4. Description of the optimal interpolation algorithm.

5. Minimum Curvature Circle Transition

To ensure path feasibility for a given UAV, the curve $(\overrightarrow{s(t)})$ must fulfill certain aerodynamic constraints. The maximum curvature (κ_{max}) is an achievable constraint by the vehicle in 2D space. Although the path generated in Section IV has continuous curvature, the curvature extrema have yet to be considered. Those piecewise paths whose curvatures exceed a specified threshold are physically unreachable. Here, a geometric path transition mechanism is established to meet these curvature upper bound constraints.

Considering the complete space of solutions, two classes of over-curved path transition are proposed: no waypoint on path and one waypoint on path. When replanning a piecewise path, it is essential to understand what kind of transition problem it presents so that it can be solved in an appropriate manner. Figure 5 shows the two classes of path transition. The first category (shown in the rectangular boxes) is where the over-curved segment is a part of the path curve between two adjacent waypoints. Similarly, the second category (shown in the elliptical boxes) represents curve segments that pass through a waypoint.

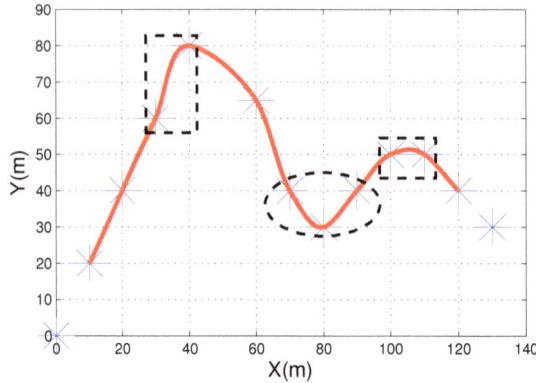

Figure 5. No waypoint on path and one waypoint on path cases: The piecewise curves are shown within the rectangular and elliptical boxes.

5.1. Analytical Solutions of the Minimum Curvature Circle Transition

It has been demonstrated that a cubic curve can be used for transitions preserving G^2 continuity and that two variables (m, θ) are required to generate the expected connections [35]. From the results of [35], it appears that the value of parameter m was determined by trial and error. Here, we want to determine the value of m for given star and end point. This can be achieved using the following theorem.

Theorem 1. *If the centers of the two minimum curvature circles are known, there is a unique curvature-constrained curve connecting two points, one located on each circle.*

Proof. When a local frame is set up, the angle between the x-axis and the line segment between the two circle centers is (for s-shape):

$$\alpha = \arctan \frac{16m^2 \cdot \frac{\sin^2 \theta}{27 \cos^2 \theta} \cdot r_0^2}{2 \cdot Pr_0 \cdot \sqrt{\frac{2\sin\theta}{3}} + h\cos\theta} \tag{11}$$

The parameters for s-shaped and c-shaped curves can then be obtained from (12) and (13), respectively.

$$\begin{cases} (8\,r_0\,\alpha\varphi - 8\,r_0\,\varphi^2)\,m^2 + 24\,r_0\,\alpha\varphi m + 54\,r_0 = 0 \\ 8\,r_0\,\varphi^2\,m^2 - (2\,r_0(9 - 12m - 2\,m^2) + \delta) = 0 \end{cases} \tag{12}$$

$$\delta = \sqrt{729\,r^2 + r_0^2\,(6 + 2m)^2(4\,m^2 + 24m - 72)}$$

$$\begin{cases} 2\,r_0\,m^2\,\varphi^3 - 3\,x_1\,\varphi^2 + (\Delta - 3\,r_0)\varphi - 3\,x_1 = 0 \\ 2\,r_0\,m^2\,\varphi^4 + (\Delta - 3\,y_1)\,\varphi^2 + 3\,r_0 - 3\,y_1 = 0 \end{cases} \tag{13}$$

This is derived by assigning $P = m \cdot \frac{1}{\cos\theta} \cdot \sqrt{\frac{8\sin\theta}{27}}$ and $h = m^2 \cdot \frac{1}{\cos^2\theta} \cdot \frac{8\sin\theta}{27} \cdot R_0$ to (12) and (13). Here, Δ is equal to $2r_0m^2 + 4mr_0 - 3r_0$.

r_0 is the radius of the minimum curvature circle; φ is equal to $\tan(\theta)$; α is the angle tangent of the x-axis and the straight line connecting two circle centers; r represents the distance between the circles' centers; and (x_1, y_1) denotes the center coordinate where endpoints are located. □

It can be seen that a feasible transition can be figured out by solving (m,φ) from Equation (13).

5.2. Transition Model for the No Waypoint on Path Case

Inspired by the approach taken in [35], where curvature-bounded connections are achieved by utilizing arcs and cubic Bezier curves, we extend the basic c and s-shaped curves to develop a series of hybrid transitions, which include $c - c, s - s, c - s$ and $s - c$ shapes. Figure 6 shows an enumeration of these shapes, where we can see that both the shape and length of the curves vary substantially between transition types. Since UAVs proceed in a predetermined direction along their trajectories, the tangent to the transition curve must be considered. Transition curves can be divided into two categories depending on whether or not the rotation orientation switches between the start and end points. To be specific, the $c, c - c$ and $s - s$ shapes guarantee that the curve goes clockwise or anticlockwise, while it reverses direction for $s, s - c$ and $c - s$ shapes.

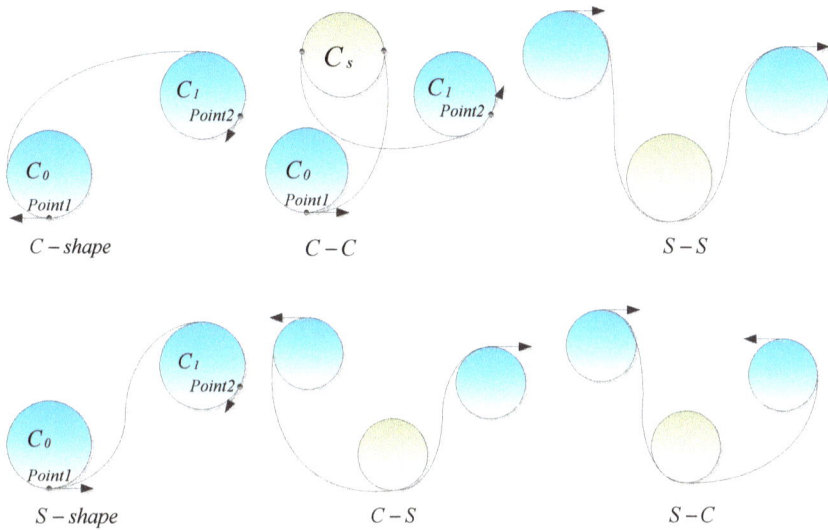

Figure 6. Transition instances: s-shape, $c - s$ and $s - c$ transitions reverse the tangent direction, while c-shape, $c - c$ and $s - s$ transitions reserve this direction.

To satisfy the maximum curvature condition, a minimum-curvature circle is employed to pass through the points whose curvature values exceed the threshold. A minimum curvature circle is defined as one whose curvature radius is in accordance with the settled curvature threshold. As the minimum curvature circle transition model is built in a local coordinate system, the global coordinates have to be converted to the local coordinates. We set the first point with the curvature threshold as the center of the local coordinates. The transformation matrix is constructed as follows.

$$[x\,y] = [x'\,y']\boldsymbol{R} + \boldsymbol{T} \tag{14}$$

The transformation matrix includes rotation and translation operations, $\boldsymbol{R}$ and $\boldsymbol{T}$, respectively. Using the inverse of this matrix, the local coordinates can be derived from the global reference system:

$$P_g = (P_l - \boldsymbol{T}) \cdot \boldsymbol{R}^{-1} \tag{15}$$

where P_g denotes the global coordinates and P_l denotes the local coordinates.

Although the points to be connected are located on the minimum curvature circles, it can be shown that both the s and c-shape transition formulas hold in a bounded interval, which is determined by Theorem 2.

Theorem 2. *The lower bound on the feasible interval is determined by setting m to one and the upper bound by the intersection of the inner or outer tangent lines and circles, respectively, for s and c-shape transitions.*

Proof. We consider an s-shape transition curve $Z(t)$ here. Set the angle of $\overrightarrow{P_2P_3}$ and $\overrightarrow{P_1P_2}$ to θ. In addition, $\overrightarrow{P_1P_2}$ and $\overrightarrow{P_2P_3}$ are parallel to the x-axis. This can be summarized as:

$$P_0 = (0,0) \quad P_1 = (g,0) \quad P_2 = (g + h\cos\theta, h\sin\theta)$$
$$P_3 = (g + h\cos\theta + k, h\sin\theta)$$

where g, h and k are the straight-line distances between the control points of Bezier curves. $P_i (i = 0,1,2,3)$ are the control points of Bezier curves. According to the results in [34], whether or not the curvature derivative of $Z(t)$ vanishes on [0,1] depends on the value of Q, where:

$$Q = 3\cos^2\theta f_1(s) + 4m^2 s f_2(s)$$

We also have:

$$Q = Am^2 + Bm + C$$

where:

$$A = 12s^4 - 4s^3 - 4s^2 + 12s,$$
$$B = -3(3s^5 - 5s^4 + 10s^3 + 10s^2 - 5s + 3)\cos^2\theta,$$
$$C = 3(s^5 - 2s^4 - s^3 - s^2 - 2s + 1)\cos^2\theta.$$

Letting $\theta \to 0$ yields $\lim_{m\to 1, s\to 0} Q = 0$, while $\exists s$, which makes $Q \leq 0$ as $m \leq 1$. Meanwhile, the value of h approaches zero when the transition curve points start at the intersection of the inner tangent lines and the circle. $\square$

From the above proof, we can see that $\theta \to 0$ and $h \to 0$ determine the boundary condition for s-shape transition.

5.3. Approach for the One Waypoint on Path Case

In addition to the approaches above, a hybrid transition mechanism has also been proposed for more complicated situations. Although the s-shape and c-shape models can solve curve transition problems between two points with controllable curvature extrema, this mechanism is essentially for determining the parameters of the cubic Bezier curves' control points using curvature monotonicity conditions. Compared with the situations studied in Section 5.2, a much more complicated application is one where one or more waypoints on the path need to be replanned. A single s-shape or c-shape solution could not satisfy replanning requirements because the transition curve control parameters are determined once the start and end points are given. In such cases, the curve shapes are fixed, and thus, the requirement to pass through the waypoints cannot be ensured. Therefore, changes are needed for each pair of points on the path curve. To solve this problem, a hybrid transition approach is proposed, which includes the following combined connections: $s - s$-, $c - c$-, $c - s$- and $s - c$-shaped curves. Using these, an auxiliary minimum curvature circle can be constructed to reach the specified waypoint. Afterwards, an s-shape or c-shape transition is employed to connect the points to this circle within the curvature threshold via piecewise arcs.

The key to finding solutions for these hybrid transitions is to estimate an appropriate position for the center of a minimum curvature circle. Results in Section 5.2 prove that a transition is valid when the start and end control points of an s or c curve fall within restricted ranges. For this reason, there is a wide choice of possible curve parameters, which give rise to different curve lengths. To optimize the path, an optimal position for the center must be found.

To determine the optimal center position of the auxiliary circle, the boundary conditions should first be established. Results in Section 5.2 show that the points of tangency on the inner tangent lines between two minimum circles are upper bounds on the end points for s-shape transition curves. The lower bounds can be determined by setting m to one. As an illustrative example, consider three pairs of points and their minimum curvature circles, as shown in Figure 7. Here, Point 1 and Point 2 are the predetermined start and end points, and P_0 is a predefined waypoint. The tangent directions at the start and end points are both counter-clockwise. Additionally, the fixed point P_0 is required to be on the sub-path. For simplicity's sake, the circle that passes through P_0 is defined as an auxiliary circle as follows.

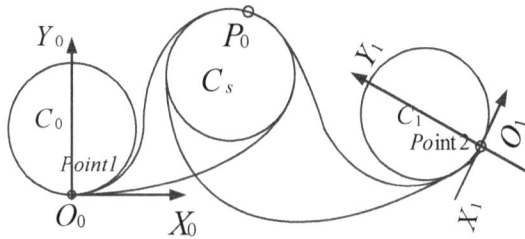

Figure 7. A minimum curvature circle transition example for the one waypoint on-line case.

By constructing an outer tangent line to C_0 (the minimum curvature circle passing through *Point1*) from P_0 as shown in Figure 8, the intersection point determines a bond on C_s. This is because the end point of the transition curve crosses the waypoint as the circle center moves in a counter-clockwise direction along its orbit, and thus, we want the UAV to move along an arc on the minimum curvature circle to travel across the waypoint, leading to an increase in path length. In addition, another bound on C_s can be obtained by drawing a line parallel to the x-axis from point P_0, which intersects the circle at B_i^1 as shown in Figure 8. As the circle center moves clockwise, the horizontal velocity component simultaneously reverses direction. Consequently, bounds on the circle center for an $s-s$ transition are given by the blue dotted line. Auxiliary circle center bounds for other hybrid transitions can be acquired likewise.

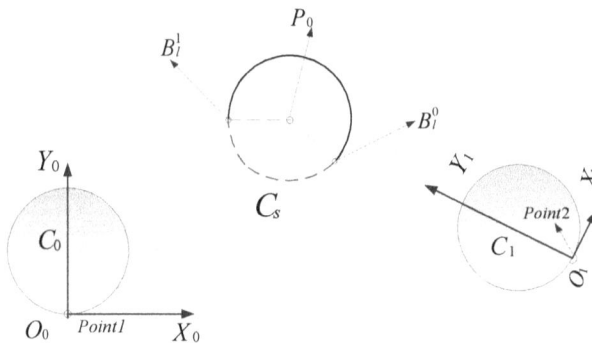

Figure 8. Circle center trajectory bounds for the $s-s$ transition.

Compared with conventional search algorithms, binary search has the advantages of low time complexity and no need for any other search region interventions. Thus, it is used to determine the position for the center of the auxiliary circle. The iterative process is summarized in Algorithm 1.

Algorithm 1 Binary searching algorithm for optimal circle center.

Input: Lower and upper bound
Output: Optimal circle center
1: Initialize the circle center (x,y);
2: Determine the starting position $C_{s,s}$ and the ending position $C_{s,f}$;
3: Let $C_{s,s}$ and $C_{s,g}$ be the initial and final poses of the circle center trajectory, and compute the center (x_0,y_0) and (x_1,y_1);
4: **while** $\left((x,y) \in \left[C_{s,s} C_{s,f} \right] \right)$;
5:    **if** $\left(\left| \sum_{i=0}^{2} \ell_i(x,y) - \sum_{i=0}^{2} \ell_i(x_0,y_0) \right| > 0 \right)$ **then**;
6:      $(x,y) = ((x,y) + (x_0,y_0))/2$;
7:    **else**
8:      **Break**;
9:    **end if**
10: **end while**

The above process consists of two phases: specification of the search region and iterative search. Utilizing the auxiliary circle center bounds determined by the previous analysis, we create an objective function that optimizes for curve length. As a result, the iteration terminates when the minimum length is obtained. Binary search is well known to have a time complexity of $O(log(n))$.

5.4. Novelty of the Proposed Transition Method

Although constructing curves with monotonic curvature using arcs and Bezier curves is quite standard in math, our method was unique as we apply the simple method in UAV path planning. The novelty is exhibited as follows: (1) on the condition that the re-planned curves had monotonic curvature, arrival of intermediate waypoints was achieved by using a minimum curvature circle; (2) $s − s$-, $s − c$-, $c − c$- and $c − s$-shaped curves were designed to connect adjacent waypoints, ensuring a UAV flied along the predefined direction; (3) the shortest replanning path was determined using the binary searching method on the basis of the auxiliary circle center's boundary condition.

6. Experiments and Discussion

To evaluate the performance of the proposed algorithm, simulated experiments were carried out to generate continuous-curvature constrained paths. To demonstrate the superiority of the proposed algorithm, typical path planning methods including the G^2 Bezier curve and Dubins curves were compared with the proposed algorithms.

6.1. Continuous-Curvature Bounded Catmull-Rom Path Generation

To verify the effectiveness and robustness of the proposed method, the parameters of a physical fixed-wing UAV were used for the numerical simulations.

The parameters for each experiment are listed in Table 1. It should be noted that the turning acceleration, fixed as 3 g, was determined by the aircraft itself.

Table 1. Parameters for the numerical experiments.

Turn Acceleration (m· s^{-2})	R (m)	κ	V_{max} (m/s)
29.42	250	0.004	85.76
29.42	66.667	0.015	44.3

Firstly, the key-point shift algorithm incorporated with the c-shaped transition mechanism was used to generate a feasible path as shown in Figure 9a. The corresponding curvature of the path curve is as displayed in Figure 9b. From the results, we can see that the curvature of the designed path is continuous and under upper bound constraints. The hybrid transitions included the $s-s$, $c-c$, $s-c$ and $c-s$ cases. Figure 10 shows the results of using $s-s$ and $c-c$ transitions for local replanning. The $s-c$ and $c-s$ solutions were scrapped due to the tangent directions of the start and end waypoints. The limiting case for an $s-s$ transition is that two circles are tangent to each other, while a $c-c$ transition only requires the circle centers to be outside each other's circumference. The center trajectories of the curvature circles for the intermediate waypoints are shown as black arcs in Figure 10b,c. Binary search was used to determine the appropriate circle center positions. Since transition paths are composed of circle arcs and Bezier curves with monotonic curvature, the curvatures of both the $s-s$ and $c-c$ transitions are bounded by the maximum value κ_{max}.

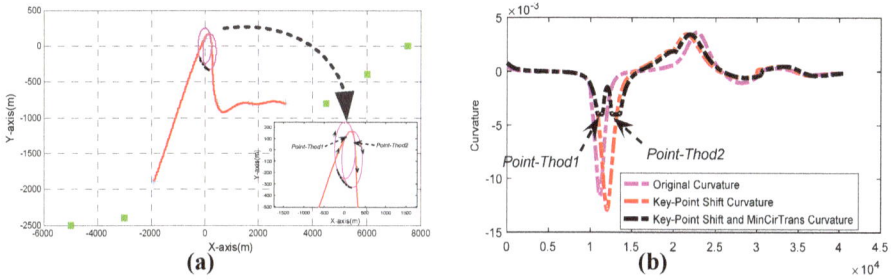

Figure 9. Path generation: (**a**) c-shaped transition (**b**) Curvature value of path curves.

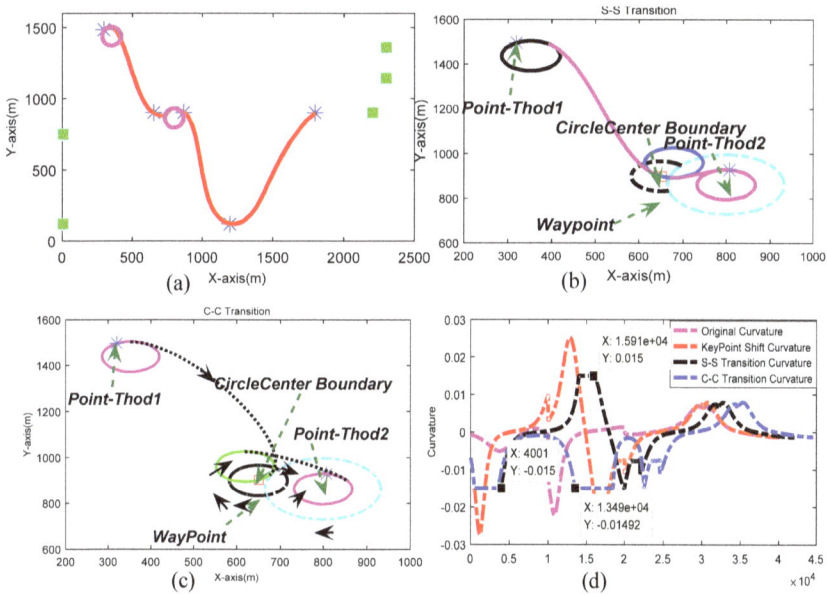

Figure 10. Combined transition mechanisms: (**a**) Final path (**b**) $s-s$ transition (**c**) $c-c$ transition (**d**) Curvature value of corresponding curves.

273

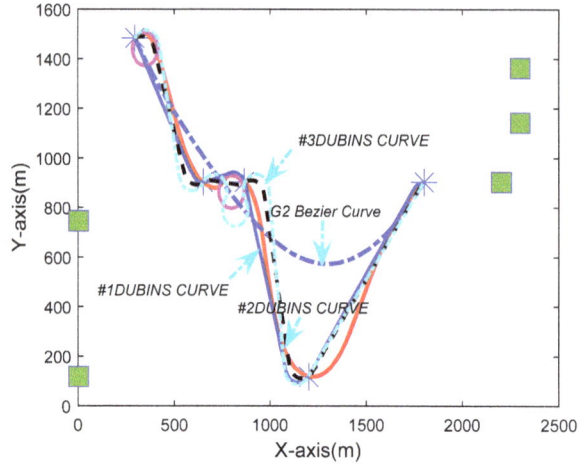

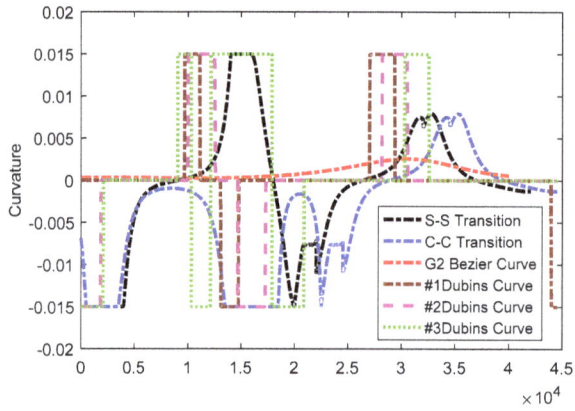

Figure 11. Performance Comparison of different path planning results: (**a**) Curve shape of G^2 Bezier, Dubins, $s - s$ and $c - c$ curves; (**b**) Comparisons of corresponding curvature value.

6.2. Path Smoothing Algorithm Comparison

To evaluate the effectiveness of the proposed path planning method, we compared the performance of the three different curves: G^2 Catmull-Rom, Dubins, and G^2 Bezier. Figure 11a presents the predefined waypoints and paths generated by the three kinds of methods. Three Dubins curves with different steering angles are given in Table 2, while the path lengths of three algorithms are listed in Table 3. The shortest distances from the designed path to all of the control points are also given in Table 3 to compare the paths produced by the three methods. The path curvatures are shown in Figure 11b, where it can be seen that only the proposed method generated a path whose curvature was both continuous and bounded.

Based on the results given in Table 3 and Figure 11b, we can see that the G^2 Catmull-Rom method can generate a continuous curvature, locally-adjustable path. Although the curvature of the G^2 Bezier curve was continuous and under the specified limit, the path merely approximated the control polygon, meaning that it would have missed some control points. Additionally, the Dubins curve paths travel across all of the waypoints, but have discontinuous curvature values at the junction points. We employed quintic Bezier curves to connect all of the waypoints and developed a minimum curvature circle transition method to solve the over-curvature problem. Thus, both the curvature continuity constraint and the curvature upper bound were satisfied. Table 3 does, however, indicate that the designed path was longer than the G^2 Bezier curve path. This is mainly because approximating splines were adopted between sequential waypoints to achieve an interpolating curve as a whole result. Meanwhile, due to the tangent direction at each waypoint, a *c-c* transition needs circular arcs as sub-paths to adapt the flight direction, which leads to even longer paths than using *s-s* transitions.

Furthermore, we can see from Figures 9b and 11b that the curvature extremes varied with the predefined maximum flying speed. This is because the curvature is inversely proportional to a UAV's velocity with fixed lateral acceleration.

Table 2. Steering angles of the Dubins curves.

	$\varphi1$	$\varphi2$	$\varphi3$	$\varphi4$	$\varphi5$
#1 Dubins	−1.027	0	−1.17	0.915	0
#2 Dubins	0.5236	0.5236	0.5236	0.523	0
#3 Dubins	1.0472	1.0472	1.0472	1.0472	1.0472

Table 3. Comparison of the three types of methods.

	G^2 Bezier	#1 Dubins	#2 Dubins	#3 Dubins	$s-s$	$c-c$
$\lvert\kappa_{max}\rvert$	0.0025	0.015	0.015	0.015	0.015	0.015
Path Length	2012.9	2859.1	2933.5	3509.1	3483.5	2775.7
Point 1	0	0	0	0	0	0
Point 2	73	0	0	0	0	0
Point 3	85	0	0	0	0	0
Point 4	459	0	0	0	0	0
Point 5	0	0	0	0	0	0

Further, we have also tested our methodology using fixed-wing UAV models in STK, which is a physics-based software package from Analytical Graphics, Inc (220 Valley Creek Blvd. Exton, PA USA). The turn acceleration (an inherent parameter of the UAV) is 29.42 m/s^2, and the cruise velocity is fixed as 160 km/h. To simulate the real flying conditions, we chose an area from Google Earth, and for the verification of the algorithm, the elevation map of the same region obtained from GlobalMapper was used. The experiments were carried out by using a topology map as shown in Figure 12, the longitude of which ranges from 121°50′5.99″ W–122°5′52.79″ W, and the latitude ranges from 46°42′3.6″ N–46°48′32.4″ N.

In order to observe the effect of the generated path, the path was discretized into points that were embedded into the UAV control system. Figure 13 shows the path generated by the proposed algorithm and the realistic trajectory. The result comparisons are demonstrated in Figure 14. The blue markers represent the path generated by our method, while the red ones denote the realistic trajectory. From the comparisons, we can see that the motion trajectory of the UAV coincides with the trajectory of planning. A remarkable detail was shown circled in maroon. It can be seen that the pre-designed path had the shape of a bump, but the UAV actually flies along a circle. This is because an s-s-shaped transition was used in our method, while the UAV followed a c-c-shaped transition. In addition, we can also see the phenomenon of path deviation in some piecewise paths. However, these path biases were mainly caused by two reasons: resolution restriction on Google Earth and GPS positioning error. Google Earth provided DEM of an area with 0.5–1-m resolution, while the positioning error of commercial GPS can be 5–20 m. This consequently resulted in path errors under those allowed.

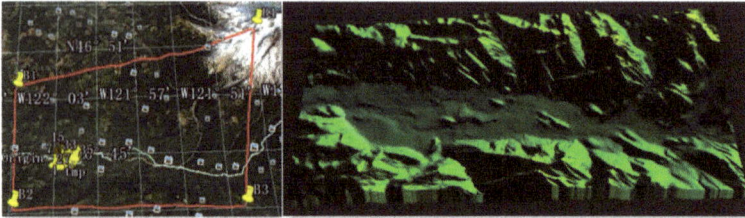

Figure 12. Areas in the experiments.

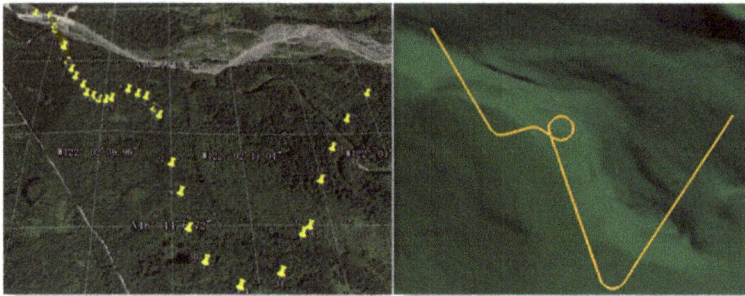

Figure 13. Pre-designed path and motion trajectory.

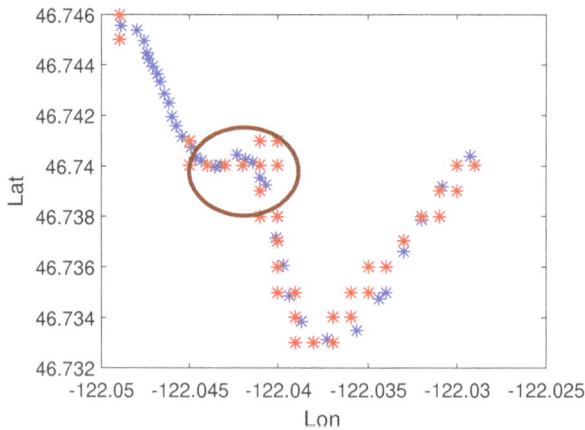

Figure 14. Comparisons between the designed and actual paths: the blue markers are designed path and the red ones represent the actual path.

6.3. Discussion

To solve the UAV path planning and optimization problem, the following requirements must be satisfied: path feasibility and path accuracy. Paths generated by the proposed method and by using other curves, including Dubins and G^2 Bezier curves, were compared in terms of path feasibility and accuracy in Section 6. Figure 11b demonstrates that the curvatures of the paths designed by our method were continuous and bounded, indicating their superior path feasibility. From Figure 11a, we can see that the designed path passes through all of the predefined waypoints, demonstrating our method's superiority in path accuracy.

7. Conclusions

In this paper, a continuous-curvature path planning algorithm was proposed, which allowed the generated path to pass through all of the sequential waypoints while satisfying the curvature upper bound constraint. A curvature continuity condition has been derived for parametric Catmull-Rom curves. Based on this condition, an optimal interpolation algorithm and a key-point shift algorithm were developed to achieve continuous curvature. Furthermore, a minimum curvature circle transition method was designed to replan the over-curved sub-paths using circular arcs and Bezier curves. In particular, a transition selection mechanism was established in accordance with the distribution of waypoint. Numerical results demonstrated that the algorithm was capable of generating a continuous path meeting the curvature upper bound constraint.

This work is just the beginning of the implementation of this approach in the navigation of UAVs during autonomous missions for ground observation. In future works, we will develop a path-following algorithm incorporate in our problems, and we will analyze the path under external disturbances (e.g., wind field). We are interested in extending our results to 3D environments.

Acknowledgments: This work was supported by the National Natural Science Foundation of China (61571334) and the State 863 project of China (2014AA09A512).

Author Contributions: The work presented here was carried out in collaboration among all authors. All authors have contributed to, seen and approved the manuscript.

Conflicts of Interest: The authors declare no conflict of interest.

References

1. Beard, R.W.; McLain, T.W.; Goodrich, M.A.; Anderson, E.P. Coordinated target assignment and intercept for unmanned air vehicles. *IEEE Trans. Robot. Autom.* **2002**, *18*, 911–912.
2. Oh, J.S.; Choi, Y.H.; Park, J.B.; Zheng, Y.F. Complete coverage navigation of cleaning robots using triangular-cell-based map. *IEEE Trans. Ind. Electron.* **2004**, *51*, 718–726.
3. Park, S.; Hashimoto, S. Autonomous mobile robot navigation using passive RFID in indoor environment. *IEEE Trans. Ind. Electron.* **2009**, *56*, 2366–2373.
4. Samaniego, R.; Lopez, J.; Vazquez, F. Path Planning for Non-Circular, Non-Holonomic Robots in Highly Cluttered Environments. *Sensors* **2017**, *17*, 1876.
5. Hwang, K.-S.; Ju, M.-Y. A propagating interface model strategy for global trajectory planning among moving obstacles. *IEEE Trans. Ind. Electron.* **2002**, *49*, 1313–1322.
6. Qu, Y.H.; Pan, Q. Flight path planning of uav based on heuristically search and genetic algorithms. In Proceedings of the IECON 2005, 31st Annual Conference of IEEE Industrial Electronics Society, Raleigh, NC, USA, 6–10 November 2005; pp. 2750–2757.
7. De Filippis, L.; Guglieri, G. Path planning strategies for uavs in 3d environments. *J. Intell. Robot. Syst.* **2012**, *65*, 247–264.
8. Pettersson, P.O.; Doherty, P. Probabilistic roadmap based path planning for an autonomous unmanned helicopter. *J. Intell. Fuzzy Syst.* **2006**, *17*, 395–405.
9. Tang, Y.; Gao, H.; Kurths, J.; Fang, J. Evolutionary pinning control and its application in UAV coordination. *IEEE Trans. Ind. Inform.* **2009**, *8*, 828–838.
10. Tsai, C.C.; Huang, H.C.; Chan, C.-K. Parallel elite genetic algorithm and its application to global path planning for autonomous robot navigation. *IEEE Trans. Ind. Electron.* **2011**, *58*, 4813–4821.
11. Wai, R.-J.; Liu, C.-M. Design of dynamic petri recurrent fuzzy neural network and its application to path-tracking control of nonholonomic mobile robot. *IEEE Trans. Ind. Electron.* **2009**, *56*, 2667–2683.
12. Shau, S.J.; Yu, H.L. Integrated Flight Path Planning System and Flight Control System for Unmanned Helicopters. *Sensors* **2011**, *11*, 7502–7529.
13. Richards, A.; Schouwenaars, T.; How, J.P.; Feron, E. Spacecraft trajectory planning with avoidance constraints using mixed-integer linear programming. *J. Guid. Control Dyn.* **2002**, *56*, 755–764.
14. Kuwata, Y.; How, J.P. Cooperative distributed robust trajectory optimization using receding horizon milp. *IEEE Trans. Control Syst. Technol.* **2011**, *19*, 423–431.

15. Magid, E.; Keren, D.; Rivlin, E.; Yavneh, I. Spline-based robot navigation. In Proceedings of the 2006 IEEE/RSJ International Conference on Intelligent Robots and Systems, Beijing, China, 9–15 October 2006; Volume 18, pp. 2296–2301.

16. Dubins, L.E. On curves of minimal length with a constraint on average curvature, and with prescribed initial and terminal positions and tangents. *Am. J. Math.* **1957**, *79*, 497–516.

17. Wang, Y.; Wang, S.; Tan, M.; Zhou, C.; Wei, Q. Real-time dynamic dubins-helix method for 3-D trajectory smoothing. *IEEE Trans. Control Syst. Technol.* **2015**, *23*, 730–736.

18. Sahingoz, O.K. Generation of bezier curve-based flyable trajectories for multi-uav systems with parallel genetic algorithm. *J. Intell. Robot. Syst.* **2014**, *74*, 499–511.

19. Rathinam, S.; Sengupta, R.; Darbha, S. A resource allocation algorithm for multivehicle systems with nonholonomic constraints. *IEEE Trans. Autom. Sci. Eng.* **2007**, *4*, 98–104.

20. Wang, Y.; Wang, S.; Tan, M. Path generation of autonomous approach to a moving ship for unmanned vehicles. *IEEE Trans. Ind. Electron.* **2015**, *62*, 5619–5629.

21. Fraichard, T.; Scheuer, A. From reeds and shepp's to continuous curvature paths. *IEEE Trans. Robot.* **2004**, *20*, 1025–1035.

22. Kito, T.; Ota, J.; Katsuki, R.; Mizuta, T.; Arai, T.; Ueyama, T.; Nishiyama, T. Smooth path planning by using visibility graph-like method. In Proceedings of the 2003 IEEE International Conference on Robotics and Automation (ICRA'03), Taipei, Taiwan, 14–19 September 2003; Volume 3, pp. 3770–3775.

23. Neto, A.A.; Campos, M.F. A path planning algorithm for uavs with limited climb angle. In Proceedings of the 2009 IEEE/RSJ International Conference on Intelligent Robots and Systems (IROS 2009), St. Louis, MO, USA, 10–15 October 2009; pp. 3894–3899.

24. Yang, K.; Sukkarieh, S. An analytical continuous-curvature path smoothing algorithm. *IEEE Trans. Robot.* **2010**, *26*, 561–568.

25. Ghilardelli, F.; Lini, G.; Piazzi, A. Path generation using η^4 splines for a truck and trailer vehicle. *IEEE Trans. Autom. Sci. Eng.* **2014**, *11*, 187–203.

26. Piazzi, A.; Visioli, A. Global minimum-jerk trajectory planning of robot manipulators. *IEEE Trans. Ind. Electron.* **2000**, *47*, 140–149.

27. Cai, W.; Zhang, M.; Zheng, Y.R. Task Assignment and Path Planning for Multiple Autonomous Underwater Vehicles Using 3D Dubins Curves. *Sensors* **2017**, *17*, 1607.

28. Hwang, J.H.; Arkin, R.C.; Kwon, D.S. Mobile robots at your fingertip: Bezier curve on-line trajectory generation for supervisory control. In Proceedings of the 2003 IEEE/RSJ International Conference on Intelligent Robots and Systems, Las Vegas, NV, USA, 27–31 October 2003; Volume 2, pp. 1444–1449.

29. Hota, S.; Ghose, D. Optimal path planning for an aerial vehicle in 3D space. In Proceedings of the 2010 49th IEEE Conference on Decision and Control (CDC), Atlanta, GA, USA, 15–17 December 2010; pp. 4902–4907.

30. Shanmugavel, M.; Tsourdos, A.; Zbikowski, R.; White, B. 3D path planning for multiple uavs using pythagorean hodograph curves. In Proceedings of the AIAA Guidance, Navigation and Control Conference and Exhibit, Hilton Head Island, SC, USA, 20–23 August 2007.

31. Machmudah, A.; Parman, S.; Zainuddin, A. UAV bezier curve maneuver planning using genetic algorithm. In Proceedings of the 12th Annual Conference Companion on Genetic and Evolutionary Computation, Portland, OR, USA, 7–11 July 2010; pp. 2019–2022.

32. Connors, J.; Elkaim, G. Analysis of a spline based, obstacle avoiding path planning algorithm. In Proceedings of the VTC2007-Spring, IEEE 65th Vehicular Technology Conference, Dublin, Ireland, 22–25 April 2007; pp. 2565–2569.

33. Walton, D.; Meek, D. Curvature extrema of planar parametric polynomial cubic curves. *J. Comput. Appl. Math.* **2001**, *134*, 69–83.

34. Khatib, M.; Jaouni, H.; Chatila, R.; Laumond, J.P. Dynamic path modification for car-like nonholonomic mobile robots. In Proceedings of the 1997 IEEE International Conference on Robotics and Automation, Albuquerque, NM, USA, 25–25 April 1997; Volume 4, pp. 2920–2925.

35. Habib, Z.; Sakai, M. G^2 cubic transition between two circles with shape control. *J. Comput. Appl. Math.* **2009**, *223*, 134–144.

36. Barsky, B.A.; DeRose, T.D. Geometric continuity of parametric curves: Three equivalent characterizations. *IEEE Comput. Graph. Appl.* **1989**, *9*, 60–69.

37. DeRose, A.D. Geometric Continuity: A Parametrization Independent Measure of Continuity for Computer Aided Geometric Design. Ph.D. Thesis, EECS Department, University of California, Berkeley, CA, USA, 1985.
38. DeRose, T.D.; Barsky, B.A. Geometric continuity, shape parameters, and geometric constructions for catmull-rom splines. *ACM Trans. Graph. (TOG)* **1988**, *7*, 1–44.

sensors

MDPI

Article

Saliency Detection and Deep Learning-Based Wildfire Identification in UAV Imagery

Yi Zhao [1,*], Jiale Ma [2], Xiaohui Li [1] and Jie Zhang [3]

[1]   AInML Lab, School of Electronics and Control Engineering, Chang'an University, Xi'an 710064, China; xiaohui.li@chd.edu.cn
[2]   School of Automation, Southeast University, Nanjing 210009, China; jiale.ma@yahoo.com
[3]   Shengyao Intelligence Technology Co. Ltd., Shanghai 201112, China; zhangj@hitrobotgroup.com
*   Correspondence: z1@chd.edu.cn; Tel.: +86-158-2930-2501 or +86-029-8233-4543

Received: 8 December 2017; Accepted: 19 February 2018; Published: 27 February 2018

Abstract: An unmanned aerial vehicle (UAV) equipped with global positioning systems (GPS) can provide direct georeferenced imagery, mapping an area with high resolution. So far, the major difficulty in wildfire image classification is the lack of unified identification marks, the fire features of color, shape, texture (smoke, flame, or both) and background can vary significantly from one scene to another. Deep learning (e.g., DCNN for Deep Convolutional Neural Network) is very effective in high-level feature learning, however, a substantial amount of training images dataset is obligatory in optimizing its weights value and coefficients. In this work, we proposed a new saliency detection algorithm for fast location and segmentation of core fire area in aerial images. As the proposed method can effectively avoid feature loss caused by direct resizing; it is used in data augmentation and formation of a standard fire image dataset 'UAV_Fire'. A 15-layered self-learning DCNN architecture named 'Fire_Net' is then presented as a self-learning fire feature exactor and classifier. We evaluated different architectures and several key parameters (drop out ratio, batch size, etc.) of the DCNN model regarding its validation accuracy. The proposed architecture outperformed previous methods by achieving an overall accuracy of 98%. Furthermore, 'Fire_Net' guaranteed an average processing speed of 41.5 ms per image for real-time wildfire inspection. To demonstrate its practical utility, Fire_Net is tested on 40 sampled images in wildfire news reports and all of them have been accurately identified.

Keywords: UAV; wildfire; deep learning; saliency detection

1. Introduction

Wildfire is a natural disaster, causing irreparable damage to local ecosystem. Sudden and uncontrollable wildfires can be a real threat to residents' lives. Statistics from National Interagency Fire Center (NIFC) in the USA show that the burned area doubled from 1990 to 2015 in the USA. Recent wildfires in northern California (reported by CNN) have already resulted in more than 40 deaths and 50 missing. More than 200,000 local residents have been evacuated under emergency. The wildfires occur 220,000 times per year globally, the annual burned area is over 6 million hectares. Accurate and early detection of wildfire is therefore of great importance [1].

Traditional wildfire detection is mostly based on human observation from watchtowers. Subject to the spatiotemporal limitation, it is inefficient. Unmanned aerial vehicles (UAVs or drones) have been increasingly used as surveillance tools [2]. UAVs equipped with global positioning systems (GPS) and ultrahigh resolution digital camera can provide HD aerial images with precise location information. Our previous work demonstrated that well organized UAV swarm is rapid, efficient and low-cost in conducting complicated agricultural applications [3]. Recent works also suggest that UAVs are quickly emerging implement in a variety of monitoring tasks [4–6]. The authors of this paper (Jie Zhang, Shanghai Shengyao Intelligence Technology Co. Ltd., Shanghai, China) developed a specialized UAV system (Figure 1) for forest and wildfire monitoring. The technique parameters of the forest monitoring UAV are presented in Table 1 below.

Figure 1. Forest Monitoring UAV PHECDA II developed by author of this paper.

Table 1. Technical parameters of the forest monitoring UAV PHECDA II.

Technical Parameters	Value
Wing span	2.7 m
Takeoff weight	18 to 25 kg (full load)
Maximum flight altitude	4000 m
Average cruising speed	90 km/h
Max flight speed	120 km/h
Endurance	1 h (pure battery powered)/6 h (gas–electric hybrid powered)
Digital camera equiped	FOV 94″ 20 mm F2.8 4K-12 million pixels
Positioning system	GPS/GLONASS dual-module

The main issue of this UAV monitoring system is that the whole system is operated by observers in surveillance center. Limited by the wireless transmission distance or the blind zone of GPRS/GSM network, real-time video transmission from the UAV can hardly be guaranteed. The surveyors often wait for the return of the UAV and examine its stored video. It is time consuming, labor intensive, and inefficient. Thus, an autonomous fire localization and identification system for this UAV is right now the top demand.

Early reported works on fire detection are generally based on sensor techniques. For example, heat and smoke detectors are mostly employed. These methods work properly in indoor environments. However, when applied to large outdoor areas, environmental factors can significantly impact their performance. Later, sensor network-based methods are reported effective in prediction rather than detection of wildfire [7], this is mainly because these methods depend on restricted data of relative temperature rise or wind speed in calculating the probability and intensity of fire. However, due to the limited measuring distance of each sensor, fine grained wildfire mapping over a large geographical area demands very dense deployment of sensors. It is difficult to implement in practice. Advanced sensor technology has been applied in fire detection: infrared sensors are used to capture the thermal radiation of fire, light detection and ranging system (LIDAR) [8] is employed to detect smoke by examining the backscattered laser light. Yet, these optical systems are also found sensitive to the varying atmospheric and environmental conditions: clouds, light reflections, and dust particles may lead to false alarms.

Satellite imagery is a common method for wildfire detection [9], but the long scan period and low flexibility make it difficult for early fire detection. Infrared (IR) thermographic cameras are used to generate thermal image of an area [10]. It can obtain reliable heat distribution data for fire detection. Yet most aerial IR thermographic imaging systems work on the wave band from 0.75 to 100 μm. They detect much less environmental information on this band. This information can also be very important, especially when flammable and combustible materials are presented. Also, limited by Nyquist Theorem, the recorded thermal image has lower spatial resolution than visible spectrum cameras. Moreover, thermographic systems are quite expensive with high maintenance costs.

Admittedly, visible spectrum CCD/CMOS imaging sensors are less sensitive to heat flux. Still they are capable of recording high resolution image of fire, smoke along with the surroundings. They are much cheaper than IR cameras and other type of sophisticated sensor system. Environmental changes have less influence on their performance. Additionally, the visible video-image-based methods combine well with existing monitoring systems: most UAVs and watchtowers have the visible spectrum camera equipped. These are the main reasons that many works using visible spectrum video and image to provide solid fire detection results [11–13].

Previous image processing-based fire detection research mostly relies on color and texture models. Chen proposed empirical models with experimental thresholds in detect flame pixels. Vipin proposed a fire pixel classification method using rule-based models in the RGB and YCbCr color space [14]. Angayarkkani and Radhakrishnan developed fuzzy rules in the YCbCr color space for fire image segmentation and detections [15]. Yuan developed a set of forest fire tracking algorithms including median filtering, color space conversion, and Otsu threshold [16]. These works are based on image processing techniques with handcrafted feature extractors, the results highly depend on the accuracy of the manually selected parameters.

Statistical and machine learning methods have been reported. Gaussian mixture model (GMM) is used in flame detection [17]. However, the use of empirical value of mixture number may not lead to the best results. SVM classifier is applied in [18]. It is noted that when commonly used feature descriptors like scale-invariant feature transform (SIFT) histogram of oriented gradients (HOG) are employed with these classifiers, the false alarm rate is not low enough.

Indeed, previous machine vision-based fire detection research are divided into two categories: flame detection and smoke detection [12]. Subject to the methods employed, these works are reliable on specific scene. Many are tested on short-range-shooting video or images with low infestation background. The volume of existing test benchmark is very small [19]. Up to this point, the fundamental difficulty has not yet been well addressed: how to build an adaptive classifier to identify complex fire features of different color, form, and texture from a varying cluttered background. Therefore, the objective of this paper is to provide an optimal wildfire feature learning model.

Deep convolutional neural networks (DCNN) have been reported to achieve state-of-art performance on a wide range of image recognition tasks [13], its architecture and learning scheme lead to an efficient extractor of complex, upper-level features which are highly robust to input transformations [20]. However, from the survey on UAV-based wildfire applications, deep learning technique has rarely been used in wildfire detections [21]. Until recently, deep learning has been reported in fire recognition in conference publications. Kim proposed an eight-layered CNN model in fire image classification [22]. The training image dataset are manually cropped and resized. Furthermore, the effect of key parameters and coefficients like dropout ratio, batch size, and learning rate have not been discussed.

Problem Description

In this work, we try to present a complete fire localization and identification solution. To achieve this, there are some practical and technical problems must be solved:

1. Localization of fire area: First, unlike most contour prominent objects (humans, animals, cars, planes, buildings, etc.) wildfires have irregular form with very vague contours, their shape can vary dramatically with time. Yet, most machine vision-based detection methods are applied on objects with clear contour. Therefore, instead of find the whole fire boundary, localization of the core burring area can be more practical. Also, it is of great importance for further firefighting operation. As discussed in the previous section of this paper, most fire detection methods are tested on shoot-range shooting video or images. The effectiveness of these methods in localization of core fire area in high altitude aerial photographs remains uncertain.
2. Fixed training image size: Most deep learning models require input training image of fixed size (e.g., 128 × 128, 224 × 224). Yet, the most web-based image sources are of different sizes. In prevalent works, the reformatting of these images requires direct manual manipulation. Handcraft operations are slow, expensive, and inefficient. Worse still, wrapping, cropping, and direct resizing (ex., Gaussian pyramid and Laplacian pyramid method) may cause detail and feature loss or image blurring. In any case, this could lead to a poor DCNN training results [23]. Although method like Spatial pyramid pooling (SPP) [23] is used for DCNN to deal with varied size input images. Some disadvantages are notable: first, SPP is a high computational cost operation, according to the study in [24], the processing speed of SPP-net is considered unsuitable for real-time recognition. Secondly, the convolution layers preceding SPP cannot be updated during its fine tuning. This can be a major limit for a deeper network. Furthermore, high resolution aerial photography is often taken with a wide field of vision, these images usually contain much more background clutter information, which leads to high data redundancy, long calculating times, and poorer classification performance.
3. Limited amount of aerial view wildfire images: Generally, a DCNN requires a substantial amount of data to fully optimize the network's parameters and weight values during its training procedure [25]. Insufficient training data will lead to overfitting and poor classification performance. The amount of wildfire aerial images available online is still very limited, thus a data augmentation necessary.

To address these problems, we proposed a new saliency detection-based segmentation method [26]. The method is tested effective in localization and extraction of the core fire area. In this way, most of the fire features can be conserved without severe feature loss. In addition, the method can be used to crop multiple fire regions into different fire images so that data augmentation can be achieved.

The rest of this paper is organized as follows. In Section 2, we present the new saliency detection method in localization and segmentation of core fire regions. In Section 3, the DCNN architecture is introduced in detail. The results and discussions are presented in Sections 4 and 5 concludes the paper.

2. Saliency Segmentation and Image Dataset Formation

In this work, the original wildfire images were obtained from image searching engine of Google, Baidu, and the database of AInML lab in Chang'An University, The data source contains over 1500 images. These images are of varied size from 300 × 200 to 4000 × 3000 pixels. As our proposed DCNN architecture is implemented on Caffe [27]. The input images should be normalized into a standard LMDB database file. Thus, the image must be formatted to fixed size (e.g., 227 × 227 or 128 × 128). Therefore, a technical issue is raised: what measures should be taken to extract maximum fire features into the resized image dataset?

As discussed in introduction, most prevalent methods use direct resizing or handcraft warping and cropping. These methods have some known disadvantages like detail loss and geometric distortion [23]. UAV imagery is usually taken with a wide-angle digital camera, the captured image contains highly redundant clutter information. When direct resizing is applied, some of the important fire features may be submerged into the background. For example, In Figure 2, the original image contains both flame and smoke features; however, after the resizing, the flame features have been submerged into the smoke.

original image directly resize into 128*128 sized image

Figure 2. Direct resizing leads to detail loss.

In this work, we developed a new algorithm of wildfire localization and segmentation algorithm by combining saliency detection and logistic regression classifier (Figure 3). It can quickly locate the core fire region out of a complex background. An example of the method is illustrated in the Figure 4.

The algorithm can be divided into two phases: region proposal and region selection. In the first phase, saliency detection method is used to extract the region of interest (ROI, possible fire region in our case), we calculate the color and texture features of the ROIs. In the second phase, two logistic regression classifiers are used to determine whether the feature vector of ROIs belong to flame or smoke, if positive, we segment these regions.

Algorithm: Saliency detection-based Fire region localization and segmentation
Step 1. Region of Interest (ROI) proposal-
Apply the Bayes-based saliency detection to original image *I*
The banalization of the saliency map into image mask I_m
Mask the original image to extract the possible fire regions R_1 to R_n
Step 2. Region selection using machine learning
Calculate the color moment and the Image energy and entropy of each ROI (vector X_1 to X_n)
Using two independently trained Logistic regression classifiers (*Lr_flame(X) and Lr_smoke(X)*) to determine whether the ROI's feature vector corresponds to either flame or smoke, if positive, go to (6), else, cancel the ROI.
Segment the fire region with MBR [30] and fixed size bounding and resizing

Figure 3. Saliency detection and logistic regression-based segmentation algorithm.

Figure 4. Example of the proposed fire localization and segmentation method.

Previous works suggests that color and texture are two effective and efficient features in image-based fire detection, therefore, we employ the value of color moment [28] as color feature descriptor while the image's angular second moment (ASM) energy and entropy [29] as texture descriptors. The two groups of values are combined into the feature vector. The use of this feature vector can largely reduce the computational complexity.

The implementation of the algorithm are as follows.

2.1. Saliency Detection

Inspired by the primate's vision system, saliency detection is highly effective in focusing on the core object from a complicated scene [31]. Early works on saliency detection have some common disadvantages like low resolution and blurred region boundary. Rahtu demonstrated in his work that the combination of saliency measure and conditional random field (CRF) model is valid in segmenting both still image and scenes from video sequences [26].

Rahtu's Bayesian method uses a slide window to calculate the conditional probability of each pixel as its estimated saliency value, if the object's feature strongly differs from the background, it obtained a high saliency value. The mechanism of the method is presented as follows: a slide window W is composed of both inner kernel K and border B. A pixel x in W can be either salient H_1 *(in K)* or non-salient H_0 *(in B)*. Accordingly, the probability for the each cases is $P(H_1)$ and $P(H_0)$, suppose the kernel K contains the salient object, its feature distribution is therefore differs from the distribution of border B. Based on this assumption, the conditional feature distribution $p(F(x)|H_1)$ and $p(F(x)|H_0)$ can be estimated with the feature value $F(x)$ computed at x. By Bayes' definition of $P(A|B)$, we have the expression of $P(H_1|F(x))$ written as

$$P(H_1|F(x)) = \frac{p(F(x)|H_1)P(H_1)}{p(F(x)|H_0)P(H_0) + p(F(x)|H_1)P(H_1)} \qquad (1)$$

Based on Equation (1), for each pixel in K, we use this estimated probability as its saliency level S(x)

$$S(x) = P(H_1|F(x))$$ (2)

This value shows the contrast level of the feature values in K and B. If the contrast between K and B is stronger, the saliency level of pixels in K is higher. In order to improve the robustness of the method, a Gaussian kernel is applied to smooth the histograms.

$$\hat{p}(F(x)|H_0) = N(g(F) * h_B(F))$$ (3)

$$\hat{p}(F(x)|H_1) = N(g(F) * h_K(F))$$ (4)

where N is the normalization operation, $h_K(F)$ and $h_B(F)$ are the histogram respectively in K and B, $g(F)$ is the Gaussian smoothing function. The authors of this paper applied this method in segmenting pests from its background, and we enhanced the segmentation details by adding edge-aware weightings [32].

2.2. Color Moment, ASM Energy, and Entropy

Color moment provides a measurement for color feature, it uses three central moments of an image's color distribution. They are mean, standard deviation, and skewness. For example, an RGB image has three color channels, thus, three moments multiplied by three color channels makes nine moments in total. We define the ith color channel at the jth image pixel as P_{ij}. The three color moments can be defined as

$$E_i = \sum_N^{j=1} \frac{1}{N} p_{ij}$$ (5)

$$\sigma_i = \sqrt{\frac{1}{N} \sum_N^{j=1} (p_{ij} - E_i)^2}$$ (6)

$$s_i = \sqrt{\frac{1}{N} \sum_N^{j=1} (p_{ij} - E_i)^3}$$ (7)

where E is the mean value, it is the average color value, σ stands for the standard deviation, it is calculated by the square root of color-distribution's variance, S is the skewness, and it represents the asymmetry in the distribution.

The ASM energy and entropy are used to describe the texture feature of an image. They are expressed mathematically below

$$ASM_{Energy} = \sum_{i,j} (P(i,j \mid d,\theta))^2$$ (8)

$$ASM_{Entropy} = \sum_{i,j} (P(i,j \mid d, \theta) \log(P(i,j \mid d, \theta)))$$ (9)

where P is the element of the ASM matrix, it stands for the occurrence probability that a gray value i and j of a pair pixels separated by distance d with an angle θ. The ASM energy (Equation (8)) is the square sum of all element values in ASM matrix. It reflects the uniformity of the gray level distribution and the roughness of the texture. If all the values of the matrix are nearly equal, the energy value is small; otherwise, the energy value is high. Similarly, ASM entropy is a measurement of non-uniformity or complexity of the texture in an image. A more dispersed distribution of ASM elements value leads to high entropy.

2.3. Logistic Regression Classifier

Logistic regression is one of the most employed binary classification method. As introduced in the previous section, the input of the classifier is the extracted ROI's feature vector, expressed as $X = [x_0, x_1, x_2, x_3 \cdots\cdots x_{10}]$. It is a vector of eleven dimensions, in which, x_0 to x_8 are the color moment, x_9 is the ASM energy and x_{10} the ASM entropy. Accordingly, the trained weight's vector should also be 11 dimensions as $W = [w_0, w_1, w_2, w_3 ...... w_{10}]$. We chose sigmoid as classifier function, the form of the classifier is

$$h_W(X) = \sigma(W^T X) = \frac{1}{1 + e^{-(W^T X)}} \tag{10}$$

Given an input feature vector X and the trained weight vector of W, the output value of this equation will situate between 0 and 1, if this value is above 0.5, we consider the feature vector belongs to positive class, otherwise, if the value is below 0.5, it is then negative. As the classification results are directly related to the weight vector, the aim of logistic regression is to find the optimal W in Equation (10). To achieve this, a gradient rise method is applied. First, based on Equation (10), the correspondent probability can be written as

$$P(y|X; W) = (h_W(X))^y (1 - h_W(X))^{1-y} \tag{11}$$

Followed by a maximum likelihood estimation on W

$$L(W) = \prod_{i=1}^{m} P\left(y^{(i)} \middle| X^{(i)}; W\right) = \prod_{i=1}^{m} (h_W(X^{(i)}))^{y^{(i)}} (1 - h_W(X^{(i)}))^{1-y(i)} \tag{12}$$

where i is the sample number of training feature vectors. For the convenience of calculation, Equation (12) is usually used in Logarithmic form. To gain the maximum value of L, we use the gradient ascent algorithm to update the value of W

$$W_j = W_j + \alpha \, \nabla(\log L(w)) + \lambda X_j^{(i)} = W_j + \alpha \sum_{i=1}^{m} (y^{(i)} - h_W(X^{(i)})) X_j^{(i)} + \lambda X_j^{(i)} \tag{13}$$

where α is the learning rate, $\nabla(\log L(w))$ is the gradient of $\log L(w)$, and it equals to $\sum_{i=1}^{m} (y^{(i)} - h_W(X^{(i)})) X_j^{(i)}$. $\lambda X_j^{(i)}$ is the disturbance term we add to avoid the overfitting problem.

In this work, two regression classifiers for flame and smoke feature are trained with two sets of 150 sampled flame and smoke image blocks, respectively. Stochastic gradient ascent method is applied during the training, thus the weights get updated in every training epoch. The training rate α is 0.001, and the coefficient of disturbance is set to 0.00001.

The reason for using two binary classifiers instead of one multiple-labeled classifier (e.g., Softmax classifier) is that we consider the feature of flame and smoke are not mutually independent. Notably, the gray level texture features of flame and smoke are often mixed. In this case, two independent binary logistic regression classifiers can be more effective.

As shown in Figure 4, the fire region shows the highest saliency value. After extraction by the saliency map, the core fire region's image feature vector is calculated. The vector is then examined by the trained logistic regression classifier. With the positive results, minimum bounding rectangle (MBR) [30] method is used to locate the core fire region's geometric center and crop the fire region into a standard sized image. This method keeps the full feature of fire (both flame and smoke) without distortion. This method has been applied to form the standard image dataset 'UAV_Fire'. Furthermore, as demonstrated in Figure 4, this method can locate the core flame zone, which is considered highly practical in UAV wildfire inspection. More results can be found in the Section 4 of this paper.

3. The DCNN Model

In this work, we try to provide a more complete assessment of DCNN's self-learning capability by comparing different architectures and evaluating the effects of the key parameters. DCNN is used as a classifier to determine whether wildfire happens in the UAV imagery. We developed a 15-layered DCNN model called 'Fire_Net'. This section provides a precise and detailed description of the proposed DCNN model.

3.1. Overall Architecture

The overall architecture is illustrated in Figure 5. The left part is the image saliency segmentation and image cropping stage. The right part is the core DCNN model. Inspired by Alexnet (eight-layered DCNN) [25], we proposed a deeper 15-layered DCNN model 'Fire_Net'. Its structure is detailed in Table 2. With its self-learning structure, it can extract wildfire features from low to high level. The first 12 layers consist of 8 convolutional layers (Conv1–Conv8) and 4 max pooling (Pool1–Pool4). The convolutional layers extract the fire features from low to high level. The max pooling layers are used to capture deformable parts and reduce the dimension of the convolutional output. The last three fully connected layers (FC1, FC2, and IP3) capture complex co-occurrence statistics. A final classification layer summarize the previous abstracted high level features for the recognition of a given image. This architecture is appropriate for learning complex wildfire features from the training dataset. The complete schematic of the DCNN model can be found in Reference 1.

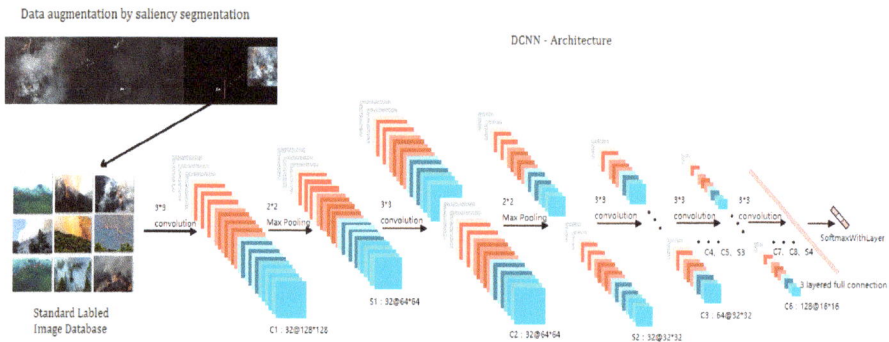

Figure 5. The overall architecture of 'Fire_Net'.

Table 2. Definition of layers in 'Fire_Net'.

Layer	Definition	Feature Maps	Kernel Size	Step
1	Conv1	32	3×3	1
2	Pool1	32	2×2	2
3	Conv2	32	3×3	1
4	Pool2	32	2×2	2
5	Conv3	64	3×3	1
6	Conv4	64	3×3	1
7	Conv5	128	3×3	1
8	Pool3	128	2×2	2
9	Conv6	128	3×3	1
10	Conv7	256	3×3	1
11	Conv8	256	3×3	1
12	Pool4	256	2×2	2
13	FC1(Full-Connection)	Number of neurons: 256		
14	FC2(Full-Connection)	Number of neurons: 256		
15	IP3(Softmax)	Output layer neurons: 2		

3.2. Fire_Net: The DCNN Model in Detail

- Input Layer: the input data are the saliency segmented image database in LMDB format, the standard size is 128 × 128 pixels. In the input layer, the color of the range of three channels of RGB is normalized to 0–1, so the proportion of 0.00390625 (1/256) is applied to the original value. During training, the batch size is 50 images, and the batch size in the validation phase is 10 images.

- Convolution layers: In Fire_Net, eight convolutional layers (Conv1–Conv8) are integrated, convolutional kernel size is defined 3 × 3, and sliding step length is 1. The weight initialization of parameter uses the Xavier way [33] and the bias is initialized with an experienced value. A rectified linear units (ReLU) activation function [34] follows each convolutional layer. The output of the ReLU function is then send to next convolution layer or pooling layer. The main reason of using ReLU instead of Tanh-Sigmoid as activation function is discussed as follows. When the input is very small or very large, the gradient of the sigmoid function can be near zero. This is called the 'saturation' effect. Comparatively, the ReLU function can be much faster in gradient descent-based training because it is a non-saturating function.

- Pooling layer: the objective of pooling layer is to reduce the data size and prevent over fitting. The kernel size of all pooling in this project is 2 × 2, the sliding step length is 2, and the max-pooling method [35] is applied.

- Full connection layer: FC1 and FC2 are the hidden layers, the number of neurons are 256 per layer, the weight initialization used Xavier mode [33]. Each neuron in the hidden layer uses a dropout technique to reduce the dependency between neurons, thus, the network's overfitting can be reduced [36]. The activation function in full connection layer is also ReLU.

- Classification layer: the classification layer contains the output layer of the full connection neural network, a Softmax and cross entropy function. The number of the output layer neurons is two. Following the output layer neuron, a Softmax function [37] is applied, in this work, the Fire_Net model should be able to identify if the wildfire occurs. Thus, it becomes a binary classification problem (fire or non-fire).

$$\text{Softmax} = \frac{\exp(a_i)}{\sum_j \exp(a_j)}, \tag{14}$$

where i corresponds to the number of classes, in this case, i equals to either 0 or 1 since it is a binary classification problem. The output of the Softmax function corresponds to the probability of the output class. The function is a monotonic function, thus, the output increases as the input increases. If the test image shows a dense wildfire feature, the output of the Softmax therefore corresponds to a higher value. The Softmax then adopted a cross-entropy loss regression to choose the outputs with the highest probability as the classification result.

3.3. Network Training

Fire_Net's training is operated on a GPU of NVidia GeForce 840M. It is a low power 64-bit Maxwell architectured graphic processor for mobile applications. Thus, it is more appropriate to be used on a UAV. The optimal configuration of the Fire_Net's training parameters are presented in the following section. As Caffe supports breakpoints observation, we observe the training error after each iteration in the terminal. Also, we save the training state and weight parameters at every 5000 iterations.

During the training, labeled images of size 128 × 128 pixels are given to the DCNN. The errors between actual output value and the desired value (label) are calculated. A stochastic gradient descent (SGD) method [24] is applied to train the DCNN. The SGD method optimizes coefficients and weights by computing derivatives after back-propagating the errors through the previous layers. This learning procedure is iterated until the training error is considered negligible.

4. Results and Discussion

4.1. Saliency Segmentation and Data Augmentation

As introduced in Section 2 of this paper, we used saliency detection-based segmentation method to locate the core fire region and segment them into standard sized images for Fire_Net training. The result is shown in Figure 6 below.

Figure 6. The saliency detection-based image segmentation: (**1**) Original image; (**2**) Saliency detection results; (**3**) ROI selection with logistic regression classifier and segmentation; (**4**) Segmented standard sized image for Fire_Net training. (**a**) Image with both flame and smoke features; (**b**) Image with only smoke feature; (**c**) Image with multiple fire regions.

As shown in Figure 6, the proposed saliency methods can efficiently locate the core fire regions in aerial images, even very tiny ignition zone (see Figure 6(a-3,c-3)) are well located. The segmented image contains a maximum fire feature without severe distortion or feature loss. Furthermore, in both Figure 6a,c, we used the proposed method to segment the original image into independent fire images and they all contain full fire features. This is how the saliency segmentation technique is adopted in augmentation of the original training image dataset. The augmentation of data helps to overcome the overfitting effect due to smaller training samples. More examples of test results are presented in the Figure 7.

The performance of the method is shown in Table 3. We tested the method on a set of selected 450 fire images with both flame and smoke features from the original image database. TP, FN, TN, FP refer to true positive, false negative, true negative, and false positive respectively. TPR is the true positive rate while FPR is false positive rate. It is found that the method achieved high accuracy in locating core fire areas in aerial images.

Finally, the segmentation method is applied on 1105 images with both flame and smoke features from the original image dataset, the rest of the images are not applied with this method. Because either these images are small in size or the flame and smoke features already covered most of the area. In this case, segmentation is not necessary. After data augmentation, the new image dataset contain over 3500 images. The detail of the augmented image database is presented in Table 4 below.

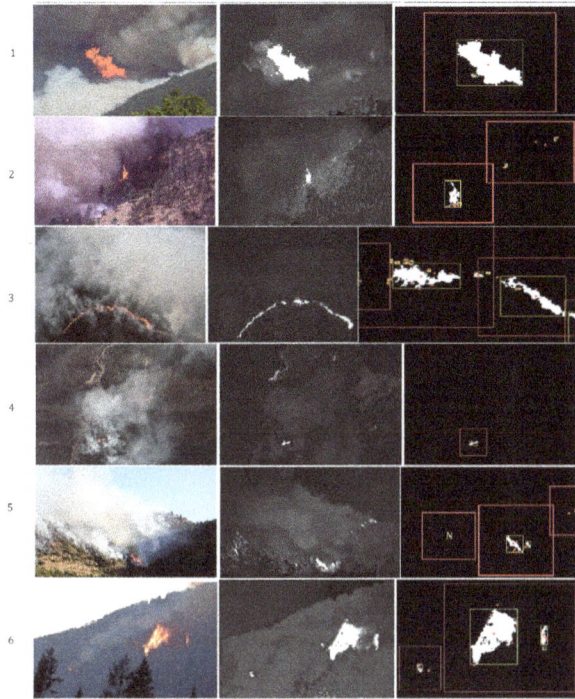

Figure 7. Examples of fire localization and segmentation method on several complex scenes.

Table 3. Performance of the proposed localization and segmentation methods.

Image Type	190 Aerial View Fire Images Total ROIs Detected 384						260 Normal View Images Total ROIs Detected 581					
	TP	FN	FP	TN	TPR	FPR	TP	FN	FP	TN	TPR	FPR
Performance	269	12	5	98	95.7%	4.9%	421	38	9	113	91.7%	7.37%

Table 4. Description of original and augmented image database 'UAV_Fire'.

Image Type	Original Image Database			Augmented Image Database 'UAV_Fire'		
	UAV/Aerial/ Remote-Sensing	Ordinary View	Total	UAV/Aerial/ Remote-Sensing	Ordinary View	Total
Fire	197	435	632	462	1097	1559
No-fire	250	658	908	677	1325	2002
total	447	1093	1540	1139	2422	3561

4.2. Fire_Net Parameter Optimization

The Fire_Net DCNN model includes a set of parameters such as the convolutional kernel size, the number of kernels, learning rate, dropout ratio, batch size, etc. Most of the parameters are initialized with experienced values from previous works [38]. After a full parameter tuning test, we find that some of the experienced value (input image size of 128 × 128, 128 convolution kernels size of 3 × 3, learning rate of 0.0001, convolution stride and pad of 1, max-pooling kernel size of 2, etc.) have indeed achieved optimum results in term of validation and test accuracy. Some of the experienced parameters does not lead to optimal results. Particularly, we find the value of dropout ratio and batch size are more sensitive regarding the model's accuracy. Thus, we present the results of the test on the two parameters in Table 5 below.

Table 5. Different dropout ratio and batch size effects on validation accuracy.

Fire_Net Training Parameters			Different Batch Size Effects		Different Dropout Ratio Effects	
Learning Rate	Momentum	Max-Iteration	Batch Size	Validation Accuracy	Dropout Ratio	Validation Accuracy
0.0001	0.9	20000	16	0.952	0.2	0.95
0.0001	0.9	20000	32	0.958	0.3	0.946
0.0001	0.9	20000	48	0.969	0.4	0.967
0.0001	0.9	20000	50	0.972	0.5	0.98
0.0001	0.9	20000	64	0.97	0.6	0.972
0.0001	0.9	20000	128	–	0.7	0.955

As shown in Table 5, the optimal coefficient combination in Fire_Net is the batch size of 50 and the dropout ration of 0.5.

4.3. Model and Method Comparasion

In order to further test the proposed method, we explored several types of DCNN architectures based on Fire_Net. Furthermore, we used both original image dataset and the augmented image dataset to evaluate the effects of the data augmentation to the model's accuracy. The results are presented in Table 6 and the training–validation curve of Model 1–Model 3 is presented in Figure 8.

Both Figure 8 and Table 6 suggest that Fire_Net has an optimal performance in consideration of both speed and accuracy. In Figure 8, Fire_Net's training and validation loss reduces to the smallest loss value. In Table 6, it is noted that the 20-layered Model 4 has also reached validation accuracy of 0.98, the same of Fire_Net, however, the average processing time for Model 4 is 1.63 times longer than Fire_Net. This is mainly due to the computation time cost in its added layers. Thus, Fire_Net is considered as a more efficient DCNN architecture in Fire recognition in terms of both validation accuracy and processing time. A more notable results in Table 6 is that we find all type of models trained by our augmented image database have shown better performance compared to those trained with original images. This is mainly because the deep architecture's coefficients are optimized with sufficient training images while models trained with original image database suffers from lower training data quality. Furthermore, augmented database give the architecture more learning samples so that the coefficient and weights value become more optimal after the training. Original image dataset contains smaller amount of training image which could be easier to have an overfitting problem.

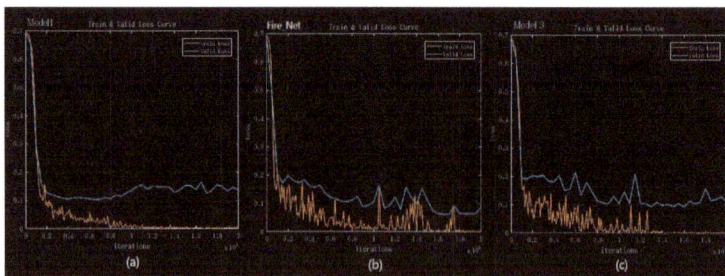

Figure 8. Training and validation loss curve of the three models: (**a**) Model 1; (**b**) Fire_Net; (**c**) Model 3. The curve is computed over an epoch of 20,000 iterations. The training loss is marked every 100 iterations while the validation loss is recorded every 500 iterations.

Table 6. Effects of model architecture on performance and runtime.

Model Type Index	Architecture	Average Processing Speed	Training Dataset	Validation Accuracy
Model 1	13-layered architecture: 6xConv + 4xPooling: Conv1-Conv2-Pool1-Conv3-Pool2-Conv4 -Conv5-Pool3-Conv6-Pool4-FCx2-Softmax	31.9ms	Original image dataset	0.952
			Saliency-based augmented dataset	0.971
Model 2	Fire_Net	41.5ms	Original image dataset	0.974
			Saliency-based augmented dataset	0.98
Model 3	15-layered architecture: 8xConv + 4xPooling: Conv1-Conv2-Pool1-Conv3-Conv4-Pool2- Conv5-Conv6-Conv7-Pool3-Conv8-Pool4 -FCx2-Softmax	38.6ms	Original image dataset	0.972
			Saliency-based augmented dataset	0.978
Model 4	20-layered architecture: 11xConv + 6xPooling: Conv1-Conv2-Conv3-Pool1-Conv4-Conv5- Pool2-Conv6-Pool3-Conv7-Conv8-Conv9- Pool4-Conv10-Pool5-Conv11-Pool6-FCx2- Softmax	67.8ms	Original image dataset	0.977
			Saliency-based augmented dataset	0.98

A standard test image dataset is used to evaluate Fire_Net's recognition accuracy and generalization performance. Given an input test image, the convolution-pooling layers extract its high-level features and the full connection layer will determine whether the abstracted features correspondent to fire or not.

We visualized the feature maps after Conv1, Conv2, and Conv3 layers in Figure 9. It is well observed that the lower layers captures the low level features (texture, edge, or color), the higher layers response to higher level sparse features by eliminating irrelevant content. Some examples of Fire_Net recognition results are presented in Figure 10.

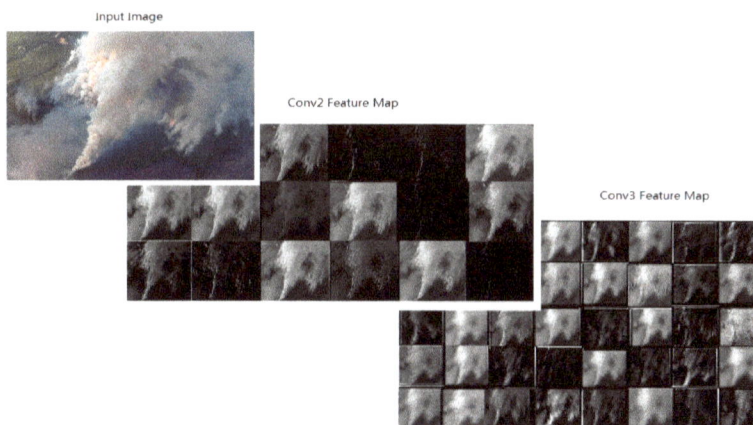

Figure 9. Visualization of feature maps after layer Conv2 and layer Conv3 in Fire_Net.

a. Correctly classified forest fire image samples with both flame and smoke features

b. Correctly classified forest fire image with mainly smoke feature

c. Correctly classified non fire image samples

d. Wrongly classified image samples

Figure 10. Samples of Fire_Net classification results. (**a**) Fire image of full fire features; (**b**) fire image with mainly smoke features; (**c**) no-fire images; (**d**) wrongly-classified images.

It is shown in Figure 10 that normal aerial images of wildfire (part a, part b) and negative image samples (part c) are correctly classified. The effectiveness of the model is justified. Still, it is admitted that Fire_Net worked inadequately on a rare case of 'forest mist' images (wrongly classified samples in part d). Fire_Net wrongly classified these forest mist images into fire case. The main reason is that the mist features are highly similar to smoke features in both color and texture. Images like 30 and 31 contain a very strong feature of smoke, even the human vision system can mistake it for fire. In these cases, sensor techniques are needed.

The following table shows the performance of Fire_Net on the test set of 512 images. The test set is composed by 256 fire image and 256 no-fire images. These images are randomly collected from online resources, they are not used in the training or validation phase. These images are not processed by our segmentation method. TPR is true positive rate while TNR is the true negative rate. It is exhibited in Table 7 that Fire_Net has a very low false negative rate.

Table 7. Performance of Fire Net on test dataset.

TP	FN	TN	FP	TPR	TNR	FNR	General Accuracy
253	3	249	7	98.8%	97.2%	0.12%	98% (502/512)

A comparison of our models with prevalent machine vision classification methods is made. Since these methods are all trained and evaluated with the same training and testing data source of

Fire_Net ('UAV_fire' as training dataset, the 512 randomly collected images as test dataset), the general classification accuracy can be a direct indicator of their performance.

The classification results by different methods are summarized in Table 8, it is shown that the 'HOG + SVM' solution is inadequate in fire detection. The HOG features are highly related with the contour and shape of the object while fire and smoke has no regular shape or contours. Deep belief net with a back-propagation neural network combines the advantage of unsupervised learning and supervised learning, unlike CNN, DBN cannot take a multi-dimensional image as input, images are turned into vectors to fit its input format, the image's structural information is lost, which limited the image feature learning ability of DBN. The rest of the methods are different models based on CNN. Fire_Net outperformed the previous deep learning models.

Table 8. Comparison of Fire_Net with other methods.

Models	Classifier	Accuracy
HOG + SVM	SVM	42.9%
Deep belief net + neural net	BPNN	87.2%
Kim's CNN model[22]	Softmax	92.8%
Eight-layer CNN + Fisher vector	SVM	94.7%
AlexNet	Softmax	97.1%
Fire_Net	Softmax	98.0%

We further evaluated Fire_Net's generalization performance on a set of 40 sampled image from wildfire news reports. These samples are aerial images of Bilahe forest fire, in Noth China 2017(a); Mongolia Forest fire, in China, 2017(b); Envisat satellite image of wildfire in Southern California, USA. 2007(c1); Kern county wildfire, CA, US, 2016(c2); Var wildfire, south France, 2017(d); La Tuna and Napa Valley wildfire, CA, US, 2017(e); etc. All these samples are correctly classified as fire images: the output of Fire_Net ranges from 0.9765 to 1 for fire class, while for no-fire class, it is from 1.0064×10^{-8} to 0.0235. This indicates that strong fire features are presented in these images (see Figure 11).

Figure 11. Examples of Fire_Net correctly-classified fire images from news reports. (**a**) Bilahe Forest fire, 2017; (**b**) Mongolia Forest fire, 2017; (**c**) South California wildfire and kern county wildfire, 2016; (**d**) Var wildfire, 2017; (**e**) La tuna and Napa valley wildfire, 2017.

5. Conclusions

In this paper, we demonstrated the effectiveness of using saliency detection and deep convolutional neural network in localization and recognition of wildfire in aerial images. The saliency detection method is used to locate core fire area and extract fire regions into multiple fire images. The proposed technique prevented severe feature loss due to direct resizing. Also, this technique significantly enriched the volume of the database. We also proposed a DCNN architecture named

'Fire_Net'. It obtained satisfactory classification results. The pipeline of adopting saliency detection and deep learning in wildfire identification have not yet been reported previously, it is firstly proposed and verified in this work. Still, our method can be improved further. Regarding the wrongly classified forest mist images, IR sensors could be implemented in the decision making layers (full connection layers) of Fire_Net so that both image and sensor data can be used to determine whether fire occurs.

Acknowledgments: This research was partly supported by Fundamental Research Funds for the Central Universities (Chang'An University No. 310832161011 and No. 310832171001) and the Project of International Cooperation and Exchanges of Shaanxi Province (2016KW-035).

Author Contributions: Y.Z. conceived the pipeline, performed the experiments and wrote the paper; J.M. performed part of the experiments and analyzed the data; X.L. optimized the network architecture; J.Z provide the hardware and experiment equipment. All authors reviewed the manuscript.

Conflicts of Interest: The authors declare no conflict of interest.

Appendix A

The complete structure of Fire_Net:

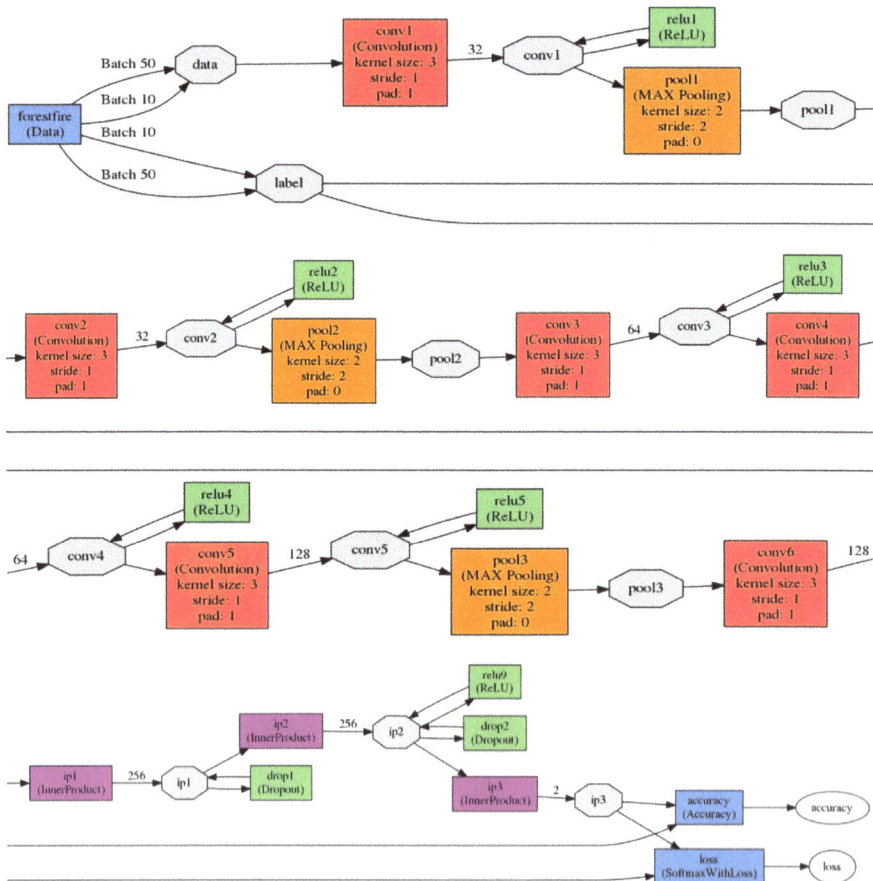

References

1. Skala, K.; Dubravić, A. Integrated System for Forest Fire Early Detection and Management. *Period. Biol.* **2008**, *110*, 205–211.
2. Colomina, I.; Molina, P. Unmanned aerial systems for photogrammetry and remote sensing: A review. *ISPRS J. Photogramm.* **2014**, *92*, 79–97. [CrossRef]
3. Li, X.; Zhao, Y.; Zhang, J. A Hybrid PSO Algorithm Based Flight Path Optimization for Multiple Agricultural UAVs. In Proceedings of the International Conference on TOOLS with Artificial Intelligence, San Jose, CA, USA, 6–8 November 2016; pp. 691–697.
4. Trasviñamoreno, C.A.; Blasco, R.; Marco, Á.; Casas, R.; Trasviñacastro, A. Unmanned Aerial Vehicle Based Wireless Sensor Network for Marine-Coastal Environment Monitoring. *Sensors* **2017**, *17*, 460. [CrossRef] [PubMed]
5. Gonzalez, L.F.; Montes, G.; Puig, E.; Johnson, S.; Mengersen, K.; Gaston, K.J. Unmanned Aerial Vehicles (UAVs) and Artificial Intelligence Revolutionizing Wildlife Monitoring and Conservation. *Sensors* **2016**, *16*, 97. [CrossRef] [PubMed]
6. Lelong, C.C.D.; Burger, P.; Jubelin, G.; Bruno, R.; Sylvain, L.; Frederic, B. Assessment of Unmanned Aerial Vehicles Imagery for Quantitative Monitoring of Wheat Crop in Small Plots. *Sensors* **2008**, *8*, 3557–3585. [CrossRef] [PubMed]
7. Hefeeda, M.; Bagheri, M. Forest Fire Modeling and Early Detection using Wireless Sensor Networks. *Ad Hoc Sens. Wirel. Netw.* **2009**, *7*, 169–224.
8. Wulder, M.; White, A.; Alvarez, J.C.; Han, F.; Rogan, T.; Hawkes, B. Characterizing boreal forest wildfire with multi-temporal landsat and lidar data. *Remote Sens. Environ.* **2009**, *113*, 1540–1555. [CrossRef]
9. Li, Z.; Nadon, S.; Cihlar, J. Satellite detection of Canadian boreal forest fires: Development and application of the algorithm. *Int. J. Remote Sens.* **2000**, *21*, 3057–3069. [CrossRef]
10. Katayama, H.; Naitoh, M.; Suganuma, M.; Harada, M.; Okamura, Y.; Tange, Y. Development of the Compact InfraRed Camera (CIRC) for wildfire detection. *SPIE Opt. Eng. Appl.* **2009**, *8*, 745806–745814.
11. Jakovcevic, T.; Braovic, M.; Stipanicev, D.; Krstinic, D. Review of wildfire smoke detection techniques based on visible spectrum video analysis. In Proceedings of the International Symposium on Image & Signal Processing & Analysis, Dubrovnik, Croatia, 4–6 September 2011; pp. 480–484.
12. Çetin, A.E.; Dimitropoulos, K.; Gouverneur, B.; Grammalidis, N.; Günay, O.; Habiboglu, Y.H. Video fire detection—Review. *Digit. Signal Process.* **2013**, *23*, 1827–1843. [CrossRef]
13. Lecun, Y.; Bengio, Y.; Hinton, G. Deep learning. *Nature* **2015**, *527*, 436–444. [CrossRef] [PubMed]
14. Vipin, V. Image processing based forest fire detection. *Int. J. Emerg. Technol. Adv. Eng.* **2012**, *2*, 87–95.
15. Angayarkkani, K.; Radhakrishnan, N. An Intelligent System for Effective Forest Fire Detection Using Spatial Data. *Int. J. Comput. Sci. Inf. Secur.* **2010**, *7*, 202–208.
16. Yuan, C.; Liu, Z.; Zhang, Y. UAV-based forest fire detection and tracking using image processing techniques. In Proceedings of the International Conference on Unmanned Aircraft Systems IEEE, Denver, CO, USA, 9–12 June 2015; pp. 639–643.
17. Töreyin, B.U.; Dedeoğlu, Y.; Güdükbay, U.; Cetin, A.E. Computer vision based method for real-time fire and flame detection. *Pattern Recognit. Lett.* **2006**, *27*, 49–58. [CrossRef]
18. Byoung, C.K.; Kwang-Ho, C.; Jae-Yeal, N. Fire detection based on vision sensor and support vector machines. *Fire Saf. J.* **2009**, *44*, 322–329.
19. Toulouse, T.; Rossi, L.; Akhloufi, M.; Celik, T. Benchmarking of wildland fire colour segmentation algorithms. *Image Process. IET* **2015**, *9*, 1064–1072. [CrossRef]
20. Girshick, R.; Donahue, J.; Darrell, T.; Malik, J. Rich feature hierarchies for accurate object detection and semantic segmentation. In Proceedings of the Conference on Computer Vision and Pattern Recognition, Columbus, OH, USA, 23–28 June 2014; pp. 580–587.
21. Yuan, C.; Zhang, Y.; Liu, Z. A survey on technologies for automatic forest fire monitoring, detection. *Can. J. For. Res.* **2015**, *45*, 783–792. [CrossRef]
22. Kim, S.; Lee, W.; Park, Y.S.; Lee, H.W.; Lee, Y.T. Forest fire monitoring system based on aerial image. In Proceedings of the International Conference on Information and Communication Technologies for Disaster Management, Vienna, Austria, 13–15 December 2017; pp. 1–6.

23. He, K.; Zhang, X.; Ren, S.; Sun, J. Spatial Pyramid Pooling in Deep Convolutional Networks for Visual Recognition. *IEEE Trans. Pattern Anal. Mach. Intell.* **2014**, *37*, 1904–1916. [CrossRef] [PubMed]

24. Girshick, R. Fast R-CNN. In Proceedings of the IEEE International Conference on Computer Vision IEEE, San Tiago, Chile, 7–13 December 2015; pp. 1440–1448.

25. Krizhevsky, A.; Sutskever, I.; Hinton, G.E. ImageNet classification with deep convolutional neural networks. In Proceedings of the International Conference on Neural Information Processing Systems, Lake Tahoe, CA, USA, 3–8 December 2012; pp. 1097–1105.

26. Rahtu, E. Segmenting Salient Objects from Images and Videos. In Proceedings of the 11th European Conference on Computer Vision, Heraklion, Greece, 5–11 September 2010; pp. 366–379.

27. Jia, Y.; Shellamer, E.; Donahue, J.; Karayev, S.; Long, J.; Girshick, R.; Guadarrama, S.; Darrell, T. Caffe: Convolutional Architecture for Fast Feature Embedding. In Proceedings of the ACM International Conference on Multimedia, Orlando, FL, USA, 3–7 November 2014; pp. 675–678.

28. Huang, Z.C.; Chan, P.P.K.; Ng, W.W.Y.; Yeung, D.S. Content-based image retrieval using color moment and Gabor texture feature. In Proceedings of the International Conference on Machine Learning and Cybernetics, Qingdao, China, 11–14 July 2010; pp. 719–724.

29. Ortiz, A.; Górriz, J.M.; Ramírez, J.; Salas-González, D.; Llamas-Elvira, J.M. Two fully-unsupervised methods for MR brain image segmentation using SOM-based strategies. *Appl. Soft Comput. J.* **2013**, *13*, 2668–2682. [CrossRef]

30. Kwak, E.; Habib, A. Automatic representation and reconstruction of DBM from LiDAR data using Recursive Minimum Bounding Rectangle. *J. Photogramm. Remote Sens.* **2014**, *93*, 171–191. [CrossRef]

31. Itti, L.; Koch, C.; Niebur, E. A Model of Saliency-Based Visual Attention for Rapid Scene Analysis. *IEEE Trans. Pattern Anal. Mach. Intell.* **1998**, *20*, 1254–1259. [CrossRef]

32. Zhao, Y.; Li, X. Edge-Aware Weighting Enhanced Saliency Segmentation of Pests Images. In Proceedings of the International Conference on Computational Science and Computational Intelligence IEEE, Las Vegas, NV, USA, 15–17 December 2016; pp. 847–851.

33. Glorot, X.; Bengio, Y. Understanding the difficulty of training deep feedforward neural networks. *J. Mach. Learn. Res.* **2010**, *9*, 249–256.

34. Nair, V.; Hinton, G.E. Rectified Linear Units Improve Restricted Boltzmann Machines. In Proceedings of the International Conference on Machine Learning, Haifa, Israel, 21–24 June 2010; pp. 807–814.

35. Giusti, A.; Dan, C.C.; Masci, J.; Gambardella, L.M.; Schmidhuber, J. Fast image scanning with deep max-pooling convolutional neural networks. In Proceedings of the IEEE International Conference on Image Processing, Melbourne, Australia, 15–18 September 2013; pp. 4034–4038.

36. Srivastava, N.; Hinton, G.; Krizhevsky, A.; Sutskever, I.; Salakhutdinov, R. Dropout: A simple way to prevent neural networks from overfitting. *J. Mach. Learn. Res.* **2014**, *15*, 1929–1958.

37. Tüske, Z.; Tahir, M.A.; Schlüter, R.; Ney, H. Integrating Gaussian mixtures into deep neural networks: Softmax layer with hidden variables. In Proceedings of the IEEE International Conference on Acoustics, Speech and Signal Processing, Brisbane, Australia, 19–24 April 2015; pp. 4285–4289.

38. Ren, S.; He, K.; Girshick, R.; Sun, J. Faster R-CNN: Towards Real-Time Object Detection with Region Proposal Networks. *IEEE Trans. Pattern Anal. Mach. Intell.* **2015**, *39*, 1137–1149. [CrossRef] [PubMed]

sensors

MDPI

Article

Experimental Study of Multispectral Characteristics of an Unmanned Aerial Vehicle at Different Observation Angles

Haijing Zheng [1], Tingzhu Bai [1,*], Quanxi Wang [2], Fengmei Cao [1], Long Shao [1] and Zhaotian Sun [1]

[1] School of Optics and Photonics, Beijing Institute of Technology, Beijing 100081, China;
    zhjsea08@163.com (H.Z.); liuba@bit.edu.cn (F.C.); shaolong@bit.edu.cn (L.S.); 2120150546@bit.edu.cn (Z.S.)
[2] System Division, Navy Equipment Research Institute, Beijing 100161, China; ccwangok@sina.cn
*   Correspondence: tzhbai@bit.edu.cn; Tel.: +86-010-6891-2748

Received: 31 October 2017; Accepted: 29 January 2018; Published: 1 February 2018

Abstract: This study investigates multispectral characteristics of an unmanned aerial vehicle (UAV) at different observation angles by experiment. The UAV and its engine are tested on the ground in the cruise state. Spectral radiation intensities at different observation angles are obtained in the infrared band of 0.9–15 μm by a spectral radiometer. Meanwhile, infrared images are captured separately by long-wavelength infrared (LWIR), mid-wavelength infrared (MWIR), and short-wavelength infrared (SWIR) cameras. Additionally, orientation maps of the radiation area and radiance are obtained. The results suggest that the spectral radiation intensity of the UAV is determined by its exhaust plume and that the main infrared emission bands occur at 2.7 μm and 4.3 μm. At observation angles in the range of 0°–90°, the radiation area of the UAV in MWIR band is greatest; however, at angles greater than 90°, the radiation area in the SWIR band is greatest. In addition, the radiance of the UAV at an angle of 0° is strongest. These conclusions can guide IR stealth technique development for UAVs.

Keywords: UAV; infrared radiation; exhaust plume; spectra

1. Introduction

With their advantages of high flexibility, low cost, high survival rate, and low operational requirements, unmanned aerial vehicles (UAVs) are widely used in civil fields such as urban management, agricultural monitoring, environmental protection, disaster relief, and film and television photography [1–4]. In the military fields of surveillance, monitoring, electronic confrontation, and damage assessment, UAVs play an important role as well [5,6]. The target drone is a kind of UAVs that has important applications in the military field [7,8]. As the imaginary target of various artillery or missile systems, the target drone should have some a certain stealth performance. Furthermore, infrared (IR) radiation features is the major factor for stealth performance. For this reason, studies on the infrared (IR) radiation characteristics of the target drone are of great significance to IR stealth techniques regarding target drone's body, engine, and fuel formulation design and to IR warning system development. Generally, the power plant of a UAV with long endurance is commonly a turbo fan engine. For the target drone and other small high-speed aircraft, such as unmanned reconnaissance aircraft, a gas turbine engine is used as its propeller.

A number of IR radiative sources impinge on UAV elements such as its surface, engine, and exhaust plume. The IR radiation emitted from the exhaust plume is one of the main analysis targets in IR steath technology. Many studies have investigated the IR radiation of both an aircraft and its exhaust plume by numerical simulations and/or experimental methods. The IR characteristics of a typical commercial quad-rotor UAV were investigated through thermal imaging with an IR camera by Gong et al. [9]. Blunck et al. measured the narrowband radiation intensity of exhaust plumes exiting from a converging nozzle with varying Reynolds numbers, Mach numbers, temperatures, and species compositions corresponding to low fuel-to-air equivalence ratios [10]. For the test conditions studied, the intensity emanating from diametric paths first decayed linearly with increasing distance from the nozzle exit and then decayed exponentially with an inflexion point near the end of the plume core. In 2014, a concise band selection method employing the multispectral signatures of stealth aircraft was proposed by Liu et al. [11]. The stability of the selected bands was tested under varying environments. The key step in the model was to select two or more optimal bands that could clearly signify the radiation difference between the target and its background. Two narrow bands of 2.86–3.3 μm and 4.17–4.55 μm were ultimately selected after detailed analyses on contrast characteristics between the target and background. In 2016, Huang studied the plume IR radiation of a long-endurance UAV [12] and found that the IR radiation energy of the plume was distributed mainly at the IR wavelengths of 2.7 μm and 4.5 μm. In 2017, a multi-scale narrowband correlated-k distribution (MSNBCK) model has been developed by Zhou et al. [13] to simulate IR radiation from the exhaust system of a typical aircraft engine. In the model, an approximate approach significantly reduces the required computing power by converting the exponential increase of computations required with the increase of participating gas species to a linear increase. The results indicate that a wall's IR emission should be considered in both 3–5 μm and 8–14 μm ranges while the IR emission of a gas plays an important role only in the 3–5 μm band. They also found that carbon dioxide's IR emission is much more significant than that of water vapor in both the 3–5 μm and 8–14 μm bands. In particular, in the 3–5 μm band, water vapor's IR signal can even be neglected compared with that of carbon dioxide. In 2017, Retief et al. measured a plume in short-wavelength infrared (SWIR) (1.1–2.5 μm), mid-wavelength infrared (MWIR) (2.5–7 μm), and long-wavelength infrared (LWIR) (7–15 μm) cameras and a spectral radiometer covering the entire mid, long, and upper part of the short-wavelength infrared bands [14]. The experimental results showed that the MWIR band was the most versatile regarding plume observations. The fuel consumption of an aircraft and the observation configuration were found to be the main influences on the detected emission of an aircraft plume.

Few studies have concentrated on the IR features of a target drone and the IR difference between a UAV fuselage and its engine. The research presented here will be helpful for the IR stealth performance of the target drone. Nevertheless, the methodologies used in previous studies are the references to this research. In this paper, the measurements of both a target drone and its gas turbine engine in different IR bands within the 0.9–15 μm band are shown. SWIR, MWIR, and LWIR cameras are used to obtain IR images of the targets. The IR spectrum is obtained by a spectral radiometer cooled by liquid nitrogen. By rotating the UAV or its engine, the observation angle between the target and the observer changes from 0° to 180°. Subsequently, IR characteristics of the complete UAV and its engine at different observation angles are recorded by the instruments mentioned above. The complete UAV and its engine operating in cruise on the ground are tested separately. The results show that the difference in IR characteristics between the complete UAV and its engine are very important for detecting a UAV when other IR sources are present, such as IR jamming. The results are also significant for the IR stealth design of UAVs.

2. Materials and Methods

2.1. Measuremental Layout

The tested UAV was a target drone in the research and development phase. Its power plant was a turbine engine with a shrinkable nozzle with inlet and outlet diameters of 96.5 and 72.1 mm, respectively, as shown in Figures 1 and 2. Beneath the nozzle on the wall was a thermal resistor that was used to measure temperature. The engine burned aviation kerosene as its fuel.

Figure 1. The unmanned aerial vehicle (UAV).

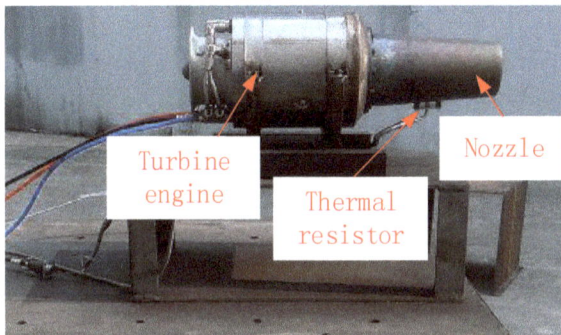

Figure 2. The turbine engine.

The experimental layout is shown in Figure 3. In the test, the instruments and the target (plume or complete UAV) were mounted orthogonal to the plume centerline. The target could be rotated from 0° to 180°. An observation angle of 90° meant that the viewing direction was perpendicular to the direction of the plume flow. An observation angle of 0° meant that the exhaust plume flowed directly toward the observer.

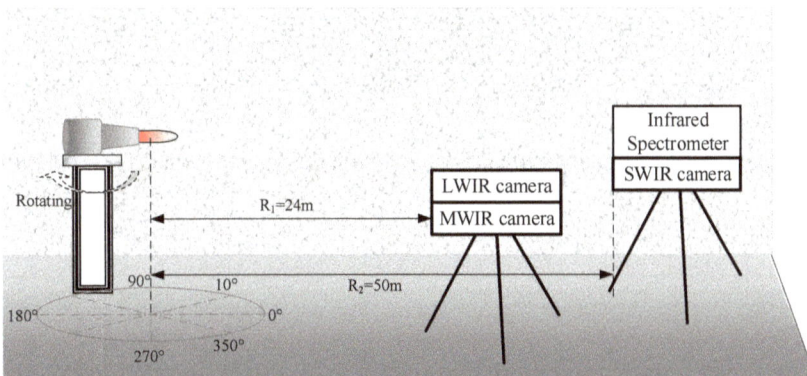

Figure 3. Experiment layout.

2.2. Radiation Measurements

A liquid-nitrogen-cooled MR170 portable IR spectral radiometer with a spectral range of 740–5000 cm^{-1} (2–13.5 μm), and a spectral resolution of 1 cm^{-1} was used in the test to obtain the spectrum. This instrument's field of view (FoV) covered the whole area of the plume and also the complete UAV.

Integrated radiation intensity was calculated by

$$I_{\Delta\lambda} = \int_{\lambda_1}^{\lambda_2} I_\lambda d\lambda \tag{1}$$

where the subscript $\Delta\lambda$ denotes the wave band, λ_1 and λ_2 are the band boundary respectively, and I is the radiation intensity.

Different bands IR images were recorded by three different IR cameras. The wavelength ranges of the LWIR, MWIR, and SWIR cameras were 7.7–9.3 μm, 3.0–4.8 μm, and 0.9–1.7 μm, respectively. The instrument specifications used in this experiment are summarized in Table 1 below.

Table 1. Instrument specifications.

Instrument	Model	Wavelength Range
Infrared Spectral Radiometer	AnalyzeIIT MR170	2–13.5 μm
LWIR Camera	FLIR SC7300L	7.7–9.3 μm
MWIR Camera	FLIR SC7300M	3.0–4.8 μm
SWIR Camera	Xenics Bobcat-640	0.9–1.7 μm

The radiation area of the target was obtained through processing the data. For calculating the radiation area of the plume or the complete UAV, a blue frame was marked manually in the IR images, illustrated in Figure 4.

Figure 4. Blue frame illustration.

The height and width of the blue frame were calculated by

$$h = eLN_h/f', w = eLN_w/f', \tag{2}$$

where h is the object height, w is the object width, e is the pixel size, L is the operating distance, f' is the focal length of the camera, N_h is the pixel number of the blue frame's height, and N_w is the pixel number of the blue frame's width. The radiation area was then calculated as the product $h \times w$.

2.3. Surrounding Radiation and Calibration

Since the atmosphere surrounds the UAV and IR detectors, the IR characteristics of the atmosphere play an important role in dictating the IR signatures as perceived by the detectors. The atmosphere is a good absorber of IR radiation. Radiant flux from the target aircraft is selectively absorbed by several atmosphere gases and scattered away by small particles suspended in atmosphere. Generally, sky radiance, cloud reflecting radiance, and other background radiance are also should be taken into account. In this work, the test was implemented indoors. Hence, the sky radiance and cloud radiance could be neglected. To further eliminate the effect of background radiation, a black curtain was arranged at the end of the test site.

These cameras were calibrated using an extended area blackbody. The blackbody was placed at the same distance from the camera as the plume centerline and was surrounded by the same atmosphere with the UAV. This approach was used to account for atmosphere attenuation of the IR radiation.

2.4. Operating Conditions

The rotation speed and temperature of the engine were measured during the experiment as shown in Figure 5. As the turbine engine began to operate, the rotation speed and temperature increased rapidly. At the cruise stage, the rotation speed and temperature stabilized. Once operating stably, the IR spectral radiometer and IR cameras were used to obtain information from the targets at a time of about 400 s. Afterward, the engine's rotation speed was reduced in consideration of safety, and the whole target was rotated by some angle to achieve the purpose of changing the observation angle. Subsequently, the engine was returned to the cruise stage, and instruments obtained information at the new observation angle.

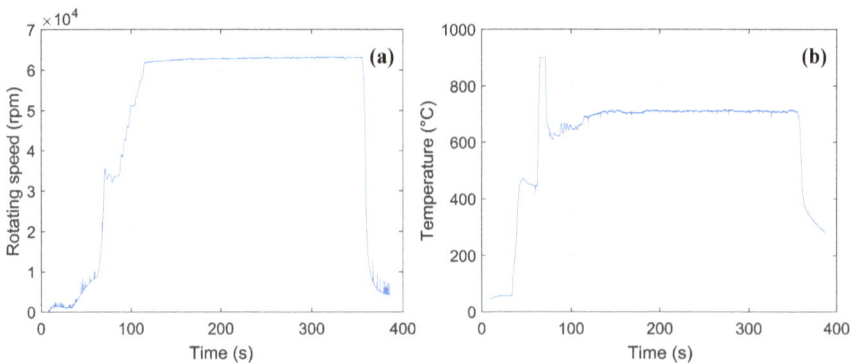

Figure 5. Engine rotation speed and temperature. (**a**) Time-dependent rotating speed of the engine; (**b**) Time-dependent temperature of the outlet wall.

3. Results

3.1. Turbine Engine

3.1.1. Spectral Radiometer Recording

Figure 6 shows the spectral radiation intensity of the engine in its cruise state at different observation angles: $0°, 15°, 30°, 45°, 60°, 90°, 120°, 150°,$ and $180°$. Several points should be mentioned in regard to this figure.

1.   All spectra were continuous, with spectral bands and lines.

2. All spectra had three highlighted bands: 2.7 μm, 4.3 μm, and 6.3 μm.
3. The values of radiation intensity at 2.7 μm and 4.3 μm changed with the observation angle, but that at 6.3 μm remained unchanged.

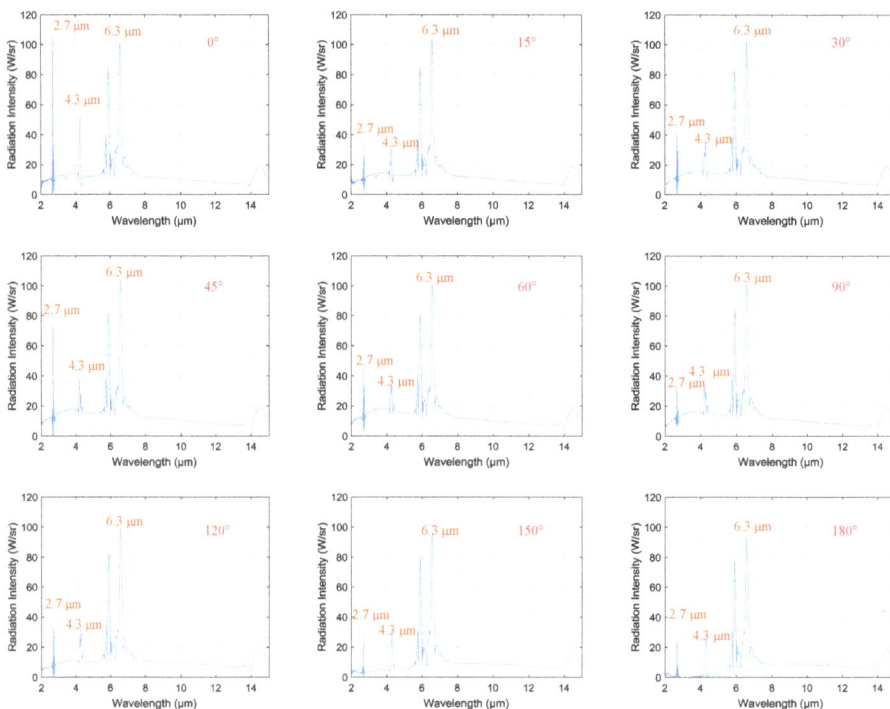

Figure 6. Spectral radiation intensity of the engine at different observation angles.

Regarding gas molecules, each spectral band involved a change in the vibrational energy of the molecules. Lines in the spectral bands resulted from simultaneous changes in rotational energy [15]. Hence, the spectral bands and lines of spectra in Figure 6 were produced by hot gas emissions in the exhaust plume. The continuous spectra were caused by solid matter, most probably carbon soot [14]. We know that water vapor's main emission bands occur at 2.7 μm and 6.3 μm [16], while those of carbon dioxide occur at 2.7 μm and 4.3 μm [17] within the range of 2–15 μm. Therefore, the emission bands of 2.7 μm, 4.3 μm, and 6.3 μm result from hot water vapor and carbon dioxide emissions. However, the values of radiation intensity at 2.7 μm and 4.3 μm changed with the observation angle and that at 6.3 μm remained unchanged. Therefore, emitters producing 2.7 μm and 4.3 μm bands were in the exhaust plume and emitters producing the 6.3 μm band were in the background. Considering the substantial combustion product already expelled into the air, the 6.3 μm band may have been caused by water vapor in the background.

The main bands of 2.7 μm and 4.3 μm are resulted from exhaust plume emissions and the 6.3 μm band is resulted from background emissions. According to Rao [18], IR radiation emitted from aircraft exhaust plumes arises primarily from carbon dioxide and carbon monoxide in the exhaust. The experimental results for spectral radiation intensity agree with Rao's conclusion.

Sensors **2018**, *18*, 428

3.1.2. IR Camera Recording

Figure 7 shows images of the engine operating in the cruise stage captured by the LWIR camera at different observation angles. The target was relatively easy to discriminate from the background. In Figure 7, the bright area can be identified as the engine, but exhaust plume structures are difficult to distinguish.

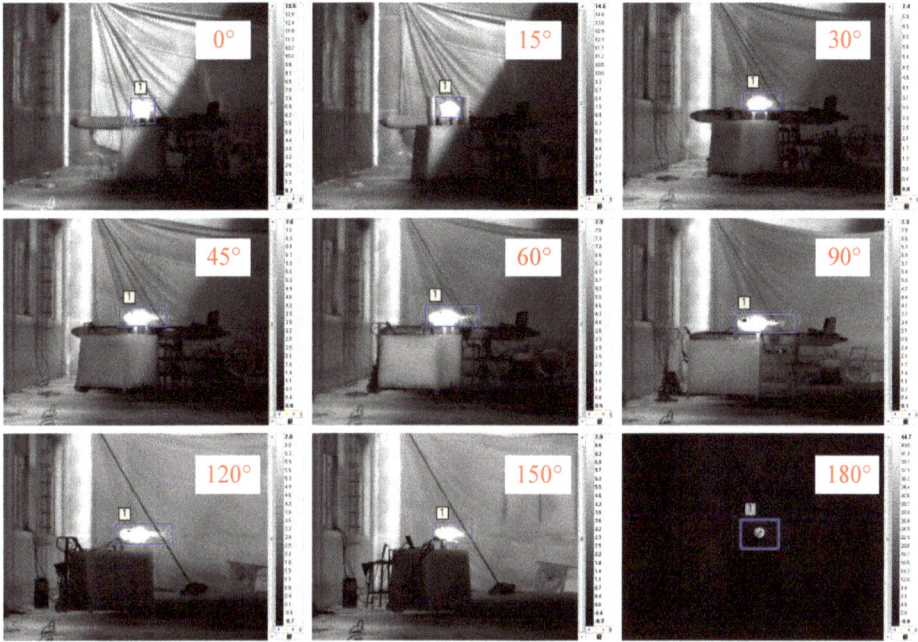

Figure 7. Engine images captured by the long-wavelength infrared (LWIR) camera at different observation angles.

Although the target was only the engine, the body of the drone may be recognized at some observation angles. The cause of this problem was that the turbine engine was disassembled but must remain connected to the UAV. During the test, this problem was not noticed at first. Afterward, from the observation angle of 120°, the body of the drone was removed. Nevertheless, there was a minimal effect on the radiance because of the blue frame.

Figure 8 shows images of the engine operating in the cruise stage captured by the MWIR camera at different observation angles. The target was easy to discern relative to the background, and the exhaust plume structures can be distinguished clearly in the images at observation angles of 45°, 60°, 90°, 120°, and 150°. At an observation angle of 0°, the exhaust plume presented as a disc. In contrast, at an observation angle of 180°, the exhaust plume could scarcely be seen because it was blocked by the cold parts of the engine.

Figure 9 shows images of the engine operating in the cruise stage captured by the SWIR camera at different observation angles. The target was difficult to discern from its background. At an observation angle of 180°, the exhaust plume could not be seen.

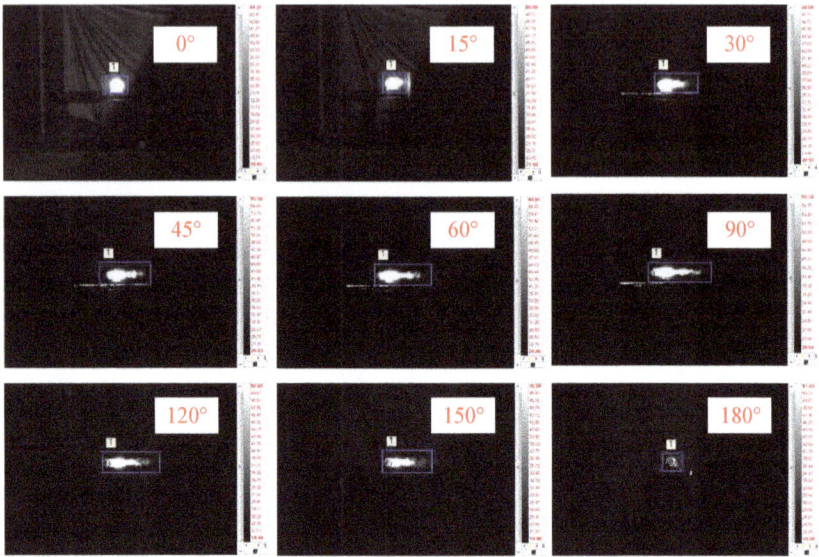

Figure 8. Engine images captured by the mid-wavelength infrared (MWIR) camera at different observation angles.

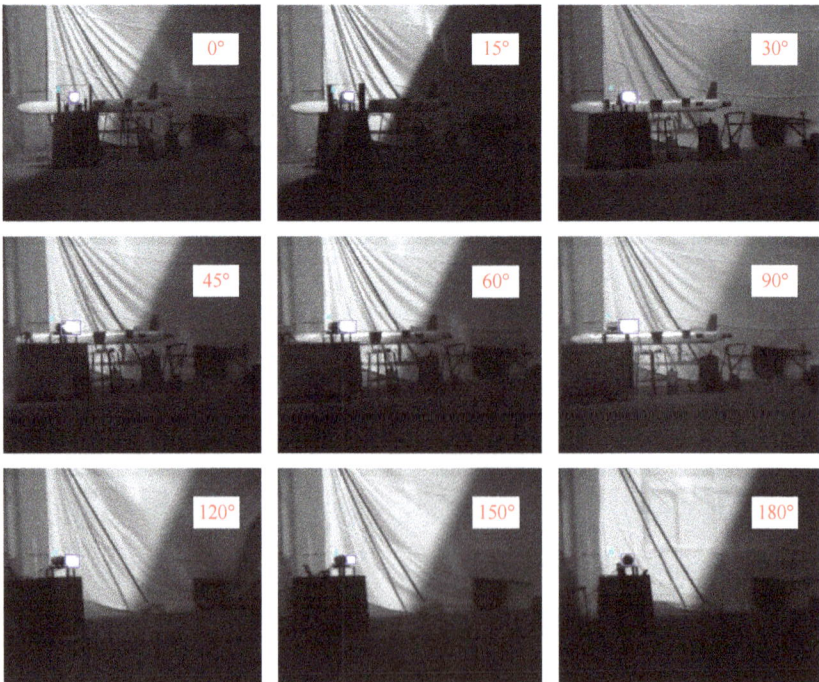

Figure 9. Engine images captured by the short-wavelength infrared (SWIR) camera at different observation angles.

From Figure 10, the exhaust plume along with the hot engine is relatively obvious in the MWIR band because it has the largest radiation area. The tendencies of the radiation area similarly change with the observation angles in the three IR bands. The radiation area increased with the observation angle from 0° to 90°. At an observation angle of 90°, the radiation area was largest. However, the further increase of the observation angle resulted in a rapid decrease in radiation area. The least radiation area occurred at observation angle of 180°, mostly because the hot engine and exhaust plume were blocked in that configuration by cold parts of the engine.

The radiance in different IR bands at different observation angles were also calculated, as shown in Figure 11. The exhaust plume and hot engine emitted the largest amount of radiance in the LWIR band though the smallest radiation area. Nevertheless, in LWIR band, radiance at an observation angle of 30° was the largest, almost 60 W/(Sr*m²). With an increasing observation angle, the radiance decreased. The least radiance in both the LWIR band and the MWIR band occurred at an observation angle of 180°. In LWIR band, the least radiance was 18 W/(Sr*m²), while in the MWIR band, it was almost zero.

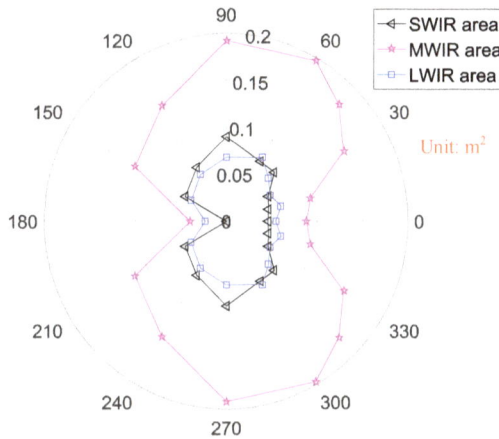

Figure 10. Orientation map of the engine's radiation areas in different infrared (IR) bands.

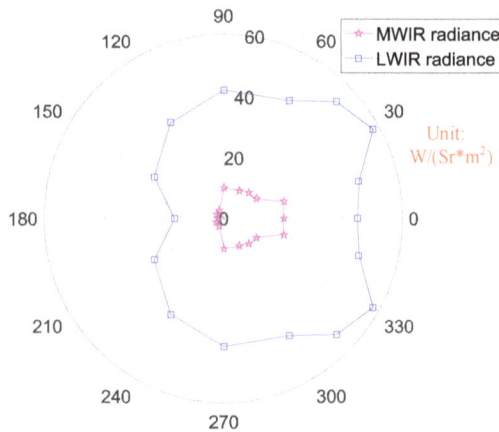

Figure 11. Orientation map of the engine's radiance in different IR bands.

3.2. Complete UAV

3.2.1. Spectral Radiometer Recording

The behavior of spectral radiation intensity for the complete UAV was the same as that for the engine which is shown in Figure 12. The values of radiation intensity in the 2.7 and 4.3 μm bands changed with the observation angle, but that in the 6.3 μm band remained unchanged. The only difference was in the absolute value of radiation intensity. Overall, the complete UAV's radiation intensity value was about three times that of the engine's. Take the observation angle of 90° for example: at an observation angle of 90° with a wavelength of 6.3 μm, the radiation intensity of the complete UAV was 298 W/sr, while the radiation intensity of the engine was 101 W/sr.

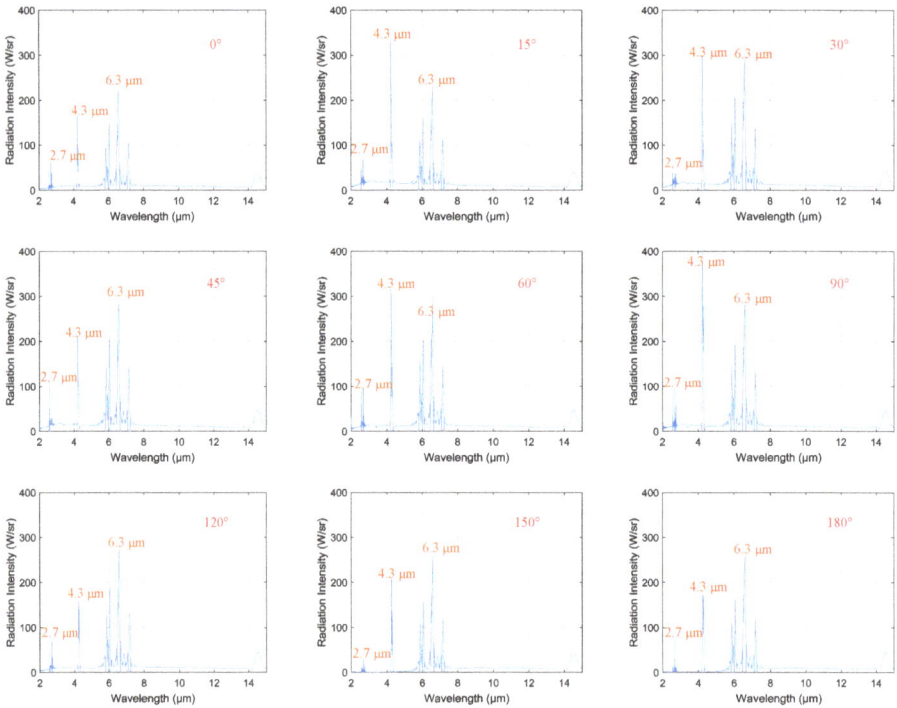

Figure 12. Spectral radiation intensity of the UAV at different observation angles.

3.2.2. IR Camera Recordings

Figure 13 shows images of the whole UAV operating in the cruise stage captured by the LWIR camera at different observation angles. In the LWIR band, the body of the UAV emitted some radiation; however, compared to the engine and the exhaust plume, the body's radiation level was too low.

Figure 14 shows images of the whole UAV operating in the cruise stage captured by the MWIR camera at different observation angles. Only the hot engine and exhaust plume could be seen at observation angles from 0° to 150 °. At 180°, owing to the large body size of the UAV, the hot engine and the exhaust plume did not appear.

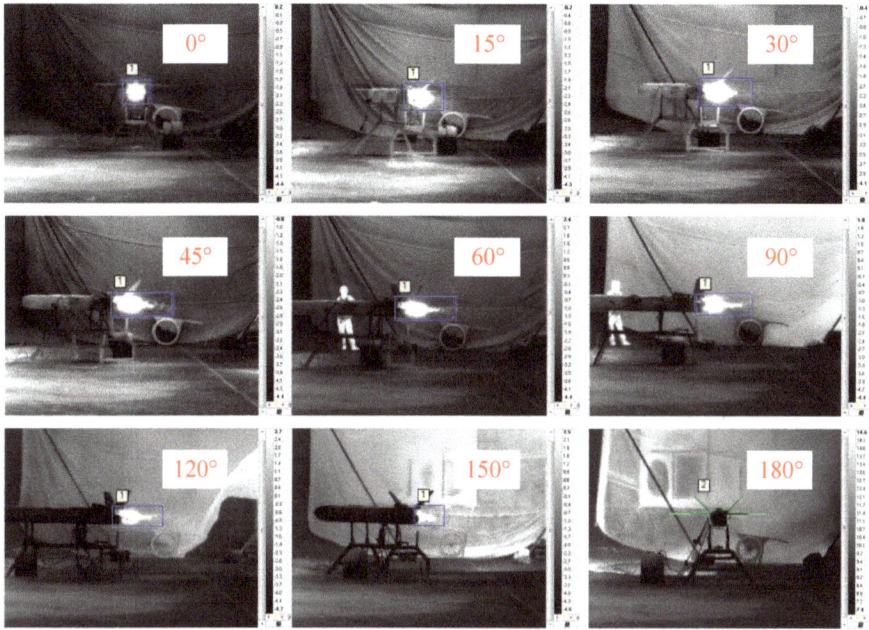

Figure 13. UAV images captured by the LWIR camera at different observation angles.

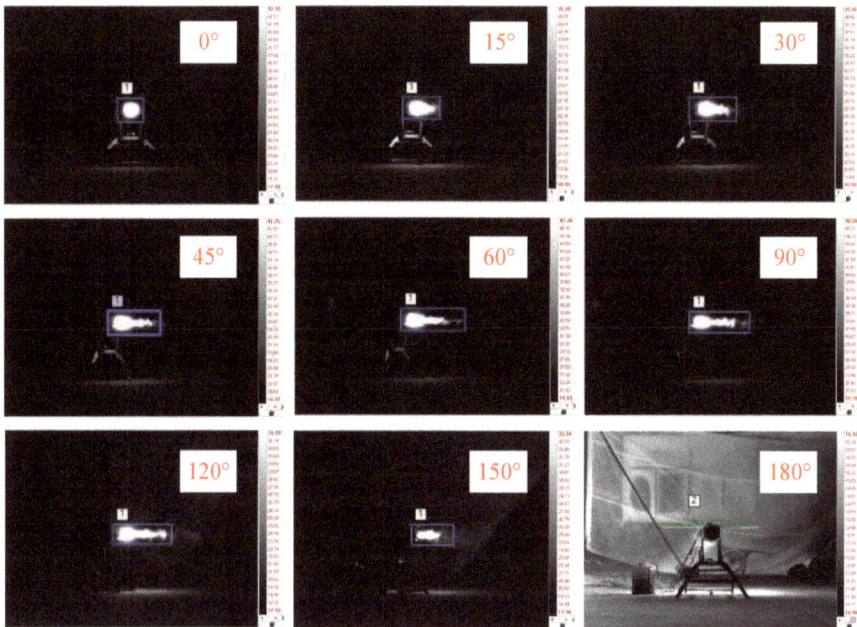

Figure 14. UAV images captured by the MWIR camera at different observation angles.

Figure 15 shows images of the complete UAV operating in the cruise stage captured by the SWIR camera at different observation angles. In the SWIR band, the exhaust plume was not obvious, but the body's appearance was still relatively distinct.

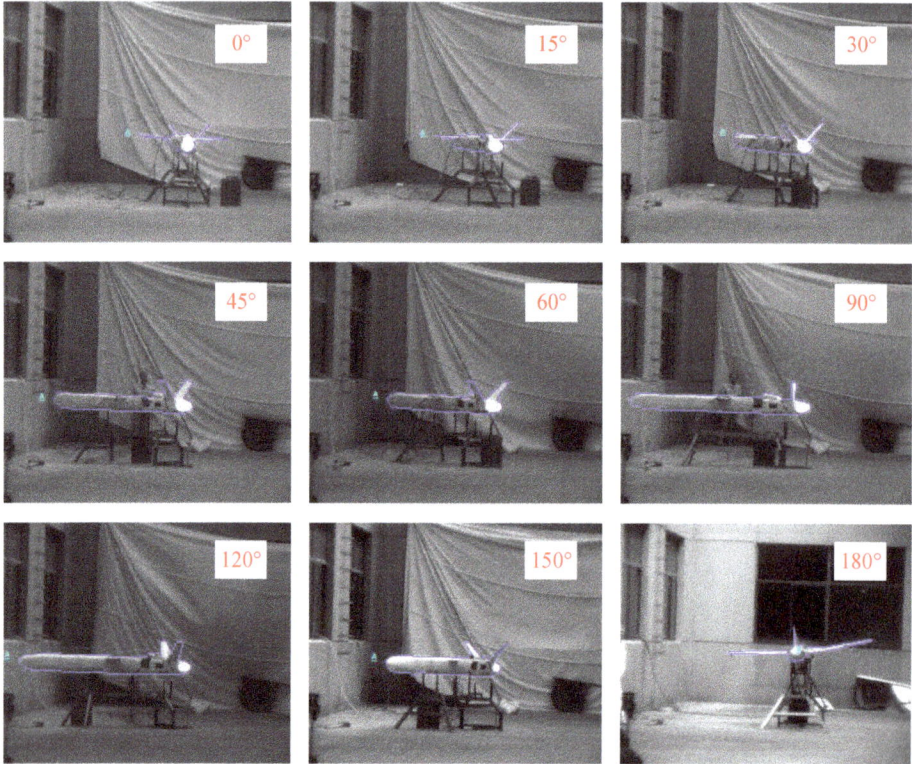

Figure 15. UAV images captured by the SWIR camera at different observation angles.

The orientation map of the UAV's radiation area in different IR bands is shown in Figure 16. Because of the obvious body in the SWIR band, the radiation area in SWIR band was larger than that in the MWIR band at observation angles from 120° to 150°. But the largest radiation area in SWIR band was 0.14 m^2 at observation angle of 120° because of the very large body of the UAV, while that in MWIR band was 0.08 m^2 at 60°. There is the least radiation area in the LWIR band.

The radiance of the complete UAV in different IR bands at different observation angles was calculated; the results are shown in Figure 17. The radiance values at observation angles from 0° to 90° were relatively close, about 42 W/(Sr*m^2). The maximum radiance occurred at an observation angle of 0°, 45 W/(Sr*m^2). When the observation angle was greater than 90°, the radiance value diminished rapidly from 40 W/(Sr*m^2).

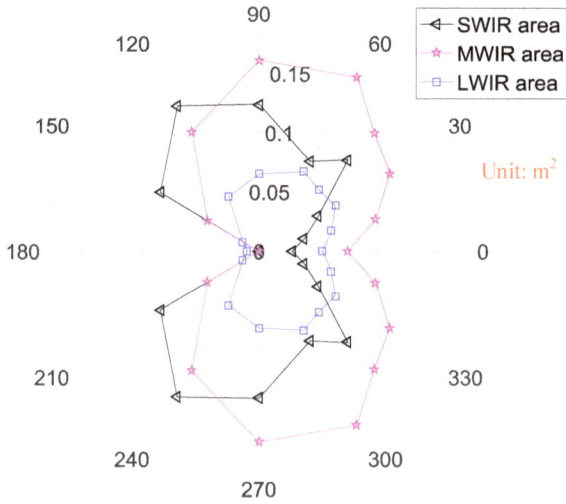

Figure 16. Orientation map of the UAV's radiation areas in different IR bands.

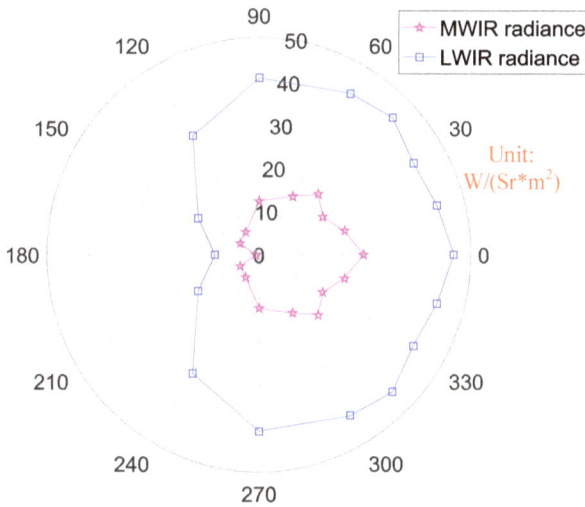

Figure 17. Orientation map of the UAV's radiance in different IR bands.

3.3. Comparison

Figure 18 shows the difference of the levels measured in terms of LWIR radiance orientation maps between the turbine engine and the complete UAV. The LWIR radiances of the turbine engine and UAV were roughly the same at different observation angles, except 30°–45°. At an observation angle of 30°, the engine's radiance in LWIR band was 58 W/(Sr*m^2), while the complete UAV's radiance in LWIR band was only 41 W/(Sr*m^2). It is worthwhile to note that the largest radiance of the engine in LWIR band is at an observation angle of 30°, that is 58 W/(Sr*m^2). In other words, under the current structure design of the UAV, the LWIR characteristics of the engine and its exhaust plume were mainly eliminated.

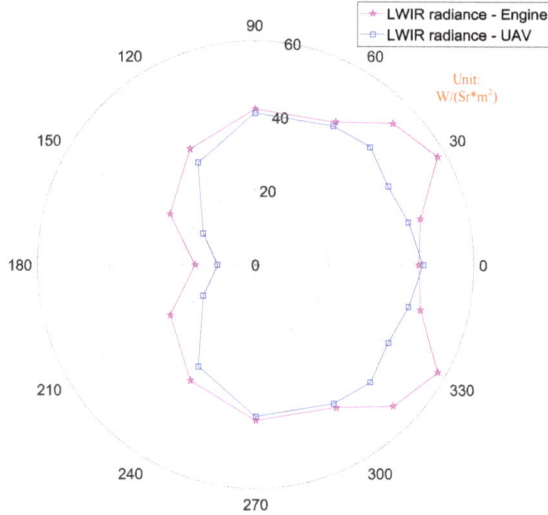

Figure 18. Comparison of LWIR radiance between engine and UAV.

Figure 19 shows the difference of the levels measured in terms of MWIR radiance orientation maps between the turbine engine and the complete UAV. For both the engine and the complete UAV, the radiance values in MWIR band were half that in LWIR band. In MWIR band, the largest radiance of the UAV was at 0°, which was 18 W/(Sr*m^2). As the observation angle increasing, radiances of the engine and the UAV continuously decreased. The main difference of the MWIR radiance values between the engine and the UAV was from 0° to 30°. At other angle of 30°–180°, the radiance values were almost the same. This difference suggests that the fuselage only affects the MWIR radiance of engine and its exhaust plume from observation angles of 0–30°.

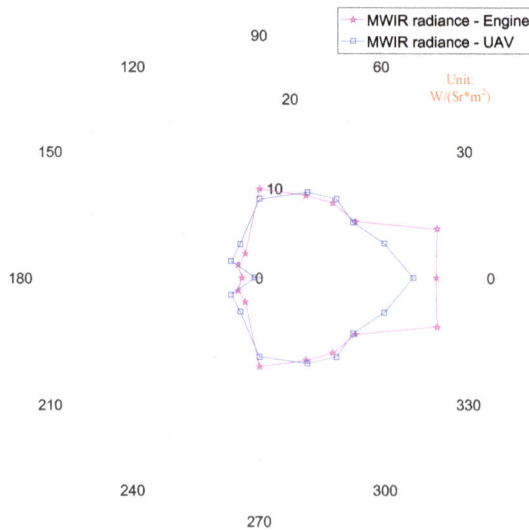

Figure 19. Comparison of MWIR radiance between engine and UAV.

4. Discussion and Conclusions

In this work, the IR radiation characteristics of a complete target drone and its engine were studied experimentally. The UAV and its engine were tested in the cruise state on the ground. The spectral radiation intensity for the UAV and its turbine engine at different observation angles were obtained by a spectral radiometer, and IR images were captured by LWIR, MWIR, and SWIR band cameras separately. The orientation maps of the radiation area in different IR bands were calculated, and the orientation maps of radiance in different IR bands were obtained. From the results, the following conclusions can be drawn.

- The spectral radiation features of the UAV in IR bands of 2–15 μm is determined by its exhaust plume; specifically, by the combustion products. According to Cain, fuel chemistry affects selectivity for specific decomposition pathways, and unburned fuel components are observed in the engine exhaust plume during operation with all fuels [19]. Except for the 2.7, 4.3, and 6.3 μm bands, other lines in Figures 5 and 11 may result from unburned fuel. Further research must be conducted to study the influence of other material in the exhaust plume on spectral radiation characteristics.
- IR radiation emitted from the UAV is primarily from carbon dioxide and water vapor. The main IR emission bands are at 2.7 and 4.3 μm. According to Rao [18], IR radiation emitted from a gas turbine engine exhaust plumes arises primarily from carbon dioxide and carbon monoxide. And Liu [11] found that 2.86–3.3 μm and 4.17–4.55 μm were the main emission bands determined by the two main constituents of the plume gas, namely CO_2 and H_2O. These conclusions are in good agreement with the current experimental results.
- At observation angles from 0–90°, the radiation area of the target drone in the MWIR band was the largest. However, at observation angles greater than 90°, the radiation area of the UAV in the SWIR band was the largest. This is important for the IR detection of target drones. When aiming at a target drone in transit, the observation angle is less than 90°. In this case, an MWIR detector may be the best choice. However, in the case of aiming at a target drone flying toward the observer, the observation angle is greater than 90°. In this case, an SWIR detector may be the best choice. But it should be noted that this conclusion has a precondition that the test is implemented on the ground indoors. During flight conditions, the drone interacts with its environment and undergoes solar illumination or background radiation from the sky and the ground. Further research under these circumstances is required.
- In the IR band, the radiance of the target drone at an observation angle of 0° is the strongest; however, regarding the engine, the radiance at an observation angle of 45° was strongest in the LWIR and MWIR bands. The radiance values of the engine at observation angles from 0° to 45° are relatively close. However, as the engine is embedded in a UAV, at an observation angle of 45°, the body of the UAV blocks most of the radiance emitted from the exhaust plume and the hot engine.

This paper provided additional and improved insight into the emission propensity in a target drone powered by a turbine engine. These conclusions are of great significance for IR stealth techniques regarding target drone's body, engine, and fuel formulation design and for IR warning system development.

Author Contributions: Tingzhu Bai, Quanxi Wang and Fengmei Cao conceived and designed the experiments; Long Shao and Zhaotian Sun performed the experiments; Haijing Zheng analyzed the data; Haijing Zheng wrote the paper.

Conflicts of Interest: The authors declare no conflict of interest.

References

1. Reddy, K.; Poondla, A. Performance analysis of solar powered unmanned aerial vehicle. *Renew. Energy* **2017**, *104*, 20–29.
2. Kumar, G.; Sepat, S.; Bansal, S. Review paper of solar powered UAV. *Int. J. Sci. Eng. Res.* **2015**, *6*, 41–44.
3. Yu, X.; Qing, L.; Liu, X.; Liu, X.; Wang, Y. A physical-based atmospheric correction algorithm of unmanned aerial vehicles images and its utility analysis. *Int. J. Remote Sens.* **2017**, *38*, 3101–3112. [CrossRef]
4. Lee, J.; Yu, K. Optimal path planning of solar-powered UAV using gravitational potential energy. *IEEE Trans. Aerosp. Electron. Syst.* **2017**, *53*, 1442–1451. [CrossRef]
5. Crusiol, L.G.T.; Nanni, M.R.; Silva, G.F.C.; Furlanetto, R.H.; Gualberto, A.; Gasparotto, A.; Paula, M. Semi professional digital camera calibration techniques for Vis/NIR spectral data acquisition from an unmanned aerial vehicle. *Int. J. Remote Sens.* **2017**, *38*, 2717–2736. [CrossRef]
6. Hu, S.; Goldman, G.H.; Borel-Donohue, C.C. Detection of unmanned aerial vehicles using a visible camera system. *Appl. Opt.* **2017**, *56*, B214–B221. [CrossRef] [PubMed]
7. Park, Y.; Nguyen, K.; Kweon, J.; Choi, J. Structural analysis of a composite target-drone. *Int. J. Aeronaut. Space Sci.* **2011**, *12*, 84–91. [CrossRef]
8. Li, Q.; Xie, W.; Luo, C. Identification of aircraft target based on multifractal spectrum features. In Proceedings of the 2012 IEEE 11th International Conference on Signal Processing, Beijing, China, 21–25 October 2012.
9. Gong, M.; Guo, R.; He, S.; Wang, W. IR radiation characteristics and operating range research for a quad-rotor unmanned aircraft vehicle. *Appl. Opt.* **2016**, *55*, 8757–8762. [CrossRef] [PubMed]
10. Blunck, D.L.; Gore, J.P. Study of narrowband radiation intensity measurements from subsonic exhaust plumes. *J. Propuls. Power* **2011**, *27*, 227–235. [CrossRef]
11. Liu, F.; Shao, X.; Han, P.; Bin, X.; Cui, Y. Detection of infrared stealth aircraft through their multispectral signatures. *Opt. Eng.* **2014**, *53*, 094101. [CrossRef]
12. Huang, Z.; Li, X.; Feng, Y. Numerical calculation of the plume infrared radiation of a long-endurance UAV. *Proc. SPIE* **2016**, *10141*, 101410.
13. Zhou, Y.; Wang, Q.; Li, T. A new model to simulate infrared radiation from an aircraft exhaust system. *Chin. J. Aeronaut.* **2017**, *30*, 651–662. [CrossRef]
14. Retief, S.J.P.; Dreyer, M.M.; Brink, C. Infrared recordings for characterizing an aircraft plume. *Proc. SPIE* **2014**, *9257*, 92570C-1.
15. Goody, R.M.; Yung, Y.L. Vibration-rotation spectra of gaseous molecules. In *Atmospheric Radiation Theoretical Basis*; Oxford University Press: New York, NY, USA, 1989; pp. 67–121.
16. Alberti, M.; Weber, R.; Mancini, M.; Fateev, A.; Clausen, S. Validation of HITEMP-2010 for carbon dioxide and water vapour at high temperatures and atmospheric pressures in 450–7600 cm^{-1} spectral range. *J. Quant. Spectrosc. Radiat.* **2015**, *157*, 14–33. [CrossRef]
17. Alberti, M.; Weber, R.; Mancini, M.; Modest, M.F. Comparison of models for predicting band emissivity of carbon dioxide and water vapour at high temperatures. *Int. J. Heat Mass Transf.* **2013**, *64*, 910–925. [CrossRef]
18. Cain, J.; DeWitt, M.J.; Blunck, D.; Corporan, E.; Striebich, R.; Anneken, D.; Klingshirn, C.; Roquenore, W.M.; Wal, R.V. Characterization of gaseous and particulate emissions from a turboshaft engine burning conventional, alternative, and surrogate fuels. *Energy Fuels* **2013**, *27*, 2290–2302. [CrossRef]
19. Rao, G.; Mahulikar, S. Aircraft powerplant and plume infrared signature modelling and analysis. In Proceedings of the 43rd AIAA Aerospace Sciences Meeting and Exhibit, Reno, NV, USA, 10–13 January 2005. [CrossRef]

sensors

MDPI

Article

Robust Vehicle Detection in Aerial Images Based on Cascaded Convolutional Neural Networks

Jiandan Zhong [1,2,3,*], **Tao Lei** [1] **and Guangle Yao** [1,2,3]

[1] Institute of Optics and Electronics, Chinese Academy of Sciences, No. 1, Guangdian Avenue, Chengdu 610209, China; taoleiyan@ioe.ac.cn (T.L.); guangle.yao@std.uestc.edu.cn (G.Y.)
[2] School of Optoelectronic Information, University of Electronic Science and Technology of China, No. 4, Section 2, North Jianshe Road, Chengdu 610054, China
[3] University of Chinese Academy of Sciences, 19 A Yuquan Rd, Shijingshan District, Beijing 100039, China
* Correspondence: jdzhong@std.uestc.edu.cn; Tel.: +86-138-8049-5638

Received: 13 October 2017; Accepted: 22 November 2017; Published: 24 November 2017

Abstract: Vehicle detection in aerial images is an important and challenging task. Traditionally, many target detection models based on sliding-window fashion were developed and achieved acceptable performance, but these models are time-consuming in the detection phase. Recently, with the great success of convolutional neural networks (CNNs) in computer vision, many state-of-the-art detectors have been designed based on deep CNNs. However, these CNN-based detectors are inefficient when applied in aerial image data due to the fact that the existing CNN-based models struggle with small-size object detection and precise localization. To improve the detection accuracy without decreasing speed, we propose a CNN-based detection model combining two independent convolutional neural networks, where the first network is applied to generate a set of vehicle-like regions from multi-feature maps of different hierarchies and scales. Because the multi-feature maps combine the advantage of the deep and shallow convolutional layer, the first network performs well on locating the small targets in aerial image data. Then, the generated candidate regions are fed into the second network for feature extraction and decision making. Comprehensive experiments are conducted on the Vehicle Detection in Aerial Imagery (VEDAI) dataset and Munich vehicle dataset. The proposed cascaded detection model yields high performance, not only in detection accuracy but also in detection speed.

Keywords: vehicle detection; convolutional neural network; aerial image; deep learning

1. Introduction

Vehicle detection in aerial images is an important task in various fields, such as: remote sensing, intelligent transportation and military reconnaissance. With the great development of Unmanned Aerial Vehicle (UAV) technologies, aerial images are captured conveniently and flexibly in this way. For the growing aerial imagery data, vehicle detection has become a challenge, attracting extensive attention recently. As a fundamental task in computer vision, vehicle detection is widely studied in some practical applications, such as traffic monitoring [1,2] and safety assistant driving [3,4], but for aerial images, it is still a tough problem due to the obscurity, relatively small size of the targets and cluttered backgrounds. Additionally, other objects such as big containers and road marks always show a similar appearance to vehicles, which will cause false detection or accuracy loss. Furthermore, in a detection model, not only detection accuracy is demanded, but also good detection speed.

In last decade, target detection technology has developed greatly, and can be roughly divided into three stages. In the first stage, the combination of hand-crafted features and discriminative classifiers were utilized to detect targets. On the one hand, some classical method like Histogram of Oriented Gradient (HOG) [5] and Scale-Invariant Feature Transform (SIFT) [6] were designed for feature extraction. On the other hand, the discriminative classifiers like Support Vector Machine (SVM) [7] and Ada-Boost [8] were adopted for classification. Felzenszwalb et al. [9] proposed a deformable parts model (DPM), which employs various trained components to detect targets from an image pyramid in sliding-window fashion. Although DPM is an excellent detector, the sliding-window strategy is time consuming in the detection phase. In the second stage, the sliding-window method was replaced with a region proposal way [10–12]. It means that the detectors don't need to detect the targets from the image pyramid, but from thousands of candidate target-like regions. This is a very efficient way to reduce the detection time. For example, the candidate regions of an image (of size 400×500) is about 10^3, which is much less than the search space (about $10^4 \sim 10^5$) of the image pyramid with a sliding-window way. The third stage started in 2012, when Krizhevsky et al. [13] applied the convolutional neural networks (CNNs) method in an image classification challenge (ILSVRC2012) and obtained striking results [14], which turned CNN-based methods into the mainstream in the field of computer vision. Recently, Girshick et al. [15] and Sermanet et al. [16] proposed efficient detection models based on CNNs. Especially, the method described in [15], called Regions with CNN (R-CNN), has become the baseline for the detection framework. The workflow of R-CNN is mainly divided into two steps: (a) it employs the region proposal method discussed in [10] to generate a set of candidate regions, and then (b) these regions are warped into a fixed size and fed into a CNN to extract the deep features. From the extensive experimental results, the CNN features show more discriminative capability than the traditional hand-crafted features. It is noteworthy that the region proposal method [10] always takes several seconds on an image of medium size (e.g., 500×300 pixels), and the CNN features in the different regions would be extracted repeatedly. Then, improved methods named SPP-Net [17] and Fast R-CNN [18] were proposed to accelerate the detection speed. In Fast R-CNN, a region of interesting (ROI) strategy was used to deal with the problem of repeated CNN feature extraction, which speeds up the CNN feature extraction procedure significantly. Another main bottleneck of R-CNN is the computational costs in the region proposal procedure. Ren et al. [19] proposed a CNN-based architecture called Region Proposal Network (RPN) to replace the method described in [10]. They combined RPN with Fast R-CNN and trained a unified detection model, which achieves state-of-the-art performance on PASCAL 2007/2012 and MS COCO datasets [20]. The detection speed reached 5 fps with a VGG-16 network [21].

Although the works [17–19] show promising results in target detection, they are not suitable for aerial images. The first reason is that the vehicles in the aerial image are relatively small in size (the average size of a vehicle is 40×20 pixels), and due to the coarseness of the feature map (output of the deep convolutional layer of the CNN), RPN has the poor localization performance for small targets. Moreover, the detection models [17–19] are designed for multi-category detection, but for the specific category "vehicle", they perform poorly due to the false positives. The second reason is that the vehicles always appear as vehicle roofs in aerial images, which has similarity with other background targets. This would cause accuracy loss without specific training. Furthermore, unlike the large scale public datasets (such as: ImageNet [14] and MS COCO [20]) comprising millions of images, the training data of available annotated aerial image datasets (for vehicles) is insufficient.

In this paper, we propose a cascaded CNN model to detect vehicles in aerial imagery data, which maintains high detection accuracy and fast speed. The framework of our model, shown in Figure 1, comprises two CNN-based networks. The first network is called the vehicle-regions proposal network (VPN) which aims to generate the vehicle-like regions. The second part is the vehicle detection network (VDN) which performs decision making for the regions generated by the first network. The workflow of the detection phase is divided into three steps: (1) an input image is put into the VPN to generate candidate regions, (2) the generated regions incorporating the input image are fed into VDN to extract each region's feature and predict the confidence score, (3) the regions with high score (greater than a threshold) are output as detections. Compared with the work [19], our model has three main differences: (1) unlike the work [19] that trains a unified network, we train two independent networks. It means that we do not share the convolutional layers of two networks, which avoids re-training the unshared layers of two networks; (2) the feature maps output from the deep convolutional layers (of CNN) can detect the target with high recall but poor localization performance, while the feature map from the shallow layers have better localization performance but obtain a reduced recall [22]. To take advantage of both, we combine the feature maps of the shallow layers and deep layers together to generate the vehicle-like regions in various scales and hierarchies. In this way, our method obtains finer and more accurate vehicle-like regions than RPN; (3) the VDN is trained as a specific category detector which is applied to detecting multi-type vehicles.

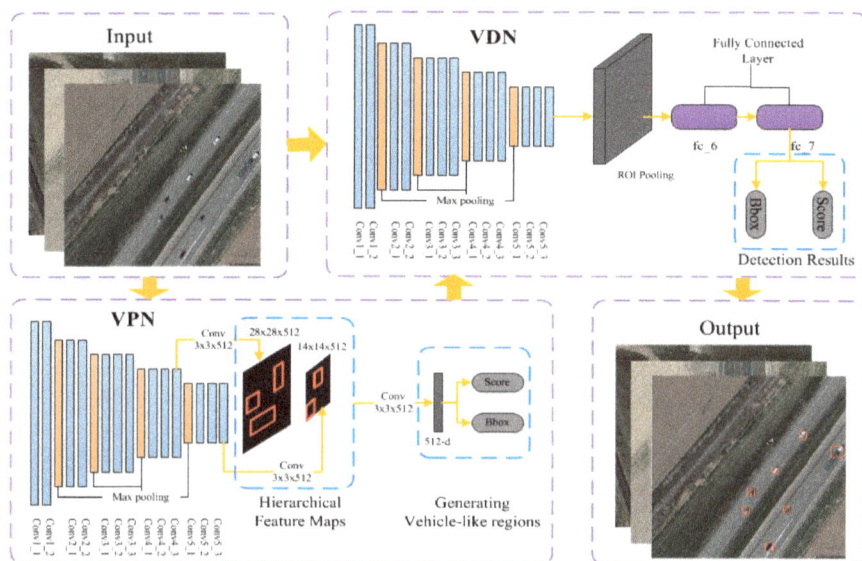

Figure 1. Framework of the proposed model.

Additionally, the original annotations of aerial image data are not suitable for VDN due to the fact that the bounding boxes of targets are annotated with various orientations. In this paper, the target bounding box is transformed into a vertical or horizontal format. To avoid overfitting in such a deep network, the training data are augmented by flipping and rotating operations.

The contributions of this paper are:

- A fast and accurate detection model is designed for vehicle detection in aerial images, which is different from the traditional sliding-window-based model and the recent CNN-based model. Our model is a cascaded architecture which incorporates two independent CNNs: the first is employed to generate vehicle-like regions, and the second is a specific-category detector which makes a final decision.
- The VPN is proposed to extract vehicle-like regions. Unlike the RPN that uses only one feature map, the proposed VPN combines multi-feature maps of different size and hierarchy for generating better vehicle-like regions. Actually, the proposed VPN takes effect on other categories as well, especially for the small targets in aerial image.
- A category-specific detector named VDN is developed, which can detect the various types of vehicles in aerial images. Additionally, unlike the Faster R-CNN which employs two-stage alternative training to share some convolutional layers, our VDN and VPN are trained independently once to increase training efficiency. This also avoids re-training the unshared layers that exist in the two networks. The VDN can be easily transferred to other target detection tasks.

- An augmented dataset is built for vehicle detection in aerial images. To make the training data fit for our CNN-based model, we re-annotated the available public dataset. To avoid the overfitting, we performed data augmentation in two operations.

The rest of this paper is organized as follows: in Section 2, we describe the related work about the region proposal method, CNN-based detectors and the related detectors designed for aerial image data. The preliminary theories and analysis of data augmentation, VPN and VDN are introduced in Section 3. In Section 4, we show the evaluation results on the VEADI and Munich vehicle datasets. We conclude this paper and propose some future work in Section 5.

2. Related Work

In this section we review the recent methodologies related to target detection. Moreover, some recent vehicle detection methods are introduced as well.

2.1. Region Proposal Method

In many target detection approaches a small number of candidate regions which cover all the objects in an image is proposed, and extensive studies on region proposal methods can be found in [10–12,19,23–28]. Carreira et al. [23] proposed a rough segmentation method to generate candidate regions, which has been shown to be effective. Promising results were obtained by the method of estimating the objectness score on an image [11,12,24]. Uijlings et al. [10] proposed the Selective Search (SS) way, which generates regions with better objectness based on its hierarchical segmentation and grouping strategies. Additionally, the works [29–32] adopted the method of super-pixels segmentation to generate image regions. In particular, Achanta et al. [29] proposed a simple and efficient method called simple linear iterative clustering (SLIC), which performs well in image segmentation. However, the candidate regions generated by a segmentation method cannot be directly fed into CNN for feature extraction because the segmented regions are polygonal regions which should be converted into rectangles first. Recently, the use of CNN-based methods to generate the candidate regions has become a trend. Deepbox [28] trained a slight CNN model and learned to re-rank candidate regions generated by [24]. Ren et al. employed RPN [19] and Fast R-CNN [18] to train a unified detection model. Through this two-stage alternative training, this model yields state-of-the-art performance.

2.2. Target Detection with the CNN-Based Models

By virtue of its powerful feature extraction capability, CNN has been widely used in target detection. References [15,16,33] are the pioneering works of employing CNN to deal with target detection tasks. Although these works perform well in detection accuracy, they are time consuming in feature extraction. Then, shared convolution computation has attracted more attention, and the methods in references [17,18,34] were proposed to deal with this problem. The SPP-Net [17] and Fast R-CNN [18] proposed further improvements on [15], which showed compelling accuracy and speed. To obtain more efficient and accurate localization of targets, more and more works have employed CNN-based models to generate proposal regions. RPN [19] and MultiBox [35] are two representative works. Moreover, Redmon et al. [36] presented a proposal-free framework named You Only Look Once (YOLO), which directly predicts bounding boxes and evaluate probabilities without proposing candidate regions. In practice, the region-based models like [19] outperform YOLO with respect to the detection accuracy.

2.3. Vehicle Detection in Aerial Imagery

Detecting vehicles in aerial imagery data is an interesting topic nowadays. Xu et al. [37] proposed a hybrid method which adopted the traditional hand-crafted features (HOG) and linear SVM. For vehicles on a highway, this method yields high performance, however, this method uses a lot of road-line information as auxiliary. Nassim et al. [38] proposed a deep learning method to detect vehicles in the aerial images captured by UAVs, where they first segment the regions of interest in the image and then feed them into a CNN model for feature extraction. The final decision was made by a SVM. Qu et al. [39] combined the region proposal method in [11] with SPP-Net [17] to build a vehicle detection model. The works [38,39] both employed the CNN as a feature extractor. Tang et al. [40] proposed the Hyper Region Proposal Network (HRPN) to localize the vehicle-like regions, and utilized hard negative examples to improve the detection accuracy. Deng et al. [41] modified RPN and Fast R-CNN to build a unified CNN-based model for vehicle detection. In fact, the works [40,41] concatenated multiple convolution layers into one hyper-feature map, but the multi-hierarchy and scale information concepts were not adopted.

3. Overview of the Proposed Model

The proposed vehicle detection model (shown in Figure 1) consists of two cascaded CNNs: a vehicle-regions proposal network (VPN) and a vehicle detection network (VDN), which are trained independently during the training phase. The VPN aims to generate candidate vehicle-like regions accurately at first. Afterwards, these vehicle-like regions are fed into the VDN to make inference. Moreover, to avoid overfitting, we augment the original dataset artificially.

3.1. Training Data Augmentation

The VEDAI [42] and Munich vehicle datasets [43] are adopt to evaluate the performance of detection model. The VEDAI dataset includes about 1240 images with two kinds of resolutions: 1024×1024 and 512×512 pixels. The training data is relatively small, which is just applicable for many situations (such as vehicle detection in urban, country road, crop and residential areas) and may not be able to meet the needs of a larger range practical application (e.g., to detect the vehicles are partially covered by vehicle-like regions with trees or artificial structures). Therefore, the experimental design of this study makes it difficult to comment on the feasibility of large-scale implementation. Additionally, it is very inefficient to directly use CNN-based models for target detection in the image with such a large scale of resolution (5616×3744 pixels). For this reason an input image will be resized by the designed CNN model (the shorter side of the image will be resized to 600 pixels for convenience). For large size images, this will cause an accuracy loss. Hence, the images in Munich vehicle dataset are cropped to a size of 702×468 pixels for training and testing.

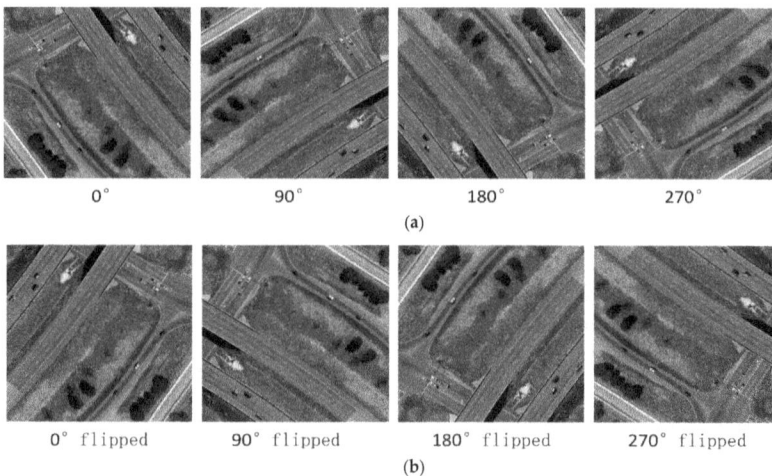

$0°$ $90°$ $180°$ $270°$

(a)

$0°$ flipped $90°$ flipped $180°$ flipped $270°$ flipped

(b)

Figure 2. (a) Training images are rotated with four angles in clockwise; (b) The flip operation of the training images.

Additionally, due to the lack of training data, we augment the training data by two operations: rotation and flip (described in Figure 2). For each training image, we rotate it with four angles ($0°$, $90°$, $180°$ and $270°$) in a clockwise direction. Further, we flip the rotated images as well (shown in Figure 2b). Another problem is that the original annotation information of these datasets is not suitable for CNN-based models, because the bounding boxes of targets are rotated with various angles. We adjust the coordinates of bounding box according to the steps below:

(1)   Obtaining the original four coordinates of bounding box: $[x_{lt}, y_{lt}]$, $[x_{rt}, y_{rt}]$, $[x_{rb}, y_{rb}]$ and $[x_{lb}, y_{lb}]$;
(2)   Calculating the *height*: $h = max(y_{lt}, y_{rt}, y_{rb}, y_{lb}) - min(y_{lt}, y_{rt}, y_{rb}, y_{lb})$;
(3)   Calculating the *width*: $w = max(x_{lt}, x_{rt}, x_{rb}, x_{lb}) - min(x_{lt}, x_{rt}, x_{rb}, x_{lb})$;
(4)   Updating the left-top coordinate as $[min(x_{lt}, x_{rt}, x_{rb}, x_{lb}), min(y_{lt}, y_{rt}, y_{rb}, y_{lb})]$;
(5)   Using the height, width and left-top coordinate to update other coordinates.

Figure 3 gives examples of the original annotation and the updated annotation.

The Original Annotation   The Updated Annotation

Figure 3. Examples of the original and updated annotations.

3.2. Vehicle-Regions Proposal Network

The proposed VPN takes an image as input and outputs a set of vehicle-like regions with the corresponding objectness scores. RPN [19] adopts the feature map of the deep convolutional layer to generate candidate regions. To improve this framework, references [41,44] concatenated multiple convolutional layers and built a hyper-feature map. Enlightened by these works [19,41,44], we combine deep and shallow convolutional layers to construct a hierarchical structure which comprises coarse and fine feature maps with various sizes and scales. In our VPN, the region proposals are generated from each feature map. As a result, more accurate regions are proposed than by using the methods of [19,41,44], which adopt only one feature map. The detailed description of VPN is provided below.

3.2.1. Overview of the Architecture

The architecture of VPN is based on the VGG-16 model [21], which is a deep CNN including 13 convolutional layers and three fully connected layers (shown in Figure 4a). The original VGG-16 is an excellent model that is usually applied in image classification. Firstly, it generates a deep feature map by 13 convolutional layers. Then, the deep feature map is fed into the three fully connected layers to form a 4096-d (dimension) feature vector. Lastly, the feature vector is input into a soft-max for classification. However, VPN is used to deal with region-proposal task, which aims to not only predict the position of candidate regions, but also evaluate their objectness scores. Therefore, we reserve the 13 convolutional layers to generate multi feature maps, and make further modifications. Specifically, we modify this model by two strategies: (1) deleting the last three fully connected layers (from fc_6 to fc_8) and Soft-Max layer; (2) adding two small networks behind conv4_3 and conv5_3 respectively to generate candidate regions. The outputs of each small network are fed into two sibling fully connected layers for predicting bounding box and evaluating objectness score. Figure 4b illustrates the modifications and process of VPN.

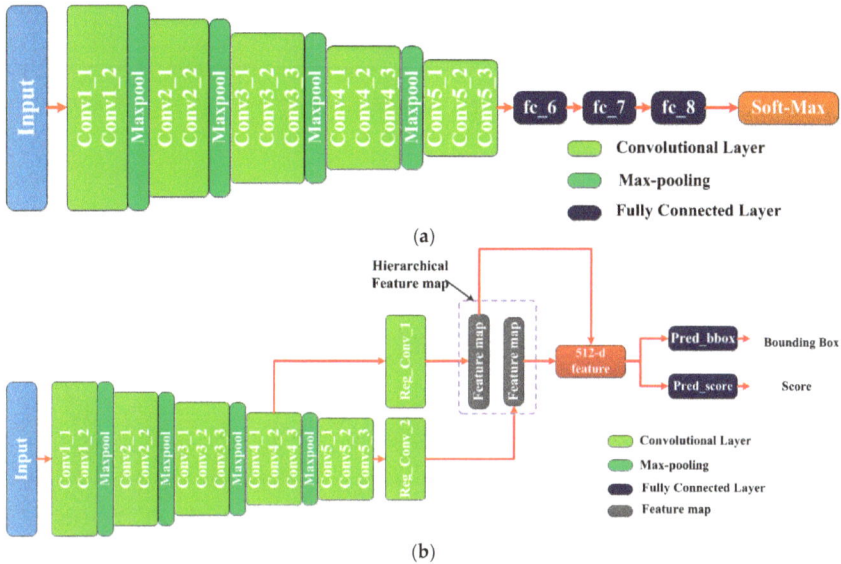

(a)

(b)

Figure 4. (**a**) The architecture of VGG-16 model; (**b**) The architecture of VPN.

Detailed descriptions of each layer are presented below:

Input data: this model requires RGB images (of any size) as the input.

Conv1 layers: Conv1 layers include two convolution layers (*conv1_1* and *conv1_2*), and the rectified linear units are configured after each convolutional layer. 64 kernels of sizes 3×3 are adopted for each layer.

Conv2 layers: configurations of Conv2 layers are almost as same as Conv1 layers'. The only difference is that Conv2 layers adopt 128 kernels of sizes 3×3.

Conv3, Conv4 and Conv5 layers include three convolutional layers, and the rectified linear units are configured after each convolutional layer. 256, 512 and 512 kernels (of size 3×3) are adopted respectively.

Pooling layers: this model adopts four pooling layers which are placed between the aforementioned Conv layers. The pooling layers are configured as max pooling with kernel of size 2×2.

Reg_Conv_1 layer and Reg_Conv_2 take *conv4_3* and *conv5_3* as the input respectively. Then, 512 kernels (of size 3×3) are adopted to generate two feature maps with different size.

Feature map: the hierarchical feature map architecture combines the output of the shallow convolutional layer and the deep convolutional layer. Because the shallower layers are better for localization and deeper layers are better for classification, the hierarchical feature map architecture integrates the advantages of both. Especially for small vehicles in aerial images, it shows better performance. In the hierarchical feature map architecture, a window of size $3 \times 3 \times 512$ is slid to generate the vehicle-like regions. At each position, a 512-d (dimension) feature is extracted and fed into two sibling fully connected layers. The *pred_bbox layer* is used to predict the bounding box and the *pred_score layer* outputs a discrete probability distribution over two categories (vehicle-like region or background).

Following the anchor scheme in [19], this network predicts multiple regions associated with the different aspect ratios and scales at each sliding-window position. According to the average size of a vehicle (which is about 20×40 pixels), three aspect ratios (1:2, 1:1, 2:1) and four scales (16^2, 32^2, 48^2, 64^2) are set for vehicle-like regions. Hence, each sliding-window position generates 12 types of regions. We assign a positive label to the regions which have higher intersection-over-union (*IoU*) overlap ratio (which is greater than 0.7) with a ground-truth bounding box. Inversely, we assign a negative label to the regions which have lower *IoU* ratio (between 0.1 and 0.3) with ground-truth. The definition of *IoU* is seen as below (Equation (1)):

$$IoU_{ratio} = \frac{A_{reg} \cap A_{gt}}{A_{reg} \cup A_{gt}} \tag{1}$$

where, A_{reg} and A_{gt} represent the bounding-box area of candidate regions and ground truth respectively.

3.2.2. Loss Function

A multi-task loss function L (shown in Equation (2)) is employed to jointly train for classification and bounding-box regression:

$$L(p^t, l^t) = L_{cls}(p^t, p^g) + \lambda * p^g * L_{br}(l^t, l^g) \tag{2}$$

For the *pred_score* layer, p^t is the predicted probability of region being an object. The ground-truth label p^g is 1 if the region is positive, and is 0 if the region is negative. L_{cls} is log loss over two categories (vehicle-like region and background).

The *pred_bbox layer* outputs a vector representing the four parameterized coordinates ($x, y\ w, h$) of the predicted bounding box. x, y, w, and h denote the box's center coordinates and its width and height. l^g and l^t represent the ground-truth bounding box and predicted bounding box respectively. And L_{br} adopts smooth L1 loss function [18] defined in Equations (3) and (4). The parameter λ is the balancing parameter, and it is set to 10:

$$L_{br}(l^t, l^g) = S_{L1}(l^t - l^g) \tag{3}$$

$$S_{L1}(x) = \left\{ \begin{array}{ll} 0.5x^2 & \text{if} |x| < 1 \\ |x| - 0.5 & \text{otherwise} \end{array} \right\} \tag{4}$$

3.2.3. Training

The VPN is trained by the method of stochastic gradient descent (SGD) [45]. In the experiments, we initialize our model by a pre-trained VGG-16 weights which is previously trained on ILSVRC [14]. Because that the weights of new added convolutional layers should be initialized firstly, we initialize them by zero-mean Gaussian distribution with a 0.01 standard deviation, which is a widely used initialization way for CNN model in Caffe—deep learning framework [46]. Specifically, the initializations are configured in the model file (a 'prototext' file to describe the structure of the model). During training, a mini-batch is generated from one image, and it is set to 256. We keep the ratio of positive and negative examples to 1:1. If there are fewer than 128 positive examples in an image, we pad the mini-batch with negative ones. After the training process, VPN can generate a set of candidate regions; actually, there is no need to feed all of the regions to VDN. The works [10–12] have proven that top 2000 candidate regions almost cover all objects in the images. The RPN performs better than the traditional works [10] by adopting the top 300 candidate regions. As an improved version of RPN, the VPN also adopts top 300 highly overlapped candidate regions and feeds them into VDN for the further inference.

3.3. Vehicle Detection Network

Vehicle detection network takes the generated vehicle-like regions and image as the input and outputs a set of detections. The details of VDN are described as below.

3.3.1. Overview of the Architecture

The architecture of VDN is also based on the VGG-16 model. Because the sizes and scales of the candidate regions are different, in order to extract the fixed-length feature vector from each region, the ROI polling layer [18] and two fully connected layers (fc_6 and fc_7) are adopted. Additionally, as a detection model, VDN is required to output the vehicle's bounding box of and evaluate its confidence score. Two sibling fully connected layers are added behind fc_7 layer. Figure 5 illustrates the architecture of VDN.

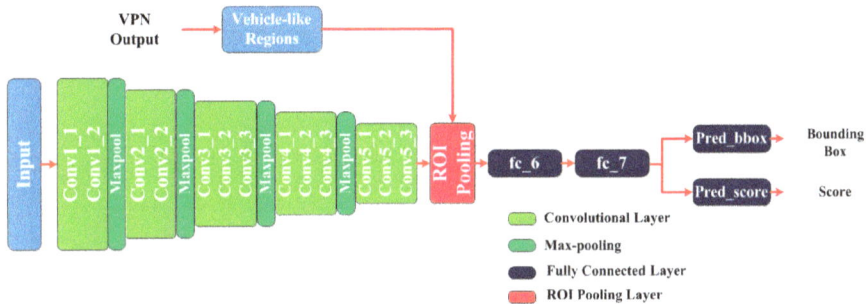

Figure 5. The architecture of VDN.

Input data: this model requires two kinds of input data. One input is the same RBG image as the input of VPN. Another input is a set of candidate regions generated by VPN, which are directly mapped into ROI pooling layer.

Convolutional layers: the convolutional layers from Conv1 to Conv5 take identical settings as VPN.

ROI pooling layer: because the generated vehicle-like regions have various sizes, this layer extracts a fixed-length feature vector for each vehicle-like region. Specifically, this layer works by dividing the ROI (region of interest) window into a 6 × 6 grid of sub-windows and then max-pooling the values in each sub-window into the corresponding output grid cell [18]. Pooling is applied independently to each feature map channel, as in standard max pooling. The generated feature is the input of fc_6.

Fc_6 is a fully connected layer that outputs a 4096-d feature vector. Fc_7 takes the same settings as fc_6 and it is branched into two sibling fully connected layers, named *pred_bbox* and *pred_score* respectively. The *pred_bbox* layer predicts the bounding box of vehicle, and the output of *pred_score* layer is the corresponding confidence score.

3.3.2. Loss Function and Training

The output of VDN and VPN is similar; therefore, the multi-task loss function L as given by Equation (2) is adopted to jointly train this network for vehicle classification and bounding-box regression. Moreover, the pre-trained VGG-16 weights are adopted as well. The training parameters and settings are similar to VPN.

4. Experiment and Results

We report the experimental results on two benchmark datasets: the VEDAI dataset [42] and the Munich vehicle dataset [43]. The performance of our detection model is compared with other methods on two aspects: detection accuracy and detection speed. Detailed evaluation metrics are described in Section 4.1. All methods in the experiments were programmed based on Matlab 2014a and Caffe deep learning framework [46]. All experiments were run on a desktop computer equipped with an Intel Core i7 5930k CPU (6 Core, 3.5 GHz), 64 GB memory, a NVIDIA Titan X GPU (with 12 GB video memory) and Ubuntu 14.04 OS.

4.1. Evaluation Metrics

We employ the widely used four metrics including: the precision-recall curve (PRC) [47], average precision (AP), recall rate and F1-Score [48] to quantitatively evaluate the performance of our model. The definition of F1-score is shown in Equation (5):

$$F1\_Score = \frac{2 * recall * precision}{recall + precision} \tag{5}$$

where, recall and precision are calculated by Equations (6) and (7):

$$Recall = \frac{True\ Positive}{True\ Positive + False\ Negative} \tag{6}$$

$$Precision = \frac{True\ Positive}{True\ Positive + False\ Positive} \tag{7}$$

Recall and precision evaluate the correctly identified positive detections and true positive detections respectively. The AP is defined as the area under the PRC, which is a comprehensive indicator of precision and recall rate. To sum up, F1-Score and AP are two key criteria to reveal the performance of detectors. The higher the F1-Score and AP score, the better the performance. In the experiments, the detections with IoU_{ratio} value greater than 0.5 was defined as true, otherwise, it was false.

4.2. VEDAI Dataset

VEDAI is a public dataset providing various types of vehicle in the images which were taken during spring 2012 in Utah, USA. The images comprise different backgrounds such as road, desert, rural and urban areas (shown in Figure 6). This dataset provides images with two different sizes, which are referred as VEDAI 512 (512 × 512 pixels) and VEDAI 1024 (1024 × 1024 pixels) respectively. VEDAI 1024 has a ground sampling distance of 12.5 cm/pixel, and the VEDAI 512 comprises the downscaled images of VEDAI 1024 and has a ground sampling distance of 25 cm/pixel.

This dataset contains nine different classes of vehicles, there are 'car', 'pick-up', truck', 'plane', 'boat', 'camping car', 'tractor', 'van', and the 'other' category. There is an average of 5.5 vehicles per image, and they occupy about 0.7% of the total pixels of the images. The statistical data of each class is described in Table 1. Due to the scarcity of samples, we discard some categories (such as 'boat', 'plane' and 'tractor') in the experiments.

In the training stage, we adopted 996 images from VEDAI 1024 and augmented them according to the descriptions in Section 3.1. Each input image was resized such that its shorter side has 600 pixels. Moreover, for both networks (VPN and VDN), the training parameters were equivalent. We applied a weight decay of 0.0005 and a momentum of 0.9. There were 40,000 iterations in total during the whole training process, and the learning rate was set as 0.001 for the first 30,000 iterations, and 0.0001 for the next 10,000 iterations.

Figure 6. Examples from the VEDAI dataset.

Table 1. The statistical data of VEDAI.

Classes	Tag	Number
Car	car	1340
Pick-up	pic	950
Truck	tru	300
Plane	pla	47
Boat	boa	170
Camping car	cam	390
Tractor	tra	190
Vans	van	100
Other	oth	200

In the test stage, about 240 images (rest images of the dataset) with different size were selected to evaluate the performance. Our model was compared with super-pixels segmentation based methods (such as SLIC [29]) and recent CNN-based detectors, including: Faster R-CNN with Z&F model [49], Faster R-CNN with VGG-16 model and Fast R-CNN with VGG-16model. For the SLIC based methods, we first segmented the image into 768 regions by SLIC, and then converted the generated polygonal regions into approximate rectangular regions. The converted regions were fed into VGG-16 and Z&F model respectively. These two models were referred as: SLIC with VGG-16 and SLIC with Z&F. As the comparison results in Table 2 illustrate, for VEDAI 1024, our detection model outperforms the super-pixels segmentation based methods and recent CNN-based detectors, which obtains the best AP (54.6%) and F1-score (0.305). Especially, the AP outperforms the second best detector by 12.5 percentage points. And the recall rate also reaches a comparable level with Faster R-CNN (VGG 16). For VEDAI 512, our model obtains the best AP and F1-Score as well. Figure 7a,b show the PRC of the various models on VEDAI 1024 and VEDAI 512, respectively. Compared with other models, our model shows significant improvement.

Table 2. Comparison results of various detection models on VEDAI.

Detection Model	Image Size	Recall Rate	AP	F1-Score
Faster R-CNN (Z&F)	1024 × 1024	63.5%	30.8%	0.229
Faster R-CNN (VGG-16)	1024 × 1024	**73.9%**	42.1%	0.232
Fast R-CNN (VGG-16)	1024 × 1024	72.2%	39.8%	0.216
SLIC with Z&F	1024 × 1024	58.3%	25.4%	0.066
SLIC with VGG-16	1024 × 1024	58.8%	23.2%	0.064
Our Model	1024 × 1024	72.3%	**54.6%**	**0.320**
Faster R-CNN (Z&F)	512 × 512	60.9%	32.0%	0.212
Faster R-CNN (VGG-16)	512 × 512	**71.4%**	40.9%	0.225
Fast R-CNN (VGG-16)	512 × 512	69.4%	37.3%	0.224
Our Model	512 × 512	69.7%	**50.2%**	**0.305**

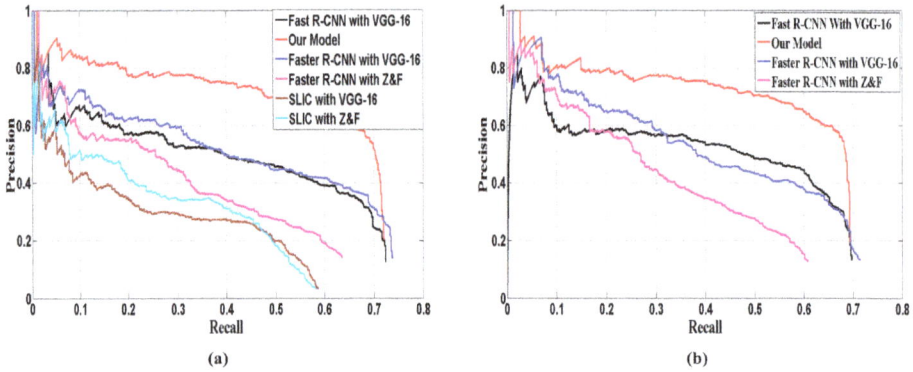

Figure 7. Precision-recall curve of four models: (**a**) VEDAI 1024 (**b**) VEDAI 512.

The performance of VPN determines the results of detection model, to evaluate the localization performance of VPN; we compared it with other RPN-based region proposal methods. Reference [19] designed the RPN based on Z&F and VGG-16 model respectively. We adopted the recall-IoU curve (shown in Figure 8) for evaluation.

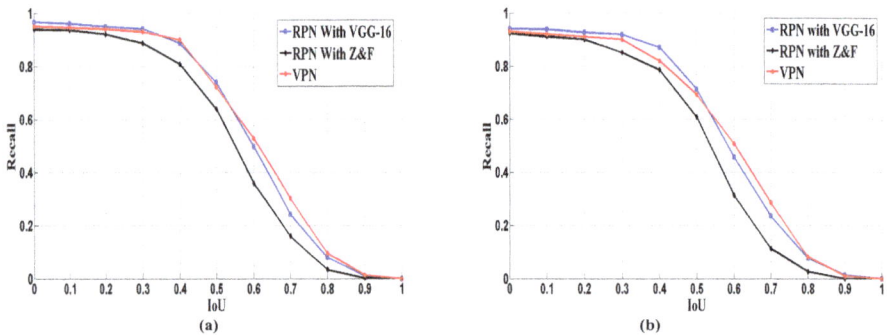

Figure 8. Recall vs. *IoU* curve of three CNN-based models: (**a**) VEDAI 1024 (**b**) VEDAI 512.

As the results in Figure 8 show, our model obtains a comparable recall rate to Faster R-CNN (with VGG-16). When the IoU_{ratio} is greater than 0.5, our model achieves the best performance. Additionally, we evaluated the detection speed of different detection models by *fps* (frames per second). Table 3 illustrates the detection time and training time of each detection model. From the aspect of detection time, our model, SLIC based models and other two Faster R-CNNs achieve comparable detection speed. The Fast R-CNN that uses the Selective Search [10] scheme for region proposal performs poorly, and its detection speed is much slower than the speed of the other five. The Faster R-CNN (with Z&F model) adopts a simple and shallow CNN, so it achieves the fastest detection speed. However, it obtains the lower detection accuracy (30.8% and 32%). The SLIC based models perform well on detection speed, which are benefit for the segmentation speed of SLIC algorithm, but they obtain the lowest detection accuracy (23.2%). This may be caused by the inaccurate segmentation and the conversion of segmented regions. The detection speed of our model is a little slower than Faster R-CNN with VGG-16, because the proposed VPN is a hierarchy architecture, which spends a little more time on generating more but accurate candidate regions. Actually, this gap is very small in practical application. Hence, we made the trade-off between detection speed and accuracy. For the training time, Fast RCNN and SLIC based models perform well, because training CNN is time consuming and they just adopt one CNN for feature extraction, the rest models employ two CNNs for region proposal and feature extraction respectively. Our model is better than the Faster RCNNs, because the Faster RCNNs are alternatively trained twice, but we train each CNN (VPN and VDN) only once. In practical application, detection time is considered more. Due to the fact that detection systems always adopt the trained model and no extra training cost during the detection phase.

Table 3. Comparison of detection time (fps: frames per second) and training time (h: hours).

Detection Model	Image Size	Detection Time	Training Time
Faster R-CNN (Z&F)	1024 × 1024	**5.8 fps**	28.4 h
Faster R-CNN (VGG-16)	1024 × 1024	5.4 fps	28.5 h
Fast R-CNN (VGG-16)	1024 × 1024	0.4 fps	8.2 h
SLIC with Z&F	1024 × 1024	5.6 fps	7.9 h
SLIC with VGG-16	1024 × 1024	4.9 fps	8.2 h
Our Model	1024 × 1024	4.5 fps	10.7 h
Faster R-CNN (Z&F)	512 × 512	**6.3 fps**	28.3 h
Faster R-CNN (VGG-16)	512 × 512	5.6 fps	28.6 h
Fast R-CNN (VGG-16)	512 × 512	0.4 fps	8.1 h
Our Model	512 × 512	4.6 fps	10.6 h

Figure 9 shows some detection examples of VEDAI 1024. Figure 9a,c,e,g,i,k is the input images, and the ground truths are annotated by yellow boxes. Figure 9b,d,f,h,j,l is the detections annotated by red boxes.

Figure 9. (a–l) some detection examples of VEDAI dataset.

4.3. Munich Vehicle Dataset

The Munich vehicle dataset is an aerial imagery dataset captured by the DLR 3 K camera system [50] over the area of Munich, Germany. It comprises of 20 aerial images which were mainly taken from urban and residential areas. The original images in this dataset were taken at the height of 1 km above the ground with the resolution of 5616 × 3744 pixels, and the approximate ground sampling distance is 13 cm/pixel. Training and testing set include 10 images respectively.

We performed our model on the testing set and compared the performance with other two RPN-based models (Faster R-CNN with VGG-16 and Faster R-CNN with Z&F). In the training process, we firstly cropped the original images into the size of 702 × 468; in this way, then collected 640 training images from Munich dataset. Secondly, we combined the training set of VEDAI 1024 and these cropped images to form a joint training set. During training, we used the same parameters and settings as that were adopted in VEDAI dataset.

In testing phase, each testing image was cropped into 702 × 468 pixels as well. Hence, 640 cropped images were employed as the testing set. As the evaluation results showed in Table 4, our model obtains the best detection accuracy. Especially, the AP outperforms other two models by approximate 20 and 10 percentage points. The detection speed also achieves a comparable level with that of others.

Table 4. Comparison results of various detection models on Munich Vehicle dataset.

Detection Model	Recall Rate	AP	F1-Score	Detection Time (fps)
Faster R-CNN (Z&F)	66.8%	53.9%	0.657	**5.2**
Faster R-CNN (VGG-16)	78.3%	64.8%	0.779	4.9
Our Model	**80.3%**	**73.7%**	**0.782**	3.2

In addition, the precision-recall curve and recall-IoU curve are showed in Figure 10a,b. Figure 11 gives some examples of the detection on the Munich vehicle dataset. Figure 11a,c,e,g,i,k is the input images, and the ground truths are annotated by yellow boxes. Figure 11b,d,f,h,j,l is the detect results, and the detections are annotated by red boxes.

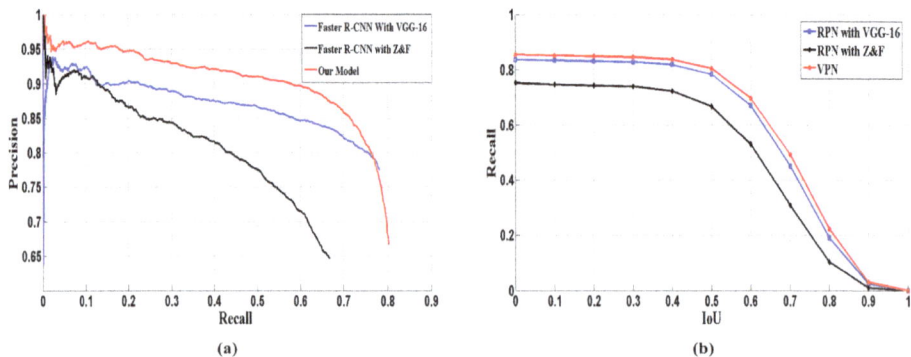

(a)

(b)

Figure 10. Comparisons of three detection models (**a**) precision-recall curve (**b**) recall vs. IoU curve.

Figure 11. (a–l) some detection examples of VEDAI dataset.

5. Conclusions

In this paper, we propose a fast and accurate vehicle detection model for aerial images. Unlike the traditional sliding-window-based detection models and recent CNN-based models, our detector is a cascaded CNNs architecture that combines two CNNs (VPN and VDN) for generating candidate regions and making decisions, respectively. The proposed VPN is based on a VGG-16 model; taking advantage of the shallow and deep feature map, we build hierarchical feature maps. Compared with other CNN-based region proposal methods (such as RPN with VGG-16, RPN with Z&F), the VPN generates more accurate candidate regions, especially for the small vehicles in aerial images. Moreover, we trained a category-specific detection network called VDN, which is combined with VPN and obtained high performance. From the extensive experimental results presented in Section 4, the proposed model outperforms the state-of-the-art detection model [18,19] in detection accuracy, and the detection speed achieves a comparable level.

Although our model has obtained favorable performance on vehicle detection in aerial image data, it still has some limitations. One limitation is in hard example detection, for example, when some vehicles in aerial images are partially occluded by other objects or extremely small vehicles. Moreover, to distinguish some intra-class vehicles is also difficult, such as camping cars and big vans. In the future work, we focus on the further optimization of VPN. Firstly, a deeper CNN model will be adopted and built finer architecture of feature maps. Moreover, to reduce the time cost of region proposal stage, we will try to improve the performance of the efficient super-pixel segmentation method like SLIC, which shows advantages in speed of generating regions, but the capability of generating accurate candidate regions should be improved. Multi-GPUs should be adopted collaboratively in the region proposal stage.

Acknowledgments: This work was supported by Youth Innovation Promotion Association, CAS (Grant No. 2016336). The authors would appreciate the anonymous reviewers for their valuable comments and suggestions for improving this paper.

Author Contributions: Jiandan Zhong proposed the original idea and wrote this paper; Tao Lei gave many valuable suggestions and revised the paper; Guangle Yao designed a part of experiments and revised the paper.

Conflicts of Interest: The authors declare no conflict of interest.

References

1. Tang, Y.; Zhang, C.; Gu, R.; Li, P.; Yang, B. Vehicle detection and recognition for intelligent traffic surveillance system. *Multimedia Tools Appl.* **2017**, *76*, 5817–5832. [CrossRef]
2. Wen, X.; Shao, L.; Fang, W.; Xue, Y. Efficient Feature Selection and Classification for Vehicle Detection. *IEEE Trans. Circuits Syst. Video Technol.* **2015**, *25*, 508–517.
3. Xu, H.; Zhou, Z.; Sheng, B.; Ma, L. Fast vehicle detection based on feature and real-time prediction. In Proceedings of the IEEE International Symposium on Circuits & Systems, Beijing, China, 19–23 May 2013; pp. 2860–2863.
4. Gu, Q.; Yang, J.; Zhai, Y.; Kong, L. Vision-based multi-scaled vehicle detection and distance relevant mix tracking for driver assistance system. *Opt. Rev.* **2015**, *22*, 197–209. [CrossRef]
5. Dalal, N.; Triggs, B. Histograms of oriented gradients for human detection. In Proceedings of the IEEE Conference on Computer Vision and Pattern Recognition, San Diego, CA, USA, 20–25 June 2005; pp. 886–893.
6. Lowe, D. Distinctive image features from scale-invariant keypoints. *Int. J. Comput. Vis.* **2004**, *60*, 91–110. [CrossRef]
7. Chang, C.; Lin, C. LIBSVM: A library for support vector machines. *ACM Trans. Intell. Syst. Technol.* **2011**, *2*, 389–396. [CrossRef]
8. Viola, P.; Jones, M. Rapid object detection using a boosted cascade of simple features. In Proceedings of the IEEE Conference on Computer Vision and Pattern Recognition, Kauai, HI, USA, 8–14 December 2001; pp. 511–518.
9. Felzenszwalb, P.; Girshick, R.; McAllester, D.; Ramanan, D. Object detection with discriminatively trained part based models. *IEEE Trans. Pattern Anal. Mach. Intell.* **2010**, *32*, 1627–1645. [CrossRef] [PubMed]
10. Uijlings, J.; Van de Sande, K.; Gevers, T.; Smeulders, A. Selective search for object recognition. *Int. J. Comput. Vis.* **2013**, *104*, 154–171. [CrossRef]
11. Cheng, M.; Zhang, Z.; Lin, W.; Torr, P. BING: Binarized Normed Gradients for Objectness Estimation at 300 fps. In Proceedings of the IEEE Conference on Computer Vision and Pattern Recognition, Columbus, OH, USA, 23–28 June 2014; pp. 3286–3293.
12. Alexe, B.; Deselaers, T.; Ferrari, V. Measuring the objectness of image windows. *IEEE Trans. Pattern Anal. Mach. Intell.* **2012**, *54*, 2189–2202. [CrossRef] [PubMed]
13. Krizhevsky, A.; Sutskever, I.; Hinton, G. ImageNet classification with deep convolutional neural networks. In Proceedings of the 25th International Conference on Neural Information Processing System, Lake Tahoe, NV, USA, 3–6 December 2012; pp. 1097–1105.
14. Deng, J.; Berg, A.; Satheesh, S.; Su, H.; Khosla, A.; Li, F. ImageNet Large Scale Visual Recognition Competition 2012 (ILSVRC2012). Available online: http://www.image-net.org/challenges/LSVRC/2012 (accessed on 10 July 2017).
15. Girshick, R.; Donahue, J.; Darrell, T.; Malik, J. Rich feature hierarchies for accurate object detection and semantic segmentation. In Proceedings of the IEEE Conference on Computer Vision and Pattern Recognition, Columbus, OH, USA, 23–28 June 2014; pp. 580–587.
16. Sermanet, P.; Eigen, D.; Zhang, X.; Mathieu, M.; Fergus, R.; LeCun, Y. OverFeat: Integrated Recognition, Localization and Detection using Convolutional Networks. *arXiv* **2013**, arXiv:1312.6229.
17. He, K.; Zhang, X.; Ren, S.; Sun, J. Spatial Pyramid Pooling in Deep Convolutional Networks for Visual Recognition. *IEEE Trans. Pattern Anal. Mach. Intell.* **2015**, *37*, 1904–1916. [CrossRef] [PubMed]
18. Girshick, R. Fast R-CNN. In Proceedings of the IEEE International Conference on Computer Vision, Santiago, Chile, 7–13 December 2015; pp. 1440–1448.
19. Ren, S.; He, K.; Girshick, R.; Sun, J. Faster R-CNN: Towards Real-Time Object Detection with Region Proposal Networks. *IEEE Trans. Pattern Anal. Mach. Intell.* **2017**, *39*, 1137–1149. [CrossRef] [PubMed]
20. Lin, T.; Maire, M.; Belongie, S.; Hays, J.; Perona, P.; Ramanan, D.; Dollar, P.; Zitnick, L. Microsoft COCO: Common Objects in Context. In Proceedings of the European Conference on Computer Vision, Zurich, Switzerland, 6–12 September 2014; pp. 740–755.
21. Simonyan, K.; Zisserman, A. Very Deep Convolutional Networks for Large-Scale Image Recognition. *arXiv* **2014**, arXiv:1409.1556.

22. Ghodrati, A.; Pedersoli, M.; Tuytelaars, T.; Diba, A.; Gool, L. Deepproposal: Hunting objects by cascading deep convolutional layers. In Proceedings of the IEEE International Conference on Computer Vision, Santiago, Chile, 7–13 December 2015; pp. 2578–2586.
23. Carreira, J.; Sminchisescu, C. CPMC: Automatic object segmentation using constrained parametric min-cuts. *IEEE Trans. Pattern Anal. Mach. Intell.* **2012**, *34*, 1312–1328. [CrossRef] [PubMed]
24. Zitnick, C.; Dollar, P. Edge boxes: Locating object proposals from edges. In Proceedings of the European Conference on Computer Vision, Zurich, Switzerland, 6–12 September 2014; pp. 391–405.
25. Hosang, J.; Benenson, R.; Dollar, P.; Schiele, B. What makes for effective detection proposals? *IEEE Trans. Pattern Anal. Mach. Intell.* **2016**, *38*, 814–830. [CrossRef] [PubMed]
26. Chavali, N.; Agrawal, H.; Mahendru, A.; Batra, D. Object-Proposal Evaluation Protocol is 'Gameable'. In Proceedings of the IEEE International Conference on Computer Vision and Pattern Recognition, Las Vegas, NV, USA, 27–30 June 2016; pp. 2578–2586.
27. Arbeláez, P.; Pont-Tuset, J.; Barron, J.; Marques, F.; Malik, J. Multiscale combinatorial grouping. In Proceedings of the IEEE International Conference on Computer Vision and Pattern Recognition, Columbus, OH, USA, 23–28 June 2014; pp. 328–335.
28. Kuo, W.; Hariharan, B.; Malik, J. Deepbox: Learning objectness with convolutional networks. In Proceedings of the IEEE International Conference on Computer Vision, Santiago, Chile, 7–13 December 2015; pp. 2479–2487.
29. Achanta, R.; Shaji, A.; Smith, K.; Lucchi, A.; Fua, P. SLIC superpixels compared to state-of-the-art superpixel methods. *IEEE Trans. Pattern Anal. Mach. Intell.* **2012**, *34*, 2274–2282. [CrossRef] [PubMed]
30. Vedaldi, A.; Soatto, S. Quick shift and kernel methods for mode seeking. In Proceedings of the European Conference on Computer Vision, Marseille, France, 12–18 October 2008; pp. 705–718.
31. Veksler, O.; Boykov, Y.; Mehrani, P. Superpixels and supervoxels in an energy optimization framework. In Proceedings of the European Conference on Computer Vision, Heraklion, Crete, Greece, 5–11 September 2010; pp. 211–224.
32. Bergh, M.; Boix, X.; Roig, G.; Capitani, B.; Gool, L. SEEDS: Superpixels Extracted via Energy-Driven Sampling. *Int. J. Comput. Vis.* **2013**, *7578*, 1–17.
33. Zhang, Y.; Sohn, K.; Villegas, R.; Pan, G.; Lee, H. Improving object detection with deep convolutional networks via bayesian optimization and structured prediction. In Proceedings of the IEEE Conference on Computer Vision and Pattern Recognition, Boston, MA, USA, 7–12 June 2015; pp. 249–258.
34. Long, J.; Shelhamer, E.; Darrell, T. Fully convolutional networks for semantic segmentation. In Proceedings of the IEEE Conference on Computer Vision and Pattern Recognition, Boston, MA, USA, 7–12 June 2015; pp. 3431–3440.
35. Erhan, D.; Szegedy, C.; Toshev, A.; Anguelov, D. Scalable object detection using deep neural networks. In Proceedings of the IEEE International Conference on Computer Vision and Pattern Recognition, Columbus, OH, USA, 23–28 June 2014; pp. 2155–2162.
36. Redmon, J.; Divvala, S.; Girshick, R.; Farhadi, A. You only look once: Unified, real-time object detection. In Proceedings of the IEEE International Conference on Computer Vision and Pattern Recognition, Las Vegas, NV, USA, 27–30 June 2016; pp. 779–788.
37. Xu, Y.; Yu, G.; Wang, Y.; Wu, X.; Ma, Y. A Hybrid Vehicle Detection Method Based on Viola-Jones and HOG + SVM from UAV Images. *Sensors* **2016**, *16*, 1325. [CrossRef] [PubMed]
38. Ammour, N.; Alhichri, H.; Bazi, Y.; Benjdira, B.; Alajlan, N. Deep Learning Approach for Car Detection in UAV Imagery. *Remote Sens.* **2017**, *9*, 312. [CrossRef]
39. Qu, T.; Zhang, Q.; Sun, S. Vehicle detection from high-resolution aerial images using spatial pyramid pooling-based deep convolutional neural networks. *Multimedia Tools Appl.* **2016**, *76*, 21651–21663. [CrossRef]
40. Tang, T.; Zhou, S.; Deng, Z.; Zou, H.; Lei, L. Vehicle Detection in Aerial Images Based on Region Convolutional Neural Networks and Hard Negative Example Mining. *Sensors* **2017**, *17*, 336. [CrossRef] [PubMed]
41. Deng, Z.; Sun, H.; Zhou, S.; Zhao, J.; Zou, H. Toward Fast and Accurate Vehicle Detection in Aerial Images Using Coupled Region-Based Convolutional Neural Networks. *IEEE J. Sel. Top. Appl. Earth Obs. Remote Sens.* **2017**, *10*, 3652–3664. [CrossRef]
42. Razakarivony, S.; Jurie, F. Vehicle detection in aerial imagery: A small target detection benchmark. *J. Vis. Commun. Image Represent.* **2016**, *34*, 187–203. [CrossRef]

43. Liu, K.; Mattyus, G. Fast Multiclass Vehicle Detection on Aerial Images. *IEEE Geosci. Remote Sens. Lett.* **2015**, *12*, 1938–1942.

44. Kong, T.; Yao, A.; Chen, Y.; Sun, F. Hypernet: Towards accurate region proposal generation and joint object detection. In Proceedings of the IEEE International Conference on Computer Vision and Pattern Recognition, Las Vegas, NV, USA, 27–30 June 2016; pp. 845–853.

45. LeCun, Y.; Boser, B.; Denker, J.; Henderson, D.; Howard, R.; Hubbard, W.; Jackel, L. Backpropagation applied to handwritten zip code recognition. *Neural Comput.* **1989**, *4*, 541–551. [CrossRef]

46. Jia, Y.; Shelhamer, E.; Donahue, J.; Karayev, S.; Long, J. Caffe: Convolutional Architecture for Fast Feature Embedding. In Proceedings of the 22nd ACM International Conference on Multimedia, Orlando, FL, USA, 3–7 November 2014; pp. 675–678.

47. Everingham, M.; Gool, L.; Williams, C.; Winn, J.; Zisserman, A. The Pascal Visual Object Classes (VOC) Challenge. *Int. J. Comput. Vis.* **2010**, *88*, 303–338. [CrossRef]

48. Lipton, Z.; Elkan, C.; Naryanaswamy, B. Optimal Thresholding of Classifiers to Maximize F1 Measure. In Proceedings of the European Conference on Machine Learning and Knowledge Discovery in Databases, Nancy, France, 15–19 September 2014; pp. 225–239.

49. Zeiler, M.; Fergus, R. Visualizing and Understanding Convolutional Networks. In Proceedings of the European Conference on Computer Vision, Zurich, Switzerland, 6–12 September 2014; pp. 818–833.

50. Leitloff, J.; Rosenbaum, D.; Kurz, F.; Meynberg, O.; Reinartz, P. An operational system for estimating road traffic information from aerial images. *Remote Sens.* **2014**, *6*, 11315–11341. [CrossRef]

MDPI

St. Alban-Anlage 66

4052 Basel

Switzerland

Tel. +41 61 683 77 34

Fax +41 61 302 89 18

www.mdpi.com

Sensors Editorial Office

E-mail: sensors@mdpi.com

www.mdpi.com/journal/sensors